PHYSICS OF MATERIALS

PHYSICS OF MATERIALS

Yves Quéré

École Polytechnique, Paris–Palaiseau, France
Académie des Sciences, Paris, France

Translated by

Stephen S. Wilson, *FIL*

With the support of the French Ministry of Culture
and of the École Polytechnique

CRC Press
Taylor & Francis Group
Boca Raton London New York

CRC Press is an imprint of the
Taylor & Francis Group, an **informa** business

CRC Press
Taylor & Francis Group
6000 Broken Sound Parkway NW, Suite 300
Boca Raton, FL 33487-2742

© 1998 by Taylor & Francis Group, LLC
CRC Press is an imprint of Taylor & Francis Group, an Informa business

No claim to original U.S. Government works

ISBN 13: 978-90-5699-119-7 (pbk)
ISBN 13: 978-90-5699-118-0 (hbk)

Visit the Taylor & Francis Web site at
http://www.taylorandfrancis.com

and the CRC Press Web site at
http://www.crcpress.com

CONTENTS

PREFACE ix

CHAPTER I
MADELUNG OR ROCK SALT 1
  1. Coulomb attraction 2
  2. Interionic repulsion 3
  3. Van der Waals attraction 5
  4. Stable structure of halides. Some remarks 7
  5. Compressibility of halides 8

CHAPTER II
DRUDE OR 'ARCHAEO-METAL' 11
  1. The situation in 1900 11
  2. Drude's model 14
  3. Where Drude's model goes wrong 29

CHAPTER III
SOMMERFELD OR 'GOOD' METAL 31
  1. Free electrons in one dimension 31
  2. Free electrons in three dimensions 43
  3. Brushing the dust off Drude 53

CHAPTER IV
SOME PROPERTIES OF 'GOOD' METALS 67
  1. Cohesion of the 'good' metal 67
  2. A 'good' metal: aluminium 76

CHAPTER V
PACKINGS: ORDER OR DISORDER 83
  1. Amorphous packings 83
  2. Crystalline packings 95
  3. Neither glasses nor crystals 109

CHAPTER VI
BLOCH AND BRILLOUIN OR THE ELECTRONS IN A CRYSTAL 115
  1. The potential taken as a perturbation 116
  2. Bloch functions 122
  3. Brillouin planes and zones 124
  4. Forbidden bands (continued) 126
  5. Energy gap 129
  6. Dynamics of electrons 138
  7. Bloch, Anderson, Mott *et al.* 147

CHAPTER VII
SURFACES AND POINT DEFECTS 155
   1. Free surfaces 155
   2. Point defects 161

CHAPTER VIII
DIFFUSION AND PRECIPITATION 173
   1. Self-diffusion 173
   2. Heterodiffusion 181
   3. Precipitation 186

CHAPTER IX
DISLOCATIONS AND BOUNDARIES 201
   1. Genesis of a concept 201
   2. Elastic model of dislocations 205
   3. Dislocations in crystals 212
   4. Sub-boundaries and boundaries 217

CHAPTER X
DEFORMATION AND FRACTURE 223
   1. Generalities 223
   2. Deformation at low temperatures 226
   3. Deformation at high temperatures 239
   4. Brittle fracture 246

ANNEX 1
DIFFRACTION BY A CRYSTAL 255

ANNEX 2
BLOCH'S THEOREM 259

ANNEX 3
BOLTZMANN'S EQUATION. ELECTRICAL CONDUCTIVITY 261
   1. Generalities 261
   2. Electrical conductivity of free electrons 262
   3. Electrical conductivity of almost-free electrons 264

ANNEX 4
TIGHT BINDING METHOD 267
   1. Approximate form of the Bloch functions 267
   2. Linear combinations of atomic orbitals (LCAO) 268
   3. Energies of LCAO 269

ANNEX 5
BURGERS VECTOR 271
   1. Definitions 271
   2. Nodes 272

PROBLEMS WITH SOLUTIONS 275

PROBLEM 1
LITHIUM PLATELETS 277

PROBLEM 2
INSTABILITY OF CAROTENE (OR WHY CARROTS ARE RED) 281

PROBLEM 3
ONE-DIMENSIONAL A–B COMPOUNDS 289

PROBLEM 4
EXTENDED STATES AND LOCALISED STATES ON A CHAIN 295

PROBLEM 5
INSULATOR–METAL TRANSITION (FROM HYDROGEN TO DOPED SEMICONDUCTORS) 305

PROBLEM 6
BAND STRUCTURE OF A SUPERCONDUCTING OXIDE 325

PROBLEM 7
MAGNETIC PROPERTIES OF A CRYSTAL 339

PROBLEM 8
FERROELECTRICITY OF BARIUM TITANATE 345

PROBLEM 9
ORDER AND DISORDER IN ALLOYS 355

PROBLEM 10
CRYSTALLINE SURFACES AND STEPS 365

PROBLEM 11
COLOUR CENTRES IN IONIC CRYSTALS 381

PROBLEM 12
DIFFUSION OF HYDROGEN ON INTERSTELLAR DUSTS 391

PROBLEM 13
DIFFUSION IN THE PRESENCE OF A FORCE (FROM PERRIN GRANULES TO COTTRELL CLOUDS) 399

PROBLEM 14
OXIDATION OF METALS 409

PROBLEM 15
SNOEK EFFECT AND MARTENSITIC STEELS 415

PROBLEM 16
SWELLING OF NUCLEAR FUELS   427

PROBLEM 17
EVOLUTION OF IRRADIATED STEELS   439

PROBLEM 18
ATOMIC VIBRATIONS IN SOLIDS   453

PROBLEM 19
STRUCTURE AND ELASTICITY OF POLYMERS   461

COMPLEMENTARY READING   471

SUBJECT INDEX   473

AUTHOR INDEX   479

TABLE OF CONSTANTS   483

ENERGY   EQUIVALENCIES   483

PREFACE

The solids which play a significant role in the evolution of our environment and our civilisations are called *materials*. The word is therefore sufficiently general that it can be applied to solids as different as stainless steels or semiconductors, the bones of the skeleton or photographic emulsions, polymers or the rocks of the Earth's mantle.

Thus, the science of materials is at once that of the chemist who invents, develops and characterises materials, that of the mechanician who establishes the laws of macroscopic behaviour, that of the physicist who attempts to understand this behaviour at the atomic scales, that of the mathematician who carries out modelling, that of the engineer who uses materials, and also that of the biologist or geologist to whom materials are a familiar landscape. Thus, also, for the student of materials science, it is like being in a good restaurant, faced with a copious and varied menu, in which the entrées are, of course, presented separately from the desserts or the wines, but everything is necessary and everything is ultimately linked together to form a good meal.

Entrée? Dessert? Roast? physics has a choice place in the banquet. While it is certainly not sufficient, it is strictly necessary if one wishes to *understand* what a material is, what its properties are and how it reacts to the many external demands. From this point of view, the physics of materials becomes one of the most interdisciplinary areas of physics and one of the closest to applied science. Using the relatively simple ingredients of electromagnetism, quantum mechanics, statistical thermodynamics, elasticity and optics, it constitutes an excellent recapitulative theme for a course at the end of the undergraduate level.

This manual, which constitutes one of the courses offered to the students (*élèves*) of the *École Polytechnique*[1] is at this level. After having all taken, in Physics, a course in Quantum Mechanics[2] (abbreviated to Q.M. in the text) and a course in Statistical Physics[3] (Stat. Ph.), the *élèves* then have a choice between several optional courses (Astrophysics, Quantum Optics, ...) including this course on the Physics of Materials. These optional courses supplement numerous other courses (Mathematics, Computer Science, Mechanics, Chemistry, Biology, Economic Sciences and also Philosophy, Languages, ...) and are

---

[1] Created in 1794, at the very heart of the French Revolution, the *École Polytechnique* was assigned the double mission, something quite new for the period, of recruiting its students strictly according to their scientific talents (independently of their social origin) and centering its system of education on a global vision of the Sciences, and on mathematics and physics in particular, rather than on the Humanities or the Practical Arts. This emphasis on Science was in the spirit of the time, at this, the end of the Age of Enlightenment, when, in Paris, Voltaire, Lavoisier, Condorcet, Benjamin Franklin, Coulomb, Carnot and many others were discoursing endlessly on the importance of the Sciences as an element of the universal culture, with the promise of benefits for mankind.

This principle of a generalist Scientific education, marked in the first decades of the *École* by men such as Monge, Lagrange, Laplace, Ampère, Poisson, Fourier, Cauchy, Coriolis, Fresnel, Bravais, Poincaré, Becquerel, ...has been broadly retained. Thus, to this day, the *École* continues to recruit almost 450 *élèves* (there were 394 in 1794) each year in France and abroad, via a rigorous competition, and provide them with a two-year multidisciplinary scientific education. It is after these two years that they specialise and move either into a more precise area of science, in which they write a thesis, either at the *École* itself or outside, or into careers in engineering or in management.

[2] J. L. Basdevant, Mécanique quantique, Ellipses, Paris, 1986.

[3] R. Balian, Du microscopique au macroscopique, Ellipses, Paris, 1982. From microphysics to macrophysics, Springer-Verlag, 1991–2 (English translation). E. Brézin, Physique statistique, École Polytechnique, 1993.

strictly limited to a set number of hours. They are thus forced to make difficult selections and therefore aim to provide an introduction to a number of key points in the subject rather than a global presentation.

In this case, this consists of an elementary description of the electric and mechanical properties of solids, presented at an undergraduate level and limited to the following: a substrate of **free electrons**, an introduction to **band structures**, a little about **atomic diffusion** and the **evolution of alloys**, and the rudiments of **plasticity** and **fracture**. This is very classical knowledge, which is probably sufficient for the student seeking only a first (and possibly last) contact with solids and is certainly necessary for anyone wishing to go further in Solid-State Physics or Materials Science. This selection sidelines many important subjects (Magnetism, Crystalline Symmetries, Dielectric Properties, Rheology of Polymers, Composite Materials, or Semiconductor Physics which is the subject of a separate course[4]), some of which are broached in the form of **problems** (see later). These omissions are a sign of the vastness of the subject area: materials science is a limitless field of knowledge, techniques and applications, bringing together chemists, physicists, and others named above, and also archaeologists, anthropologists, artisans, artists .... Its history merges with the history of mankind and its diversity espouses the profusion of human talents.

The content of a course is one thing, how it is approached is another. The route taken here is heavily intuitive. Since the beginning of time, materials science has built on the astute empiricism and industrious intuition of smiths and potters. It has now, God be thanked, adopted the scientific method for its own, but it is necessary for it not to challenge or forget its past: it never progresses as well as when it is able to marry rigour and astuteness, logic and intuition.

Following the exposition which constitutes the first part of the book, a large second part is devoted to **problems** and their solutions. These are the subjects of tests set for the *élèves* of recent years in the final examination. They lasted three or four hours each, as appropriate. In addition to being able to use them to test his or her assimilation of general notions, the reader will find that they deal with or touch on some of the subjects omitted in the first part.

The passages in **small print** concern matters (calculations, applications, examples), which are not needed for a general understanding on a first reading. In particular, they include the proofs of a number of propositions previously introduced in a qualitative or intuitive manner.

It is a pleasure for me here to thank my colleagues at the *École Polytechnique* for numerous stimulating exchanges, and, in particular, Henri Alloul, Roger Balian, Jean-Louis Basdevant, Edouard Brézin, Claudine Hermann, Guy Laval, Jean Salençon, Bernard Sapoval, Jean-Claude Tolédano, Libero Zuppiroli and Claude Weisbuch. I am also grateful to those who allowed me to use some of the present problems which they drew up: Henri Alloul, Roger Balian, Jean-Louis Basdevant, Maurice Bernard, Jean-Noël Chazalviel, Maurice Guéron, Roland Omnès, Pierre Petiau, Ionel Solomon and Alfred Vidal-Madjar.

I should also like to thank those who have provided me with, sometimes unpublished, documents, and, in particular, Manuel Allain, Loïc Boulanger, Alain Bourret, Christian Colliex, Catherine Corbel, Omourtague Dimitrov, Jean-Claude Jousset,

---

[4] B. **Sapoval** and C. **Hermann**, *Physics of Semiconductors* (abbreviated to S.C.Ph in the text), Springer-Verlag, 1993.

René Kormann, Daniel Lesueur, Alfred Manuel, Jean Mory, Hubert Pascard, Lawrence Slifkin, Tiziana Stoto, Junzo Takahashi and Michel Treilleux.

I am grateful to those who taught me, the Professors Jacques Friedel and André Guinier at the *Université de Paris-Sud* and Paul Lacombe at the *École des Mines de Paris*. Neither can I forget those whom I have had the pleasure of working with in founding the French summer schools in Materials Science, Yves Adda and Jean-Michel Dupouy at the *Commissariat à l'Énergie Atomique*, and Jean Philibert at the *Institut de recherche de la sidérugie* and at the *Université de Paris-Sud*.

Finally, working with Dr. S. S. Wilson who translated and typeset this book has been for me both enjoyable and stimulating.

The search for new superconductors.
After Van Gogh.

CHAPTER I

MADELUNG OR ROCK SALT

Remembrance of Things Past (Vol. 1)
Marcel Proust

Back to Methuselah
George Bernard Shaw

Around 1900 most questions about the nature of *chemical bonding* and *a fortiori* about the *cohesion of solids* were unanswered. This was even true for the 'simple' case of the hydrogen molecule which was to remain a mystery until 1927 (Heitler and London)! All attempts to develop a convincing theory of the cohesion of solids such as *carbon* (graphite or diamond), *copper* or *ice* were doomed to failure whilst the essential tool of quantum mechanics was missing.

However, the cohesion of one type of solid, namely **ionic solids**, was correctly understood from 1910. Here, we shall describe the inductive and often intuitive approach which enabled Madelung to develop an initial theory of *alkali halides*[1]; in this chapter we shall use this family as a prototype for ionic crystals. We note immediately that a proper description of alkali halides does not *only* end with the structure of rock salt. In fact, it provides a good initial approach to all ionic crystals and, importantly, to most **minerals**.

Alkali halides are stoichiometric substances (rock salt: NaCl; sylvine: KCl ...) consisting of equal proportions of atoms of a *monovalent metal* and a *halogen*. Once the chemical composition of these solids is known, the only two ingredients needed for further progress are *Mendeleyev's periodic table* and *Coulomb's law*.

The *first* reminds us that monovalent metals follow directly after noble gases in the classification whilst halogens precede them. Thus, it is tempting to imagine that each metal atom may give up an electron to a halogen atom. The species formed in this way are metallic *cations* (A^+) and halogenic *anions* (B^-) which both have the electronic structure of a noble gas.

The *second* provides us with the key to cohesion. If each cation (*resp.* anion) tends to prefer to surround itself with anions (*resp.* cations) the Coulomb attraction summed over the pairs A^+B^- is likely to be greater than the repulsion summed over the pairs A^+A^+ and B^-B^-.

To show that this is actually the case, we must imagine[2] a structure which is plausible and which favours this preference for surrounding ions. The simplest approach

---

[1] E **Madelung**, *Physik Z.* **11**, 898 (1910) and **19**, 524 (1918).

[2] In 1910, 'imagine' was the right word, since the structure of rock salt was not determined (by **Bragg**) until 1912 by X-ray diffraction.

(in Madelung's words, '*das Näturlichste*') involves forming neutral rows of regularly spaced ions with alternating signs and assembling these rows in parallel prismatic 'tubes' with a square base (see Figure 1). In this arrangement, we recognise the so-called face-centred-cubic NaCl structure (see Figure 23, p. 105). These tubes may also be based not on squares but on regular centred hexagons. This leads to the so-called CsCl cubic structure (see Figure 4, p. 7) in which the axis of the tubes is a diagonal of the cube. We shall return to these various packings later (Chapter V).

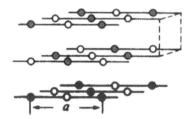

Figure 1   Packing of Na^+ (o) and Cl^- (•) ions into the solid NaCl, as described by Madelung. Rows of alternating ions, which are globally neutral, are grouped into tubes with a square base. It is easy to find a cube with centred faces and edge a (see p. 104).

1. COULOMB ATTRACTION

In each of these two structures, a cation (*resp.* anion) is surrounded by n_1 anions (*resp.* cations) at a distance r, then by n_2 cations (*resp.* anions) at a distance r_2, etc .... For charged ions (charge e), the Coulomb energy (in CGS units) of such a crystal containing $2N$ ions is

$$E_c = N \left(-n_1 \frac{e^2}{r} + n_2 \frac{e^2}{r_2} - \dots \right).$$

Since r is the smallest possible distance between two ions, we may write $r_2 = \alpha_2 r \dots$ $r_i = \alpha_i r$ ($\alpha > 1$). Thus, we write

$$\boxed{E_c = -N \frac{e^2}{r} M} \tag{1}$$

where the number

$$\boxed{M = \sum_i \frac{(\mp) n_i}{\alpha_i}} \tag{2}$$

is called the **Madelung constant**. This constant is characteristic for the crystalline structure. Notably, it is independent of the interionic distances. It is always positive. Thus, around a given ion, it is the Coulomb interaction with the *first layer* of ions with the opposite sign which determines this sign: **it is an attraction**, since the energy E_c of (1) is negative.

Thus, for the NaCl structure, the number n_1 of 'first' neighbours at distance r is 6. The number n_2 of 'second' neighbours at distance $r_2 = r\sqrt{2}$ is 12 etc ... and the Madelung constant is given by

$$M = \frac{6}{1} - \frac{12}{\sqrt{2}} + \frac{8}{\sqrt{3}} - \frac{6}{2} + \frac{24}{\sqrt{5}} - \ldots = 6 - 8.485 + 4.620 \ldots$$

which converges slowly and has sign fixed by that of the first term.

Exercise   Consider an infinite chain of alternating $+$ and $-$ equidistant ions: $\ldots - + - + - + \ldots$. Show that the Madelung constant of this chain is equal to $2 \ln 2$.

The Madelung constant defined above in terms of the distance r (cation–anion) may also be defined in terms of the cube root of the molecular volume (constant M') or in terms of the edge of the cube in the case of cubic crystals (constant M''). The values of these constants, calculated numerically (as $i \to \infty$), are given below for various structures:

Table 1   Values of the Madelung constants for four different crystalline structures. The constants M' and M'' are defined in the text.

	M	M'	M''
Sodium chloride (NaCl) structure	1.7476	2.2018	3.4951
Caesium chloride (CsCl) structure	1.7627	2.0354	2.0354
Blende (ZnS) structure	1.6381	2.3831	3.7829
Wurtzite (ZnS) structure	1.641	2.386	

2. INTERIONIC REPULSION

If Coulomb attraction (see above) were the only force present, the crystal would collapse, since r would tend to zero in (1) – like the specific volume – so as to decrease the energy. This is not the case and we know that this is a fundamental consequence of **Pauli's principle**. We shall return to this aspect in Chapter IV. For Madelung, this could still only amount to a *'non-penetrability'* of ions, with which he associated a **repulsion** term, defined, in a phenomenological fashion, to be equal to

$$\boxed{E_R = Br^{\cdot} \qquad n > 1} \qquad (3)$$

where B is a positive constant. The important thing here is that *this term counts for little in the total energy at the equilibrium distance* r_0. Thus, for NaCl we have $r_0 = 2.81$ Å and the Coulomb energy E_c (1) per NaCl molecule is equal to:

$$E_c = -\frac{(4.8 \times 10^{-10})^2}{2.81 \times 10^{-8}} 1.7476 \ \text{erg}$$
$$= -8.9 \ \text{eV}.$$

This value is quite close to the experimental value of the binding energy -7.9 eV; thus, in the crystal, *almost all the binding is due to the Coulomb energy* E_c *calculated from* (1).

Let us define this **binding energy** E_b. This is the energy needed to dissociate the crystal into Na^+ and Cl^- *ions*. It is not the **cohesive energy** E_{coh} which is equal (up to sign) to the **sublimation energy** E_s which corresponds to dissociation into Na and Cl *atoms*. Since the creation of Na^+ from Na requires the **ionisation energy** E_i and since the creation of Cl^- from Cl generates the **affinity energy** E_a, we simply have (see accompanying diagram)

$$E_b = E_{coh} - E_i - E_a.$$

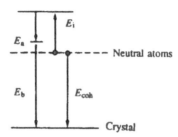

The three quantities on the right-hand side are known experimentally ($E_{coh} = -6.5$ eV, $E_i = 5.14$ eV for sodium and $E_a = -3.71$ eV for chlorine) and we deduce that E_b has value

$$E_b = -7.9 \text{ eV}.$$

Expressed as a function of the ionic distance $Na^+ - Cl^-$, r, the total energy of the crystal is simply ((1) and (3)):

$$E(r) = -N\frac{e^2}{r}M + \frac{B}{r^n} \qquad M \text{ and } B > 0 \text{ and } n > 1. \qquad (4)$$

The distance r_0 for which attraction and repulsion are balanced is given by $(dE/dr)_{r_0} = 0$; in other words,

$$r_0 = \left(\frac{nB}{Ne^2M}\right)^{1/(n-1)}.$$

Since E_b and E_c are, respectively, the values of the total energy (4) for $r = r_0$ and the Coulomb energy (1), it is easy to show that

$$n = \frac{E_c}{E_c - E_b}. \qquad (5)$$

Thus, for the example of NaCl ($E_c = -8.9$ eV and $E_b = -7.9$ eV), the exponent n of the repulsion potential (3) turns out to be 8.9. This high value explains the abrupt nature of the repulsion term. The two contributions E_c and E_R to the total energy are shown in Figure 2 for the case of NaCl.

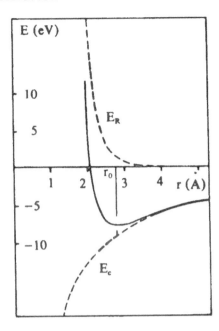

Figure 2 Energy of the NaCl crystal. E_c: Coulomb energy (1). E_R: repulsive energy (3). The values of B and n are adapted to the experimental values of the binding energy and the interionic distance r_0. E: the total energy (continuous line).

3. VAN DER WAALS ATTRACTION

Let us consider two neighbouring ions A^+ and B^-. That there might exist an attraction between these *other than* – and independent of – *the Coulomb attraction* is strongly hinted at in the case of *atoms of noble gases*, of which, moreover, they have the electronic structure. In fact, there must exist an attraction between these neutral objects (atoms) since they may condense into liquids or even solids. For example, *neon* is solid up to $T = 24$ K and krypton up to 117 K at normal pressure. It was through studying the interaction of neutral molecules in gases under high pressure that Van der Waals (1873) deduced the existence of weak, short-range attractive forces between these molecules. These forces were only correctly described in *quantum* terms at a much later date (London 1930); however, we can now give a *classical* semi-quantitative description based on the example of two (neutral) atoms of a single noble gas.

Let $\bar{r}$ be the vector joining the centres of these two atoms denoted by (1) and (2). For each atom, the mean dipole moment taken over a sufficiently long time is zero. On the other hand, *at a given instant* the orbital movement of the electrons gives rise to a *dipole moment* which is generally non zero. Suppose that $\overline{m}_1$ is the instantaneous dipole moment of the atom (1). This creates an electric field $\bar{\varepsilon}$ at the centre of the atom (2) which induces an instantaneous dipole moment $\overline{m}_2$ proportional to the *polarisability* of the atom in question:

$$\overline{m}_2 = P\overline{\varepsilon}.$$

The two dipoles ($\overline{m}_1$) and ($\overline{m}_2$) interact and we see (Figure 3) that, taking into account the structure of the lines of the force field $\overline{\varepsilon}$, *this interaction tends*, in all cases, *to decrease the energy*.

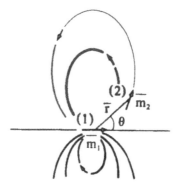

Figure 3   Van der Waals–London interaction. The atom (1) with dipole moment $\overline{m}_1$ induces a dipolar moment $\overline{m}_2$ on the atom (2) (position $\overline{r}$) parallel to the lines of force of the dipolar field. For all values of $\overline{r}$, the interaction of the dipoles ($\overline{m}_1$) and ($\overline{m}_2$) is attractive.

We recall, more precisely, that the radial and normal projections of the field $\overline{\varepsilon}$ at the point $\overline{r}$ (see, for example, Figure 3) are (in CGS):

$$\varepsilon_r = 2m\frac{\cos\theta}{r^3} \qquad \text{where } m = |\overline{m}_1| \text{ and } r = |\overline{r}|$$

$$\varepsilon_H = m\frac{\sin\theta}{r^3}$$

and that the energy $E(\overline{r})$

$$E(\overline{r}) = P\left(2m^2\frac{\cos^2\theta}{r^6} - m^2\frac{\sin^2\theta}{r^6} - \frac{3}{r^5}(mr\cos\theta)\left(2mr\frac{\cos\theta}{r^3}\right)\right)$$

$$= -\frac{Pm^2}{r^6}(4\cos^2\theta + \sin^2\theta), \tag{6}$$

is always negative.

This short-range attraction between neutral atoms (which varies as r^{-6}, see (6)) gives rise to the **Van der Waals–London** forces. It originates in the instantaneous and local fluctuations of the dipole moments of the atoms or ions.

The Van der Waals–London forces generally have a modest intensity, as is shown by the mediocre cohesion of solids in which they are the only active forces (noble gases). In the case of ionic solids, the contribution corresponding to the cohesion energy is *small* compared with the Coulomb energy (around 1 to 2% of the latter in alkali halides).

However, it is not unimportant since, in fact, it may determine the crystalline structure adopted by the halide (see the remarks below).

4. STABLE STRUCTURE OF HALIDES. SOME REMARKS

The crystalline structure chosen from amongst the various possibilities (notably, NaCl and CsCl structures) is that which minimises the free energy, or the energy at $T = 0$ K. The Coulomb energy (1) depends only upon the Madelung constant: in this respect, the CsCl structure seems slightly more favourable than the NaCl structure (difference of 1.5%, see Table 1). On the other hand, the number of first-neighbour atoms is greater for the CsCl structure (eight, see Figure 4, against six for NaCl, see Figure 1 or Figure 22, p. 105), whence the repulsive part of the energy, which is directly associated with the interaction of the first neighbours, is also greater. This accrued repulsion counterbalances the effect of the Madelung constant and may even, in general, have the edge since most alkali halides adopt the NaCl structure.

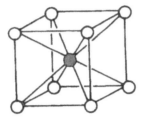

Figure 4   CsCl structure. o: Cs$^+$ (or Cl$^-$). •: Cl$^-$ (or Cs$^+$).

However minimal it is the Van der Waals–London contribution may tip the balance described above, particularly when the polarisation of ions increases as they increase in size, as shown in Table 2.

Table 2   Electronic polarisability P of ions of alkali halides, in (Å^3). After **J R Tessman** et al. *Phys. Rev.* **92**, 890 (1953). Using (6), it can be shown that the polarisability is homogeneous to a single volume.

F$^-$	0.65	Li$^+$	0.03
Cl$^-$	2.97	Na$^+$	0.41
Br$^-$	4.17	K$^+$	1.33
I$^-$	6.44	Cs$^+$	3.34

We note the large value of P_{Cs^+} compared with that of P_{Na^+}. The Van der Waals–London energy between Cs$^+$ and Cl$^-$ is greater than that between Na$^+$ and Cl$^-$. Since it partly wipes out the effect of repulsion on the eight first neighbours of the CsCl structure, it favours the stability of the structure over that of the NaCl structure. The crystals CsCl, CsBr and CsI have the CsCl structure. CsF (for which the anion F$^-$ is the smallest and least polarisable of all the halogen ions) retains the NaCl structure.

5. COMPRESSIBILITY OF HALIDES

The compressibility χ is defined, for a variation in the hydrostatic pressure p, by

$$\chi = -\frac{1}{V}\left(\frac{\partial V}{\partial p}\right)_T$$

where V is the volume. We immediately have:

$$\frac{1}{V\chi} = \left(\frac{\partial^2 E}{\partial V^2}\right)_{V=V_0} \tag{7}$$

where V_0, the volume at equilibrium, is defined by

$$\left(\frac{\partial E}{\partial V}\right)_{V=V_0} = 0. \tag{8}$$

Let us apply these equations to the case of CsCl, reducing the energy to that of the volume V of a molecule. Then we have

$$E(r) = -\frac{e^2}{r}M + \frac{b}{r^n}$$

where $b = B/N$ and V (the volume of the cube of Figure 4) is given by $V = \left(2/\sqrt{3}\right)^3 r^3$.

Expanding expressions (7) and (8), it is easy to show that

$$b = \frac{Me^2}{n}r_0^{n-1} \tag{9}$$

whence

$$n = \frac{8\sqrt{3}r_0^4}{Me^2\chi} + 1 \tag{10}$$

and

$$E_b = E(r_0) = -\frac{Me^2}{r_0}\left(1 - \frac{1}{n}\right) \tag{11}$$

where this final expression (11) is simply (5).

Table 3   Calculated and measured cohesive energies (in eV) and the exponent n of the repulsion potential for various alkali halides.

Solid	Structure	n (10)	E_{coh} (theor.)	E_{coh} (exper.)
CsCl	CsCl	10.5	−6.46	−6.74
CsBr	CsCl	11.0	−6.24	−6.46
CsI	CsCl	12.0	−5.92	−6.32
NaCl	NaCl	8.7	−7.84	−7.92
NaBr	NaCl	8.5	−7.47	−7.53
NaI	NaCl	9.5	−7.00	−7.23

Experimental determination of r_0 and χ then allows us to calculate the values of n and E_b. Moreover, since we know E_a and E_i (see p. 4), we may determine E_{coh} which can be compared with measured cohesive energies. Table 3 shows that there is a satisfactory agreement between this elementary theory and experiments for a number of halides with the CsCl and NaCl structures.

N.B. For the latter, it can be shown that the results (9), (10) and (11) remain valid provided $8\sqrt{3}$ is replaced by 18 in (10).

It is clear that the calculation *underestimates* the cohesive energy, partly because the Van der Waals–London energy is not taken into account. However, it leads to a good approximation of E_{coh} and provides a correct description of the variations of this inside a series of a given cation (or anion).

However, while the results are satisfactory, this can scarcely be said of the approach of Section I.2. The potential in terms of r^{-n} is an *ad hoc* creation which is only used to prevent the solid from collapsing. Thus, we shall certainly need to return to the physical origin of these interionic repulsions, which we shall do in the case of metallic solids (Chapters III and IV).

We shall again find the same aura of 'archaism', but also the same inductive progression, the same qualities of intuition and, to a certain extent, the same success in the next chapter which describes the initial theory of the metallic state. In this case, we shall study and explain not the energy and the cohesion of the solid but its conductivity and its optical properties.

CHAPTER II

DRUDE OR 'ARCHAEO-METAL'

Remembrance of Things Past (Vol. 2)
Marcel Proust

The Box of Delights
John Edward Masefield

Transparent Things
Vladimir Nabukov

1. THE SITUATION IN 1900

At the beginning of this century, no serious theory of the the metallic state exists. The most important facts known to craftsmen (points i) and ii)) or from experiments (points **iii**) to **viii**)) around 1900 may be summarised as follows.

i) Metals – and, very generally, their alloys – are **plastic solids**. The transition from the Stone Age to the Iron Age established this fundamental discovery. Little by little, man has learnt that, in general, the plasticity *increases with temperature* (the art of the blacksmith who heats his workpiece in order to deform it) and decreases in *alloys* (the art of the jeweller who prefers to use a gold alloy rather than pure gold which is too soft).

ii) Metals are good **thermal conductors**. Thermal conductivity κ (defined by Fourier's law, see Figure 1) varies greatly from one metal to another (by factors of $\simeq 10$ between *silver* and *lead* and $\simeq 100$ between *silver* and *uranium*). It tends to decrease in alloys. It varies little with temperature (Figure 2).

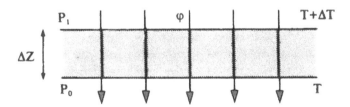

Figure 1   Fourier's law. The heat flux φ between the walls P_1 and P_0 is proportional to the temperature gradient $\Delta T / \Delta z$ ($\simeq dT/dz$): $\varphi = -\kappa dt/dz$. κ is the **thermal conductivity** of the medium between P_0 and P_1.

iii) Metals are good **electrical conductors**. They are prototypes of conductors in electrostatics and are notably characterised by the impossibility of creating a non-zero electric field within them *at equilibrium*, other than over distances of the order of interatomic distances:

$$\overline{\mathcal{E}} = 0.$$

iv) The **electrical conductivity** σ ($\sigma = \rho^{-1}$, where ρ is the electrical **resistivity**) varies with temperature. It decreases as the temperature increases. In many cases, the resistivity varies linearly with T (absolute temperature), except at very low temperatures when the variation is often in T^5. If we neglect this area, the linearity with T often approximates to a proportionality (Figure 2 and Figure 22, p. 59):

$$\rho \simeq AT. \tag{1}$$

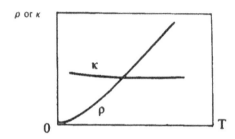

Figure 2 General behaviour of the electrical resistivity ρ and the thermal conductivity κ of metals as a function of temperature.

v) The conductivity σ varies greatly from one metal to another. It was soon seen ($\simeq$ 1850) that **metals which are good** (*resp.* bad) **electrical conductors are also good thermal conductors**: silver, copper … (*resp.* lead, uranium …).

   More precisely, when measuring the thermal and electrical conductivity (κ and σ, respectively) of various metals at standard temperature, Wiedemann and Franz (1853) observed that the ratio κ/σ is constant. Thus, κ_{Ag}/σ_{Ag} is equal to κ_{Pb}/σ_{Pb} although κ (and σ) are around 10 times larger for silver than for lead. This factor may be as large as $\simeq$ 100 if we compare silver with plutonium (see Table 1).

   Taking these measurements again at different temperatures, Wiedemann and Franz observed that at each temperature, the ratio κ/σ remains independent of the metal used and that, moreover, the value found is practically proportional to T. In other words, writing

$$\boxed{\frac{\kappa}{\sigma} = LT} \tag{2}$$

(**Wiedemann–Franz law**), one defines a quantity L (called the *Lorenz number*) which varies very little *either* from one metal to another *or* from one temperature to another. This can be seen in Table 1.

For a large range of applications, it is much more important to know the value of the thermal conductivity than that of the electrical conductivity (for example, for alloys of heat exchangers, for nuclear fuels and their cladding ...), although the latter is much easier to measure than the former. The Wiedemann–Franz law enables us to obtain a low-cost, first-order estimate of κ (for example, in order to compare a series of alloys with each other). For this, it is sufficient to measure σ (easy measurement with Wheatstone bridge) and deduce κ using a 'reasonable' value for L (for example, 2.3×10^{-8} W.Ω K^{-2} for copper alloys).

Table 1 Some values of the thermal conductivity κ (in W.cm^{-1} K^{-1}) and the Lorenz number $L = \kappa/\sigma T$ (in 10^{-8} W.Ω K^{-2}).

	273 K		373 K	
	κ	L	κ	L
Li	0.71	2.22	0.73	2.43
Al	2.37	2.14	2.30	2.19
Fe	0.80	2.61	0.73	2.88
Cu	3.85	2.20	3.82	2.29
Ag	4.18	2.31	4.17	2.38
Pb	0.38	2.64	0.35	2.53
Bi	0.09	3.53	0.08	3.35
Pu	0.065	2.50		
'Nichrome' steel	0.03	1.45		

vi) Metals are generally **paramagnetic**.

We recall that a substance placed in a magnetic field $\overline{H}$ acquires an induced magnetism $\overline{M}$ parallel to $\overline{H}$ and that its *magnetic susceptibility* is defined by

$$\chi = \lim_{H \to 0} \overline{M}/\overline{H}. \tag{3}$$

The substance is said to be paramagnetic (*resp.* diamagnetic) if χ is positive (*resp.* negative). The (paramagnetic) susceptibility of metals general *varies little with temperature*.

This paramagnetism is different from that of ionic salts for which χ varies as T^{-1} (Curie's law). This is called *Pauli paramagnetism*. We shall return to this type of paramagnetism later (see p. 64).

Some metals including Fe, Ni and Co (Gd (gadolinium) and Dy (dysprosium) were added to this list later), have a non-zero spontaneous magnetisation in a zero field. These are the *ferromagnets*. This property disappears at the Curie temperature T_c ($T_c = 1043$ K for iron, 631 K for nickel) above which these metals become paramagnetic (see also **Problem 7**).

vii) The **specific heat** of metals is, at first sight, not very different from that of insulators. It increases rapidly with temperature from zero and stabilises at a value close to 6 cal/mole.K (Dulong and Petit's law) or $3N_0 k$/mole (N_0 is the Avogadro constant and k is the Boltzmann constant).

viii) Most metals are found towards the left in **Mendeleyev's periodic table** (1869).

2. DRUDE'S MODEL

It was the discovery of the electron (J J Thomson, 1897) which made it possible to lay the foundations of an initial theory of metals.

Drude immediately perceived the role played by electrons in metals and soon proposed[1] a phenomenological description of conduction. For this, he introduced a number of simple concepts (free electron, relaxation time, ...) which (naturally after some refinement and deeper investigation) have remained useful and important tools in later theories of the metallic state. On the other hand, Drude's use of inappropriate statistics imposed a fundamental limitation on the model which was not removed for a quarter of a century (Chapter III). We also note that this model does not touch on the plastic properties (point i) of the previous list) at all. This came half a century later (Chapter X).

2.1 Free-electron approximation

The position of metals in the periodic table (point viii)) suggests that these are substances in which some electrons outside the atom to which they are only loosely bonded are liable to leave the latter from the moment when the solid forms from isolated atoms. These electrons, of which there are z ($z = 1$ for sodium, 3 for aluminium ...) are the **valence electrons**.

The agglomeration of N atoms, which gives birth to a metallic solid then ends with a crystal of ions each with charge $z+$ (for example, Na^+ in the case of sodium, Figure 3(a)) and a 'gas' of Nz electrons released from their atom of origin. As we shall see, it is this electron gas which guarantees the stability of the structure.

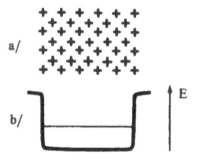

Figure 3   An aggregate (for example, crystalline, see Chapter V) of + ions (a) creates a potential well for the electrons (b).

Like Drude, we make the following additional simplifications.

i)   The repulsive Coulomb *mutual interaction of the electrons is negligible*. We shall see the (relatively minor) consequences of this approximation later (p. 49). In return, this approximation provides for a radical simplification of the calculations.

[1] *Ann. der Physik* **1**, 566 and **3**, 369 (1900).

ii) The interaction of each of the Nz valence electrons with the ions is schematised as follows: the presence of the N positive ions creates an *attractive potential* for the electrons in the spatial region occupied by the solid: an electron trying to escape from the metal is subject to an attractive potential from the latter due to the excess of *a single* positive charge created in this way. In other words, the electrons are trapped in a **potential well** the form of which follows that of the solid (Figure 3(*b*)).

Apart from this *global* action of the ions, their *local* action on the electrons is taken as *negligible*. In other words, in the metal, the valence electrons are subjected to a *constant potential* which we may take to be equal to zero:

$$V(r) = \begin{cases} 0 & \text{inside the metal} \\ V_0 \quad (V_0 > 0) & \text{outside.} \end{cases} \tag{4}$$

These electrons, considered without mutual interactions and without interaction with the ions other than that imposed by the walls of the potential well (the free surfaces) are called **free electrons**. The free surfaces form a barrier which internally reflects the free electrons colliding with it. For this 'gas' of free electrons, the surfaces of the metal form the rigid walls of an empty *box* in which the electrons evolve freely (Figure 4).

The existence of free electrons in metals received direct confirmation in 1916 in an experiment by Tolman. A metal wire is wound on a coil which may turn around its axis. Its two ends are connected to a galvanometer G by revolving contacts (T) (Figure 5). If the coil, which is initially rotating at a high speed, is stopped suddenly, at the moment of stopping the galvanometer records an instantaneous passage of electric current: the wire constitutes a box (in the sense of Figure 4(*b*)) within which the free electrons are carried along in the rotational movement. When the rotation is abruptly stopped the free electrons continue their movement by inertia. It is this transitory movement of charges which is observed on the galvanometer.

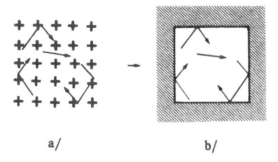

a/ b/

Figure 4 The aggregate of ions (*a*) constitutes a closed box for the electrons (*b*).

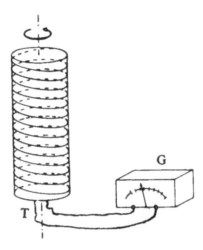

Figure 5 **Evidence for free electrons.** When the rotation of the cylinder carrying the conductor is stopped, a current passes through the galvanometer G.

iii) Drude added a further postulate to these simple and reasonable assumptions which later turned out to be fundamentally flawed. He considered the electrons as discernible particles which, like the atoms or molecules of a perfect gas, obeyed the *Maxwell–Boltzmann statistics* (see (Stat. Ph.)). Thus, Drude applied results from the kinetic theory of gases to this free-electron gas. We shall clear up this misunderstanding in the next chapter.

2.2 Electrical conductivity of a metal

The assumptions i) and ii) (above) imply that each electron has a rectilinear movement in the metallic box which is only perturbed when the electron rebounds from the free surface. All directions of movement are equally likely and this movement of electrons does not generate a *current*.

In the presence of an electric field $\overline{\mathcal{E}}$, each electron is subjected to a force $\overline{f} = -e\overline{\mathcal{E}}$ ($-e$ is the charge of the electron; *here, and throughout what follows, e is a positive quantity*). This force carries all the electrons along in the direction opposite to $\overline{\mathcal{E}}$. The previous equiprobability of all directions is violated and a current $\overline{i}$ appears.

Suppose that, **in a unit volume**, N electrons all move with the same velocity $\overline{v}_e$. In a time dt, $N S dt$ electrons cross a surface S perpendicular to $\overline{v}_e$. Denoting the *current density vector* (this is a vector parallel to the flow of the charges with modulus equal to the number of charges crossing a unit surface in a unit time interval), we may write:

$$\overline{i} = -Ne\overline{v}_e. \tag{5}$$

The previous force $\overline{f}$, which is constant like $\overline{\mathcal{E}}$, gives the electrons a **drift velocity** $\overline{v}_e$, the modulus of which increases linearly with time. Thus, according to (5), this apparently results in an electrical current which increases with time; however, this manifestly contradicts experience and is absurd (i tending to infinity).

N.B. Do not confuse the *drift velocity* v_e and the intrinsic velocity $\bar{v}$ of electrons. The former adds to the latter and is generally much weaker. This is the case for a cloud in which the water molecules (or drops) i) have (large) natural velocities v_i distributed isotropically, to which the wind may add a drift velocity $\bar{v}_e$, which is then that of the cloud.

1 – Relaxation time, collisions

It was in order to resolve this difficulty that Drude introduced an additional assumption. To prevent $\bar{v}_e$ from increasing indefinitely, he assumed that the electrons are subjected to a *relaxation* mechanism in the metal. The drift velocity still increases in proportion to t because of the force $\bar{f}$ over a time Δt_1, but it falls abruptly to zero and begins to increase again during time Δt_2, etc ....

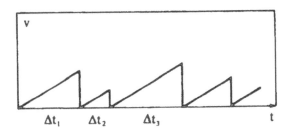

Figure 6   Velocity of a free electron as a function of time, in Drude's theory.

If this is the case, the drift velocity has mean value

$$\bar{v}_e = -\frac{1}{m}e\bar{\mathcal{E}}\tau \tag{6}$$

where τ is the mean value of the time intervals Δt_i and m is the mass of the electrons. The time τ is called the **relaxation time**. It is defined below from the *exponential distribution*.

Introduced in this way, relaxation and relaxation time are only *ad hoc* concepts which enable $\bar{v}_e$ (whence also $\bar{i}$) to have constant values. The underlying physical idea is that of *collision*: at the end of each period Δt_i, the electron is involved in a collision. These collisions which Drude considered as collisions with *ions* of the crystal must obey the following conditions for (6) to be valid:

(i)   They must give rise to an **exponential distribution**.

Let us briefly use this example of random collisions to recall some of the characteristics of this distribution.

The *probability of collision* for an electron is assumed to be *independent of time*. During the period dt, the probability that this electron is involved in a collision is taken to be proportional to dt, namely dt/τ. The constant τ which characterises the type of collision (and notably the effective cross section) is called the **relaxation time**. Under these conditions, the probability that at time $t = 0$ an arbitrary electron is not involved in a collision in the *next* interval t is

$$p(t) = \lim_{dt \to 0} \left(1 - \frac{dt}{\tau} \right)^{t/dt} = e^{-t/\tau}. \tag{7}$$

Clearly, the probability that this same electron has not been involved in a collision during the *previous* period t (the interval between $-t$ and 0) has the same value

$$p(-t) = e^{-t/\tau}. \tag{8}$$

The probability that an electron has not been involved in a collision in a period t and that it is involved in a collision between times t and $t + \delta t$ is $p(t) \times dt/\tau$ or $P(t)dt$, where

$$P(t) = \frac{1}{\tau}e^{-t/\tau}. \tag{9}$$

The probability density $P(t)$, called the *exponential distribution*, may be used to calculate the *mean time between collisions* (or flight time):

$$\langle t \rangle = \int_0^\infty t P(t) \, dt = \tau. \tag{10}$$

Thus, the relaxation time is just the **mean flight time** of the electrons. Similarly, it is possible to calculate $\langle t^2 \rangle$

$$\langle t^2 \rangle = \int_0^\infty t^2 P(t) \, dt = 2\tau^2. \tag{11}$$

From (10), τ is the mean time between two collisions; thus, the number of collisions per unit time and per unit volume is

$$n_{coll} = N/\tau. \tag{12}$$

But τ is also the mean time (averaged over all the electrons) between an arbitrary time (for example, $t = 0$) and the next collision (see (7)). It is also the mean time (averaged over all the electrons) between the last collision and this arbitrary time (see (8)). It follows that, at an arbitrary given instant (for example, $t = 0$), the *mean time* (averaged over all the electrons) *between the last collision and the next is 2τ.*

This result does not contradict (10). In fact, the probability that at a given instant, the flight time of a given electron is t (to within δt) is proportional to t. More precisely, the distribution of the flight times of all the electrons *at this instant* is $t/\tau P(t)$, or $t/\tau^2 \exp(-t/\tau)$. This is no longer an exponential distribution and the mean flight time for all the electrons is

$$\frac{1}{\tau^2} \int_0^\infty t^2 e^{-t/\tau} \, dt \tag{13}$$

which is equal to 2τ (see the integral (11)).

This point is illustrated in Figure 7 which shows sequences of collisions of 30 electrons. The collisions are shown by vertical dashes. The flight times obey an exponential law with relaxation time $\tau = 5$. It is clear from the figure that the *ongoing* flight times at a given time t_0 (bold lines) do not obey an exponential law. The mean of these flight times (at $t = t_0$) for the figure is $9.8 \simeq 2\tau$.

It is actually this assumption of an exponential distribution which justifies the calculation of formula (6). In fact, the average velocity $\overline{v}_e$ resulting from the acceleration $\overline{\gamma} = -e\overline{\mathcal{E}}/m$ is, at all times, equal to $\frac{1}{2}\overline{\gamma} \times 2\tau$ (see (13)) or $-e\overline{\mathcal{E}}/m\tau$.

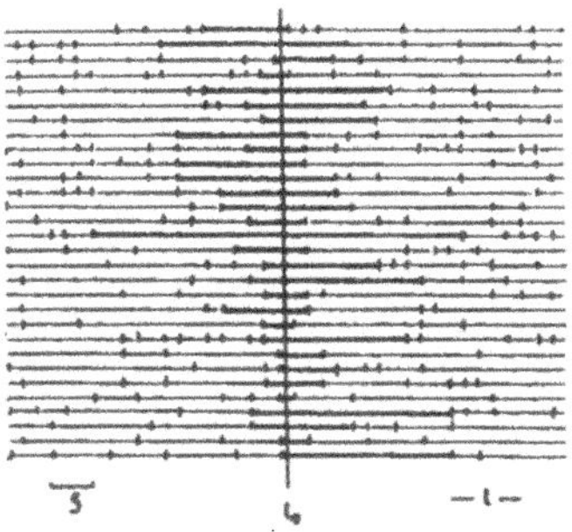

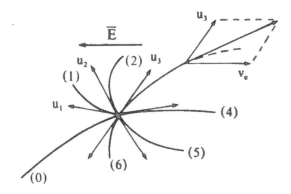

Figure 7 Electron flight times. Along the time axis, we have marked the instants of the collisions amongst some 30 electrons (small vertical dashes). The flight times were taken randomly in an exponential distribution with mean time $\tau = 5$. At time t_0 the *ongoing* flight times are shown by bold lines. The average of these *ongoing* flight times here is equal to 9.8 ($\simeq 2\tau$).

(ii) The collisions must **return the drift velocity to zero.**

(iii) Collisions **must not favour any direction**; in other words, the electron must exit the collision with a velocity in an arbitrary direction. This is the price one has to pay to ensure that the velocity $\bar{u}$ of the electrons on emerging from collisions (the *velocity of emergence*) does not contribute to the drift velocity $\bar{v}_e$. In Figure 8, a collision of an incident electron (0) gives rise to various equiprobable trajectories (1), (2), ... differing in their velocities of emergence $\bar{u}_1$, $\bar{u}_2$ ... (their moduli are assumed to be equal).

(iv) Collisions must occur with **a probability which is independent of both the position and the velocity** of the electron. In other words, the relaxation time (the inverse of which is the probability of the occurrence of a collision in unit time) must characterise *the metal* (its nature, its temperature, its structure ...), but not the dynamic state of the electron.

Figure 8 After the collision involving the electron (0), the trajectories of emergence (1), (2) ... are assumed to be equiprobable.

2 – Ohm's law

Let us return to the expression (6). Together with (5), this enables us to see that the establishment of the electric field $\overline{\mathcal{E}}$ is accompanied by an electric current with density vector $\overline{i}$ such that

$$\overline{i} = \frac{Ne^2\tau}{m}\overline{\mathcal{E}}. \tag{14}$$

Here, we have rediscovered Ohm's law, which says that the current is proportional to the electric field. The constant of proportionality is the **electrical conductivity** σ:

$$\boxed{\sigma = \frac{Ne^2\tau}{m}} \tag{15}$$

which is a scalar quantity in this instance[2]. Its inverse $\rho = 1/\sigma$ is the **resistivity**.

We may use (15) to obtain an experimental order of magnitude for the time τ in a metal such as aluminium:

$$\tau = \frac{m}{\rho Ne^2}$$

At standard temperature, ρ is of the order of 2×10^{-8} Ωm and we have $N = 18 \times 10^{28}$ m^{-3}. Whence

$$\tau \simeq \frac{9.1 \times 10^{-31}}{2 \times 10^{-8} \times 18 \times 10^{28}(1.6 \times 10^{-19})^2} \simeq 10^{-14} \text{ s.}$$

At very low temperatures (for example, $T = 4.2$ K), and for very pure aluminium, the previous value of ρ may have to be divided by a factor 10^4 (see Figure 2). Under these conditions, we have $\tau \simeq 10^{-10}$ s.

Such considerable variations of ρ for the same metal show clearly that the initial idea of simple collisions with ions is not sufficient to describe relaxation. In fact, correct determination of the relaxation time (or rather of the different relaxation times, since there may be several superimposed mechanisms for a given metal: *electron–crystal, electron–defect, electron–foreign atom relaxations* and, of course, to a lesser extent, *electron–electron relaxation*) is still a non-trivial objective of research in solid state physics.

3 – Joule's law

The assumption (p. 19) that the drift energy $\frac{1}{2}mv_e^2$ is lost by the electron at the time of the collision implies that it is transferred to the solid: this is the **Joule effect**.

For a flight time prior to collision equal to t, this energy is:

$$\Delta W = \frac{1}{2m}(e\mathcal{E}t)^2 \qquad \text{see (6),}$$

which has mean value (see (11)):

$$\langle \Delta W \rangle = \frac{1}{m}(e\mathcal{E}\tau)^2.$$

---

[2] It loses this quality later (see Annex 3, p. 261).

The number of collisions per unit volume per unit time is N/τ (see (12)) and the energy transferred is

$$w = \frac{N}{m}(e\mathcal{E})^2\tau \quad \text{(see (15))}$$
$$w = \sigma\mathcal{E}^2 = i^2/\sigma = \rho i^2 \quad (i: \text{current density; see (5))}.$$

If the metal consists of a cylinder of length ℓ and cross-section s lying in the field $\mathcal{E}$ parallel to its length, it is traversed by a current $I = is$ and the energy released per unit time is

$$W = \ell sw = \rho\ell si^2 \quad \text{or}$$
$$W = RI^2$$

where $R = \rho\ell/s$ is the electrical *resistance* of the metal cylinder. Here, we have rediscovered *Joule's law*.

2.3 Thermal conductivity. Wiedemann–Franz law

The phenomenological character of the previous model – and notably the introduction of the *ad hoc* parameter τ – will not have escaped the reader. It is tempting to eliminate this irritating parameter from two physical quantities which both depend on it. In this respect, the Wiedemann–Franz law (W.F.) pointed Drude in the right direction (look again at point v), p. 12). The parallelism between electrical and thermal conductivity which appears in W.F. strongly suggests the idea that the *agent carrying the electrical current* is also the *heat-carrying agent*. In the language of Drude, the thermal conductivity would then be that of a free-electron gas, treated in the framework of the kinetic theory of gases.

In this theory, the conductivity of a gas with specific heat for a constant volume C_v is given by:

$$\boxed{\kappa = \frac{1}{3}C_v\underline{v}\Lambda} \tag{16}$$

where $\underline{v}$ is the root mean square velocity $\left(\underline{v} = \sqrt{\langle v^2\rangle}\right)$ and Λ is the **mean free path** of the molecules. This classical formula tells us that the transportation of energy from a hot wall P_1 to a cold wall P_0 (Figure 1) is all the *more efficient* when the molecules transport *more energy* (term C_v), with a *greater velocity* (term $\underline{v}$) over a *larger distance* without collisions (term Λ). Initially, for simplicity, we assimilate $\underline{v}$ in the average velocity $\bar{v}$ and recall (see (Stat. Ph.)) that we have $\frac{1}{2}m\langle v^2\rangle = \frac{3}{2}kT$ and that for a monatomic gas, $C_v = \frac{3}{2}Nk$. Formula (16) becomes:

$$\kappa = \frac{3}{2m}Nk^2T\frac{\Lambda}{\bar{v}}. \tag{17}$$

We now apply this brutally to the free-electron gas. If we assume that in the thermal conductivity experiment (see Figure 1) the electrons are involved in the same collisions (in the sense of Section 2.2.1, p. 17) as in an experiment on electrical conductivity[3], *in*

---

[3] In both cases, the external perturbation (thermal gradient or electric field) is considered to be weak.

other words, that the relaxation time τ *is the same in both cases,* then we simply have $\Lambda = v \cdot \tau$ and (17) becomes

$$\kappa = \frac{3}{2m} N k^2 T \tau.$$

The sought-for elimination of τ from κ and σ (formula (15)) is then immediate and gives:

$$\boxed{\frac{\kappa}{\sigma} = \frac{3}{2} \left(\frac{k}{e} \right)^2 T.} \tag{18}$$

This success is notable since the experimental law of Wiedemann–Franz (see (2)) is rediscovered and appears to be understood. The theoretical constant $\frac{3}{2}(k/e)^2$ is equal to 1.2×10^{-8} W.Ω K^{-2}, which is smaller than, but of the same order of magnitude as, the experimental values of $L = \kappa/\sigma T$ (see Table 1, p. 13).

We note that in his original article, Drude wrote $\sigma = Ne^2\tau/2m$ (see (15)), whence $\kappa/\sigma = 3(\kappa/e)^2 T$, which accords better with experiments than (18). The error came from an erroneous evaluation of the velocity $\bar{v}_e$ (see (6)), which was divided by 2 under the pretext that the average velocity during a flight of duration τ was $-e\mathcal{E}/m\tau/2$.

2.4 Optical properties of the electron gas

1 – Plasma oscillations

We now study the propagation of electromagnetic radiation in Drude's electron gas (free electrons with density N per unit volume) and the resulting oscillations of the plasma of free electrons.

For this, we need to solve the Maxwell equations, namely, here:

$$\boxed{\operatorname{div}\overline{\mathcal{E}} = 0 \quad \operatorname{div}\overline{B} = 0 \quad \operatorname{curl}\overline{\mathcal{E}} = -\frac{1}{K}\frac{\partial \overline{B}}{\partial t} \quad \operatorname{curl}\overline{B} = \frac{\mu_0}{K}\overline{j} + \frac{\varepsilon_0\mu_0}{K}\frac{\partial \overline{\mathcal{E}}}{\partial t}} \tag{19}$$

where $K^2 = c^2\varepsilon_0\mu_0$ (in MKS, $K = 1$; in CGS, $K = c$, $4\pi\varepsilon_0 = 1$ and $\mu_0 = 4\pi$). $\overline{j}$ is the current vector. We calculated it previously for the static case ($\overline{\mathcal{E}} = $ constant) and we now wish to evaluate it for an oscillating electric field.

Assuming the relaxation time τ, the equations of motion of an arbitrary electron with coordinates x, y, z are of the type:

$$m\ddot{x} = -m\dot{x}/\tau - e\mathcal{E}_x \tag{20}$$

where the first term on the right-hand side describes the friction due to the relaxation (see (6)).

The reader will have noticed that in writing (20), we neglect the Laplace force $-e\bar{v} \wedge \overline{B}$ in comparison with the electrostatic force $-e\overline{\mathcal{E}}$. This approximation is reasonable since, we recall that for an electromagnetic wave $(\overline{\mathcal{E}}_0, \overline{B}_0)$ propagating in the vacuum we have $(\mathcal{E}_0/B_0) = c$. Thus, the two previous forces are in the ratio v/c. But, for a current $j = 1$ A mm^{-2}, we have $v_e = J/Ne \simeq 10^6/(5 \times 10^{28}) \times (1.6 \times 10^{-19}) \simeq 10^{-4}$m s$^{-1} \simeq 0.1$ mm s^{-1}, a velocity which is completely negligible in comparison with c (v_e is the drift velocity, see (5)).

We have also neglected the movement of the ions, the presence of which enabled us to write $\rho = 0$ (ρ is the charge density) in the first Maxwell equation.

We look for a stationary solution

$$x = \mathrm{Re}\,(x_\omega e^{-i\omega t})$$

resulting from a monochromatic excitation:

$$\mathcal{E}_x = \mathrm{Re}\,(\mathcal{E}_{\omega x} e^{-i\omega t}).$$

Equation (20) then becomes

$$-m\omega^2 x_\omega = \frac{i\omega m x_\omega}{\tau} - e\mathcal{E}_{\omega x} \qquad \text{or}$$

$$x_\omega = \frac{e}{m}\,\frac{1}{\omega^2 + (i\omega/\tau)}\,\mathcal{E}_{\omega x}.$$

Similarly, the current $j_x = -Ne\dot{x}$ may be written as

$$j_x = \mathrm{Re}\,(j_{\omega x} e^{-i\omega t}) \qquad \text{where}$$

$$j_{\omega x} = i\omega N e x_\omega$$

$$= \frac{Ne^2\tau}{m}\,\frac{1}{1 - i\omega\tau}\,\mathcal{E}_{\omega x}.$$

This formula defines a **complex conductivity** σ_ω such that

$$\overline{j}_\omega = \sigma_\omega \overline{\mathcal{E}}_\omega \qquad \text{where} \tag{21}$$

$$\boxed{\sigma_\omega = \sigma_0(1 - i\omega\tau)^{-1}} \tag{22}$$

where σ_0 is the conductivity at frequency zero given earlier (15).

The monochromatic solutions of Maxwell's equations (19) are obtained by setting

$$\overline{\mathcal{E}} = \mathrm{Re}\,(\overline{\mathcal{E}}_\omega e^{-i\omega t}) \quad \overline{B} = \mathrm{Re}\,(\overline{B}_\omega e^{-i\omega t}) \quad \text{and} \quad \overline{j} = \mathrm{Re}\,(\overline{j}_\omega e^{-i\omega t}).$$

Eliminating $\overline{B}_\omega$, we find

$$\overline{\Delta}\,\overline{\mathcal{E}}_\omega = -i\omega(\mu_0 \overline{j}_\omega - i\omega\varepsilon_0\mu_0\overline{\mathcal{E}}_\omega)$$

or, using (21) and (22), and taking account of the fact that $\mathrm{div}\,\overline{\mathcal{E}} = 0$:

$$\overline{\Delta}\,\overline{\mathcal{E}}_\omega = -\frac{\omega^2}{c^2}\left[1 + \frac{i}{\varepsilon_0}\,\frac{Ne^2\tau}{m\omega}\left(\frac{1}{1 - i\omega\tau}\right)\right]\overline{\mathcal{E}}_\omega$$

$$= -\frac{\omega^2}{c^2}\varepsilon_\omega\overline{\mathcal{E}}_\omega. \tag{23}$$

Here we recognise a classical *equation for propagation* in a medium with a **complex dielectric constant** equal to the contents of the square brackets: $\varepsilon_\omega = [\ ]$.

If we have:

$$\omega\tau \gg 1, \tag{24}$$

the dielectric constant may simply be written as

$$\varepsilon(\omega) = 1 - \frac{\omega_p^2}{\omega^2}$$

where (in CGS)

$$\omega_p^2 = \frac{4\pi N e^2}{m}. \tag{25}$$

The corresponding frequency $\nu_p = \omega_p/2\pi$ is called the **plasmon frequency**. According as the frequency of the electromagnetic wave is greater or less than ν_p, the dielectric constant ε (which is real provided the relaxation time τ satisfies condition (24)) is positive or negative and the solutions of (23) propagate or (conversely) are damped in the free-electron gas. Thus, we expect metals to be *opaque to light for wavelengths greater than λ_p (= c/ν_p) and transparent for $\lambda < \lambda_p$.*

In fact, this transparency appears in metals for wavelengths smaller than λ_0 (experimental), which will be compared in Table 2 with the values which may be calculated from (25), namely $\lambda_p = (2\pi c/e)(m/4\pi N)^{1/2}$ for a number of metals.

Table 2   Plasmon wavelengths λ_p for a number of metals compared with the wavelengths λ_0 below which the latter become transparent.

	λ_p (nm)	λ_0 (nm)
Li	150	$\simeq 200$
Na	200	$\simeq 210$
K	280	$\simeq 310$
Ag	140	$\simeq 160$

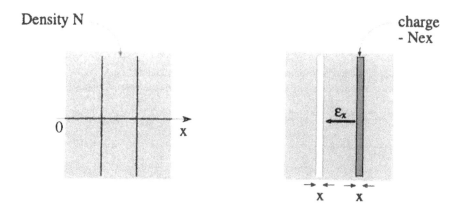

Density N   charge
                                                                              - Nex

Figure 9   Illustration of the oscillation of plasmons in a free-electron gas.

2 – Physical meaning of the plasmon frequency ν_p.

Suppose we have the following fluctuation in a metal: the free electrons situated in a planar section with unit area parallel to yOz are displaced globally by a quantity x parallel to Ox (Figure 9). The charges $+Nex$ and $-Nex$ created in this way generate an electric field $\mathcal{E}_x = 4\pi Nex$ which subjects all the electrons in the section to a return force $-e\mathcal{E}_x = -4\pi N e^2 x$. Thus, these electrons are subject to the oscillatory movement of frequency

$$\nu_p = \frac{1}{2\pi} \sqrt{\frac{4\pi N e^2}{m}} \qquad (26)$$

implied by the equation

$$m \frac{d^2 x}{dt^2} = -4\pi N e^2 x.$$

This frequency is just the plasmon frequency introduced earlier (25). It is characteristic of the metal considered *via* the quantity N: thus, the electrons are subject to oscillations at the frequency ν_p even for fluctuations having a density of a form less stylised than that of Figure 9.

This **plasma oscillation** (a plasma is a medium containing equal concentrations of positive and negative charges, at least some of which are mobile) constitutes a collective excitation (longitudinal in this case) of the electron gas created by a local fluctuation. It is harmonic and the *plasmon* is a quantised plasma oscillation with energy $\hbar\omega_p$. The plasmon excitation may be observed in an electron microscope, a device in which the monokinetic electrons (for example, with energy 100 keV) crossing thin samples ($\simeq$ 100 nm) lose their energy. Measurement of these energy losses actually reveals a discrete structure: they are integer multiples of an elementary energy which we associate with the excitation of a *single* plasmon (see Figure 10). Plasmon energies measured in this way are recorded in Table 3, in which they are compared with those calculated using formula (26).

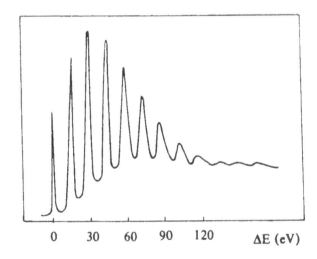

Figure 10 **Plasmon excitations.** Spectra of the energy losses ΔE (in eV) of electrons with energy 50 keV crossing a sheet of aluminium of thickness 240 nm. The excitations of plasmons with energy 14.37 eV are observed. After P Trebbia and C Colliex.

Table 3 **Measured and calculated (formula (26)) plasmon energies for various metals (in eV).**

	Na	Mg	Al
$\hbar\omega_p$ (measured)	5.71	10.6	14.4
$\hbar\omega_p$ (calculated)	5.95	10.9	15.8

Exercise   Between the altitudes of $\simeq 200$ and 400 km, our atmosphere is a plasma consisting of ionised gas (ionised by solar radiation) and free electrons: this is the *ionosphere*. The free-electron density (which varies, notably between day and night) is of the order of 10^6 cm^{-3}. Show that the ionosphere is *transparent* to *television* waves (frequency $\nu \simeq 100$ MHz) and *opaque* to *radio* waves ($\nu = 100$–1000 kHz), which are reflected like visible light on a metal.

Show also that, given that the **solar** atmosphere decreases as the distance from the surface of the Sun increases, in detecting centimetre or decimetre (*resp.* metre or decametre) EM waves, we observe internal regions or the *chromosphere* (*resp.* external regions or the *corona*) of this atmosphere.

In all these cases (metals, ionosphere, solar atmosphere) the ionised medium which constitutes the plasma behaves like a *high-pass filter*, the cut-off frequency of which depends solely on the density (variation in $N^{1/2}$; see (25)).

3 – Absorption and reflection

In studying the optical properties of the electron gas, we neglected the effects due to the polarisability of the ions. This results from the deformation of the orbitals of electrons of deep layers of each atom when they are subjected to the electric field. Because of the inertia of the electrons, the polarisation induced by an oscillating electric field generally depends on the frequency. We may write:

$$\overline{P}_\omega(t) = \mathrm{Re}\left(\overline{P}_\omega e^{-i\omega t}\right) \quad \text{and} \quad \overline{P}_\omega = \chi_\omega \overline{E}_\omega.$$

where χ_ω is the complex *electrical susceptibility* analogous to the complex conductivity (see (22)). We introduce the *complex dielectric constant*:

$$\varepsilon_\omega = \varepsilon_0(1 + \chi_\omega).$$

Taking these polarisable centres and the currents due to the free electrons into account, the Maxwell equations (19), with ε_0 replaced by ε_ω give:

$$\overline{\Delta}\,\overline{E}_\omega = -\mu_0 i\omega(\sigma_\omega - i\omega\varepsilon_\omega)\overline{E}_\omega.$$

In searching for a solution of the type $\overline{E}_\omega = \overline{E}_{\omega_0} \exp(i\overline{k} \cdot \overline{r})$, we obtain the condition

$$k^2 = i\omega\mu_0\sigma + \omega^2\mu_0\varepsilon_\omega$$

or

$$k = \frac{\omega}{c}\left(\frac{\varepsilon_\omega}{\varepsilon_0} + i\frac{\sigma}{\varepsilon_0\omega}\right)^{1/2}.$$

By analogy with insulators for which we have $\sigma = 0$, whence

$$k = \frac{\omega}{c}\left(\frac{\varepsilon}{\varepsilon_0}\right)^{1/2} = \frac{\omega}{c}n,$$

we define in a general fashion, a *complex refractive index*

$$n = n_r + in_i = \left(\frac{\varepsilon}{\varepsilon_0} + i\frac{\sigma}{\varepsilon_0\omega}\right)^{1/2}$$

which enables us to describe the optical properties of the medium in a convenient way.

If we fix the direction Oz along $\overline{k}$, the expression for the wave takes the form:

$$\overline{\mathcal{E}} = \overline{\mathcal{E}}_0 \exp\left(i\left(\frac{\omega}{c}(n_r + in_i)z - \omega t\right)\right)$$

or

$$\overline{\mathcal{E}} = \overline{\mathcal{E}}_0 \exp\left(i\omega\left(\frac{n_r z}{c} - t\right)\right) \times \exp\left(-\frac{n_i \omega z}{c}\right)$$

where the first exponential describes the *propagation* of a wave with velocity c/n_r and the second describes its *damping*. The latter, which is due to **absorption** and **reflection**, is characterised by the *damping coefficient* η which is the inverse of the damping in terms of energy ($\propto \overline{\mathcal{E}}^2$):

$$\eta = \frac{2n_i \omega}{c}.$$

η (and thus n_i) may be determined experimentally by measuring the *transmission* of waves (for example, light waves) in very thin sheets of metal.

A more detailed study of the wave transmission enables us to calculate the *reflection coefficient* (which, for normal incidence, is the fraction of the incident energy reflected). We have:

$$R = \left|\frac{1 - n}{1 + n}\right|^2 = \frac{(n_r - 1)^2 + n_i^2}{(n_r + 1)^2 + n_i^2}.$$

Taken to its end, the calculation of R (Drude, 1902; Zener, 1933) provides for a correct description of the broad lines of the optical behaviour of a free-electron gas: as the frequency ν increases, such a gas passes successively from a mode of predominant *absorption* to a mode of *reflection*, before entering the region of *transparency* at $\nu = \nu_p$ (see above).

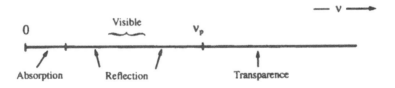

Here, we rediscover the patent fact that, for metals, reflection is the dominant mode for frequencies in the visible area.

4 – Example of an industrial application of the above properties

One important objective of energy-saving programmes, is to reduce the consumption expended on heating houses and buildings. In fact, in a country such as France, this takes up a quarter of the total energy (or $\simeq$ 50 MTEP – million tonnes equivalent petroleum – out of a total of 206 MTEP in 1990). Important work is being carried out to produce 'selective' windows. The idea is simple: it involves considering glazing which, in Winter, in reasonably sunny countries (such as the southern half of France) allows solar radiation to enter freely into buildings but does not allow internal radiation to escape, and if possible reflects it. The internal radiation is found at wavelengths of around 10 μm (*infrared*), while solar radiation is largely biased towards the visible (0.3–0.8 μm) or even the *ultraviolet*. Thus, windows coated with a thin film of a solid with an appropriate plasmon frequency ν_p (or wavelength λ_p) may fulfil this role, if the coating is *transparent* in the visible and *a fortiori* in the ultraviolet and *reflecting* in the infrared.

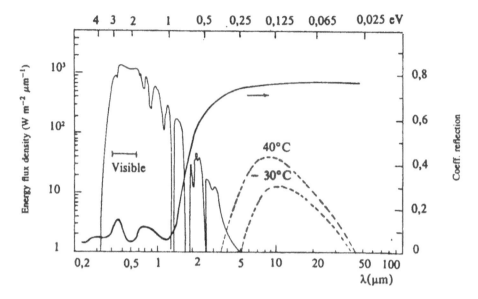

Figure 11   Reflection coefficient of a silver-coated window (composite film TiO_2–Ag–TiO_2) as a function of the wavelength λ or the radiation frequency ν (bold line). For comparison, the figure includes a standard spectrum of solar radiation together with the spectra of black-body radiation at 40° and −30° C. After C M Lambert *Sol. Energy Mater.* **6**, 1 (1981).

A solid such as SnO_2 meets these requirements well. It is a semiconductor which at room temperature contains $N \simeq 2 \times 10^{20}$ electrons cm^{-3}, whence λ_p ($= 2\pi c/\omega_p$, see (25)) $\simeq 2$ μm. It is transparent in a thin film ($\simeq 100$ nm) in the visible and opaque to infrared radiation. Good results are also obtained with a metal such as *silver*. For the thicknesses at which this is transparent in the visible ($\simeq 20$ nm), formula (25) no longer applies as such and the cut-off wavelength λ_0 (see Table 2) is appreciably greater ($\lambda_0 \simeq 2$ μm). Figure 11 shows the optical characteristics of a window covered with a film of silver. Its reflection coefficient largely satisfies the problem in hand: it is equal to $\simeq 0.8$ in the infrared, but is less than or equal to $\simeq 0.1$ below λ_p, i.e. in the area of sunlight. Installation of such silver-coated windows could decrease energy consumption by 10 to 40%, according to the surface area glazed, the hours of sunshine . . . .

In reality, in the example of Figure 11 the silver film (20 nm) is sandwiched between two films of TiO_2 (each 20 nm), one of which ensures the adhesion to the window glass and the other protects the silver against atmospheric corrosion (oxygen, sulphur . . .).

It is easy to imagine the technical difficulties involved in the industrial manufacturing of large surface areas of such films, with a thickness which must be constant to within 2 nm (for fear of causing irisations which make the product unsellable) and which must be *stable* for decades under all weather conditions (the building guarantee itself already covers 10 years). If we add that once the *physical problem* has been solved in the laboratory and the *manufacturing problems* have been overcome in the factory there still remains the important question of *sales* (which involves the question of *costs* and that of the *tastes* of the purchasers and, in particular, in this case of the architects), the reader will appreciate the dimension and diversity of the parameters to be mastered for a successful launching of a new material!

In any case, it is clear that, even for a traditional industry outside the high-technology sector which is given over to mass production of low-price materials, physics, and, more generally, research activities (experimental or theoretical) constitute an ingredient which, although certainly not sufficient, is *increasingly necessary*, above all at the initial stage, if one wishes to design and develop high-performance materials.

3. WHERE DRUDE'S MODEL GOES WRONG

The successes of Section II.2 appear remarkable. Notably, they provide us with a solid anchor for the concepts of *free electrons* and *relaxation time*. They speak strongly in favour of a *thermal conductivity* of metals which is *principally electronic*.

However, **they are based on a false postulate** (see point **iii**), p. 16). It is well known (see (Stat. Ph.)) that electrons obey the Fermi–Dirac rather than the Maxwell–Boltzmann statistics. We shall return to this point (pp. 42, 53) and we shall see that changing these statistics without modifying the general ideas of Section II.2.3 introduces two corrections into the calculation of the thermal conductivity which, miraculously, cancel each other out. This will lead us back to a scarcely modified version of formula (18).

Such miracles are rare, and we should not be surprised that in other areas, Drude's model has rapidly reached a dead end.

This is the case, for example, for the *specific heat* of electrons. We recalled (p. 21) that this should be equal to a constant $\frac{3}{2}Nk$. This is in flagrant disagreement with experiments. The specific heat of solids (and that of metals in particular) decreases rapidly as the temperature decreases (point **vii**), p. 13) until it reaches zero at $T = 0$. In addition, the theory of Einstein (1907), in which the specific heat is attributed to *atomic* (or ionic) *vibrations* and all contributions from electrons are dispensed with, very properly takes account of experimental results including, notably, Dulong and Petit's law. Thus, the free-electron gas must play a minimum role in the total specific heat.

This difficulty, along with others, such as the silence of Drude's model as far as magnetic properties are concerned, (point **vi**), p. 13) point to the limits of this model and the need for more adequate postulates. It is of course quantum mechanics which will provide these.

CHAPTER III

SOMMERFELD OR 'GOOD' METAL

Twenty Years After
Alexandre Dumas

1. FREE ELECTRONS IN ONE DIMENSION

1.1 Reminders

In the first instance, we discuss the case of electrons constrained to move along a line.

An electron is located on a segment of length L on which it is free to 'slide' but which it cannot leave. It is assumed that the *potential* is *constant* (and equal to zero) and that it abruptly takes on a high value (for example, infinity) at the extremities. This is the one-dimensional, infinitely deep potential well (see Q.M.), which is the simplest model for representing an electron tied to a long molecule and moving freely along this molecule.

It is also the first picture which one may have of a 'free' electron in **certain** so-called *one-dimensional* **organic crystal conductors**. In these crystals, cyclic molecules are stacked on top of each other (like plates) and each of these 'piles' constitutes a molecular line. The agglomeration of these lines forms a crystal in which each line has a length equal to that of the crystal (for example, several millimetres). Electrons may move along these lines with, in certain cases, a very small probability of jumping from one line to a neighbouring line. These electrons give rise to a notable electrical conductivity in the direction of the lines; thus, at room temperature the resistivity of the compound TTF–TCNQ[1] is $\sim 10^3$ $\mu\Omega$ cm along the lines (around 10^3 times that of copper). It is $\simeq 120$ times greater transversally to the lines than longitudinally.

The **classical** equivalent of this movement is that of a bead constrained to move without

---

[1] A compound in which flat molecules of tetrathiofulvalene (with 4 atoms of hydrogen, 6 of carbon and 4 of sulphur) with formula are stacked together, forming lines (or chains) parallel to chains of molecules of tetracyanoquinodimethane (6C and 4N) with formula

The TCNQ molecules give up electrons to the TTF molecules. It is these electrons which move along the TTF chains.

friction along a rod and striking a buffer at the two extremities (Figure 1). If the potential energy (constant) is taken to be equal to zero, the energy of the bead is purely kinetic. It may take all values $E = p^2/2m$ (where $p = mv$). The velocity v is uniform and changes sign with each rebound. The *positional probability density* $g(x)$ ($g(x)dx$ is the probability of finding the bead between x and $x + dx$) is constant along the x-axis and equal to $1/L$. By striking each buffer with frequency $\nu = v/2L$, the bead produces a *force* $\nu \Delta p = v/2L \times 2mv = 2E/L$ at the buffer, in the same way that molecules striking a wall create a *pressure* there.

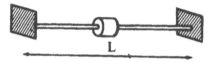

Figure 1   Bead constrained to slide without friction along a horizontal rod of length L.

The **quantum** solution of this problem is trivial (see Q.M.). The Schrödinger equation

$$H\Psi = E\Psi \qquad \text{where}$$

$$H = -\frac{-\hbar^2}{2m}\Delta + V \quad \text{or in this case} \tag{1}$$

$$H = -\frac{\hbar^2}{2m}\frac{d^2}{dx^2}$$

can be solved immediately. There may be two different expressions for the solution, according to the boundary conditions adopted.

i)   **Either** one places importance on the **extremal** effects (at $x = 0$ and $x = L$) and constrains the wavefunction, which is continuous and zero in the exterior, to be zero at the extremities:

$$\Psi(0) = \Psi(L) = 0, \tag{2}$$

which leads to the following solutions of equation (1):

$$\Psi_n(x,t) = \sqrt{\frac{2}{L}}\sin n\frac{\pi}{L}x\, e^{-i\frac{E_n t}{\hbar}} \quad 0 < x < L \quad \Psi_n(x,t) = 0 \quad x < 0 \quad \text{and} \quad x > L$$

$$\tag{3}$$

where n is a positive integer and the eigenvalue of the **energy** E_n is

$$E_n = E_0 n^2 \quad \text{where} \quad E_0 = \frac{\hbar^2}{2m}\frac{\pi^2}{L^2}. \tag{4}$$

The Ψ_n form an orthonormal basis ($\langle\Psi_i|\Psi_j\rangle = \int \Psi_i^*\Psi_j\, dx = \delta_{ij}$) of *stationary states*. The positional probability densities $P_n(x,t) = \Psi_n^*\Psi_n$ are time independent. Unlike the classical case, they are not spatially uniform.

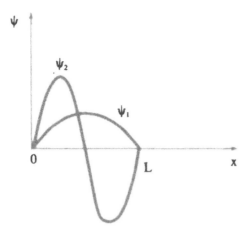

Figure 2 Eigenfunctions ψ_1 and ψ_2 of equation (1).

By way of example, Figure 2 shows graphs of the space functions $\psi(x) =$ Re Ψ for $n = 1$ and $n = 2$ against space.

Any state (not an eigenstate of H) of the electron is a linear combination of the preceding stationary states:

$$\Psi(x, t) = \sum_n \lambda_n \Psi_n(x, t) \quad \text{where} \quad \sum_n \lambda_n^* \lambda_n = 1.$$

In such a state, the density $P(x, t)$ varies with time.

Exercise Calculate the function $P(x, t)$ for the state

$$\Psi(x, t) = \frac{1}{\sqrt{2}} \left(\Psi_1(x, t) + \Psi_2(x, t) \right).$$

Observe that P varies with t and that the position of its maximum oscillates with frequency ω, to be determined. Draw the graph of P at the four successive instants $t = 0, \pi/2\omega, \pi/\omega$ and $3\pi/2\omega$.

ii) **Or** one places importance on the idea of **propagation** of the electron.

As we have seen, the solutions (3) represent stationary wave packets. These packets are actually superpositions of plane waves $\exp(ikx)$ stretching throughout space with 'weights' $f(k)dk$ such that $f(k)$ is the Fourier transform of $\psi(x)$. It is these superpositions which enable us to 'trap' the packet in the well. The function $f(k)$ is symmetrical: the weight of each wave $\exp(ikx)$ is equal to that of the wave $\exp(-ikx)$: there is *no propagation* with these solutions (3).

Exercise Calculate the function $f(k)$ corresponding to the wave packet $\psi_1(x)$.

To avoid this disadvantage, we may view the segment $(0, L)$ as part of a line of the same type, but infinite. On this infinite line, a general solution of (1) is simply:

$$\Psi(x, t) = e^{ikx}e^{-i(E/\hbar)t} \quad \text{where} \quad E = \frac{\hbar^2}{2m}k^2.$$ (5)

The solution (5) represents a plane wave which **propagates** in the direction of the vector $\bar{k}$. The price to be paid for this advantage is the loss of all information about the form of ψ at the extremities of the segment. This loss is clearly not serious if we neglect the influence of the extremities[2] or if we are interested, more particularly, in states with high n. In all cases, this amounts to assuming that the properties of the segment $(0, L)$, at least those which are linked to the presence of the free electron, are unchanged if we close the segment on itself in the form of a loop (Figure 3). If this is actually the case, we do not expect any discontinuity of the properties at the point of the solder and the corresponding boundary condition is:

$$\Psi(0) = \Psi(L) \qquad \Psi'(0) = \Psi'(L)$$ (6)

or, more generally, on the infinite line which enabled us to obtain (5):

$$\Psi(x) = \Psi(x + L) \qquad \Psi'(x) = \Psi'(x + L).$$

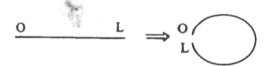

Figure 3   Closure of the line OL on itself.

Since it involves plane waves, this periodic condition or **Born–von Karman condition** (B.v.K.) reduces to

$$\Psi(x) = \Psi(x + L).$$ (6')

It quantises wavevectors and energies in the form

$$k_n = \frac{2\pi}{L}n \qquad E_n = \frac{\hbar^2}{2m}\left(\frac{2\pi}{L}\right)^2 n^2.$$ (7)

where n is a positive or negative integer.

The vectors $\bar{k}$ are the vectors joining an arbitrary origin Ω to regularly spaced points (or 'nodes') $M_1, M_2, \ldots$ with abscissae $n \times 2\pi/L$ on an axis parallel to Ox (Figure 4). They are proportional to the velocity $\bar{v}$ since the functions (5) are eigenfunctions of the momentum operator $\hat{p} = m\hat{v} = -i\hbar\partial/\partial x$:

$$\bar{v} = \frac{\hbar}{m}\bar{k}.$$ (7')

[2] We shall return to this point later (p. 155) when we study the free surfaces of a metal.

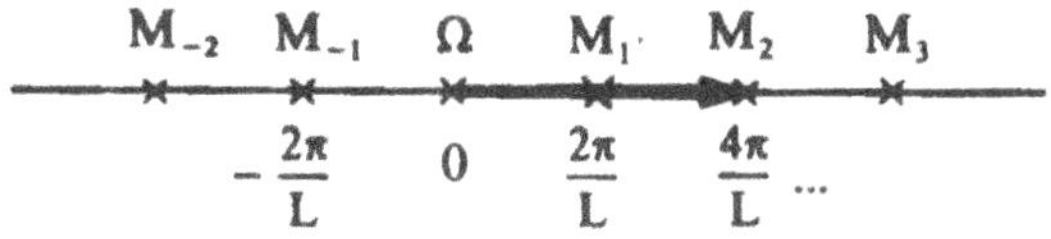

Figure 4 The wavevectors of (7) are the vectors $\bar{k}_n = \overline{\Omega M_n}$. Here, the vector $\bar{k}_2$ is shown.

If, finally, we normalise the planar functions (5) in the interior of the segment (O, L), we obtain the following orthonormal basis:

$$\Psi(x, t) = \frac{1}{\sqrt{L}} e^{ik_n x} e^{-i(E_n/\hbar)t}.$$ (8)

Any periodic wavefunction defined on the interior of the segment may be extended to a periodic function of period L, which in turn may be expanded in a Fourier series on the basis (8).

1.2 Density of states

The set of energy levels E_n (either (4) or (7)) forms a **band** of possible **energies** for the electron. The first levels (n small) are very close to one another. Thus, in the case of (4), for $L = 1$ mm $\simeq 2 \times 10^7$ a.u. (see Inset), the quantity E_0 amounts to $5/\left(4 \times 10^{14}\right)$ a.u. or 3×10^{-13} eV and we have $E_2 - E_1 = 3E_0 \simeq 10^{-12}$ eV. As n increases, these levels become more spaced out and their *density* decreases. Thus, we define the **density of states** to be the function $n(E)$ such that

$$n(E)\, dE$$ (9)

is the number of possible energy eigenstates (including possible degeneracies, *except those associated with spin*) lying between E and $E + dE$. Remaining with the formulation **i)** and taking into account (4), we immediately find that

$$n(E) = \frac{1}{\sqrt{E_0}} \frac{1}{\sqrt{E}}.$$ (10)

In the formulation **ii)**, p. 34, it is noticeable that the energy levels $(7) - 4E_0 n^2 = E_0(2n)^2$ instead of $E_0 n^2$ – are a factor of $1/2$ less numerous than in **i)**. But this factor is compensated for by the degeneracy of order two which now arises since the functions (8) ψ_n and ψ_{-n} have the same eigenvalue E_n. *Thus, the function $n(E)$ gives the same expression* (10) *in both cases.*

The density of states does not depend on which boundary conditions, i) or ii) are adopted.

The system of atomic units (a.u.)

This is the system based on the units

- m (mass of the electron at rest) = 1
- $|e|$ (charge of the electron) = 1
- $\hbar$ (Planck's constant/2π) =1
- $4\pi\varepsilon_0 = 1$

This system is not legal but it is convenient.

We recall that (see Q.M.), the radius of the first Bohr orbit in the hydrogen atom (0.53 Å) is equal to $\hbar^2/me^2$ and that the energy of the ground state (-13.6 eV) is equal to $-me^4/2\hbar^2$. We immediately deduce that:

- *unit length* in a.u. = 0.53 Å= 0.53×10^{-10} m
- *unit energy* in a.u. = 27.2 eV.

Moreover, writing $\hbar\omega = E$ and knowing that $\hbar = 1.05 \times 10^{-34}$ J.s., we have:

- *unit frequency* in a.u. = 4.13×10^{16} s^{-1}
- *unit time* in a.u. = 0.24×10^{-16} s
- *unit velocity* in a.u. = $0.53 \times 10^{-10} \times 4.13 \times 10^{16}$
  $$\simeq 2.2 \times 10^6 \text{ m s}^{-1}.$$

In particular, c (velocity of light) = 137 a.u.

1.3 Systems of N independent electrons

Suppose that we now have not *one* but N electrons on the segment (O, L). These electrons are assumed to be *free* on the line and *mutually independent*. More precisely, we shall initially neglect their Coulomb repulsion, but we shall not forget the Pauli principle. Finally, we assume that the segment is electrically *neutral*: the N electrons are compensated by N positive ionic charges assumed to be uniformly distributed along the segment.

Denoting the coordinates relative to the electrons [1], [2], ... by $x_1, x_2, \ldots$, the Schrödinger equation (1) becomes

$$H\Psi = -\frac{\hbar^2}{2m}\left(\frac{d^2}{dx_1^2} + \frac{d^2}{dx_2^2} + \ldots\right)\psi = E\Psi \tag{11}$$

the general solution of which is a product of functions for a *single* electron of type (3) or (8), according to the boundary conditions chosen. Thus, taking the condition $\Psi(O) = \Psi(L) = 0$, the **space function** is:

$$\psi(x_1 \ldots x_N) = \varphi_i(x_1)\varphi_j(x_2)\ldots\varphi_m(x_N) \tag{12}$$

where we have set

$$\varphi_n(x) = \sqrt{\frac{2}{L}} \sin n\frac{\pi}{L}x. \tag{13}$$

The solution (12) gives the following eigenvalue of the energy in (11):

$$E = E_i + E_j + \ldots + E_m.$$

This solution satisfies equation (11) but not the Pauli principle which excludes any solution which is not antisymmetric as far as the **exchange** of two electrons is concerned. Since this exchange concerns both spin variables s_i and positional variables x_i, we must introduce the **spin state** of the electron. Suppose that α and β denote the two possible spin states. In what follows, the notation $\varphi_i(x_1)\alpha(1)$ (tensor product) will describe a space state φ_i (for example, (13)) for the electron [1] associated with the spin state $|+\rangle$ (see Q.M.) with $\langle\alpha|\alpha\rangle = \langle\beta|\beta\rangle = 1$ and $\langle\alpha|\beta\rangle = 0$.

The introduction of the spin variables leads us to rewrite the solution (12) in the form

$$\Psi(x_1, 1, x_2, 2, \ldots, x_N, N) = \varphi_i(x_1)\gamma_i(1)\ldots\varphi_m(x_N)\gamma_m(N) \tag{14}$$

where γ denotes either α or β; or, in more abbreviated notation:

$$\Psi(x_1 \ldots N) = \psi_1(x_1\,1)\ldots\psi_N(x_N\,N), \tag{15}$$

where now $\psi_1(x_1\,1)$ denotes the tensor product $\varphi_i(x_1)\gamma_i(1)$.

Thus, for $N = 2$ one of the possible states is

$$\psi(x_1, 1, x_2, 2) = \varphi_3(x_1)\alpha(1)\varphi_1(x_2)\alpha(2). \tag{16}$$

In this state, the electrons [1] and [2] have, respectively, the functions φ_3 and φ_1 for space functions (see (13)) and are both in the spin state $|+\rangle$, which is represented (Figure 5) by a small upwards arrow: they are said to have *parallel spins*.

The energy eigenvalue is $E = E_3 + E_1$ $(= 10E_0)$.

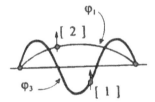

Figure 5 **Two electrons [1] and [2] in a potential well.** [1] lies on an orbital φ_3 with a spin state $|+\rangle$, [2] lies on an orbital φ_1 with a spin state $|+\rangle$.

Here again, states such as (14)–(16) do not satisfy the Pauli principle. It is easy to construct the following state by *linear combination* of degenerate states of type (15), in the form of a determinant (**Slater determinant**):

$$\Psi = \frac{1}{\sqrt{N!}} \begin{vmatrix} \psi_1(x_1, 1) & \psi_1(x_2, 2)\ldots & \psi_1(x_N, N) \\ \vdots & & \\ \psi_1(x_1, 1)\ldots & & \psi_N(x_N, N) \end{vmatrix} \tag{17}$$

This state is **antisymmetric** since exchanging two electrons i and j inverts two columns of the matrix and thus inverts its sign.

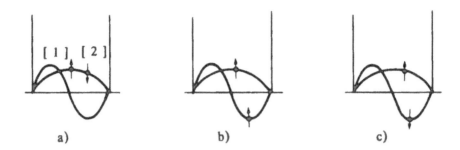

Figure 6 Two electrons [1] and [2] in space states φ_1 and φ_2, showing the various possibilities for the spin: (a) same space state; antiparallel spins; (b) different space states; parallel spins; (c) different space states; antiparallel spins.

Returning to the example of two electrons and considering the space functions φ_1 and φ_2, we see that the following states can be constructed:

$$\psi_0 = \frac{1}{\sqrt{2}} \begin{vmatrix} \varphi_1(x_1)\alpha(1) & \varphi_1(x_2)\alpha(2) \\ \varphi_1(x_1)\beta(1) & \varphi_1(x_2)\beta(2) \end{vmatrix} \tag{18}$$

$$\psi_\alpha = \frac{1}{\sqrt{2}} \begin{vmatrix} \varphi_1(x_1)\alpha(1) & \varphi_1(x_2)\alpha(2) \\ \varphi_2(x_1)\alpha(1) & \varphi_2(x_2)\alpha(2) \end{vmatrix} \quad \psi_\beta = \frac{1}{\sqrt{2}} \begin{vmatrix} \varphi_1(x_1)\beta(1) & \varphi_1(x_2)\beta(2) \\ \varphi_2(x_1)\beta(1) & \varphi_2(x_2)\beta(2) \end{vmatrix} \tag{19}$$

$$\psi_I = \frac{1}{\sqrt{2}} \begin{vmatrix} \varphi_1(x_1)\alpha(1) & \varphi_1(x_2)\alpha(2) \\ \varphi_2(x_1)\beta(1) & \varphi_2(x_2)\beta(2) \end{vmatrix} \tag{20}$$

$$\psi_{II} = \frac{1}{\sqrt{2}} \begin{vmatrix} \varphi_1(x_1)\beta(1) & \varphi_1(x_2)\beta(2) \\ \varphi_2(x_1)\alpha(1) & \varphi_2(x_2)\alpha(2) \end{vmatrix} \tag{21}$$

It is clear that $\psi_0 (= \varphi_1(x_1)\varphi_1(x_2)[\alpha(1)\beta(2) - \beta(1)\alpha(2)])$ is the ground state for the set of the two electrons. The eigenenergy is $E = 2E_0$. The total spin s is zero: this can be verified by checking that $S^2\psi_0$ is zero, where S^2 is the operator giving 'the square of the total spin', namely $S^2 = (S_1 + S_2)^2 = S_1^2 + S_2^2 + 2(S_{1z}S_{2z} + \ldots)$. We shall say that *the spins are antiparallel* (Figure 6(a)).

The four states ψ_α, ψ_β, ψ_I and ψ_{II} are the four degenerate states corresponding to the energy $E = E_1 + E_2 (= 5E_0)$. $\psi_\alpha = [\varphi_1(x_1)\varphi_2(x_2) - \varphi_2(x_1)\varphi_1(x_2)]\alpha(1)\alpha(2)$ represents a simple spin state: the two electrons are in the state α: they have *parallel spins* (Figure 6(b)). It is possible to check that the total spin is $s = 1$, whence $S^2\psi_\alpha = s(s+1)\psi_\alpha = 2\psi_\alpha$. In the state ψ_β, the two spins are also *parallel*.

The states ψ_I and ψ_{II} do not describe quite such simple cases, but, it can be shown that the linear combination

$$\psi_z = \frac{1}{\sqrt{2}}(\psi_I + \psi_{II})$$

is the product of an antisymmetric space function by the spin function $\alpha(1)\beta(2) + \beta(1)\alpha(2)$ which is *symmetric*, as $\alpha(1)\alpha(2)$ and $\beta(1)\beta(2)$ were. Thus, the states ψ_α, ψ_z and ψ_β form the thrice degenerate subspace relative to the spin $s = 1$: these are the three **triplet** states relative to the space function $\varphi_1\varphi_2$ (Figure 6(b)); whilst the combination

$$\psi_s = \frac{1}{\sqrt{2}}(\psi_I - \psi_{II})$$

with antisymmetric spin state $\alpha(1)\beta(2) - \beta(1)\alpha(2)$ describes the **singlet state** with spin zero, $s = 0$ for the same space function $\varphi_1\varphi_2$ (antiparallel spins; Figure 6(c)).

A linear combination such as (17) clearly implies that there is no question of attributing *one* electron to *one* state. There is *exchange* and the electron [1], for example, is as much 'in' state ψ_1 as 'in' state ψ_2 .... On the other hand, each space state is associated with a spin state, where the notation ψ_1 *denotes a tensor product* φ (space function) $\otimes \gamma$ (spin state) (for example, $\varphi_1 \alpha$ in example (18)). This remark enables us to associate an energy scheme with the Slater determinant. Thus, that of Figure 7 would correspond to a determinant with four lines:

$$\psi_1 \equiv \varphi_1\alpha, \quad \psi_2 \equiv \varphi_3\alpha, \quad \psi_3 \equiv \varphi_3\beta, \quad \psi_4 \equiv \varphi_5\alpha.$$

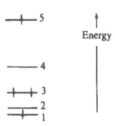

Figure 7 **Energy levels and spin states of four electrons of a linear chain.**

A *first consequence* of the Pauli principle is apparent in the determinant (17): a state constructed with two identical space and spin states (for example, φ_1 and α, respectively) cannot exist, since the corresponding Slater determinant, having two identical lines, would be zero.

A *second consequence* of this principle is the following: if two electrons are in the same spin state, *the probability of finding them at the same point in space is zero.* In fact, the same spin state for two electrons [i] and [j] implies that the spin state is symmetric in the permutation $i \longleftrightarrow j$. Thus, the *space state must be antisymmetric* in this permutation, which implies that the space function is zero for $x_i = x_j$. Each of the two electrons is thus surrounded by a region (**Pauli hole** or *exchange hole*) in which the presence of the other is excluded. It is clear that this hole will tend to decrease the Coulomb repulsion of the two electrons when we take this into account (see p. 49).

If, on the other hand, the two electrons are in different spin states, the probability that one is present is not affected by the presence of the other.

Thus, in the example of two electrons, the state ψ_0 ((18) and Figure 6(a)) may be written as:

$$\psi_0 = \frac{1}{\sqrt{2}}\varphi_1(x_1)\varphi_1(x_2)[\alpha(1)\beta(2) - \beta(1)\alpha(2)].$$

The square $\langle \psi_0 | \psi_0 \rangle$, equal to $\varphi_1^*(x_1)\varphi_1(x_1)\varphi_1^*(x_2)\varphi_1(x_2) = P(x_1x_2)$, is such that $P(x_1x_2)dx_1dx_2$ is the probability of finding the electron [1] at x_1 (up to dx_1) and the electron [2] at x_2 (up to dx_2) simultaneously. If we fix the position of the electron [1] at the point with abscissa x_0, the probability of finding electron [2] at the point x (to within dx) is

$$|\varphi_1(x)|^2 dx = \frac{2}{L} \sin^2 \frac{\pi}{L} x \, dx.$$

It is the same, whatever the value of x_0.

On the other hand, in the state ψ_α ((19) and Figure 6(b)), which may be written as:

$$\psi_\alpha = \frac{1}{\sqrt{2}} [\varphi_1(x_1)\varphi_2(x_2) - \varphi_2(x_1)\varphi_1(x_2)] \alpha(1)\alpha(2)$$

where [ ] = 0 for $x_1 = x_2$, the probability of finding the electron [2] at x (up to dx) when electron [1] is located at x_0 is equal to

$$\left\{ \frac{1}{2} \left[\varphi_1^2(x_0)\varphi_2^2(x) + \varphi_2^2(x_0)\varphi_1^2(x) \right] - \varphi_1(x_0)\varphi_2(x_0)\varphi_1(x)\varphi_2(x) \right\} dx.$$

The reader should check, for example, that if the electron [1] is at the point $x_0 = 2L/3$, then the function in the brackets becomes

$$P_2(x) = \frac{3}{8} \sin^2 \frac{\pi}{L} x + \frac{3}{8} \sin^2 \frac{2\pi}{L} x + \frac{3}{4} \sin \frac{\pi}{L} x \sin \frac{2\pi}{L} x.$$

This has the form pictured in Figure 8, which shows clearly a parabola-shaped 'hole' around the position of the electron [1].

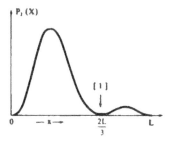

Figure 8   For two electrons with the same spin state, the probability that one is present (electron [2]) at the position of the other (electron [1]) is zero.

1.4 Influence of the temperature

If the N free electrons on the segment (O, L) are at **zero temperature**, it is immediately apparent from Figure 7 what the ground state (of minimum internal energy) of the system is: it is that of Figure 9(a), in which the states are populated with two electrons per space state as the energies increase. The energy attained when the Nth electron is positioned is called the **Fermi energy** E_F. It is defined using the density of states (see (9), p. 35), by

$$2 \int_0^{E_F} n(E) \, dE = N \tag{22}$$

or, following (10) and (4),

$$E_F = \frac{\hbar^2}{8m} \pi^2 \left(\frac{N}{L} \right)^2. \tag{23}$$

The corresponding level is the **Fermi level**.

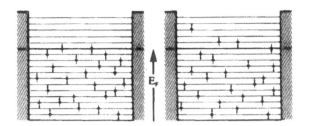

Figure 9 Fermi–Dirac distribution (*a*) at zero temperature; (*b*) at a finite temperature.

Very generally, the N electrons are provided by the atoms (or the molecules) of the segment: thus, N is proportional to L and the Fermi energy depends on the *concentration of electrons* N/L and not on the length of the segment. By way of example, if *each* atom (or molecule) in the system provides *one* electron we will have (to an order of magnitude) $N/L \simeq 1/1$ Å or 0.5 a.u. (see a.u. system, p. 36) and

$$E_F = \frac{\pi^2}{8}(0.5)^2 \text{ a.u.} \simeq 8 \text{ eV.} \tag{24}$$

This is a considerable energy in relation to thermal energies ($kT \simeq 1/40$ eV for $T = 300$ K).

The average (kinetic) energy of these N electrons is

$$\overline{E}_{\text{kin}} = \frac{\int_0^{E_F} E n(E) \, dE}{\int_0^{E_F} n(E) \, dE} = \frac{1}{3} E_F. \tag{24'}$$

The graph of the *density of states* (see (10)) is shown on Figure 10 where the hatched area (see (9)) represents the *total number of states* (or, multiplying by 2, the number of electrons, N).

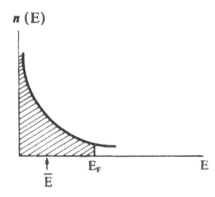

Figure 10 Density of states for a linear chain. The hatched area represents the number of states occupied.

At **non-zero temperature** T, the equilibrium distribution of the N electrons over the available levels is that which minimises the free energy (no longer the internal energy). Some electrons now occupy levels above E_F to the detriment of levels below that (Figure 9(b)); this leads to an increase in the energy but also to an increase in the entropy. The free energy is then a minimum when the probability that a level E is occupied is equal to the **Fermi–Dirac function** (see (Stat. Ph.)).

$$f(E) = 1/\exp\left((E - \mu)/kT\right) + 1 \qquad (25)$$

where μ is the *chemical potential*. This is determined by the equation

$$2 \int_0^\infty n(E) f(E)\, dE = \text{number of free electrons}, \qquad (25')$$

which generalises (22). In (25') we see that μ is equal to E_F for $T = 0$ and that it varies (but very slightly) with the temperature. In fact, the function $f(E)$ is always equal to $1/2$ for $E = \mu$ and in the limit $T \to 0$, it passes discontinuously from the value 1 to the value 0 for $E = E_F$ (Figure 9(a) and dashes on Figure 11). For all the values of T which will be of interest to us in what follows, ($T < 3000$ K; $kT < 1/4$ eV), μ differs very little from E_F. Moreover, taking into account the current orders of magnitude of E_F (see (24)), which are very much greater than kT, the function $f(E)$ differs relatively little (spreading of the order of $4kT$) from the step function $(1, E_F, 0)$ corresponding to $T = 0$. Figure 11 shows the form of $f(E)$ for a temperature T such that $kT = \mu/20$, or $T \simeq 3000$ K for $\mu = 5$ eV.

Even at these high temperatures, the proportion of thermally excited electrons remains low: it is equal to the hatched area on Figure 11 above $E = \mu \simeq E_F$, multiplied by the density of states $n(\mu)$ ($\simeq n(E_F)$, see (10)) and divided by the total number of

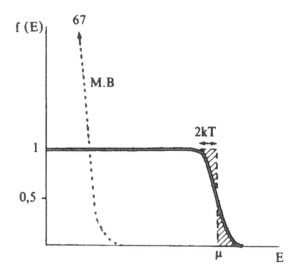

Figure 11 Fermi–Dirac curve drawn (continuous line) for $kT = \mu/20$ (extreme case corresponding to $T \simeq 3000$ K). The dashed line shows the case for $T = 0$. The dotted line shows the Maxwell–Boltzmann function corresponding to the same temperature (ordinate at the origin: $\simeq 67$).

electrons $2 \int_0^{E_F} n(E) \, dE$. This hatched area is comparable to that of a triangle with sides $1/2$ and $2kT$. Thus, the proportion of excited electrons is

$$\simeq \frac{1}{2} \times \frac{1}{2} \times 2kT \times \frac{E_F^{-1/2}}{2 \int_0^{E_F} E^{-1/2} \, dE} = \frac{1}{8} \frac{kT}{E_F}. \tag{26}$$

In the previous example ($T = 3000$ K, $E_F \simeq \mu = 5$ eV), this proportion is 0.6×10^{-2}: *the majority of the electrons are strictly unaffected by the heating from 0 K to T K*. Only a (small) proportion $\simeq kT/8E_F$ of electrons close to the Fermi level are thermally excited and their excitation energy is of the order of kT.

The radical difference between the Fermi–Dirac, and the Maxwell–Boltzmann statistics (used by Drude, see p. 16) which depend on E and T as $T^{-3/2} \exp(-E/kT)$, appears clearly in Figure 11 where the dotted line shows the tail of this distribution for the same temperature (3000 K) as the graph of the Fermi–Dirac distribution (shown by a continuous line). These dots constitute the end of an exponential distribution, the origin of which (for $E = 0$) is at the point with ordinate $\simeq 67$. The two statistics differ even more violently for less extreme temperature values (300 K, for example).

2. FREE ELECTRONS IN THREE DIMENSIONS

2.1 Generalities

We can easily generalise the above discussions, for a one-dimensional line, to three dimensions.

Here, the problem is that of a 'box' (see Figure 4(*b*), p. 15), which we may take to be a cube (edge: L) in the interior of which is confined a gas of N *free* electrons, which are initially assumed to be *independent* (in other words, without Coulomb interaction amongst themselves, but governed by the Pauli principle). The potential is taken to be constant (and zero) in the interior of the box. The x, y and z axes coincide with the edges of the cube. The number of free electrons will be assumed to be proportional to the volume of the box, which will be clearly the case whenever the latter consists of atoms which each contribute z electrons (z is the valency) to the gas.

The corresponding Schrödinger equation:

$$H\psi = -\frac{\hbar^2}{2m}(\Delta_1 + \Delta_2 + \ldots + \Delta_N)\psi = E\psi \tag{27}$$

enables us to separate the variables. Its solutions (space and spin) are products of orthonormal *space* and *spin* functions ($\bar{r}_p$ and p, respectively) for a single electron (here, the electron $[p]$).

$$\psi(\bar{r}_1, 1, \bar{r}_2, 2, \ldots, \bar{r}_N, N) = \prod_p \psi_n(\bar{r}_p, p) \quad p = 1, \ldots, N. \tag{27'}$$

As in Subsection 1.1, we may place importance on either:

i) **The surface properties**. Here, we adopt the boundary condition that the functions should be zero on the six sides of the sample:

$$\psi(0, y, z) = \psi(L, y, z) = 0 \quad (\textit{id. for } y \text{ and } z)$$

in which case, the functions ψ_n have (see (3)) the form:

$$\psi_n(\vec{r}_p, p) = \left(\frac{2}{L}\right)^{3/2} \sin n_x \frac{\pi}{L} x_p \times \sin n_y \frac{\pi}{L} y_p \times \sin n_z \frac{\pi}{L} z_p \times \gamma(p)$$

for $0 < x < L$, $0 < y < L$, $0 < z < L$ and:

$$\psi_N(\vec{r}_p) = 0 \quad \text{outside of the cube.} \tag{28}$$

Here, n_x, n_y and n_z are three positive integers and γ represents one of the two spin states α or β (see p. 37).

ii) or **the volume properties**. Here, we adopt periodic boundary conditions (Born–von Karman conditions, p. 34):

$$\psi(x, y, z) \equiv \psi(x + L, y, z) \quad (\textit{id.} \text{ for } y \text{ and } z). \tag{28'}$$

These conditions, which reduce to conceptually isolating a cube of matter with side L in an infinite substance, lead, as solutions of equation (27) to:

$$\psi_n(\vec{r}_p, p) = \frac{1}{L^{3/2}} \exp(i\vec{k}_n \cdot \vec{r}_p) \times \gamma(p) \quad \text{or more simply}$$

$$\psi_n(\vec{r}) = \frac{1}{L^{3/2}} \exp(i\vec{k}_n \cdot \vec{r}). \tag{29}$$

This is a plane wave, the wavevector $\vec{k}_n$ of which has components:

$$k_{nx} = \frac{2\pi}{L} n_x \qquad k_{ny} = \frac{2\pi}{L} n_y \qquad k_{nz} = \frac{2\pi}{L} n_z \tag{30}$$

where n_x, n_y and n_z are *positive or negative* integers which are not simultaneously zero. Here, as in (7'), the wavevector $\vec{k}_n$ is proportional to the velocity $\vec{v}$.

We introduce a cubic lattice of points with an arbitrary origin Ω and cell size $2\pi/L$. The conditions (30) reduce to saying that the vectors $\vec{k}_n$ are vectors of this lattice, which we shall sometimes call the $\vec{k}$ *lattice*. Thus, in Figure 12, which shows a projection of this lattice perpendicular to $\vec{k}_z$, the vector $\overline{\Omega M}$ is one of the vectors $\vec{k}$ enabling us to construct solutions of type (29). Its coordinates are $3 \times (2\pi/L), -(2\pi/L), 0$. In this case, we shall speak of the vector $(3, \bar{1}, 0)$ and, by extension, of the state $(3\bar{1}0)$ (pronounced: *three, minus one, zero*).

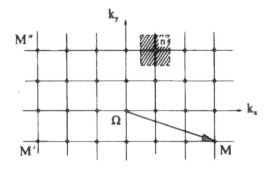

Figure 12 Projection of the $\vec{k}$ lattice onto the plane $k_z = 0$. Ω is the (arbitrary) origin of this lattice.

The eigenvalue E_n associated with the plane wave (29) is

$$\boxed{E_n = \frac{\hbar^2}{2m} |\bar{k}_n|^2}$$

(31)

$$= E_0(n_x^2 + n_y^2 + n_z^2) \quad \text{where} \quad E_0 = \frac{\hbar^2}{2m}\left(\frac{2\pi}{L}\right)^2.$$

The degeneracy which was of order *two* in the one-dimensional case (see p. 35 and Figure 4), is multiple here and, as can be seen from Figure 12, it increases with $|\bar{k}|$. In fact, each node M of the lattice corresponds to numerous nodes (M', $M'' \ldots$) such that $|\overline{\Omega M}| = |\overline{\Omega M'}| \ldots$. All these nodes lie on a single **sphere of equal energy**.

Thus, the degeneracy of the ground state is of order six: the states (1 0 0), (0 1 0), (0 0 1), ($\bar{1}$00), (0$\bar{1}$0) and (00$\bar{1}$), which are linearly independent eigenstates of (27), all have the eigenvalue $E^{(1)} = E_0$ which is the smallest possible. The second level of energy $E^{(2)} = 2E_0$ corresponds to states of the type (110). The degeneracy here is of order 12, etc ....

Each node n of the $\bar{k}$ lattice (Figure 12) corresponds to a space function $\psi_n(\bar{r}_p)$ and two states $\psi(\bar{r}_p, p)$ namely $\psi_n(\bar{r}_p)\alpha(p)$ and $\psi_n(\bar{r}_p)\beta(p)$. Since each node occupies a volume $(2\pi/L)^3$ in the $\bar{k}$ *space*, we see that this space contains:

$$\boxed{\begin{array}{cc} \dfrac{L^3}{8\pi^3} \text{ } \textit{states} \text{ (excluding spin),} & \text{or} \\[2mm] \textit{per unit volume} & \\[2mm] \dfrac{L^3}{4\pi^3} \text{ } \textit{states} \text{ (including spin).} & \end{array}}$$

(32)

(32′)

N.B. The conditions (28′) are another way of expressing the *uncertainty principle*. The uncertainty on x is L for each electron. The momentum $p_x = mv_x$ cannot be known to within better than h/L, whence (see (7′)) k_x cannot be known to within better than $2\pi/L$. The cubes with edges $2\pi/L$ of Figure 12 are the 'cubes of uncertainty' of the vectors $\bar{k}$.

2.2 Density and occupation of states

The *density of states* (see (9)) is easy to evaluate in the $\bar{k}$ space. All states with energies between E and $E + dE$ have their representative point between the spheres of radius $|\bar{k}|$ and $|\bar{k}| + |\overline{dk}|$ such that

$$E = \frac{\hbar^2}{2m} |\bar{k}|^2 \quad \text{and} \quad dE = \frac{\hbar^2}{m} |\bar{k}||\overline{dk}|.$$

The corresponding volume (hatching in Figure 13 where the cubic lattice is not shown) is $4\pi |\bar{k}|^2 |\overline{dk}|$. Taking (32) into account, we obtain a density of states:

$$\boxed{n(E) = \frac{L^3}{4\pi^2}\left(\frac{2m}{\hbar^2}\right)^{3/2} E^{1/2}.}$$

(33)

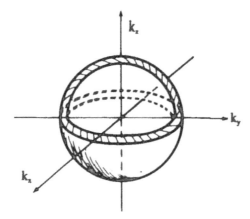

Figure 13 Two neighbouring spheres of equal energy in the $\bar{k}$ space and (hatched) the volume between these two spheres.

This density *increases with the energy* (unlike in the case of the one-dimensional metal, see (10) or Figure 10) according to a *parabolic* law.

The density n (33) is that of the space states. The total density (space + spin) is obtained by multiplying n by 2 (see (32')).

Exercise Show that the density of states for a metal of free electrons in two dimensions is constant. Hint: use the $\bar{k}$ lattice in two dimensions for the two-dimensional sample taken as a square (with side L). See also **Problem 1**.

Let us return to the problem of N electrons. The Pauli principle may be expressed in a simple way, by saying that each node of the $\bar{k}$ lattice may be assigned at most two electrons.

1 – At temperature $T = 0$ K, we populate these states beginning with the lowest energies and progressing in the $\bar{k}$ space via successive concentric spheres (of equal energy). The last sphere reached when the N electrons are assigned to a state is called the **Fermi sphere**. Its radius $k_F \left(= |\bar{k}_F|\right)$ corresponds (see (32')) to

$$\frac{4}{3}\pi k_F^3 \times \frac{L^3}{4\pi^3} = N$$

or

$$\boxed{k_F = \left(3\pi^2 \frac{N}{L^3}\right)^{1/3}.}$$
(33')

The corresponding velocity and (kinetic) energy are

$$\boxed{v_F = \frac{\hbar}{m}k_F = \frac{\hbar}{m}\left(3\pi^2 \frac{N}{L^3}\right)^{1/3} \qquad E_F = \frac{\hbar^2}{2m}\left(3\pi^2 \frac{N}{L^3}\right)^{2/3}.}$$
(34)

E_F is called the **Fermi energy**. This energy *does not depend on the size L* of the sample since the number of electrons N is actually proportional to the volume and

$$\frac{N}{L^3} = N_0 \tag{34'}$$

is a *concentration of electrons* which is a characteristic for the metal being studied. We let r_e denote the radius of an *electron sphere* defined by

$$\frac{L^3}{N} = \frac{4}{3}\pi r_e^3. \tag{34''}$$

v_F and E_F become:

$$v_F = \frac{\hbar}{m}\left(\frac{9\pi}{4}\right)^{1/3}\frac{1}{r_e}$$

$$E_F = \frac{\hbar^2}{2m}\left(\frac{9\pi}{4}\right)^{2/3}\frac{1}{r_e^2} \tag{35}$$

or, in atomic units:

$$\boxed{v_F = 1.92 r_e^{-1} \text{ a.u.} \qquad E_F = 1.84 r_e^{-2} \text{ a.u.}} \tag{35'}$$

Like Drude, we postulate that the free electrons are the z valence electrons, per atom, of the metal (*one* electron for Na, *three* electrons for Al ...), lost by each atom to the free electron gas. If N_{at} is the number of atoms located in the volume L^3, the equations $N = zN_{at}$ and (34'') enable us to calculate r_e (see Table 1).

Table 1   Values of the radius r_e (in a.u.) for a number of metals, at room temperature.

Li	Na	Cu	Ag	Al	Nb	Fe	Hg	Pb
3.25	3.93	2.67	3.02	2.07	3.07	2.12	2.65	2.30

In the two extreme cases of *sodium* and *aluminium*, we immediately find that (see p. 36):

$$
\begin{array}{llll}
\text{Na:} & v_F = & 0.49 \text{ a.u.} \simeq 1 \times 10^6 \text{ m s}^{-1}. \\
& E_F = & 0.12 \text{ a.u.} \simeq 3.2 \text{ eV}. \\
\text{Al:} & v_F = & 0.93 \text{ a.u.} \simeq 2 \times 10^6 \text{ m s}^{-1}. \\
& E_F = & 0.43 \text{ a.u.} \simeq 11.7 \text{ eV}.
\end{array} \tag{35''}
$$

We note that the velocities v_F, the velocities of the fastest electrons, of the order of one hundredth of the velocity of light, are considerable, even at $T = 0$ K.

The temperature T_F defined by $kT_F = E_F$ is called the **Fermi temperature**. For Na and Al, we have, respectively, $T_F = 11\,600 \times 3.2 = 37\,100$ K and $11\,600 \times 11.7 = 136\,000$ K.

The average kinetic energy of the N electrons (see (33)), namely $(\int_0^{E_F} En(E)\,dE)/\int_0^{E_F} n(E)\,dE)$ is:

$$\boxed{\overline{E} = \frac{3}{5}E_F = 1.1 r_e^{-2} \text{ a.u.}} \tag{36}$$

2 – At non-zero temperature T, only a small fraction ($\simeq T/T_F$) of the electrons are excited from levels located slightly *below* the Fermi level to levels located slightly *above* that level. More precisely, the fraction of the energy states E occupied is given by the Fermi–Dirac function $f(E)$.

The *density of the occupied states* $\mathcal{N}(E)$ is then simply the function

$$\mathcal{N}(E) = n(E)f(E)$$

where $n(E)$ and $f(E)$ are given in (33) and (25). This is illustrated in Figure 14, where the shaded area $\int_0^\infty \mathcal{N}(E)\,dE$ is equal to the number of electrons divided by two.

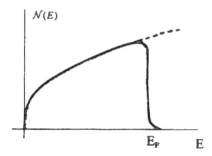

Figure 14 **Density of occupied states for free electrons in a 3-dimensional box at non-zero temperature.** The dashes show the parabola of expression (33).

N.B. 1 The energy axis (abscissa in Figure 14) is, of course, not continuous; but the separation between levels is sufficiently small that we may assume that the latter form a *quasi continuum*, which, as in Subsection 1.2, we call an **energy band**.

In fact, the difference in energy δE between a state (n_x00) and the neighbour state (n_x10) is E_0 (see (31)). Thus, we have (see (35)):

$$\delta E/E_F = E_0/E_F = \left(\frac{32\pi^2}{9}\right)^{2/3}\left(\frac{r_e}{L}\right)^2 \simeq 10\left(\frac{r_e}{L}\right)^2$$

or, for a millimetric box and a radius r_e of the order of 1 Å, a value of $\simeq 10^{-13}$.

N.B. 2 All the previous results, notably (31), (33), (34) and (36), were obtained using B.v.K. **ii)** conditions (p. 44). and functions $\exp(i\vec{k}\cdot\vec{r})$ (29). None of these results is modified if we use the conditions **i)** (p. 43) and the *sine* functions (28). In the latter case, the quantum numbers n_x, n_y and n_z are positive (more exactly, the states $(\bar{n}_x, n_y, n_z)$ and (n_x, n_y, n_z) are identical) and none of these can be zero. The corresponding $\vec{k}$ lattice constructed on the elementary cube with edge π/L (and not now $2\pi/L$) is confined to an octant of space, for example that defined by the positive k_i ($= n_i(\pi/L)$). The density of the points in this case is eight times larger than that given in (32), but, since this refers to an octant of the sphere, it reduces to the same value as the density of states (33) or the Fermi energy (34).

2.3 Interactions between electrons

Until now, we have neglected the Coulomb interaction between the electrons and also that between electrons and ions. In an elementary description in which the positive ionic charges are assumed to be uniformly distributed and the electrons are also uniformly distributed ($\psi^*\psi$ = constant), the electron–electron and ion–ion repulsion is exactly compensated by the electron–ion *attraction*. The Pauli principle brings a first infringement of this compensation[3]. The Pauli holes which make electrons with opposite spins avoid one another (see p. 39) decrease the electron–electron *repulsion*. This decrease, ΔE_p, called the **exchange term** varies as the reciprocal of the mean distance between electrons; in other words, as r_e^{-1}. To a first approximation (see below, (43) and (45)), we have

$$\Delta E_p = -0.46 r_e^{-1} \text{ a.u.} \tag{37}$$

which brings the mean energy of an electron (see (36)) to:

$$\boxed{E = 1.1 r_e^{-2} - 0.46 r_e^{-1} \text{ a.u.}} \tag{37'}$$

Let us now consider the Coulomb repulsion $e^2 / \left(|\bar{r}_i - \bar{r}_j|\right)$ between the electrons [i] and [j].

In a general way, introducing also the positive ions which constitute the box (see p. 14), we have, as Hamiltonian for the set of N electrons:

$$H = \sum_{i=1}^{N} \left(-\frac{\hbar^2}{2m} \Delta_i - ze^2 \sum_R \frac{1}{|\bar{R} - \bar{r}_i|} \right) + \frac{1}{2} \sum_{i \neq j} \frac{e^2}{|\bar{r}_i - \bar{r}_j|}.$$

The second term in the brackets represents the Coulomb interaction between the electron [i] and the set of all the z times charged ions located at the point $\bar{R}$ in the box.

The search for the eigenfunctions and eigenvalues of such a Hamiltonian is an absolutely insoluble problem.

On the other hand, it is possible to search for approximate solutions using the *variational method* (see Q.M.) which reduces to minimising the quantity $\bar{H}$ (mean value of the energy):

$$\bar{H} = \frac{(\psi, H\psi)}{(\psi, \psi)} \tag{38}$$

where the ψ are test functions which may be varied.

As test functions we may use the products of states with *one* electron (see (27')) obtained using the simplified Hamiltonian (27). In fact, we shall construct a Slater determinant similar to that previously described in (17) which enables us to begin from an antisymmetric state, in conformance with the Pauli principle. Namely

$$\psi = \frac{1}{\sqrt{N!}} \begin{vmatrix} \psi_1(\bar{r}_1, 1) & \psi_1(\bar{r}_2, 2) \dots & \psi_1(\bar{r}_N, N) \\ \vdots & & \\ \psi_N(\bar{r}_1, 1) & \dots & \psi_N(\bar{r}_N, N) \end{vmatrix} \tag{39}$$

where the states ψ_n are the plane waves together with their spin state, as given in (29). We recall that the states ψ_n are orthonormal.

---

[3] Later (see (3) and (4), p. 68) we shall take the discrete nature of ions into account.

Under these conditions, expansion of $\overline{H}$ in (38) gives:

$$\overline{H} = \sum_i \int \psi_i^*(\overline{r}) \left(-\frac{\hbar^2}{2m}\Delta_i - ze^2 \sum_{\overline{R}} \frac{1}{|\overline{R}-\overline{r}_i|} \right) \psi_i(\overline{r})\, d^3\overline{r}$$

$$+ \frac{1}{2}\sum_{ij} \int |\psi_i(\overline{r})|^2 |\psi_j(\overline{r}')|^2 \frac{e^2}{|\overline{r}-\overline{r}'|}\, d^3\overline{r}\, d^3\overline{r}'$$

$$- \frac{1}{2}\sum_{p,q} \int \delta(\gamma_p, \gamma_q)\psi_p^*(\overline{r})\psi_p(\overline{r}') \frac{e^2}{|\overline{r}-\overline{r}'|}\psi_q^*(\overline{r}')\psi_q(\overline{r})\, d^3\overline{r}\, d^3\overline{r}'. \qquad (40)$$

In this mean energy value, the first two terms ($\sum_i$ and $\frac{1}{2}\sum_{ij}$) have a simple physical significance: they integrate the kinetic energy of the *electrons*, the *electron–ion* Coulomb energy and the *electron–electron* Coulomb energy for electrons with density $|\psi_i(\overline{r})|^2$ and $|\psi_j(\overline{r}')|^2$, respectively. The last term $\sum_{p,q}$ has a more unexpected structure. It is non-zero only if $\delta = 1$ (in other words, if the two spin states γ_p and γ_q are identical) and involves the densities $\psi_p^*(\overline{r}_i)$ and $\psi_p(\overline{r}_j)$, called *exchange densities*, which associate two electrons [i] and [j] with the same space function ψ_p. The minus sign which precedes the term known as the *exchange term* is linked to the presence of a Pauli hole around the electron [i], where the probability of finding the electron [j] with the same spin state is very small. As we saw before, this hole decreases the Coulomb repulsion term (see p. 39).

Returning, by way of example, to Slater determinants for two electrons, like those on p. 38, in which we replace x by $\overline{r}$, it is easy to see that the only one which gives rise to a non-zero exchange term is the determinant ψ_a (19).

Let us return to the mean value $\overline{H}$ defined in (38) and expanded in (40). The functions ψ_i of the determinant (39) are only test functions. We shall come closer to exact solutions by varying the ψ_i and searching for a minimum of $\overline{H}$ (theorem of variations, see Q.M.). This search for a minimum ends with a series of equations (Hartree–Fock equations) which, for ψ_i are:

$$-\frac{\hbar^2}{2m}\Delta\psi_i(\overline{r}) + V^{(+)}\psi_i(\overline{r}) + V^{(-)}\psi_i(\overline{r}) - \sum_j \int \delta(\gamma_i, \gamma_j) \frac{e^2}{|\overline{r}-\overline{r}'|}\psi_j^*(\overline{r}')\psi_i(\overline{r}')\psi_j(\overline{r})d^3\overline{r}' = E_i\psi_i(\overline{r}).$$

$$(41)$$

$V^{(+)}$ and $V^{(-)}$ denote the Coulomb potentials created, respectively, by the ions and the electrons:

$$V^{(+)} = -ze^2 \sum_{\overline{R}} \frac{1}{|\overline{R}-\overline{r}|}$$

$$V^{(-)} = \int \frac{e^2\rho(\overline{r}')}{|\overline{r}'-\overline{r}|}\, d^3\overline{r}' \quad \text{where} \quad \rho(\overline{r}') = \sum_i |\psi_i(\overline{r}')|^2.$$

The extreme simplicity of the starting functions ψ (29) will enable us to find the eigenvalues of the Hartree–Fock equation (41).

In the assumption of the constant potential (Figure 3, p. 14) the positive charge of the z times charged ions, which compensates the charge of the N electrons is uniformly distributed in the space. The same applies to the electron density $\psi^*\psi d^3\overline{r} = (d^3\overline{r})/V$ (see (29)). It follows that the electron–ion and electron–electron interactions are opposing. Thus, we have $V^{(+)} + V^{(-)} = 0$ in equation (41) where now only the kinetic-energy term and the exchange term survive:

$$-\sum_j \int \delta(\gamma_i, \gamma_j)\psi_j^*(\overline{r}')\psi_i(\overline{r})\psi_j(\overline{r}) \frac{e^2}{|\overline{r}-\overline{r}'|}\, d^3\overline{r}'. \qquad (42)$$

To calculate the latter, we assume we are in the ground state at $T = 0$ K. Each space function $\psi_i(\overline{r})$ used in the Slater determinant (39) occurs in two lines of the determinant, namely the line

$\psi_i\alpha$ and the line $\psi_i\beta$. In (42), we replace $|\vec{r} - \vec{r}'|^{-1}$ by its Fourier integral[4]

$$\frac{1}{|\vec{r} - \vec{r}'|} = \frac{1}{2\pi^2} \int \frac{1}{K^2} \exp(i\overline{K} \cdot (\vec{r} - \vec{r}')) \, d^3\overline{K}.$$

The exchange term then becomes:

$$-\frac{1}{L^{3/2}} \exp(i\overline{k}_i \cdot \vec{r}) \frac{1}{L^3} \sum_j \frac{e^2}{2\pi^2} \int \frac{1}{|K|^2} \exp(i(\overline{K} + \overline{k}_j - \overline{k}_i) \cdot (\vec{r} - \vec{r}')) \, d^3\overline{K} \, d^3\vec{r}'$$

$$= \psi_i(\vec{r}) \frac{1}{L^3} 4\pi e^2 \sum_j \int \frac{1}{|K|^2} \delta(\overline{K} + \overline{k}_j - \overline{k}_i) \, d^3\overline{K}$$

$$= \psi_i(\vec{r}) \frac{e^2}{2\pi^2} \int_{k_j} \frac{d^3\overline{k}_j}{|k_j - k_i|^2}.$$

Thus, it turns out that the plane wave $1/L^{3/2} \exp(i\overline{k} \cdot \vec{r})$ is a solution of the equation (41), with, eigenvalue

$$E(k) = \frac{\hbar^2}{2m} |\overline{k}|^2 - \frac{e^2}{2\pi^2} \int_{k' < k_F} \frac{d^3\overline{k}'}{|k - k'|^2}$$

or

$$E(k) = \frac{\hbar^2}{2m} k^2 - \frac{2e^2}{\pi} k_F \, g\left(\frac{k}{k_F}\right) \tag{43}$$

where

$$g(u) = \frac{1}{2} + \frac{1 - u^2}{4u} \ln\left|\frac{1 + u}{1 - u}\right|. \tag{44}$$

The exchange term g, which is positive, is shown in Figure 15. It contributes to a *decrease* in the energy of the plane wave $(\hbar^2/2m)k^2$ evaluated in (31) without correction for electron–electron interaction. The dispersion relation $E(k)$ is no longer parabolic (Figure 16). It can be shown that, taking account of (33′) and (34″), we obtain:

$$E/E_F = (k/k_f)^2 - 0.66 r_e g(k/k_F) \text{ for } r_e \text{ in a.u.} \tag{44′}$$

This is illustrated in Figure 16 for the case of aluminium.

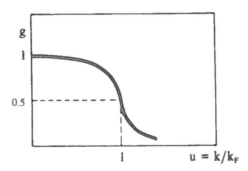

Figure 15 **Function** $g(u)$ **of expression (44).**

---

[4] This integral can be determined without difficulty, starting from the equation $\delta(\vec{r}) = (2\pi)^{-3} \int \exp(i\overline{k} \cdot \vec{r}) \, d^3\overline{k}$ together with the Poisson equation $\Delta(-(e/r)) = 4\pi e\delta(r)$.

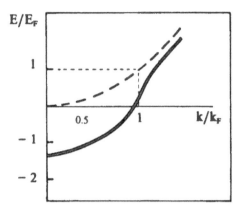

Figure 16 Value of E/E_F as a function of k/k_F for free electrons without exchange correction (dashed line) and with correction (bold line). The latter is constructed using (44') for the case of aluminium, namely $r_e = 2.07$ (see Table 1, p. 47).

The mean energy of the electrons is:

$$\overline{E} = \frac{3}{5}\frac{\hbar^2}{2m}k_F^2 - \frac{2e^2}{\pi}k_F g(\overline{k})_{k<k_F}.$$

When all the calculations are complete, we have

$$\overline{E} = \frac{3}{2}\frac{\hbar^2}{2m}k_F^2 - \frac{3}{4}\frac{e^2k_F}{\pi}.$$

The quantity k_F is still as defined by (33') and (34''). Finally (see (37'), p. 49):

$$\boxed{\overline{E} = 1.1r_e^{-2} - 0.46r_e^{-1} \quad \text{a.u.}} \tag{45}$$

The dispersion graph $E(k)$ (Figure 16) is not very different from a parabola. For k small, we may write:

$$E(k) \simeq \frac{\hbar^2}{2m^*}k^2$$

where, expanding (43), we have

$$m^* = m\left(1 + \frac{2me^2}{\hbar^2\pi k_F}\right)^{-1}$$

or

$$m^* = \frac{m}{1 + 0.33r_e} \qquad r_e \text{ in a.u.}$$

m^* is called the effective Hartree–Fock mass.

N.B. Care should be taken not to confuse this scalar m^*, defined at the bottom of the band ($|k| \rightarrow 0$), with the effective mass tensor to be described in Chapter 6.

3. BRUSHING THE DUST OFF DRUDE

The model which we have just described in Section III.2 uses certain assumptions borrowed directly from Drude (see p. 14): the metal is a *box* inside which there is a *constant potential*. The electrons, which, for the 'simple metals' such as Li, Na and Al are the *valence electrons* (for example, three per atom for Al), are *free* in the box. **On the other hand**, in introducing the Pauli principle, we have used statistics which are radically different from those used by Drude.

3.1 Specific heat of the electrons

One immediate consequence of this change in the statistics concerns the *specific heat* of the free-electron gas. We recall (p. 43) that when the temperature is raised from 0 to T K only a small proportion of the electrons close to the Fermi level are excited (see Figure 11). There are $\simeq n(E_F)kT$ of these electrons (twice the area shaded on Figure 17) and their excitation energy is of the order of $2kT$. Thus, the variation of electron energy ΔE_{el} is $\simeq 2n(E_F)k^2T^2$ and the corresponding specific heat is, for a constant volume:

$$C_v^{el} = \frac{\partial \Delta E_{el}}{\partial T} \simeq 4n(E_F)k^2T. \tag{45'}$$

This specific heat, unlike in the Maxwell–Boltzmann case (see pp. 21, 29), is **proportional to** T. On the other hand, it is small compared *both* to the specific heat generally measured due to vibrations of the ions (see **Problem 18**) *and*, of course, to that which Drude attributed to the electrons, namely $\frac{3}{2}Nk$.

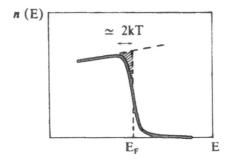

Figure 17 **Density of states in the neighbourhood of the Fermi level.** The hatched area denotes the number of pairs of thermally excited electrons.

In fact, following (33), in a.u., we may write

$$n(E_F) = \frac{L^3}{\pi^2\sqrt{2}} E_F^{1/2}.$$

Whence,

$$\frac{C_v^{el}}{C_{Drude}^{el}} = \frac{L^3}{N} \frac{1}{3\pi^2\sqrt{2}} E_F^{1/2} kT.$$

In the case of a monovalent metal $\left(L^3/N = (4/3)\pi r_e^3\right)$ with $r_e = 3$ and $E_F = 0.2$ (see (35'')), we find that $C_v^{el}/C_{Drude}^{el} = 0.6kT$ or 1×10^{-4} for $T = 30$ K and 1×10^{-3} for $T = 300$ K.

N.B. 1   A more precise calculation of C_v^{el} gives

$$\boxed{C_v^{el} = \frac{\pi^2}{3} n(E_F) k^2 T = \gamma T \qquad \gamma = \frac{\pi^2}{3} k^2 n(E_F)} \tag{46}$$

(from which we essentially rediscover (45')).

In fact, the variation of the energy ΔE_{el} in passing from 0 to T is

$$\Delta E_{el} = \int_{E_F}^{\infty} n(E) f(E)(E - E_F)\, dE + \int_0^{E_F} n(E)(1 - f(E))(E_F - E)\, dE$$

where f is the function (25) in which we set $E_F = \mu$. The specific heat is

$$C_v^{el} = \frac{\partial \Delta E_{el}}{\partial T} = \int_0^{\infty} n(E) \frac{\partial f}{\partial T} (E - E_F)\, dE$$

or, taking into account the very 'peaky' form of $\partial f/\partial T$ in the neighbourhood of $E = E_F$:

$$C_v^{el} \simeq n(E_F) \int_0^{\infty} \frac{\partial f}{\partial T} (E - E_F)\, dE$$

$$= n(E_F) k^2 T \int_{-E_F/kT}^{\infty} \frac{x^2 e^x}{(e^x + 1)^2}\, dx \quad \text{where} \quad x = \frac{E - E_F}{kT}.$$

Since the term $\exp(-E_F/kT)$ is negligible, the integration may be carried out from $-\infty$ to $+\infty$. Taking into account that $\int_{-\infty}^{\infty} x^2 e^x\, dx/(1 + e^x)^2 = \pi^2/3$, we end up with the expression (46).

N.B. 2   The specific heat due to the *vibration of the ions* varies at very low temperature like T^3: $C_v^{ion} = BT^3$, before becoming constant at high temperature (Dulong and Petit's law). Thus, at low temperature, and in the absence of other contributions, the total specific heat of the metal has the form:

$$C_v = \gamma T + BT^3.$$

At a sufficiently low temperature, $T < T_0 = (\gamma/B)^{1/2}$, the specific heat of the electrons dominates. T_0 is in general of the order of a few degrees K. Measurements in this temperature range are difficult, but feasible. The ordinate at the origin gives the value of γ and thus of $n(E_F)$. Table 2 shows some experimental values of γ determined in this way.

N.B. 3   It is worth noting here that, in the derivation of (46) illustrated in Figure 17, no reference is made to the explicit form of $n(E)$. The form (33) in $E^{1/2}$ corresponds to the simple model of free electrons. In slightly more refined descriptions (see Chapter 6), where $n(E)$ varies differently, notably in the neighbourhood of the Fermi level, this derivation remains. Thus, the coefficient γ gives access to the quantity $n(E_F)$ regardless of the form of variation of $n(E)$.

Table 2   Coefficient γ (in 10^{-4} J/mole K^2) of electron specific heat (46) for a number of metals. These values are proportional to the density of states at the Fermi level $n(E_F)$.

	Li	Na	Al	Cu	Ag	Au	V	Cr	Mn	U
γ	17.5	13.8	13.6	7.0	6.1	7.0	93	14	92	109

3.2 Electrical conductivity

We now return to the problem of *electrical conductivity* (see p. 16). Let us consider the $\overline{k}$ lattice and call 'occupied nodes' those nodes associated with a state (or a plane wave) occupied by two electrons. These occupied nodes all sit inside the Fermi sphere. In the absence of an electric field, each of them (n_x, n_y, n_z; wavevector $\overline{k}_n$; electron velocity $\overline{v}_n = \hbar k_n/m$) can be associated with the symmetric occupied node ($-n_x$, $-n_y$, $-n_z$; velocity $-\overline{v}_n$) in such a way that, globally, there is no electrical current.

The establishing of the electric field $\overline{\varepsilon}$, creates the force $\overline{F} = -e\overline{\varepsilon}$ on each electron, which modifies the wavevector $\overline{k}$ by the quantity $d\overline{k}$ such that

$$\hbar \frac{d\overline{k}}{dt} = -e\overline{\varepsilon}.$$

All the vectors $\overline{k}$ within the Fermi sphere are subject to the same variation $d\overline{k}$: all the nodes (occupied or not) of the $\overline{k}$ lattice then move at a uniform velocity in the direction opposite to $\overline{\varepsilon}$ (Figure 18), initially causing an overall uniform displacement of the Fermi sphere.

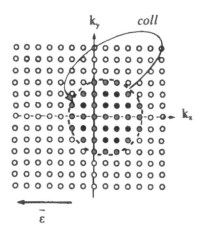

Figure 18 $\overline{k}$ lattice for a metal subjected to an electric field $\overline{\varepsilon}$. The nodes of the lattice move at a constant velocity in the direction opposite to $\overline{\varepsilon}$ (here, to the right). The Fermi sphere stabilises at a position shifted to the right, by virtue of collisions (*coll.*) *i.e.* transfers from the front face to the rear face of the sphere.

We then assume (the assumption is for the same reason as in Section III.2.2.1, p. 17) that the movement *of the sphere* should stop after an average time τ. This stabilisation of the sphere requires that those nodes which move through the front surface (here the right surface) of the sphere should return from 'occupied' to 'empty'; and vice versa on the rear (here left) surface. This 'transfer of occupation' (see *coll.* in Figure 18) takes place by means of **collisions** involving the Fermi electrons[5]. In the stationary state, reached in this way, the Fermi sphere is shifted to the right by

$$|\Delta k| = \frac{e\varepsilon\tau}{\hbar}. \tag{47}$$

Assuming that ε or τ are sufficiently small that $\Delta k \ll k_F$ (Figure 19), we see that the overwhelming majority of occupied states $\overline{k}$ are compensated by an 'opposing state' $-\overline{k}$ which does not contribute to the electrical current. The non-compensated states occupy twice the shaded volume. Thus, amongst all the free electrons, those non-compensated and defined by $(\theta, \theta + d\theta)$ occur in the proportion

$$\nu d\theta = 2 \times 2\pi k_F^2 \times 2\cos\theta\sin\theta \times \Delta k \times \frac{3}{4\pi k_F^3} \times d\theta$$

and their velocity projected onto $-\overline{\varepsilon}$ is $v_\varepsilon = \hbar k_F \cos\theta/m$.

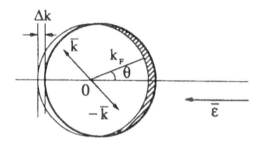

Figure 19 Fermi sphere (bold line) shifted with respect to the initial sphere (fine line) under the effect of an electric field $\overline{\varepsilon}$. In the k space, the states giving rise to electrical conduction occupy twice the shaded volume (supplementary electrons on the right, missing electrons on the left).

Denoting the number of free electrons per unit volume by N (as on p. 16), we see that the current density due to the displacement of the Fermi sphere, namely

$$i = Ne \int_0^{\pi/2} v_\varepsilon \nu d\theta$$

integrates to $i = \sigma\varepsilon$ with (taking into account (47)) electrical conductivity σ given by:

$$\boxed{\sigma = \frac{Ne^2\tau_F}{m}.} \tag{48}$$

We have rediscovered Drude's formula (15) (p. 20) *but with a very different physical meaning.* Although N (the total number of electrons) appears in (48), **only the Fermi**

---

[5] This point is handled more explicitly in the Annex (**Annex 3**, p. 261).

electrons *actually contribute to the conduction* (whence the index F of τ_F). Only these electrons may change state (transitions $\bar{k} \to \bar{k}'$) through being involved in collisions (characterised by the relaxation time τ_F). In fact, taking into account the very low excitation energies available ($\simeq kT$ or $v_F e \varepsilon \tau_F$) only **elastic** or **quasi-elastic collisions** may occur (thus, with constant energy $|\bar{k}|$). These collisions, which are impossible in the *interior* of the Fermi sphere where all the possible states are occupied, can involve only electrons **on the sphere**. Thus, the process *coll.* of Figure 18 consists of the collision $\bar{k} \to \bar{k}'$ of Figure 20, where both k and k' are $\simeq k_F$.

Figure 20 An elastic collision giving rise to the transition $\bar{k} \to \bar{k}'$ (with $k = k'$) of the wavevector. This collision corresponds to the process *coll.* in Figure 18.

In the case of an elastic collision ($k = k'$, no energy lost) the centre of the sphere is displaced (by Δk) to the right and the final state (immediately after the collision) is above the left Fermi half sphere. The corresponding electron will rejoin this half sphere by one or more **inelastic collisions** in which the excess energy will be lost to the ions. Thus, it is clear that the **Joule effect** is also due to the Fermi electrons and to these alone.

These electrons have a **mean free path** Λ defined by

$$\Lambda = v_F \tau_F$$

where τ_F is the velocity at the Fermi level, namely, following (35'):

$$\Lambda = 1.92 r_e^{-1} \tau_F \qquad \text{a.u.}$$

Thus, for aluminium, we have $\Lambda = 0.93\tau$. From the value of the resistivity, namely $\rho = 2.5 \times 10^{-8}$ Ωm at room temperature and $\rho = 0.6 \times 10^{-12}$ Ωm for very pure aluminium ($\simeq 99.999\%$) at a very low temperature (4 K), we deduce the following values of τ of 500 and 20×10^6 a.u., respectively (see p. 36). Thus, we have $\Lambda_{300K} \simeq 460$ (a.u.) $\simeq 25$ nm and $\Lambda_{4K} \simeq 18 \times 10^6$ (a.u.) $\simeq 10^6$ nm $= 1$ mm.

Let us again make our earlier assumptions (see Section II.2.2.1, p. 17) about the collisions. We consider electrons moving parallel to Ox. For each electron, the *probability that it is involved in a collision* in the segment $x, x + dx$ is dx/Λ. For a group of $n(0)$ electrons passing through $x = 0$, at the point x there remain only $n(x)$ electrons which have not been involved in a collision, where $n(x) = n(0) \exp(-x/\Lambda)$.

An easy calculation now shows that Λ is equal to $\bar{x}$, the mean distance between collisions. Let us now suppose that *two (or more) different types of collision* may occur, characterised by the probabilities dx/Λ_1, dx/Λ_2 .... Assuming that these processes are mutually independent, the probabilities add together and we may write:

$$dn(x) = -\frac{n(x)dx}{\Lambda_1} - \frac{n(x)dx}{\Lambda_2} \ldots, \quad \text{or}$$

$$n(x) = n(0) \exp\{-((1/\Lambda_1) + (1/\Lambda_2)\ldots)x\}.$$

Everything occurs as though there were only one type of collision with mean free path Λ defined by

$$\frac{1}{\Lambda} = \frac{1}{\Lambda_1} + \frac{1}{\Lambda_2} + \ldots$$

with relaxation time τ such that

$$\frac{1}{\tau} = \frac{v_F}{\Lambda} = v_F \left(\frac{1}{\Lambda_1} + \frac{1}{\Lambda_2} + \ldots \right) = \frac{1}{\tau_1} + \frac{1}{\tau_2} + \ldots.$$

In other words, the resistivity of the metal

$$\rho = \frac{1}{\sigma} = \frac{m}{Ne^2\tau}$$

is equal to a *sum of partial resistivities* characteristic of each type of collision:

$$\boxed{\rho = \rho_1 + \rho_2 + \ldots \quad \text{where} \quad \rho_i = \frac{m}{Ne^2\tau_i} = \frac{mv_F}{Ne^2\Lambda_i}.} \tag{49}$$

This additive rule for partial resistivities is called *Matthiessen's Law*.

N.B. 1 Amongst collision processes, we can at least distinguish between *collisions in the bulk* on atoms of impurities, on crystalline defects, on the elementary modes of thermal vibration of atoms called *phonons* ..., and *collisions on free surfaces*: $\rho = \rho_b + \rho_s$, where $1/\Lambda = (1/\Lambda_b) + (1/\Lambda_s)$. The path Λ_s is clearly of the order of magnitude of the smallest dimension of the sample studied (the *diameter* for a wire, the *thickness* h for a foil ...). This remark enables us to test (49) experimentally. For this, we assume conditions (h small) such that Λ_s is of the same order as Λ_b. Writing $\Lambda_s \simeq h$, for simplicity, we have

$$\rho \simeq \frac{A}{h} + \rho_0, \qquad A = \frac{mv_F}{Ne^2}$$

where $\rho_0 = A/\Lambda_b$ is the resistivity of the same metal in the bulk (in other words, such that $h \gg \Lambda_b$). From Figure 21, we note that $(\rho - \rho_0)/\rho_0$ ($\simeq \Lambda_b/h$), measured for various thicknesses of silver, varies as h^{-1}. From the experimental curve, we deduce that $\Lambda_b = 24$ nm. In fact, clearly, we have $\Lambda_s > h$ and a more elaborate calculation of the function $\rho = f(h)$ gives $\Lambda_b = 52$ nm.

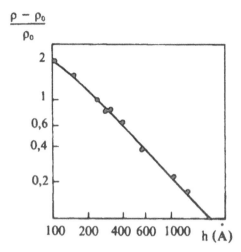

Figure 21 **Electrical resistivity** ρ **of silver versus the thickness of the sample** h. ρ_0 denotes the resistivity for an infinite thickness. The measurements were carried out at room temperature. **F W Reynolds**, *Phys. Rev.* **88**, 418 (1952).

Numerical exercise Beginning with this experimental value of Λ_b with $N = 5.85 \times 10^{28}$ m^{-3} or $r_e = 3.02$ a.u., and calculating v_F (using (35′)), determine ρ_0 for silver.

N.B. 2 Without entering into a precise description of collision mechanisms, we note that for an ideally pure metal the dominant process brings into play the interaction of the electrons with the modes of thermal vibration of the atoms. The following is a purely semi-qualitative argument: the amplitude of oscillation of the atoms is proportional to the square root of their energy, thus to $T^{1/2}$ (see **Problem 18**). The electron collision probability, proportional to the area swept by the oscillator, is proportional to the square of this amplitude, thus to T. Under these conditions, we expect τ^{-1} to vary as T (at least, if T is not very low), whence ρ should increase proportionally with the temperature (see (1), p. 12) or at least linearly with it. This is actually what we observe for a good number of metals (see Figure 22).

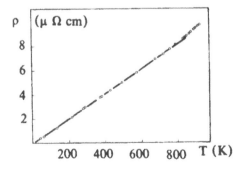

Figure 22 **Electrical resistivity of aluminium against temperature.** Note the broad linear area. After **O Dimitrov**.

3.3 Superconductivity

In certain cases, the resistivity, which decreases as T decreases, becomes **strictly zero** below a certain *critical temperature* T_c. In particular, the Joule effect becomes zero. The metal is then said to be a **superconductor**. Of the many superconducting metals, we mention mercury ($T_c = 4.15$ K) and lead ($T_c = 7.22$ K).

Of the many remarkable properties of the superconducting state, other than the disappearance of all resistivity, we mention:

i)   the appearance of a perfect diamagnetism: the magnetic field is expelled from a superconductor (*Meissner effect*);

ii)   the suppression of the superconducting state by application of a magnetic field stronger than a *critical field* H_c (see Figure 23), which, for $T < T_c$, enables us to study either the normal state (N) or the superconducting state (S), as we wish;

iii)   *the reduction of the entropy* (evaluated by measuring the specific heat of electrons) when the transition from (N) to (S) takes place at $T < T_c$;

iv)   the *decrease in thermal conductivity* on passing from (N) to (S);

v)   the relationship between T_c and the *isotopic mass M of the metal considered,* $T_c \propto M^{-1/2}$;

vi)   the fact that the 'best' superconductors are often bad normal conductors (metals or alloys with quite a high resistivity in the normal state); conversely, good conductors such as alkalis are not generally superconductors;

vii)   the fact that *magnetic impurities* (such as gadolinium) tend to decrease the value of T_c.

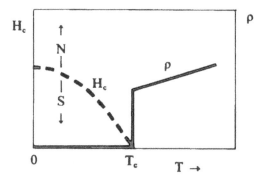

Figure 23 Critical temperature and critical field of a superconductor. The diagram relates to a so-called *type I* superconductor.

Applications of superconductivity were for a long time confined to laboratory activities such as the production of intense magnetic fields using superconducting cores. With the discovery[6] of **high-T_c superconductors** (T_c above 100 K), the possibility of numerous applications and even of a real industrial revolution has been perceived (Figure 23). These materials (oxides such as, for example, $YBa_2Cu_3O_{7-\epsilon}$) are, unfortunately, very delicate to develop and form; whence, hopes have been tempered.

[6]   J G Bednorz and K A Müller, *Physik*, B-64 (1986) 908.

We note that they illustrate point **vi)** above, in that they have very high resistivities (of the order of one ohm.cm) above T_c.

N.B. An example of the electronic structure of these oxides is given in **Problem 6**, which cannot be tackled before reading Chapter VI.

Figure 24 Levitation by the Meissner effect. The lower sample (diameter $\simeq$ 1 cm) consists of an oxide YBa$_2$Cu$_3$O$_7$ (T_c = 90 K). When cooled to 80 K by liquid nitrogen it is a superconductor. A small permanent magnet is placed on it. Since the lines of force of the latter cannot penetrate the superconductor (see point **i)**, p. 60), it 'floats' above. Possible applications of this phenomenon, for example to rail transport, are conceivable. After **H Pascard**.

There is no question of our describing the theory of the superconducting state here. We shall simply derive the phenomenological **London equation** and include a few words about the **microscopic theory** due to Bardeen, Cooper and Schrieffer (B.C.S. 1957).

We assume (Gorter and Casimir, 1934), that the state (S) consists of *two fluids*: normal electrons with a finite entropy and superconducting electrons, in the proportion $\eta = N_s/N$, which are assumed to be ordered, without dissipation and with zero entropy (see **iii)**). The fraction η decreases from 1 to 0 as T passes from 0 to T_c. The two fluids constitute two currents in parallel, such that, for $T < T_c$, only the electrons (S) contribute to the current since their resistivity is zero. Suppose that $\bar{\varepsilon}$ is the electric field, which we assume to be static at $T < T_c$. The electrons (S) are accelerated according to the equation:

$$m\frac{d\bar{v}_s}{dt} = -e\bar{\varepsilon}$$

(see (19), p. 22), with the friction term zero, or:

$$\frac{d\bar{j}}{dt} = \frac{N_s e^2}{m}\bar{\varepsilon}$$

where $\bar{j}$ is the current density. This equation, together with the third and fourth Maxwell equations (p. 22) gives:

$$\frac{d}{dt}\left(\overline{\mathrm{curl}\,\bar{j}} + \frac{N_s e^2}{mK}\bar{B}\right) = 0 \tag{50a}$$

$$\overline{\mathrm{curl}\,\bar{B}} - \frac{\mu}{K}\bar{j} = 0. \tag{50b}$$

This system of equations provides a link between *magnetic induction* and *current density*.

Since the Meissner effect (point i)) rules out the obvious solution of (50) in which $\overline{B}$ (and $\overline{j}$) is constant and non-zero, London (1935), postulates restricting equation (50a) for a superconductor to:

$$\overline{\text{curl}}\,\overline{j} + \frac{N_s e^2}{mK}\overline{B} = 0.$$

This so-called *London equation*, together with (50b), enable us to write

$$\overline{\Delta}\,\overline{B} - \lambda^{-2}\overline{B} = 0 \qquad \overline{\Delta}\overline{j} - \lambda^{-2}\overline{j} = 0$$

where we have set $\lambda^2 = mK^2/N_s e^2 \mu_0$. It follows that $\overline{B}$ is zero in the interior of the superconductor and decreases exponentially from the surface. Thus, for a planar, semi-infinite sample (surface perpendicular to Oz, with origin on the surface) placed in an external uniform magnetic field parallel to Oy, we have, for example

$$B_x = 0 \qquad B_y = B(0)e^{-z/\lambda} \qquad B_z = 0.$$

The induction decreases in the superconductor from the value $B(0)$ over a distance λ called the **penetration distance** ($\lambda \simeq 0.1\ \mu$m).

These phenomenological generalities are incorporated in the B.C.S. microscopic theory[7], only a few purely qualitative elements of which will be described here. The general idea of this is that the electrons (S) are regrouped in *pairs* (**Cooper pairs**). These transient pairs have a characteristic size called the **coherence length**. They are linked by a very weak interaction which may be loosely described as follows. An electron attracts (positive) ions of the crystal towards itself, deforming the lattice locally. The instantaneous positive fluctuation created in this way attracts another electron and links it indirectly to the first. Here, the reason for point v) above is apparent. The pairs consist of electrons with wavevectors $\overline{k}$ and $-\overline{k}$ (thus, moving at equal velocities in opposite directions) and spins $\uparrow$ and $\downarrow$. Thus, each pair is a *boson* and at low temperature, these bosons are susceptible to *Bose–Einstein condensation* (see (Stat. Ph.)). The movements of the centres of mass of these pairs are coherent: all the centres of mass have the same momentum, either *zero* (in which case, there is no current (S)), or $\overline{p}$ creating a current (S) without dissipation. It is the coherence of this movement which decreases the total entropy (point iii)). The energy Δ needed to 'break' a pair is called the **gap energy**. It may be provided by the thermal agitation (whence T_c) or by the application of a magnetic field (whence H_c, see ii)). The B.C.S. theory links Δ to T_c via $\Delta(T = 0) = 1.76kT_c$ (in other words, 1 or 2 meV).

We note, by way of a certainly provisional conclusion, that the theory of the *high-T_c superconductors* remains to be developed. Apart from their high T_c, the latter exhibit characteristics which are both **usual** (such as the existence of *pairs* of carriers, ...) and **original**. Such original characteristics include a relatively minor *isotopic effect* (see point v), above), the strong influence of the *sub-stoichiometry* in oxygen (the term ε in the formula for the oxide given on p. 61), an indifference to the addition of *magnetic impurities* (see point vii), above), the existence of an abnormally high *gap energy* (10 to 20 meV), the *bidimensional* crystallographic nature, ... Given these facts, will it be possible to retain the B.C.S. framework, even though it may have to be adapted (for example, by replacing the pair-creating *electron–ion* interaction with an *electron–plasmon* interaction, etc ...)? Or, will it be necessary to conceive a new type of superconduction in which the *magnetism* and *fluctuating valency* of copper ions (Cu^{++} and Cu^{+++}), associated with the sub-stoichiometry and the content of alkaline-earth elements (Ba ...) would play a crucial

[7] For an introduction to this theory, see, for example: **P G de Gennes**, *Superconductivity in metals and alloys*, Benjamin, 1966.

role? These questions are the root of a sizeable scientific challenge and of a possible large-scale industrial adventure in the future (see **Problem 6**).

3.4 Thermal conductivity

Finally, we return to *thermal conductivity*. We shall have recourse to the formula (see p. 21):

$$\kappa = \frac{1}{3} C_v \underline{v} \Lambda; \tag{51}$$

however, in the light of the above discussion of electrical conductivity, we shall apply it solely to the Fermi electrons, which are the only ones which may be thermally excited. As in Chapter II, we assume that the collision processes are the same for a thermal gradient as for an electric field (same values of τ or Λ). We now have $\underline{v} = v_F$, whence

$$\kappa = \frac{1}{3} C_v^{el} v_F^2 \tau_F.$$

Using (46), (33) and (34), and writing $E_F = \frac{1}{2} m v_F^2$, we obtain

$$\kappa = \frac{\pi^2}{3m} N k^2 T \tau_F$$

(where k is the Boltzmann constant), whence (see (48)):

$$\frac{\kappa}{\sigma} = \kappa \rho = \frac{\pi^2}{3} \left(\frac{k}{e} \right)^2 T. \tag{52}$$

We rediscover here the Wiedemann–Franz law (see p. 12 and p. 21) with a constant $L = (\pi^2/3)(k/e)^2 = 2.45 \times 10^{-8}$ WΩK^{-2}, called the *Lorenz constant*.

The good agreement (see Table 1, p. 13) between the predictions of this law and experimental values for the product $\kappa \rho$, unequivocally assigns *the dominant role in heat transfer* in metal to the *valence electrons*; it also provides evidence that (for ρ varying as T, see **N.B. 2** p. 59) over a large temperature range **the thermal conductivity of metals does not vary** – or varies only slightly – **with temperature**.

N.B. 1 The above agreement, which is good or excellent at medium and high temperatures, becomes sometimes frankly mediocre at low temperature. Thus, for copper, the conductivity κ, which is practically constant (and equal to 3.8 W cm^{-1} K^{-1}) from the melting point to 80 K, begins to increase as T decreases, passing through a maximum of 50 W cm^{-1} K^{-1} at 20 K. At this temperature, the measured value of L becomes around 10 times lower than the theoretical value. This disagreement is attributed to the differences between the relaxation times τ_{th} and τ_{el} relative to κ and ρ. In this case, we must, of course, write

$$\kappa \rho = \frac{\pi^2}{3} \left(\frac{k}{e} \right)^2 \frac{\tau_{th}}{\tau_{el}} T.$$

N.B. 2 The coming together of two successive errors which enabled Drude to find the result (52) correctly (up to a factor of two: see (18), p. 22) based on erroneous statistics is now clear: in his approach, C_v was constant and v^2 was proportional to T. Here, it is v ($= v_F$) which is constant and C_v which is proportional to T. C_v was overvalued by a factor $\simeq E_F/kT$ ($\simeq 200$ at room temperature), while v^2 was undervalued by the same factor.

3.5 Pauli paramagnetism

We recall that Drude's theory was unable to take account of the fact that many metals are paramagnetic and that this *Pauli paramagnetism* is characterised by a *susceptibility* χ_p which is almost *constant with temperature* (see Figure 25).

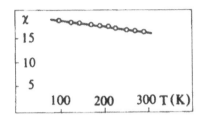

Figure 25   Paramagnetic susceptibility of aluminium χ_0 **as a function of the temperature (in** 10^{-6} **CGS). K Honda.** *Ann. der Phys.* **XXXII,** 1027 (1910).

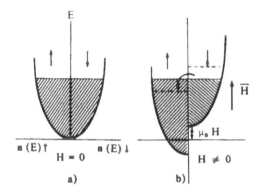

Figure 26   Evolution of two half bands (spin ↑ and ↓) in the presence of a magnetic field $\overline{H}$.

Let us consider the graph (Figure 14, p. 48) of the density of the occupied states and suppose initially that $T = 0$ K. Each space function corresponds to two spin states α and β, or ↑ and ↓. In the absence of a magnetic field, these two states have the same energy, and we may split the density $n(E)$ into two identically populated *half densities*, the first of which, $n(E)\uparrow$ corresponds to ↑ states, and the second, $n(E)\downarrow$ to ↓ states (Figure 26a).

When the magnetic field $\overline{H}$ is introduced, the spin magnetic moments may *align* with (state ↑) or *anti-align* (state ↓). In the first case, each state has its energy decreased by $\mu_B H$ (μ_B = Bohr magneton = $e\hbar e/2mc$ = 9.27×10^{-24} J/Tesla) and the half band ↑ moves downwards (decrease in energy), without deformation (**rigid band**). Similarly, the half band ↓ moves upwards by $\mu_B H$. Thus, this latter half band is depopulated for the benefit of the other, until the Fermi levels of $n(E)\uparrow$ and $n(E)\downarrow$ are identical (Figure 26b), just as though they were two communicating vessels.

The number of electrons changing the half band (in other words, changing spin) is $\Delta N = 1/2 n(E_F)\mu_B H$ (in fact, $\mu_B H$, which is equal to $\sim 6 \times 10^{-5}$ eV for $H = 1$ Tesla is always very small relative to E_F). They create a magnetisation $M = 2\Delta N \mu_B$ or

$$M = n(E_F)\mu_B^2 H$$

corresponding to a susceptibility

$$\boxed{\chi_P = n(E_F)\mu_B^2.} \tag{53}$$

This result remains true if the temperature is non-zero; thus, χ_P is *independent of the temperature*.

Indeed, since $\mu_B H$ is small in comparison to E_F, we may expand $n(E)\!\uparrow$ at a non-zero temperature in the form

$$n(E)\!\uparrow = \frac{1}{2}n(E + \mu_B H) = \frac{1}{2}n(E) + \frac{1}{2}\mu_B H n'(E)$$

and

$$n(E)\!\downarrow = \frac{1}{2}n(E - \mu_B H) = \frac{1}{2}n(E) - \frac{1}{2}\mu_B H n'(E).$$

Each half band contains $N\!\uparrow$ and $N\!\downarrow$ electrons where:

$$N\!\uparrow = \int n(E)\!\uparrow f(E)\,dE \qquad N\!\downarrow = \int n(E)\!\downarrow f(E)\,dE$$

where $f(E)$ is the Fermi–Dirac function (25).

The magnetisation (zero for $H = 0$, since then we have $N\!\uparrow = N\!\downarrow$) is

$$M = \mu_B(N\!\uparrow - N\!\downarrow)$$

or

$$M = \mu_B^2 H \int n'(E) f(E)\,dE.$$

Integration by parts gives:

$$M = -\mu_B^2 H \int n(E) f'(E)\,dE.$$

The derivative $f'(E)$ is a 'pointed' function which gets closer to the Dirac 'function' $-\delta(E - E_F)$ the lower the temperature. Thus, we have:

$$M \simeq \mu_B^2 n(E_F) H,$$

which again gives the susceptibility (53).

CHAPTER IV

SOME PROPERTIES OF 'GOOD' METALS

The Age of Bronze
Lord Byron

The properties of free electrons of metals such as those which have just been described (Chapter III) in terms of plane waves must affect the 'atomic' properties of the material. For example, the kinetic energy of the electrons (an inevitable consequence of the Pauli principle), which increases as the available volume decreases (see (36), p. 47), prevents the latter from collapsing beyond a certain limit and maintains a distance between the individual atoms.

Here, we shall use the results obtained in the previous chapter to present a simple view of certain atomic properties of 'good' metals (interatomic distance, cohesion, compressibility, thermal expansion), and show that this view is a very good approximation to reality for a common metal such as aluminium (for a definition of 'good' metals, see p. 76).

1. COHESION OF THE 'GOOD' METAL

1.1 Interatomic distance

The theory of the free electron has already provided a first cohesive equation: in (37′) (p. 49), we saw that the average energy of the electrons of a metal is composed of a *repulsive* term in r_e^{-2} and an *attractive* term in r_e^{-1}, where r_e is directly related to the dimension of the sample (for a given number of electrons N) via (34″), or more simply, to the interatomic distance.

It is clear that, since the ions did not intervene in any way, or rather, since they were replaced (see p. 49) by a uniform positive charge, we can scarcely expect (45) to provide a satisfactory description of cohesion. The latter is very sensitive to the **interaction between the electrons and the ions**. This is a difficult problem. We shall approach it below in a very simplified fashion.

For this, we describe a metal (of volume V, containing N_{at} atoms) as a packing of

67

spheres[1] of radius R defined by

$$\boxed{N_{at}\frac{4}{3}\pi R^3 = V.}$$

(1)

The radius R in the above is called the **atomic radius** and $\frac{4}{3}\pi R^3$ is the **atomic volume**.

In the case which we shall discuss from now on of a **monovalent metal** ($z = 1$), this atomic volume is equal to the electron volume ($R = r_e$; see (34″), p. 47). Each atomic sphere consists of *an ion*, likened here to a point charge $+e$ situated at its centre and *an electron charge* $-e$. Thus, each sphere is **neutral** and **does not interact** with the neighbouring spheres. Studying the energy, and, in particular, the stability, of the metal reduces to studying the energy of an atomic sphere. Finally, using space functions of type $\psi = \exp(i\vec{k} \cdot \vec{r})$ (see (29), p. 44) *reduces to assuming that the electron density is constant in the sphere* (since $\psi^*\psi = $ constant).

Under these conditions, the electrostatic energy of the *positive ion–electron cloud* ensemble is equal to

$$\boxed{E_1(R) = -\frac{9}{10}\frac{e^2}{R}\ \text{CGS}\ = -\frac{9}{10R}\ \text{a.u.}}$$

(2)

In fact, considering the Coulomb energy as a **perturbation** (see Q.M.), then, to a first order, we calculate a *cloud–ion* attraction term equal to

$$-\int_0^R \psi^* \frac{e3r^2\,dr}{R^3}\frac{e}{r}\psi = -\frac{3e^2}{2R}$$

(3)

and a *cloud–cloud* repulsion term equal to

$$\int_0^R \frac{e3r^2\,dr}{R^3}e\frac{r^3}{R^3}\frac{1}{r} = \frac{3e^2}{5R}.$$

(4)

Adding the kinetic energy ((36), p. 47) and the exchange term ((37), p. 49) to $E_1(R)$, we obtain

$$\boxed{E(R) = \frac{1.1}{R^2} - \frac{1.36}{R}\ \text{a.u..}}$$

(5)

This expression shows (Figure 1) a short-range *kinetic repulsion* proportional to $1/R^2$, which is directly associated with the Pauli principle, and a long-range *Coulomb attraction*. A stable minimum of the energy occurs for

$$R_0 = \frac{2 \times 1.1}{1.36} = 1.62\ \text{a.u.} = 0.86\ \text{Å}.$$

Table 1 shows experimental values of R_0 (deduced from the equality $(4/3)\pi R_0^3 = $ total volume/number of atoms, see (1)) for certain monovalent metals, and aluminium.

---

[1] It is clear that this packing of spheres does not fill space properly and gives rise to gaps and superpositions. As far as what follows is concerned, the error introduced remains small (see **N.B. 2**, p. 72).

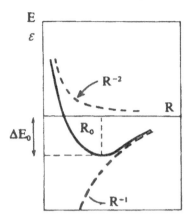

Figure 1 The two terms of (5) lead to a minimum of the energy for R (atomic radius) = R_0.

Table 1 Experimental values of the atomic radius for certain monovalent metals and aluminium.

	Li	Na	K	Rb	Cs	Cu	Ag	Al	
R_0	3.25	3.93	4.86	5.20	5.62	2.67	3.02	2.98	a.u.
	1.72	2.08	2.57	2.75	2.98	1.41	1.60	1.58	Å

We note immediately that the value found above ($R_0 = 1.62$ a.u.) is both too small and independent of the atomic species, in manifest disagreement with the increase in R_0 in the series of alkalis from Li to Cs.

These two facts point clearly to the weakest point of the above calculation, namely that we did not take into account the electron core of ions which were simply considered as point charges. In fact, the **cores** (regions with a high electron density) have a finite dimension and are difficult for free electrons to penetrate.

More exactly, in a global treatment of metal electrons, the states of valence electrons must be not only mutually orthogonal, but also orthogonal to the states of core electrons. The latter have space functions ψ_c very similar to the corresponding atomic functions (for example, 1s, 2s, 2p for Na). Thus, the space functions of valence electrons are radically modified in the core, in comparison with the planar functions $\exp(i\bar{k} \cdot \bar{r})$.

In metals, these cores are sufficiently small that they have a negligible effect on the properties studied until now (in Chapter III); however, even an elementary description of cohesion cannot ignore them.

One simple way of introducing this relative *impenetrability* of the cores involves replacing the purely ionic Coulomb potential used in (3) by an **effective potential** defined by

$$V(r) = \begin{cases} e^2 r^{-1} & \text{for } r > R_c \\ 0 & \text{for } r < R_c \end{cases} \tag{6}$$

where R_c denotes the radius of the core which is assumed to be spherical and strictly opaque (Figure 2).

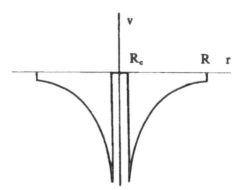

Figure 2 A simple way of simulating the electron core of the ion involves 'truncating' the Coulomb potential inside a sphere of radius R_c, where R_c is the core radius.

The integrals (3) and (4) are then replaced, respectively, by

$$\int_{R_c}^{R} \frac{3r^2\,dr}{R^3 - R_c^3} \frac{e^2}{r}$$

and

$$\int_{R_c}^{R} \frac{3r^2\,dr}{R^3 - R_c^3} \frac{r^3 - R_c^3}{R^3 - R_c^3} \frac{e^2}{r}$$

which, when expanded in terms of successive powers of $\rho_c = R_c/R$, give the Coulomb energy

$$\mathcal{E}_1 = -\frac{9}{10R} + \frac{3}{2R}\rho_c^2 - \frac{33}{10R}\rho_c^3 + \dots.$$

Thus, up to the first term in the expansion in powers of ρ_c, the Coulomb energy of the atomic sphere defined in this way is equal to

$$\mathcal{E}_1 = -\frac{9}{10R} + \frac{3R_c^2}{2R^3} \text{ a.u.}, \tag{7}$$

whence

$$\boxed{\mathcal{E}(R) = \frac{1.1}{R^2} - \frac{1.36}{R} + \frac{1.5R_c^2}{R^3} \text{ a.u.}} \tag{8}$$

The introduction of an opaque core decreases the volume available for the free electrons. Thus, it also has the effect of increasing the kinetic energy E_{kin}. We recall (see (34), p. 46) that the latter varies as $V^{-2/3}$ (where V is the available volume). Thus, we have:

$$\delta E_{kin}/E_{kin} = -\frac{2}{3}\delta V/V = \frac{2}{3}\rho_c^3.$$

This growth tends to increase the equilibrium radius R_0; however, restricting ourselves to terms of order two in ρ_c, we shall neglect this effect.

The energy $\mathcal{E}(R)$ is a minimum for

$$1.36R^2 - 2.2R - 4.5R_c^2 = 0. \tag{9}$$

The values of the *ionic radii* R_c are evaluated from the interionic distances in various ionic crystals (see Chapter I). Thus, R_c (Li) is calculated from the distances Li^+-Cl^-, Li^+-Br^- etc ... in the crystals LiCl, LiBr etc, measured by X-ray diffraction. Table 2 shows the values found in this way for the *alkali metals* and for *aluminium*.

Table 2   Ionic radii (in a.u.) of the alkali metals and of aluminium.

	Li	Na	K	Rb	Cs	Al
R_c	1.13	1.79	2.51	2.79	3.19	1.5

Then we may calculate the positive root R_0 of (9) (see Table 3). The values found in this way are in slightly better agreement with the experimental values (Table 1 and Figure 3).

Table 3   Positive root (in a.u.) of equation (9). R_c takes the values given in Table 2. For Al, see equation (11).

	Li	Na	K	Rb	Cs	Al
R_0	3.02	4.16	5.44	5.95	6.66	3.57

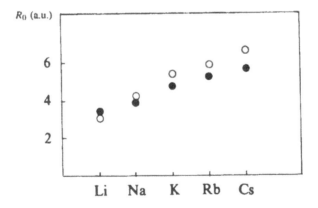

Figure 3   Experimental (black circles) and theoretical (white circles; equation (9)) values of the atomic radius, defined by (1), for the alkali metals.

N.B. 1   The above calculation overestimates R_0, except for Li. We have clearly exaggerated the 'opacity' of cores, for simplicity, by setting $V(r) = 0$ in the core. In fact, conduction electrons penetrate the cores, and it is always possible to make R_0 (theoretical) equal to R_0 (measured) by setting

$$V(r) = -V_0 \quad \text{for} \quad r < R_c \tag{10}$$

where V_0 is a tunable parameter. The corresponding effective (or 'pseudo') potential (Heine and Abarenkov) is frequently used.

N.B. 2   One may wonder about the consequences of the approximation which involves replacing the atomic volume by a *sphere* (see (1)). The former is better represented by the **Wigner–Seitz polyhedron** or *cell* (see p. 88 and Figure 4). Unlike the previous spheres, these cells side by side fill space exactly. An exact calculation of the Coulomb energy in such a cell thus leads us to replace the factor 0.9 in (2) by a number α which, as can be seen below (Table 4), is very close to 0.9 for the highly symmetric body-centred cubic (bcc), face-centred cubic (fcc) and close-packed hexagonal (hcp) structures. See Chapter V.

Table 4   Numerical constant of equation (2) for a Wigner–Seitz atomic cell.

Structure	bcc	fcc	hcp
α	0.895_{93}	0.895_{87}	0.895_{84}

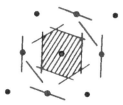

Figure 4   Wigner–Seitz cell (shaded) of a planar lattice of points.

For a *metal with valency* z, it is not difficult to check, using (45) on p. 52 and (7) together with the equation $r_e = Rz^{-1/3}$, that expression (8) becomes

$$\mathcal{E} = \frac{1.1z^{5/3}}{R^2} - (0.9z^2 + 0.46z^{5/3})\frac{1}{R} + 1.5z^2R_c^2\frac{1}{R^3}. \tag{11}$$

1.2 Cohesive energy

This is the name which we have already given (see p. 4) to the energy (referred to a single atom and with a change of sign) which a crystal must acquire for it to sublimate into separate neutral atoms; whence the alternative name **sublimation energy** E_s.

To sublimate the crystal, we may first increase the radius R to infinity (Figure 1), which costs us the energy $\Delta E_0 = \mathcal{E}(R_0)$. Doing this, we not only increase the distance between the ions but also separate each ion from its valence electron(s). Reconstituting the neutral atom at this point, we recover the **ionisation energy** E_i; whence we have

$$|\Delta E_0| = E_i + E_s \tag{12}$$

in conformance with the scheme shown (see also p. 4).

The energy ΔE_0 may be calculated from expressions (8) or (11), using calculated values of R_0 (Table 3). Table 5 shows the values obtained in this way.

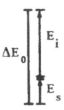

Table 5 Energies ΔE_0, E_i and E_s (see text) for certain metals (in eV).

	Li	Na	K	Rb	Cs	Al
ΔE_0 (calc.)	**−5.98**	**−5.31**	**−4.16**	**−3.84**	**−3.45**	**−50.6**
E_i (exp.)	5.36	5.12	4.34	4.18	3.87	59.9
E_s (exp.)	1.36	1.13	0.94	0.86	0.7	3.24
$E_i + E_s$	**6.72**	**6.25**	**5.28**	**5.04**	**4.57**	**63.1**

This table also shows experimental values of E_i and E_s. The theoretical values of $|\Delta E_0|$ are clearly of the order of magnitude predicted by (12), although they are systematically too small.

Here again, we pay the price for the excessive simplicity of the effective potential (6). Introducing the correction term V_0 (see (10)), which decreases the weight of the third term of (8), would increase and improve the calculated value of $|\Delta E_0|$.

1.3 Elasticity. Compressibility

The graph of $E(R)$ (Figure 1), or better still, that of $\mathcal{E}(R)$ (see (8)), provides a direct description of the **elastic behaviour** of the metal subject to a *hydrostatic stress*. The resistance to a compression is contained in the kinetic-energy term, which is itself a direct consequence of the Pauli principle. The resistance to an expansion is due to the Coulomb forces (summarised in (3)).

A solid of volume V shrinks generally in proportion to the pressure P to which it is subjected, and its **compressibility** χ at $T = 0$ K is defined by:

$$\chi = -\frac{1}{V}\left(\frac{\partial V}{\partial P}\right)_T.$$

$K = \chi^{-1}$ describes the rigidity and is called the **bulk modulus**.

Let us work with the atomic volume $(4/3)\pi R_0^3$. From $P = (1/S)(-\partial \mathcal{E}/\partial R)$ (where S is the atomic surface area, $S = 4\pi R_0^2$), it follows that

$$\chi = \frac{12\pi R_0}{\mathcal{E}''(R_0)}. \tag{13}$$

Thus, the *elastic rigidity* of a metal is *proportional to the curvature* of the graph of $\mathcal{E}(R)$ at the minimum point. Knowing $\mathcal{E}(R)$ (see (8)) and R_0 (Table 3), we calculate $\mathcal{E}''(R_0)$. The corresponding values χ_{th} are shown in Table 6, where they are compared with the experimental compressibilities.

Here again, we obtain sensible results. In particular, we rediscovered the fact that the compressibility increases strongly when one moves from Li to Cs.

Table 6 Calculated (formula (13)) and experimental (room temperature) compressibilities of alkali metals. 10^3 a.u. = 3.42×10^{-12} cm² dyne⁻¹ (see inset, p. 36).

	Li	Na	K	Rb	Cs	
χ (calc.)	1.6	5.1	14.1	20.1	30.9	10^3 a.u.
χ (exp.)	2.5	4.6	10.3	15.2	20.5	

1.4 Thermal expansion

Let us now consider two neighbouring atomic spheres[2], separated by a distance $b \simeq 2R_0$ and conceptually isolated from the rest of the metal (Figure 5). To say that the second contracts (or expands) by ΔR amounts to saying that the distance between the two decreases (or increases) by ΔR. In this case, the energy varies by $U(-\Delta R)$ (or by $U(+\Delta R)$, where U denotes the energy of the sphere, calculated from the minimum value (Figure 6). These cycles of expansion–contraction are actually constantly provided for by the *thermal vibrations* (see **Problem 18**). More simply put, we shall say that the second atom oscillates in the well with potential $U(\Delta R)$.

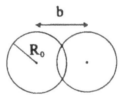

Figure 5 Adjacent atomic spheres with radius R_0, a distance b apart.

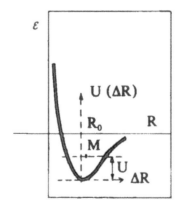

Figure 6 The potential $\mathcal{E}(R)$ (see Figure 1) is anharmonic. The midpoint M of the oscillation moves away from the origin as U (in other words, the temperature) increases.

----

[2] See footnote to p. 68.

If this well were symmetric (i.e. harmonic), the midpoint of the oscillation would be the centre of the well (R_0). The 'molecule' consisting of the two atoms would have constant size b, whatever the amplitude of the oscillation; in other words, whatever the temperature T. In fact, the well is clearly not symmetric (Figure 6) and the midpoint M of an oscillation with a given energy U drifts towards positive ΔR: the molecule grows longer: this is **thermal expansion**. To a first order, the latter is linear with temperature and is greater the more marked the asymmetry (**anharmonicity**) of the well is[3].

Let us give a simple description of this phenomenon by extending the previous 'molecule' in a straight line *of adjacent atoms*, at distance b apart. Recalling that the probability of excitation of a level U of the oscillator (Figure 6) is proportional to $\exp(-U/kT)$ and constraining the oscillations to take place on the line (of length l), we write

$$\frac{\Delta l}{l} = \frac{\overline{\Delta R}}{b}$$

where $\overline{\Delta R}$ is the mean value of the ΔR and is given by

$$\overline{\Delta R} = \frac{\int_{-\infty}^{\infty} \rho \exp(-U(\rho)/kT)\,d\rho}{\int_{-\infty}^{\infty} \exp(-U(\rho)/kT)\,d\rho}. \tag{14}$$

We have set $\Delta R = \rho$. We then write U in the form

$$U \simeq \alpha\rho^2 - \beta\rho^3 \quad \text{where} \quad \begin{cases} \alpha = (1/2)\mathcal{E}''(R_0) & (> 0) \\ \beta = (-1/6)\mathcal{E}'''(R_0) & (> 0) \end{cases}$$

knowing that $\mathcal{E}''(R_0)$ and $\mathcal{E}'''(R_0)$ may be calculated using (8) and the values of Table 3. At this level of the expansion in ρ, U is the sum of a **harmonic term** and an **anharmonic term**, which is small in comparison with the former. β is the *coefficient of anharmonicity*. Then we may expand the exponentials of (14):

$$\exp(-U/kT) \simeq \left(1 + \frac{\beta\rho^3}{kT}\right)\exp(-\alpha\rho^2/kT),$$

whence

$$\overline{\Delta R} = \frac{\beta\alpha^{-5/2}(kT)^{3/2}\int \eta^4 \exp(-\eta^2)\,d\eta}{\alpha^{-1/2}(kT)^{1/2}\int \exp(-\eta^2)\,d\eta} = \frac{3\beta kT}{4\alpha^2}.$$

Finally, we find an *extension proportional to T*, which is currently observed experimentally, and a **coefficient of linear expansion** λ (defined by $\Delta l/l = \lambda\Delta T$) equal to:

$$\lambda = \frac{3k}{4b\alpha^2}\beta \quad \text{or} \quad \lambda = \frac{-\mathcal{E}'''(R_0)}{2[\mathcal{E}''(R_0)]^2}\frac{k}{b} \tag{15}$$

which is **proportional to the coefficient of anharmonicity** β.

Formula (15) applied to *lithium* gives the following values $\alpha = 0.034$, $\beta = 0.025$, $b = (8\pi/3)^{1/3}(\sqrt{3}/2)R_0 = 1.76R_0$ (for a bcc structure), or

$$b = 5.31 \text{ a.u.} \quad \text{and} \quad \lambda = 10 \times 10^{-6} \text{ K}^{-1}$$

These are reasonable values for metals: $\lambda(Cu) = 17$, $\lambda(Ag) = 19$, $\lambda(Al) = 24 \times 10^{-6}$ K^{-1}, although the values for alkali metals are clearly underestimated: $\lambda(Li) = 45 \times 10^{-6}$ K^{-1}.

---

[3] Note (see **Problem 19**) that the mechanism (and even the sign) of the expansion of *stretched elastomers* is different.

N.B. 1   At low temperatures, the thermal expansion $\Delta l/l$ behaves like T^n ($n \simeq 3$–5) and not like T. Note that at low temperatures, the above expansion of the exponentials is no longer valid.

N.B. 2   There exist a number of exceptional cases of metals with a *negative expansion coefficient*. For example, in the face-centred cubic phase of *plutonium* (δPu), which is stable between 315 and 445° C, one finds $\lambda = -5 \times 10^{-6}$ K^{-1}: plutonium *contracts* on heating. This anomaly is attributed to the superposition of an anharmonic effect (with $\lambda > 0$, see above) and a thermally activated change in the electronic structure (transition of states f to states d), which contracts the atomic volume.

Certain magnetic alloys have been developed in such a way that their thermal expansion is practically zero, at least in a certain temperature range. These are the INVAR alloys.

N.B. 3   For metals with a *non-cubic structure*, the expansion is generally *anisotropic*. For example, in α *uranium* with an orthorhombic structure, one observes a *positive* expansion along two axes and slightly *negative* expansion along the third. Such materials, if they are polycrystalline (see later, p. 217), deteriorate rapidly when the temperature varies (in particular, during *thermal cycling*), as a result of the thermal stresses created between grains of expansions.

2. A 'GOOD' METAL: ALUMINIUM

As one might guess, the above very simple description of metals, based on the free-electron theory, cannot claim to provide a satisfactory description of all metals. In particular, it does not explain the considerable differences in conductivity, cohesion, crystal structure, magnetic properties (ferromagnetism) etc observed between metals such as lithium, tungsten or iron. To neglect the periodic crystalline structure of metals as we have done and to merely replace the electronic core of ions by a 'black' sphere must be paid for at some stage[4]. It is no accident that the metals chosen as examples in the previous sections, the *alkali metals* are those with the simplest possible atomic structure.

However, there exist metals other than alkali metals which are reasonably well described in the frame of free electrons. We call these *'good' metals*. One such is aluminium.

2.1 Aluminium is a 'good' metal

We shall see in Chapter VI that the introduction of ionic potentials transforms the *Fermi sphere* into a Fermi surface, which is the site in $\bar{k}$-space of points representing states of maximum energy. We shall speak of 'good' metals when this surface is little different from a sphere.

Among the methods used to determine this surface experimentally, one of the most direct is the study of the **annihilation of positrons**. Positrons from a radioactive source (for example, ^{64}Cu; mean energy $\simeq 0.3$ MeV) are sent onto the metal to be studied. These positrons e^+, antiparticles of the electron, are firstly slowed down by Coulomb collisions. Once *thermalised* (in other words, practically at rest), they are annihilated by the electrons of the metal, the lifetime of the positron in the metal being of the order of a hundred picoseconds. The annihilation is accompanied by the emission of two (more rarely three) photons γ. If the electron and the positron are at rest at the time

---

[4] This is what we shall do in Chapter VI.

of the annihilation, the two γ are emitted in opposite directions with energy $E = mc^2 = 511$ keV. If the electron has a momentum p ($\gg p(e_+)$), the two photons exhibit a (small) deviation from collinearity θ and a Doppler effect ΔE, such that

$$\theta = \frac{p_\perp}{mc}, \quad \Delta E = \frac{p_\parallel}{2}c$$

where $p_\perp$ and $p_\parallel$ are the components of $\overline{p}$ perpendicular and parallel to the direction $x'x$ of emission of the two photons. In the case $p_\parallel = 0$ (Figure 7), there is no Doppler effect, and we have

$$\theta = p/mc = v/c. \tag{16}$$

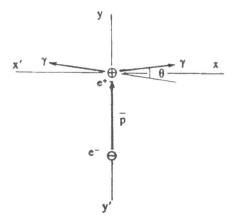

Figure 7   Annihilation of an electron $\ominus$ with momentum $\overline{p}$ with a positron $\oplus$ at rest. The two emitted photons γ form an angle θ, which depends on $|\overline{p}|$. Two counters, placed on either side of the sample at a large distance (several metres) from it are used to measure θ accurately.

For a kinetic energy of $\frac{1}{4}$ a.u. ($\simeq 7$ eV), we have

$$v = \sqrt{2E} \text{ a.u. } = \frac{1}{\sqrt{2}} = 1.5 \times 10^6 \text{ m s}^{-1}$$

(see inset p. 36), whence $\theta = 5$ mrad. Thus, measurement of the *angular correlation* between the two photons enables us to measure p.

Since, by virtue of their charge, positrons have a probability of presence which is zero on the nuclei and a maximum half-way between these, the majority are annihilated by *valence electrons*. Maximum momentum mv_F of the Fermi electrons, in an angular correlation experiment, should correspond to a maximum angle $\theta_F = v_F/c$, with all the annihilations taking place between $-\theta_F$ and $+\theta_F$.

Furthermore, turning the crystal with respect to the direction of observation $x'x$, one can check whether θ_F (whence v_F) is constant in space; in other words, if the surface on which the extremities of the vector $\overline{v}_F$ are located is actually a sphere.

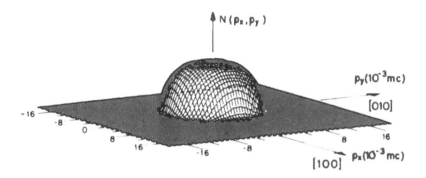

Figure 8   Positron annihilation in a single crystal of aluminium. The two emitted γ photons are detected in a direction $x'x$ (see Figure 7) in a fixed plane $p_x p_y$ of the crystal and the angle θ is measured. The angles $(-16, -8 \ldots)$ are in mrad, p_x is in the direction [100] of the crystal (see p. 99), p_y is in the direction [010].

One observes that there exists a cut-off angle ($\simeq 7$ mrad), and that, since the surface is a surface of revolution, this angle is constant in the various directions of the given plane $p_x p_y$.

A A Manuel et al. *Helv. Phys. Acta.* **52**, 255, (1979).

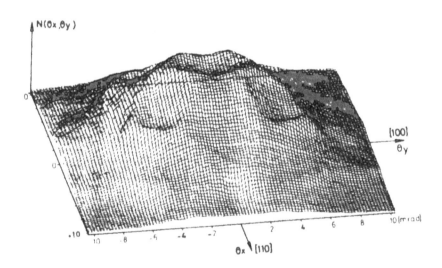

Figure 9   Positron annihilation in a single crystal of niobium. The same experimental procedure as for Figure 8. The two directions p_x and p_y are the directions [110] and [100] of the niobium crystal which is a bcc cubic crystal (see p. 106). This is far different from the symmetry of revolution observed for aluminium (Figure 8)!

A A Manuel et al. *6th Int. Conf. on Positron Annihilation (1982).*

This is effectively what is observed in aluminium. Figure 8 shows a 'surface of angular correlation', where the counting rate of the annihilation photons is shown as a function of the angle θ (measured in mrad) relative to the various directions of a plane $p_x p_y$. This plane here is parallel to a face of the elementary cube of the crystalline lattice. It can be clearly seen that *there exists a cut-off angle* θ_F and that *this angle is constant for the various directions of the plane* $p_x p_y$. Precise analysis of the results shows that this angle θ_F is equal to 6.79 ± 0.1 mrad., which is a remarkable agreement with the theoretical value of θ_F (6.78 mrad.) derived from (16) by setting $v = v_F$ (see p. 47). Using several other sections with other planes $p_x p_y$, one sees that *this angle θ_F* (thus, also v_F) *is constant for all directions*.

According to this criterion of quasi-sphericity of the Fermi surface, it appears that **aluminium is a 'good' metal**.

N.B. By way of comparison, Figure 9 shows a surface of angular correlation obtained in the same way as that of Figure 8, but for a crystal of *niobium*. It is immediately apparent that, based on the above criterion, niobium *is not* a 'good' metal.

2.2 Physical and metallurgical properties of aluminium

Aluminium is an abundant element of the Earth's crust. In particular, it is *the most abundant* metal. Isolated for the first time as a metal by Oersted in 1825, it is obtained by electrolysis of bauxite (hydrated oxide). This metal is industrially important, because of its *high conductivity* (2/3 that of copper at room temperature), its *low density* and its relatively *low price* ($\simeq$ $1500 per tonne in 1997). Its main physical characteristics are recorded below.

Table 7   Main physical properties of aluminium. a is the edge of the cubic cell at 300 K, b is the distance between adjacent atoms, r_e is the electron radius (see (34″), p. 47), R is the atomic radius (see (1)), ρ is the electrical resistivity, κ is the thermal conductivity, L is the ratio $\kappa\rho/T$ (see (52), p. 63). γ is the electron specific heat term (see (46), p. 54), χ is the compressibility (see (13)), v_F is the Fermi velocity (see (35′), p. 47) determined by positron annihilation and E_F is the Fermi energy ($= \frac{1}{2} m v_F^2$).

Atomic number	13
Atomic mass	26.98
Atomic state	$1s^2/2s^2\ 2p^6/3s^2\ 3p^1$
Valency	3
Crystal structure	fcc
Melting point	659.7° C
Density	2.7 g.cm^{-3}
a	4.05 Å, 7.64 a.u.
b	2.86 Å, 5.40 a.u.
r_e	1.10 Å, 2.07 a.u.
R	1.58 Å, 2.98 a.u.
$\rho_{300\,K}$	2.7 $\mu\Omega$ cm
$\kappa_{300\,K}$	2.37 W cm^{-1} K^{-1}
$L_{300\,K}$	2.15 WΩ K^{-2}
$\chi_{300\,K}$	1.32×10^{-12} cm^2 dyne^{-1}, 384 a.u.
γ	1.36 mJ mole^{-1} K^{-2}
χ_p	16.5×10^{-6} CGS
v_F	2.04×10^6 m s^{-1}
E_F	11.7 eV

Let us compare these experimental data with those which can be calculated from the results of this and the previous chapters.

The theoretical **Fermi velocity** $v_F = 1.92r_e^{-1}$ ((35') p. 47) equal to 2.03×10^6 m s^{-1} agrees remarkably well with the experimental value (2.04×10^6 m s^{-1}) determined by positron annihilation.

From (33) (p. 45), we deduce that $n(E_F) = L^3 \times 0.194 r_e^{-1}$ or here $n(E_F) = 62 \times 10^{23}$ a.u./mole, whence the *specific heat term* γ is given by $\gamma = 205 \times 10^{13}$ a.u. $= 0.90$ (instead of 1.36) mJ mole^{-1} K^{-2}.

A discrepancy of this magnitude is common for metals. It is often described by remarking that $n(E_F)$ (and thus γ) is proportional to the electron mass m (compare (33), p. 45 and (35), p. 47) and writing:

$$\frac{\gamma(\text{exp.})}{\gamma(\text{th.})} = \frac{m^*}{m}.$$

m^* is called the *effective thermal mass*. Here, it is equal to $1.36m/0.90 = 1.5m$. This introduction of an effective mass is linked, on the one hand, to the influence of the periodic potential of the crystal on the movement of the electrons (see Chapter VI) and, on the other hand, to the vibration modes of the atoms (*phonons*): when an electron moves, it causes a local crystalline distortion which tends to increase the apparent mass. It seems that, in the case of aluminium, it is this last effect which predominates.

The **electrical resistivity** ρ and the **thermal conductivity** κ lead to a number $L = \kappa\rho/T$ close to the theoretical Lorenz number: 2.15 (instead of 2.45) $\times 10^{-8}$ WΩ K^{-2} at 300 K (see (52), p. 63).

Moreover, since extremely pure aluminium may be obtained in the laboratory (concentration of foreign atoms less than 10^{-6}), the term ρ_i (see (49), p. 58) corresponding to the foreign atoms may be obtained to be very small. Since the thermal term ρ_{th}, decreases as T decreases (Figure 22, p. 59), the resistivity of aluminium at very low temperature (for example, 4 K) may decrease to 10^{-13} Ω m. Under these conditions, the *electron mean free path* (see p. 57) may reach a centimetre.

For aluminium, the **energy** $\mathcal{E}(R)$ (see (11)) may be written as

$$\mathcal{E}(R) = \frac{6.86}{R^2} - \frac{10.97}{R} + \frac{30.37}{R^3}$$

and we deduce $R_0 = 3.57$ a.u. (exp. val.: 2.98 a.u.),

$$\mathcal{E}(R_0) = -1.86 \text{ a.u. (exp. val.: } -2.32 \text{ a.u.)}$$

and, with the aid of (13), the **compressibility** $\chi = 336$ a.u. (exp. val.: 384 a.u.).

These theoretical values, compared with the experimental values, show differences of 17% for the interatomic distances, 25% for the cohesive energy and 15% for the compressibility. The agreement with the experimental values could be improved at little cost by replacing the potential (6) by a potential *à la* Heine and Abarenkov (see p. 72), which is less radical than (6) and tunable. This is nothing more than a numerical exercise, which does not bring additional understanding and is thus not of great interest.

It is clear that the theory of 'good' metals as described above gives a reasonable description (in comparison with its extreme simplicity) of the physical and metallurgical properties of a metal such as aluminium.

However, the absence of any reference to the **crystalline structure** of metals and, more generally, that of solids, is not tenable in the long term. The latter intervenes, often decisively, in an understanding of physical properties such as conductivity or mechanical properties.

We can no longer economise on this.

CHAPTER V

PACKINGS: ORDER OR DISORDER?

A Tangled Tale
Lewis Carroll

A Crystal Age
William Henry Hudson

The question which now arises concerns the manner in which *atoms* (or *ions* or *molecules*) pack together to form a solid.

Often (but not necessarily), packings are spatially periodic with a *long-range order*, which defines **crystals**. We met these even in Chapter 1 and quoted Madelung for whom these packings of ions seemed 'the most natural' conceivable (p. 2). Here, the term 'natural' represents an extremely subjective concept. It is also natural, perhaps even more so to anyone who has packed a jumble of identical objects – screws, marbles, sugared almonds, etc – into a bag, to state that the most probable packing does not have a long-range order. This is the case for **glasses** (or **amorphous solids**).

Let us forget the term 'natural' and look instead at nature. This exhibits an imagination far less unbridled than our own and reveals a profusion of families, including not only the two main ethnic groups mentioned above (with all their varieties, genera and types), but also many other collateral or intermediate branches. With a modicum of simplification which should not be delusive, we shall describe firstly some *amorphous packings* (typical of very different materials including wood, glass, plastics, etc) and secondly several *crystalline packings* (typical of most metals, semiconductors, rocks, etc), before considering a number of *intermediate packings* between these two extremes.

1. AMORPHOUS PACKINGS

These are characterised by the absence of a long-range order in the packing. We shall consider two very different examples: *metallic glasses* and *polymers*.

1.1 Metallic glasses

1 – General description

Discovered in the 1960s[1], these are obtained by very rapid cooling from the liquid state of certain alloys such as Pd-Si or Fe-B. Since metallic atoms (here, Pd or Fe) are very

---

[1] By **P Duwez**. Some metallic alloys may also be 'amorphised' when irradiated by fast heavy ions (**A Barbu, H Dammak, A Dunlop** and **D Lesueur**, *Mater. Res. Soc.* **20**, 29 (1995)).

much in the majority (for example, 80%) and more voluminous than metalloids (Si or B), we may conceptually reduce these solids to packings of metallic atoms, themselves similar to spheres of radius R. Finally, we regard these spheres as *hard spheres* (the case of a bag of *marbles*).

A trivial sterical constraint then means that any pair of spheres must be separated by a distance not less than $2R$. Moreover, we shall suppose that these metallic atoms tend to pack together *locally* in a compact fashion (in other words, they form regular tetrahedral aggregates of four atoms, see Figure 1), just like our marbles. In addition, the bases of two, then three,... tetrahedra may (and indeed tend to) come together during the course of the solidification, thus forming dense *aggregates* of tetrahedra. It is easy to see (if necessary, by experimenting with marbles or cardboard tetrahedra) that this aggregation cannot continue in the long term without leaving interstitial gaps here and there. In other words, it is impossible to *tile* 3D-space with regular tetrahedra, where **tiling** is the operation of packing together identical polyhedra indefinitely, without leaving spaces and without superposition (see also Subsection 3.4, p. 110).

Figure 1   Compact cluster (tetrahedral) of four atoms (or marbles).

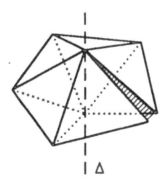

Figure 2   Five regular tetrahedra with a common edge Δ cannot all be joined together.

N.B. 1   Conversely, tiling is possible, for example, with identical *cubes* in 3D or with *rectangles* in 2D (see also p. 99).

We also note that tiling is possible in 2D with *regular hexagons*. In particular, this is the form most readily adopted by a *planar* packing of marbles. This is then *crystalline*.

N.B. 2 That it is impossible to tile 3D with regular tetrahedra, is clearly shown in Figure 2, in which tiling has been initiated around a common edge. This is a result of the fact that the dihedral angle (equal to $70°32'$) is not a submultiple of $360°$.

Here we have an example of conflict between the existence of local tendencies (in this case, the tendency towards compactness which favours regular tetrahedra) and the properties of Euclidean space when the latter cannot accommodate these tendencies: this is **geometric frustration**[2].

Let us indicate very briefly using an example how this frustration may be reduced. Thus, imagine that in 2D a certain local tendency gives rise to the formation of regular *pentagonal* objects. It is clear (Figure 3(a)) that we cannot tile the plane with these objects. However, if we give our 2D space a *constant curvature* with a suitable amplitude, this tiling becomes possible and leads to the formation of a regular dodecahedron on the sphere S_2 (Figure 3(b)). It can then be shown that *uncurving* the space (to return to the plane we started with) reduces to introducing a **defect** called a *disinclination*, the definition and an example of which will be found later (p. 207). The procedure is the same in 3D where one works on the hypersphere S_3. In this topological approach[3], the concept of *defect*, with which we shall have a great deal more to do later (Chapters VII to X) in another context, plays a critical role in the description of amorphous structures.

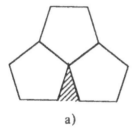
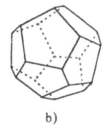

a)   b)

Figure 3 Tiling with pentagons, (a) impossible in the plane, (b) becomes possible on the sphere S_2 and forms a regular dodecahedron.

Tetrahedra may be packed together in infinitely many ways, since the manner in which the interstices are arranged and the aggregates are linked together is not fixed by any construction rule. Moreover, this is also the case during rapid cooling of our alloy, when tetrahedral aggregates form spontaneously and join together, unable to respect the symmetry of the tetrahedron on the boundaries.

In this case, the final structure exhibits:

i) an *absence* of a *long-range order* (characteristic of glasses), again referred to as *topological disorder*;

ii) the presence of a **short-range order** (SRO) tied here to the postulated packing constraints: the non-penetrability of the spheres and the tendency towards compactness, which generates *clusters of tetrahedra*;

---

[2] The notion of *frustration* was introduced (by **G Toulouse**, 1977) in the case of magnetic structures (spin glasses) the interactions of which are thwarted. Thus, in an *equilateral triangular* lattice of spins with antiferromagnetic coupling (favouring couples ↑ ↓), it is possible to satisfy at most two couples (——) but not the third (– – –):

[3] **M Kléman** and **J F Sadoc**, *J. Physique lettres*, **40**, 569 (1979).

iii) the existence of 'free volume' corresponding to the interstices of the imperfect tiling (hatching on Figure 2).

N.B. 1 Despite the infinite diversity of packings of tetrahedral clusters, experiments (for example, shaking a bag of marbles) and computer simulation lead to a very reproducible value of the 'filling factor' (the ratio of the volume of the spheres to the volume of the packing). This value is 0.637. We shall return to this factor later (p. 103).

N.B. 2 There exists a regular polyhedron, the *icosahedron*, with 20 faces and 12 vertices (see Figure 4) which enables us to pack together locally 20 tetrahedra with common faces; however, these tetrahedra *are not* exactly regular. Thus, Figures 4 and 2 are not contradictory. With one marble in the centre in contact with 12 marbles centred on the vertices, we obtain a dense aggregate of 13 marbles. The SRO obtained in this way is *approximately* good, but again, it is impossible to propagate this packing through the space in a periodic fashion.

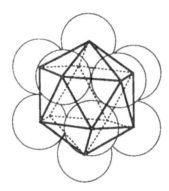

Figure 4 Icosahedral packing of 13 spheres centred on the 12 vertices and in the centre of a regular icosahedron (only the nine visible spheres are shown).

It is possible to develop aggregates of n identical atoms from vapours. If these aggregates are charged (generally +1), they may be sorted and counted using a mass spectrometer. The resulting distributions generally lead to a sequence of numbers $n_1, n_2, \ldots$ of most frequent types. In particular, one often finds (argon, bismuth, silver[4], etc) sequences (like 13, 19, 23, ...) where the signature of the icosahedron may be found. These aggregates are amorphous although the corresponding solid (argon, bismuth, etc) is crystalline. Thus, there exists a critical number n_c (which may be $\simeq 50$ or 100) above which the structure switches from amorphous to crystalline.

Exercise Place one marble in the centre and 12 marbles on the 12 vertices of an icosahedron (Figure 4), with the latter in contact with the former. Give an elementary argument to show that these 12 marbles are almost (but not exactly) in contact with one another. Determine how many of these pseudo contact points there are (see the **Exercise** on p. 104).

N.B. 3 This SRO of four atoms (marbles) forming a regular tetrahedron is created by the relatively isotropic metallic bonding. Of course, other forms of bonding result in other SROs. Thus, in the

[4] I A Harris, R S Kidwell, J A Northby, *Phys. Rev. Lett.* **53**, 2390 (1984); A Hoareau, P Mélinon, B Cabaud, *J. Phys.* **18**, 1731 (1985).

case of *silica* SiO$_2$ and more generally of *glass*, **covalent bonding** dominates. The valency four of Si implies the formation of SiO$_4$ tetrahedra, which here form the elementary 'building block' of the SRO. These tetrahedra are joined at their vertices and each oxygen common to two tetrahedra binds these in an Si-O-Si bond which is in principle rectilinear. Here, forming an amorphous structure consists of ensuring that the SiO$_4$ building blocks are as little deformed as possible (respect for the SRO) while packing them in a disorderly way, in other words, deforming the angles Si-O-Si randomly. Figure 5 shows a 2D example of this type of disorder in which *triangular* building blocks AB$_3$ have been packed together with the A-B-A angle allowed to vary.

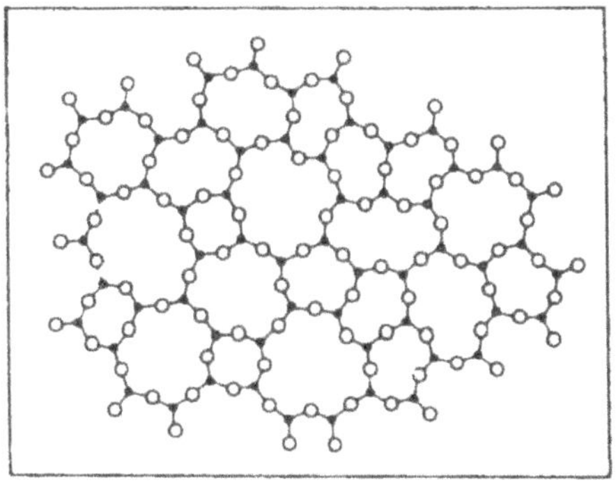

Figure 5 **Amorphous packing** (planar) of building blocks AB$_3$ (A: •; B: o). **W H Zachariasen**, *J. Am. Chem. Soc.* **54**, 3841 (1932).

Exercise Determine the formula A$_m$B$_n$ for the glass of Figure 5. Draw the distribution over p of the closed rings A$_p$B$_p$ and observe that it is centred on $p = 6$.

2 – Distribution functions

The short-range order which is generally present in glasses outlines the boundary between perfect order and total disorder: in glasses, the position of the atoms is neither fixed by a periodic rule as in a crystal, nor random as in a gas. In particular, we note that the atoms of the above metallic glass very often have near neighbours at a distance $2R$, quite often have second neighbours at a distance $4R\sqrt{2/3} \simeq 3.26R$ (twice the height of the tetrahedron of Figure 1), . . ., where these frequencies become uniform as the distances involved increase.

More generally, if we take the origin of the coordinates on a given atom, the number of atoms in an element of volume $d_3\bar{r}$ around point $\bar{r}$ (or $n_0 g(\bar{r})d_3\bar{r}$, where n_0 is the mean atomic density) enables us to define the *pair distribution function* $g(\bar{r})$ which represents the *correlations* between the atom positions. If the glass is isotropic, we have $g(\bar{r}) = g(r)$. Complete knowledge of the atomic structure implies a knowledge of the correlation functions for 2, 3, . . .,p, . . . atoms. Thus, $n_0^p g_p(\bar{r}_1 \ldots \bar{r}_p)d_3\bar{r}_1 \ldots d_3\bar{r}_p$ is the number of atoms in the elementary volumes $d_3\bar{r}_1 \ldots d_3\bar{r}_p$ around $\bar{r}_1 \ldots \bar{r}_p$. We know how to determine the correlation functions of a glass (or a liquid) experimentally by studying the *scattering* of a monochromatic wave (for example, X-ray scattering) through matter.

One example of a pair distribution function, measured by X-ray scattering, for Pd atoms in an amorphous Pd$_{80}$Si$_{20}$ alloy is shown on Figure 6. This clearly shows the absence of atoms from 0 to $\simeq 2$ Å, followed by a series of increasingly indistinct maxima, corresponding to the first, then the second, . . . layer of palladium atoms.

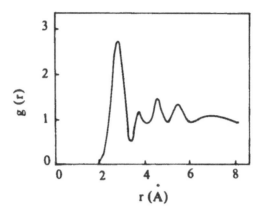

Figure 6 **Pair distribution function** $g(r)$ **in an amorphous solid.** The figure relates to palladium atoms in a $Pd_{80}Si_{20}$ alloy obtained amorphous by rapid cooling ($\simeq 10^6$ degrees s^{-1}) from the liquid phase. After D Lesueur (1973).

3. Wigner–Seitz cells

Let us consider an amorphous 3D-packing of atoms. The **Wigner–Seitz cell** (WSC), or *Voronei polyhedron*, of an atom is the polyhedron containing all points which are nearer to this atom than to any other. Its faces lie in the median planes of segments joining the atom to its neighbours. In the case of a 2D packing (Figure 7), the WSCs are polygons (see also **N.B. 2**, p. 72).

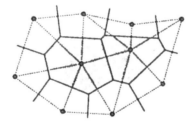

Figure 7 **Wigner–Seitz cells of a set of points** (•) **in the plane.** Points for which the corresponding cells are in contact are said to be contiguous. They are joined here by dashes.

WSCs tile the whole space. In the general case, in 3D, their faces adjoin two cells, their edges three cells and their vertices four cells (*or* two and three cells in 2D). Two atoms, the WSCs of which are in contact, are said to be **contiguous**. If the pairs of contiguous atoms are joined by rectilinear segments, we obtain a set of tetrahedra (*or* triangles in 2D) which generate another tiling of the space, the *Delaunay tiling* which is the *dual* of the Voronei tiling.

In the case of the 3D-packing of hard spheres, various numerical calculations have been used to determine the distribution of the number of faces of the WSCs and of the number of edges of these faces. The most common number of faces is 14 and the most common number of edges is five (predominance of pentagons, see Figure 8).

Figure 8 **Two examples of Wigner–Seitz cells** obtained by numerical simulation of the packing of hard spheres in 3D. **J L Finney**, Proc. Roy. Soc., A 319, 479.

1.2 Polymers

Polymers (the term plastics is also commonly used[5]) will be our second example of amorphous packings. These materials consist of extended linear molecules, which are themselves formed by laying n **monomers** end to end (here, these molecules are assumed to be identical). This laying end to end which necessitates the intervention of a catalyst is known as **polymerisation**. Thus, polyvinyl chloride (PVC) is obtained by linking n monomers of structure

$$- CH_2 - \underset{\displaystyle C\ell}{\underset{|}{CH}} -$$

to one another according to the final formula $(C_2H_3Cl)_n$.

In a given sample, these molecules have different shapes and different degrees of polymerisation n. For simplicity, we shall reduce this last variable to a unique value n_0. As far as the shape is concerned, we shall take this to be that of a chain in which the links (the monomers, of length $b \simeq 2$ to 3 Å) are connected at angles randomly selected from those permitted (or imposed) by the chemical bonding. Any correlation between the direction of the monomers is neglected[6].

1 – Shape and random walk

We consider the molecular chain of Figure 9 with the above conditions. We now suppose that the angles between the monomers are right angles, chosen amongst the six directions $(\pm x, \pm y, \pm z)$ of a reference trihedron. The shape thus obtained is strongly reminiscent of a random walk (Brownian motion) on a cubic lattice, with step length b. Thus, we know that (we shall return to this, p. 174):

---

[5] The birth of plastics may be traced back to 1830 (see **N.B. 4** p. 92), or to 1870, when a competition was organised with a view to replacing ivory which had become too expensive for large-scale manufacturing of billiard balls. This led to the birth of celluloid. The industrial boom of plastics took place only in 1920–1930. Now, one hundred million tonnes are manufactured annually: thus, for PVC alone (used for bottles, main pipes, electrical insulation, furniture, building, etc) in 1985 the figures were Spain–1.4 million tonnes, Great Britain–2.6, France–3.3, West Germany–7.6, Japan–9.2 and USA–21.7.

[6] For all the following, see **P G de Gennes**, *Scaling Concepts in Polymer Physics*, Cornell University Press, (1979).

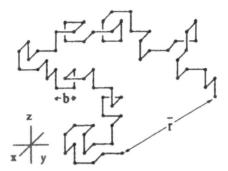

Figure 9 Polymeric chain obtain by a random walk with step length *b* on a cubic lattice.

i)   the distance *r* between the two extremities of the random walk (thus also of the molecule) is characterised by the root mean square

$$R = (\langle r^2 \rangle)^{1/2} = b n_0^{1/2}. \qquad (1)$$

Exercise The *radius of gyration* of the molecule is defined to be $R_G = \langle \rho^2 \rangle^{1/2}$ where $\langle \rho^2 \rangle$ is mean square distance between the barycentre of the molecule and the various monomers. Show that:

$$\langle \rho^2 \rangle = \frac{\langle r^2 \rangle}{6} = \frac{n_0 b^2}{6} \qquad \text{or} \qquad R_G = \frac{R}{\sqrt{6}}. \qquad (1')$$

Thus, for $n_0 = 10^4$ and $b = 3$ Å, we have $R = 300$ Å and $R_G = 122$ Å.

ii)   The probability $p(\bar{r})$ of observing a path $\bar{r}$ with uncertainty $d_3\bar{r}$ depends only on *r* and the latter is Gaussian:

$$p(r) = \left(\frac{3}{2\pi}\right)^{3/2} R^{-3} \exp\left(-\frac{3}{2}\frac{r^2}{R^2}\right). \qquad (2)$$

N.B. 1 This Gaussian variation is established later (p. 179) for a random walk in 1D when the law of $p(x)$ is then $(3/2\pi)^{1/2} X^{-1} \exp(-\frac{1}{2}(x^2/X^2))$. It is easy to check that, taking into account the independence of the movements in the three directions, passing from 1D to 3D does in fact give (2).

N.B. 2 For a molecule of given *R* (that is to say of given n_0), the function (2) describes the number $\mathcal{N}(\bar{r})$ of distinct configurations which permit it to have the extension $\bar{r}$. It follows that the *configuration entropy* (or $k \ln \mathcal{N}$) of this molecule is

$$S(\bar{r}) = S(0) - \frac{3kr^2}{2R^2} \qquad (3)$$

and thus that its free energy is

$$F(\overline{r}) = F(0) + \frac{3kTr^2}{2R^2}.\tag{4}$$

A force f applied to extremities of the molecule (assumed to be folded up, $r = 0$) produces an elongation Δr corresponding to the equilibrium $f = \partial F/\partial r$; or:

$$\Delta r = \frac{R^2}{3kT} f.\tag{5}$$

The behaviour is *elastic* and *linear* (proportionality between elongation and force); here, the elasticity (for example, that of rubber) is of entropic origin (see **Problem 19**), unlike in the case of most solids in which it results from an internal energy variation (see p. 73).

N.B. 3 If the angle between the monomers is α (some $\alpha \neq \pi/2$) it can be shown (see **Problem 19**) that formula (1) becomes

$$R = bn_0^{1/2}\left(\frac{1+\cos\alpha}{1-\cos\alpha}\right)^{1/2}.$$

iii) The probability of observing a path, whence a molecule, of 'distance' r with uncertainty dr is then $g(r)\mathrm{d}r = 4\pi r^2 p(r)$ or

$$g(r) = 6\left(\frac{3}{2\pi}\right)^{1/2} R^{-3}r^2 \exp\left(-\frac{3}{2}\frac{r^2}{R^2}\right)\tag{6}$$

(see Figure 10).

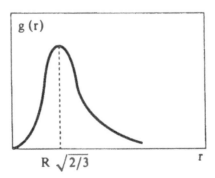

Figure 10 Distribution of the size of a polymer molecule. See (1) for definition of R.

Points i) to iii) lead us to a first conception of polymeric packing: the molecules curl up in *skeins*, the average 'size' of which (characterised by R or by ρ) varies as the square root of the degree of polymerisation (see (1) and (1′)). These skeins generally become entangled with each other, as shown in Figure 11. The complete absence of any long-range order justifies the use of the term *polymeric glass*.

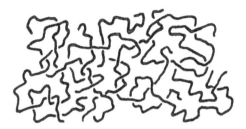

Figure 11   Tangle of polymeric skeins.

N.B. 4   It is possible to improve the stability and the rigidity of the polymeric structure by establishing bonds from skein to skein here and there by a chemical bridging (for example, with sulphur atoms: bridges – S_2 –), which decreases the term R in (5). This is the **vulcanisation** procedure, which Goodyear discovered in 1830 by adding sulphur to **latex** (the sap of the hevea, which the Indians used for boots) and thereby greatly improving the mechanical behaviour of this polymeric substance. This was **rubber** or, after further processing, **ebonite**.

More generally, any process which knits the chains together pointwise is called *reticulation*. This is frequently carried out, nowadays by irradiation of the polymer with γ photons.

2 – Excluded-volume constraint

The random-walk model comes up against an important objection: a random walk may (and often does) pass through a point which it has previously visited, although the molecule cannot superpose two monomers. Thus, the real problem is more subtle than that of Subsection 1.2.1: namely, it involves a *constrained* random walk in which the steps remain random but it is forbidden to revisit points. The term *excluded volume* is used since, within the explorable space, the molecule itself excludes a certain fraction of the volume. The term *self-avoiding* random walk (or chain) is also used.

That this constraint must modify the law (1) is evident for a 1D random walk. In this case, it forbids all backward steps and the distance between extremities is bn_0 (not $bn_0^{1/2}$). Intuitively, the constraint plays a less dramatic role in 2D and *a fortiori* in 1D, but must *increase* the size R (the molecule stretches) and modify the law (1) to

$$R = bn_0^\omega \tag{7}$$

where ω is a *critical exponent which depends on dimension d*: $\omega = \omega(d)$. ω is equal to 1 in 1D and it has recently been shown[7] that it is equal to $\frac{3}{4}$ in 3D and that (since Brownian motion is possible for any d) it is equal to $\frac{1}{2}$ for $d \geq 4$.

In 3D (usual for materials!) rigorous calculation of the exponent ω has not yet been possible. An approximate evaluation of ω ($\omega = \frac{3}{5}$) is given in **Problem 19** using the so-called **mean-field** approximation: for a given monomer, the repulsion of the monomer by the other monomers (which it is impractical to calculate) is replaced by a mean repulsion simply taken to be proportional to the local density of monomers (for a skein, this decreases from the centre outwards, according to a Gaussian law).

The previous discussion of the self-avoiding random walk may also be applied to the case of chains in solution (for example, polystyrene in ethyl acetate) and thus isolated. However, it is still a long way away from the real problem of condensed polymers (that is to say, the tangled skeins in Figure 11), since in this case, every chain must avoid not only itself but also its neighbours. In fact,

[7]   B Nienhuis, *Phys. Rev. Letters.* **49**, 1062 (1982). See also: B Duplantier, *Physica* A **163**, 158 (1990).

(the paradox is only apparent) this increase in complexity simplifies the problem: in condensed polymers measurements of R lead to $\omega \simeq \frac{1}{2}$, thus returning us to the ideal case. In fact, tangled molecules create excluded volume *uniformly* in space. Thus, the *mean field* is also uniform and the effect of the above stretching (7) tends to be softened.

3 – Tendency towards order

There exist cases (nature of the polymer, temperature, etc) in which, despite the above entropic tendency, the molecules adopt a simple form (for example, rectilinear) and regroup in an ordered (even crystalline) fashion. This tendency exists in certain textile fibres such as *nylon*, where it is apparent through the presence of *microcrystals* coexisting with the polymeric skeins and thereby contributing to the global solidity of the fibre.

4 – Polymeric gels

The polymeric chains up to this point were linear and (apart from the case of vulcanisation) independent. Once one knows how to create **branches** on certain monomers (for example, triples in the form of 'Y'), the polymer becomes a lattice of lines with multiple points (for example, triples).

Thus, we begin with a solution of monomers, the concentration c of which we increase progressively. First, numerous small isolated strands are created, with or without branches ($c \ll 1$). Then, these strands grow in size and the number of branches increases; the solution is slowly populated with polymeric *rafts*, which are still separated from each other.

We continue to increase c. Then we find a critical concentration c_p, called the **percolation** *threshold*, at which a certain number of large rafts become connected, forming a giant molecule of the size of the sample[8]. Whilst it is liquid for $n < c_p$, the solution becomes rigid for $c \geq c_p$ and the moduli of elasticity which were previously zero take non-zero values from this c_p. The solid thus formed (often at room or moderate temperature) is called a *gel* and the transformation at $c = c_p$ is called a **sol–gel transition**.

A transition of this type is responsible for the hardening of *egg white* when cooked and for the setting of milk into *cheese* or of *jellies*. This transition is also used to produce *soft contact lenses* (radicular polymerisation of methacrylate hydroxyethyl), to develop very low-density *silica gels*, etc.

5 – Fractal objects and percolation

Each of the above rafts is an object with a tangled structure and a complex shape. Its mass M does not increase as the cube of its size R. Such an object is said to be **fractal** if its mass and size are related by a power law:

$$M \propto R^d. \tag{8}$$

The exponent d, which is often non-integral and less than 3, is called the *fractal dimension* of the object. We note that the density of these objects decreases as $R^{(d-3)}$ as their size increases, which implies an increasing porosity[9].

---

[8] On the subject of percolation, one might refer to: **D Stauffer**, *Introduction to Percolation Theory*, Taylor & Francis, (1985).

[9] These objects are characterised by a shape which remains similar when they are observed at different scales. This property is also found, for example, for the successive points of a Brownian motion (as discovered by **Jean Perrin**), or in the distribution (projected onto a plane) of the galaxies (**R Balian**). See **B Mandelbrot**, *The Fractal Geometry of Nature*, W H Freeman (San Francisco) (1982), **B Sapoval**, *Universalités et fractales*, Flammarion (Paris) (1997).

Exercise We begin with a segment AB: A ————————— B, which we divide into three

equal parts and construct the line ACDEB: A___C／\̲E̲___B . Then we repeat the operation on each of the four equal segments AC, CD, DE and EB, and so on, many times. Use (8) to show that the (fractal) dimension of the object thus obtained (called a *Koch snowflake*) is 1.26. . . .

Exercise Find the fractal dimension of a polymeric chain, i) in the Brownian-motion approximation and ii) with the excluded-volume constraint (7).

Let us come back to percolation (previous page). Consider (Figure 12) a lattice of segments in the plane (for example, a square lattice, as shown by the thin lines in Figure 12(*a*), with edge *l*). Next, we distribute *n* bars of length *l* randomly amongst these segments (bold lines). According to the concentration *c* of the bars ($c = n$/number of segments in the lattice) the latter form sparse clusters (or rafts) (Figure 12(*a*)) or a continuous cluster (Figure 12(*c*)). A continuous cluster is produced at a critical concentration called the *percolation threshold*, which is well defined provided the lattice is large. In the given case, it can be shown that this concentration is equal to 0.50.

This *percolation* of bars (called *links*) must be distinguished from a *percolation of sites* in which objects (for example, chess pieces of diameter *l*) are placed on the *nodes* of the lattice (rather than on the edges). In this case, the critical concentration is equal to 0.59.

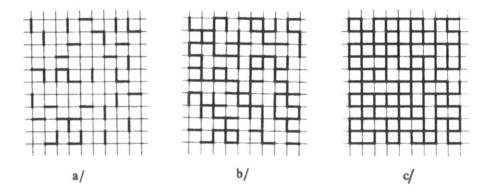

a/   b/   c/

Figure 12 Percolation of links on a 2D square lattice. (*a*) Concentration ($c = 0.20$) much lower than the percolation threshold c_p: the bars (bold) form isolated clusters. (*b*) Concentration ($c = 0.52$) just greater than c_p (= 0.50). A continuous assembly of bars appears. (*c*) Concentration ($c = 0.80$) greater than c_p. A continuous assembly of bars is formed.

In the case of the **gel** with triple points introduced above, the sol–gel transition may be described as a percolation of links over a particular lattice called a *Cayley tree*. Beginning at any point, we trace three segments (in 3D). At each extremity of these we trace two new segments, and so on (finitely) many times. Thus, we obtain a Cayley tree (Figure 13). The precipitation of the gel may then be represented as follows. We randomly distribute simple monomers on the branches of the tree and Y-shaped monomers on the forks. Then, topologically, the polymeric rafts are Cayley-tree fragments and the sol–gel transition corresponds to the percolation threshold on the tree.

Exercise On a Cayley tree with triple points, the concentration c is defined to be the proportion of 'occupied' branches. Use an argument from elementary demography to show that the percolation threshold is $c_p = 0.50$.

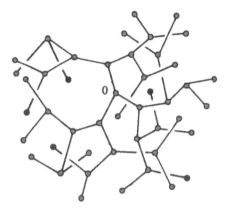

Figure 13 Cayley tree with triple points. Beginning at the origin O, this tree includes four generations of branches.

2. CRYSTALLINE PACKINGS

What is lost in terms of disorder (whence **entropy**) when passing from amorphous solids to crystals is often largely regained in terms of **energy**[10]. Many of the most common materials arise in crystalline form. This abundance, together with the simplifications resulting from the existence of a translational symmetry, explain why the following chapters (VI to X) are principally devoted to crystalline solids[11].

Here, we shall simply describe (for many students, this will probably amount to a reminder) the simplest structures, those with the greatest symmetry, which are not necessarily the most common. Thus, this paragraph makes absolutely no pretensions to cover the set of existing crystalline structures and symmetries.

In the first place, **crystallography** describes ideally periodic structures, the elements of which (atoms or molecules) are arranged in an ordering with an absolute periodicity.

This absolute state does not exist and *real* crystals are only approximately periodic. There are a number of reasons for this, some unavoidable (free surfaces, thermal vibrations), others more occasional (the existence of foreign atoms (or *impurities*) in the crystal or simply of different isotopes; the existence of local breakings of the periodicity known as **crystal defects**, etc). *Second,* crystallography must take account of all these imperfections, not only for rigour but also out of necessity. In fact, these imperfections play a major role as far as many properties of materials are concerned (in particular, *electrical* and *mechanical properties*). Two chapters (Chapters VII and IX) are devoted to the most commonly observed and the most important crystal defects.

[10] See later, **N.B.** on p. 124.

[11] In Paris, a visit to the superb collections of mineral crystals of the École des Mines (Bd. St. Michel) or the Université de Paris VI (Jussieu) is a *must*. Other magnificent collections may be seen in Washington (Smithsonian Institute), London (British Museum), Vienna, Aachen, Ouro-Prêto ...

2.1 Generalities

1 – Cell, lattice, basis

The fall of a stone may have been as important to *crystallography* as the fall of an apple to *mechanics*, in the sense of a simple, mythically crucial experiment. When, whilst visiting a friend who collected minerals, Abbot Haüy happened to drop and break a crystal of calcite ($CaCO_3$) which he was examining[12], he was struck, on picking up the fragments, by the similarities between the crystals: numerous plane faces, with sharp edges and, most of all, *equal angles* between *equivalent planes*.

Re-echoing an earlier intuition of Huyghens, Haüy postulated that these similarities were the result of the existence of a microscopic object, a sort of *elementary building block, with a characteristic parallelepiped shape, which we call a* **crystalline cell**. Then the crystal is a face-to-face packing of these building blocks, or *unit cells*, which in turn consist of atoms or molecules. Because of this, the shape of the crystal is a macroscopic reconstruction of the shape of the elementary building block[13]. This packing creates **atomic** (or molecular) **planes**, along which the crystals cleave preferentially.

Evidence of these atomic planes is provided by an experiment scarcely less simple than the 'experiment' of Haüy. These planes of atoms (more precisely of positive ions) have a positive charge. Conditions for a stable trajectory are thus afforded to a fast positive particle (proton, α particle ...) moving in a crystal, parallel to a family of planes, by two successive planes. The particle progresses between these two planes, repulsed by each alternately. It is said to be *channelled*. Under these conditions, it moves in a relatively empty space; thus, in crystal, it is subject to far less deceleration than other particles moving in random directions.

Suppose that we have a crystal of thickness h on a small *source* ($\simeq 0.5$ mm) of americium. The latter emits α particles ($\simeq 5$ MeV) by natural radioactivity. These particles penetrate the crystal. If the thickness h is sufficient, only particles which have been channelled along the atomic planes may emerge. For every family of planes, on exit, these particles constitute a planar 'curtain' of particles which can be made to materialise on a sensitive plate along the line of intersection of the atomic plane and the plate. This can be seen on Figure 14 which shows several families of atomic planes of a *gallium* crystal (orthorhombic crystal structure).

Haüy's intuition on the *existence* of cells and *crystalline planes* is fecund, but is not sufficient to account for the various crystal symmetries.

The most fundamental of these is **translational symmetry**. To describe this, we introduce a set of points called a **lattice**, points (or **nodes**) of which are defined from an arbitrary origin O by equation

$$\boxed{\overline{OM} = \bar{r}_T = m_1\bar{a}_1 + m_2\bar{a}_2 + m_2\bar{a}_3} \tag{9}$$

where $\bar{a}_1, \bar{a}_2, \bar{a}_3$ are three non-coplanar vectors, and m_1, m_2, m_3 are positive or negative integers which are not simultaneously zero. This lattice is called the **Bravais lattice** of the crystal, if from each of these points the crystal has an identical appearance in

---

[12] This does not mean that experimental clumsiness is the best way of guaranteeing scientific notoriety.

[13] **R J Haüy**, *Essai d'une théorie sur la structure des cristaux*, Paris (1784). Note, that while Haüy may be considered as the 'father' of crystallography, **Stensen** (1669), **Guglielmini** (1688) and **Romé de Lisle** (1772) had previously reported the law of equal angles.

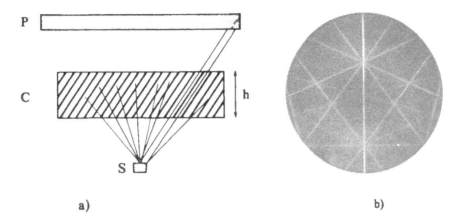

Figure 14 Direct view of an atomic order in a crystal. (*a*) α particles are emitted isotropically by an americium source S. They penetrate a crystal C, which has thickness *h* such that only *channelled* (whence, little slowed) particles may pass through it. The plate P is used to detect the impact of these at the intersection of P and the family of planes in question (only one family of planes is drawn here). (*b*) Image of the plate P after exposure of a gallium crystal. Each white line is the intersection of P with a family of crystalline planes of gallium. Thickness $h = 10 \mu$m. Distance between C and P: 10 mm. **J Takahashi** and **J Mory**, C.R. Acad. Sc. **292**, 1123, (1981).

terms of *chemical composition*, *structure* and *orientation*. In the particular case where O is situated on an atom (more exactly, at the mean point of thermal oscillation of an atom), each point M defines the position of an atom of the crystalline structure equivalent to the original atom.

Figure 15 shows a two-dimensional Bravais lattice ($a_3 = 0$ in (9)) defined on the two primitive vectors $\bar{a}_1$ and $\bar{a}_2$. It enables us to define a cell in the sense of Haüy, namely the cell $(\bar{a}_1, \bar{a}_2)$. We note immediately that this cell is not unique: the cell $(\bar{a}_1, \bar{a}_3)$ (amongst others) may be used to construct the same lattice as the cell $(\bar{a}_1, \bar{a}_2)$. They both have the same surface area. On the other hand, the cell $(\bar{a}_1, \overline{\alpha}_2)$ has

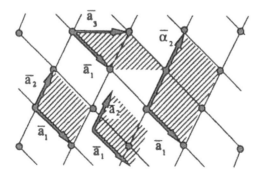

Figure 15 Some nodes of a 2-dimensional Bravais lattice. Cells of the lattice are shown in grey; these may be primitive $(\bar{a}_1, \bar{a}_2)$, $(\bar{a}'_1, \bar{a}'_2)$, or not $(\bar{a}_1, \overline{\alpha}_2)$.

double the surface area. The cells with the smallest surface area (in 3D, the smallest volume) of a Bravais lattice are called **primitive cells**. In Figure 15, $(\bar{a}_1, \bar{a}_2)$ and $(\bar{a}_1, \bar{a}_3)$ are primitive cells, $(\bar{a}_3, \bar{\alpha}_2)$ is not. Vectors $\bar{a}_1, \bar{a}_2, \bar{a}_3 \ldots$ are primitive vectors. We note that the surface area (in 3D, the volume) of the primitive cell is equal to the surface area (volume) allotted to a lattice node.

In addition to the translational symmetry (9), a Bravais lattice may have **rotational symmetries**. Thus, still in the plane, a square lattice $(a_1 = a_2, \bar{a}_1 \cdot \bar{a}_2 = 0)$ has a *4-fold* symmetry: the lattice is preserved by a rotation of $2\pi/4$ around any of its nodes (or around the centre of any of its squares). A rotation $2\pi/n$ is said to be of order n. It can be shown that in the plane and in 3D only the orders 2, 3, 4 and 6 are compatible with the translational symmetry (orders such as 5 are excluded).

We leave it as an exercise to show that, in terms of symmetries, in the plane there are five distinct types of *Bravais lattice*: the *oblique* lattice ($a_1 \neq a_2$, arbitrary angle $(\bar{a}_1|\bar{a}_2)$), the *hexagonal* lattice ($a_1 = a_2$, $(\bar{a}_1|\bar{a}_2) = \pi/3$), the *square* lattice ($a_1 = a_2$, $(\bar{a}_1|\bar{a}_2) = \pi/2$), the *rectangular* lattice ($a_1 \neq a_2$, $(\bar{a}_1|\bar{a}_2) = \pi/2$), and the *centred rectangular* lattice ($a_1 \neq a_2\sqrt{2}$, $\bar{a}_1 \cdot \bar{a}_2 = a_1^2/2$).

In 1845, Bravais showed that there are 14 types of lattice in 3D. These are the 14 *Bravais lattices*, including the *cubic* lattice, the *face-centred cubic* lattice, the *body-centred cubic* lattice, etc.

The Bravais lattice of a crystal describes the translational symmetry of the crystal. It says nothing about the nature of the crystal itself. This is defined by the *filling* (atomic or molecular) of the primitive cell. This filling (nature and position of its elements, possibly chemical bonds, spins...) is called the **basis**. The *Bravais lattice* and the *basis* together constitute the crystal:

$$\boxed{\text{(Bravais lattice)} + \text{(basis)} = \text{(crystal)}.} \tag{10}$$

Figure 16 shows an example of this 'equation' in two dimensions. The crystal (sheet of stamps) is the superposition of a Bravais lattice (rectangular) and a basis (*la Liberté*). For an arbitrary origin O (for example, the *lower left-hand corner* of the pattern or the *eye of la Liberté*, etc), the lattice 16(*a*) is a Bravais lattice, since the sheet of stamps (assumed infinite) has an identical appearance when viewed from each point of the lattice.

The choice of the *primitive cell* is arbitrary: for example, $(\bar{a}_1, \bar{a}_2)$ or $(\bar{a}_1, \bar{\alpha}_2)$. The basis may also be generated in infinitely many ways: for example, given the lattice 16(*a*), the crystal 16(*d*) may be reconstructed using either the basis 16(*c*) or the basis 16(*b*).

In some cases, it may be interesting to use a non-primitive cell, if in this way it is possible to reveal a new symmetry of the Bravais lattice. Thus, in the lattice 16(*a*), the cell defined by the basis vectors $(3\bar{a}_1, 2\bar{a}_2)$ is *square*. Under these conditions, one might prefer to use this square cell. It is self-evident that the relevant basis is then a multiple of the previous one: this new basis includes six *Libertés*. This new cell of higher symmetry is called the **unit cell**.

The basis may consist of *a single atom*. In this case, the set of atomic sites, sometimes called the *crystal lattice*, coincides with the Bravais lattice. This is the case, for example, in *aluminium*, for which the Bravais lattice (face-centred cubic) is also the crystal lattice (see Figure 21, p. 104). This is not the case in *silicon*, in which the Bravais lattice is also a face-centred cubic, but the basis includes *two* atoms, defining the *diamond structure* (see p. 107).

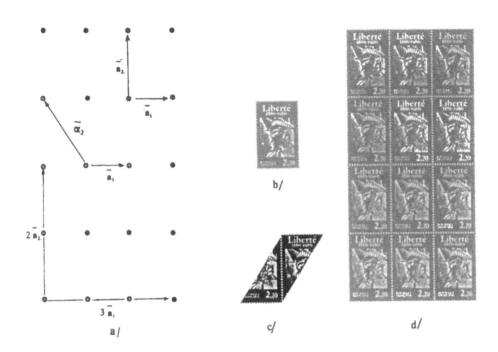

Figure 16 The planar crystal (d) is obtained by superposition of a Bravais lattice (a) and a basis, for example, (b) or (c).

Exercise Are the following objects crystals in the sense of (10)? If they are, describe the necessary conditions for this and define their Bravais lattice (in particular, its dimension) and pattern: a ream of paper, a chessboard, the arcades of the rue de Rivoli in Paris, a spider's web, a spiral spring, a sheet of corrugated iron, a pile of plates, oranges on a market stall.

2 – Rows, planes, indices

Since the crystal structure is defined by its Bravais lattice and its basis it is clear that a straight line passing through two equivalent atomic sites passes through an infinity of equivalent sites which constitute an **atomic row**. This row is parallel to a row of the Bravais lattice. A plane passing through three equivalent atomic sites defines an **atomic plane** (see Figure 14).

Once a cell (whether primitive or not) with vectors $\overline{a}_i$ has been chosen, an *atomic row* is defined by the smallest three integer coordinates of a vector of this row. For example, in a cubic lattice (Figure 17), the [110] row is a diagonal of a face of the cube. The set of equivalent rows is denoted by $\langle \ \rangle$. Thus, $\langle 110 \rangle$ represents the set of the diagonals of the faces of the cube. There are six such rows, namely: [110], [$\overline{1}$10], [101], [$\overline{1}$01], [011] and [0$\overline{1}$1]. The **Miller indices** are widely used for the *atomic planes*. The numbers n_i which denote the coordinates of the intersections of the plane in question with the three axes of the cell are inverted ($1/n_i$). The Miller indices are then the smallest three integers which are in the same ratio as the $1/n_i$. Thus, Figure 17 shows the planes (001) and (011). The set of equivalent planes is denoted by { }. There are, for example, six {110} planes in a cubic lattice.

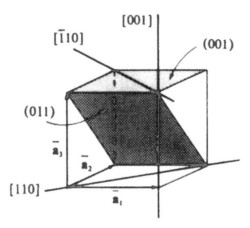

Figure 17   Cubic lattice. Indices of a number of planes and rows.

3 – Determination of crystal structures

After the first X-ray diffraction experiment with a blende crystal, interpreted by von Laue (in 1912), immediately followed by the work of the Braggs, father and son, *diffraction* is generally used to determine the structure of crystals.

We recall that the interaction of a monochromatic plane wave with a crystal of *scattering objects* (X photons or electrons incident on a crystal of electronic clouds, neutrons incident on a crystal of nuclei, etc), where the crystal is assumed to be perfect and finite (**Annex 1**, p. 255), gives rise to additive interferences in certain directions, whence to *diffracted beams*. These directions, which can be made to materialise on a photographic plate, may be conveniently determined if we introduce the **reciprocal lattice**. Let $(\overline{a}_1, \overline{a}_2, \overline{a}_3)$ be the primitive cell of the Bravais lattice of the crystal. We define a *reciprocal cell* $(\overline{a}_1^*, \overline{a}_2^*, \overline{a}_3^*)$ using the three relations

$$\overline{a}_1^* = \frac{2\pi}{v}\overline{a}_2 \wedge \overline{a}_3 \qquad \overline{a}_2^* = \dots \qquad \overline{a}_3^* = \dots \qquad \text{where} \quad v = \overline{a}_1 \cdot \overline{a}_2 \wedge \overline{a}_3$$
$$\text{or} \quad \overline{a}_i^* \overline{a}_j^* = 2\pi \delta_{ij}. \tag{11}$$

The lattice of points formed using this cell (*reciprocal lattice*) defines the 'reciprocal space' which is the *dual* space of the direct space. If $\overline{\ell}$ and $\overline{r}_T$ are translation vectors of the reciprocal lattice and of the direct lattice (see (9)), it follows immediately from (11) that

$$\exp(i\overline{\ell} \cdot \overline{r}_T) = 1. \tag{12}$$

Whence, we deduce that the *reciprocal cell of the reciprocal cell* $(\overline{a}_1^*, \overline{a}_2^*, \overline{a}_3^*)$ is the *direct cell* $(\overline{a}_1, \overline{a}_2, \overline{a}_3)$.

This reciprocal lattice is of interest because of the following property (see **Annex 1**). When a plane wave with wavevector $\overline{k}$ is incident on a crystal, the interference of the scattered waves is constructive (in other words, **diffraction occurs**) for wavevectors $\overline{k}'$ such that

$$\boxed{\begin{aligned} |k'| &= |\overline{k}| \quad \text{(elastic scattering),} \quad \textit{and} \\ \overline{\Delta k} &= \overline{k}' - \overline{k} = \overline{\ell} \end{aligned}} \tag{13}$$

where $\overline{\ell}$ is a translation vector of the reciprocal lattice. Thus, for diffraction to occur, a node of the reciprocal lattice (L on Figure 18) must be on (or very close to) the sphere passing through the origin (arbitrary) Ω of the reciprocal lattice with centre the point A such that $\overline{A\Omega} = \overline{k}$. In fact, the vector $\overline{k}' = \overline{AL}$ satisfies the condition (13), since $\overline{\Omega L}$ $(= \overline{\ell})$ is a translation vector of the reciprocal lattice. In this case, there exists energy diffracted in the direction $\overline{k}'$. The sphere defined in this way is called the **Ewald sphere**.

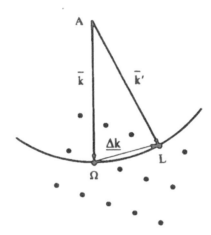

Figure 18 Ewald's construction. The incident beam with wavevector $\overline{k}$ (either all the incident beam if it is monochromatic, or the fraction having wavelength $\lambda = 2\pi/|\overline{k}|$ if it is polychromatic) is diffracted in the direction $\overline{k}'$ (where $|\overline{k}'| = |\overline{k}|$) such that $\overline{\Delta k} = \overline{k}' - \overline{k}$ is a translation vector of the reciprocal lattice (here, $\overline{\Omega L}$). The lattice of points is the reciprocal lattice. The circle represents the Ewald sphere.

In the general case, the Ewald sphere does not touch any point of the reciprocal lattice other than the origin Ω and there is no diffraction. One method introduced by von Laue to provoke diffraction involves the use of *polychromatic* radiation (for example, X-rays). Since the modulus $|\overline{k}|$ then varies continuously within a certain band, there is not one, but infinitely many Ewald spheres, all mutually tangential at point Ω. Under these conditions, a node $L_i = (\overline{\Omega L_i} = \overline{\ell}_i)$ of the reciprocal lattice is generally touched by one of these spheres, namely that with radius $|\overline{k}_i|$ and centre A_i the intersection of line $A\Omega$ with the median plane of ΩL_i. Thus, a certain number of nodes ($L_i \ldots$) give rise to a diffracted beam ($\overline{k}'_i \ldots$), where each beam has the appropriate wavelength $\lambda_i = 2\pi/|\overline{k}_i|$. The set of these beams may be traced either on a photographic plate or using counters. This enables us to reconstruct the reciprocal lattice and thus also its reciprocal, namely the direct (crystal) lattice.

In the case of **diffraction of fast electrons**, the Ewald sphere becomes almost *a plane*. For example, in electron microscopy, monokinetic electrons with energy $E = 100$ or 200 keV are currently used. The corresponding wavelength ($\lambda(\text{Å}) \simeq [150/E(eV)]^{1/2}$ or $\simeq 4 \times 10^{-2}$ Å for $E = 100$ keV) is very small in relation to the dimensions a_i of the crystalline cell. Thus, the radius of the sphere $|\overline{k}| = 2\pi/\lambda$ is very large in relation to the dimensions ($\simeq 2\pi/a_i$, see (11)) of the reciprocal cell. On the other hand, if one dimension of the crystal is small (for example, along Ox_1) (as in the case of the thin films with approximate thickness $\simeq 100$ nm studied in electron microscopy and diffraction) the nodes of the reciprocal lattice have a notable extension in direction Δk_1 (see Figure 2, p. 257) and each node becomes a *bar* centred on the theoretical position of the node. Under these conditions, despite the monochromatic nature of the incident wave, the Ewald sphere (plane) *always* touches a series of nodes (bars) of the reciprocal lattice and the diffraction diagram is always practically *a planar section of the reciprocal lattice*. Figure 19(*b*) shows such a diagram[14] in the case of a solid with two components (magnesia and sodium). This diagram is described later (see p. 107).

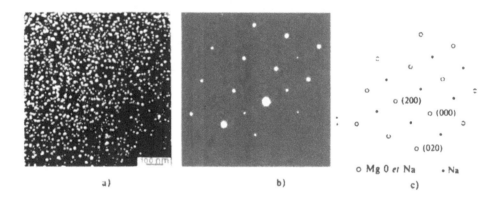

a)   b)   c)

Figure 19 **Sodium implanted in magnesia.** Sodium was introduced into magnesia (MgO) by ion implantation (see p. 107) and after thermal treatment (20 minutes at 800° C) it clustered in small colloids of diameter approximately 10 nm. (*a*) Image of these colloids from electron transmission microscopy. The sodium colloids are seen in white, the magnesia is grey. The thickness of the sample was $\simeq 100$ nm. (*b*) Diffraction diagram of the region shown in (*a*), showing the diffraction by sodium and by magnesia. (*c*) Diagram of (*b*) redrawn for interpretation (see p.107). After M Treilleux, Thesis, Lyons (1982).

2.2 Four important crystal structures

Most solids which we study in what follows crystallise into one of the four following structures: *face-centred cubic* (fcc), *close-packed hexagonal* (hcp), *body-centred cubic* (bcc) and *diamond*. Thus, we shall consider only these four cases which we shall describe succinctly. This bias towards simplicity should not give rise to any illusions: countless materials adopt more complex crystalline structures. This is the case of most intermetallic compounds, minerals, ceramics, etc. **Problem 6** gives such an example of a superconducting oxide.

[14] Here, one should beware of comparing the diagram of Figure 19 with that of Figure 14. The latter gave a *direct* image of the crystal, the former gives a *reciprocal* image.

1 – Face-centred cubic structure (fcc)

The fcc and hcp structures are obtained when packing equal spheres (for example, marbles) in a compact fashion.

To obtain a maximally compact packing of marbles, it is simplest to begin in a plane *A*. In this plane the compact packing is clearly hexagonal (touching circles shown by the thin lines in Figure 20). To begin a *second* compact packing in an adjacent plane *B* (for example, 'above'), one places a marble in contact with three arbitrary marbles in plane *A*, thus forming locally a regular tetrahedron (that of Figure 1). This first marble in plane *B* uniquely determines the complete hexagonal packing in plane *B* (shown in bold in Figure 20). There are *two possible* ways of beginning a *third* plane ('above' *B*): a marble may be placed vertically above a marble in layer *A* or not. This determines the choice between fcc and hcp packings.

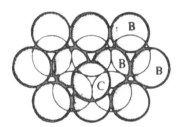

Figure 20   Close packing of identical marbles. Fine circles lie in the lower plane *A*. Bold circles lie in the middle plane *B* and the upper plane *C*. Here, the position of the marble *C* imposes a face-centred cubic packing.

If the position of this first marble in the third plane is chosen to be vertically shifted with respect to plane *A* (circle *C* in Figure 20), the three successive planes *A*, *B*, *C* form the beginnings of a dense packing which is then continued regularly: the fourth plane above *C* will be of type *A*, etc. The crystal will be formed by the stacking *ABCABCAB* .... It is easy to check that this stacking is just that of marbles of radius $a\sqrt{2}/4$ situated on an fcc crystal defined by a cube with edge *a* (Figure 21). This cube – which is the *unit cell*, see p. 98 – contains *four* marbles (or atoms). The filling factor (see **N.B. 1**, p. 86) is equal to $4 \times \frac{4}{3}\pi(a\sqrt{2}/4)^3 a^{-3} = 0.74$[15].

N.B. 1   This packing characterised by a regular tetrahedron of four marbles is a further example of the short-range order which we met previously for the amorphous packing of marbles (Figure 1). Here, the existence of a large-range order leads to a larger filling factor (see **N.B. 1**, p. 86).

N.B. 2   The following points are easy to check.

i)   The *Bravais lattice* of this structure is fcc, like the crystal lattice. The *basis* involves *one* marble.

ii)   The dense hexagonal planes of type *A*, *B*, *C* are orthogonal to the four main diagonals of the cube. These are {111} planes. By way of example, Figure 21 shows one of the four

[15] It seems obvious and is universally accepted, that this filling ratio is the highest possible when equal spheres are packed, but this has never been rigorously demonstrated up to now.

diagonals together with some of the atoms labelled *A*, *B* or *C* according to the type of plane to which they belong. Each of the three other diagonals of the cube could be used to define other dense hexagonal {111} planes, whence other packings of type *A'B'C'A'* ... identical to the first (in all (111), ($\bar{1}$11), (1$\bar{1}$1) and (11$\bar{1}$)).

iii) The **primitive cell** of the fcc Bravais lattice *is not the cube* (which contains four points instead of one for a primitive cell). We leave the determination of this primitive cell as an exercise.

iv) **The reciprocal lattice of an fcc lattice** with cube edge *a* **is a bcc lattice** with edge $4\pi/a$. To check this, apply (11) to the basis vectors of the primitive cell determined in iii) (above).

v) The most dense planes are, in decreasing order of density, {111} planes (equidistance $a\sqrt{3}/3$), the {100} planes (faces of the cube, equidistance $a/2$) and the {110} planes (equidistance $a\sqrt{2}/4$). The densest rows are the ⟨110⟩ rows (marbles aligned with the contact).

vi) The number of nearest neighbours to a given marble is 12 (six in a dense plane *B*, three in the adjacent plane *A* and three in the adjacent plane *C*). The distance of two neighbours is $a\sqrt{2}$.

If the identicalness of the two descriptions of this structure (in terms of **dense stacked planes** of type *ABCAB* ..., and in terms of **face-centred cubes**) is not clear to a reader unaccustomed to viewing things in 3D, we recommend that this reader should build a stack *ABCA* (for example, using ping-pong balls) with a view to explicit identification of the cube of Figure 21.

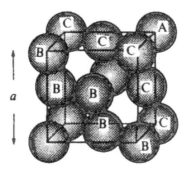

Figure 21   Face-centred cubic structure. This is also a stacking of dense hexagonal planes (*A*, *B*, *C*, ...) as in Figure 20. These planes are perpendicular to the diagonal shown by dashes. For clarity of the drawing, the spheres here have radius less than $a\sqrt{2}/4$: they are not in contact.

Exercise   In the above stacking, a given marble is surrounded by 12 marbles which are in contact with it and with each other. Contrast this shell of 12 marbles with that previously described for the icosahedral aggregate (**Exercise** of p. 86). Compare the numbers of contacts (or pseudo contacts) in the two cases and deduce that an icosahedral aggregate of 13 atoms is very likely to be more stable than an fcc aggregate of 13 atoms.

Although it is compact, the fcc packing of marbles leaves two main types of sites – called **interstitial sites** – available for inserting smaller marbles: these are *octahedral* sites (centres of the edges and centre of the cube) and *tetrahedral* sites (centres of the eight cubes of edge $a/2$). It is easy to check that the former, of which there are *four* per cell, lie at the centre of an octahedron of marbles and that the latter, of which there are *eight* per cell, lie at the centre of a regular tetrahedron of neighbouring marbles (see Figure 22).

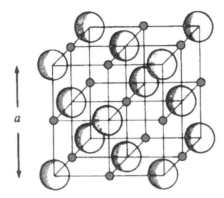

Figure 22 **Interstitial sites** in an fcc packing. 1 – White circles: fcc lattice (cell *a*). 2 – Black circles: octahedral insertions. Note that these sites also form an fcc packing identical to the previous one and shifted by $\frac{1}{2}\langle 100 \rangle$ with respect to it (NaCl structure). 3 – Centres (not shown) of eight cubes *a*/2: tetrahedral insertions.

One particular, important case of the fcc structure is the NaCl structure. This is that of most alkali halides, in which the *anions* (for example, Cl^-) form an fcc crystal where the *cations* (for example, Na^+) occupy the octahedral sites[16]. This structure is shown in Figure 22, in which the anions and the cations are indicated, respectively by white and black circles.

In this structure, the ionic $\langle 100 \rangle$ rows are alternating $(+ - + - \ldots)$, whilst the $\langle 110 \rangle$ rows are alternately positive and negative (see Figures 22, 23; and 1, p. 2).

Exercise Identify the Bravais lattice and the basis of this structure.

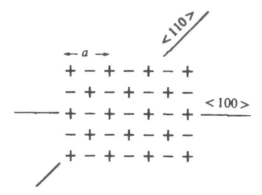

Figure 23 **NaCl structure.** Plane of type {100} of an alkali halide of NaCl structure.

[16] Or conversely.

2 – Hexagonal close-packed structure (hcp)

We return to the procedure for stacking compact planes of type $A, B, \ldots$ described above (Figure 20). When it comes to arranging the *third* plane, we may proceed in two ways. We chose one of these ways. The other way involves arranging the third plane *vertically above the first*. Thus, we form a stacking $ABABAB \ldots$. The sequence ABA generates hexagonal prisms which are generally taken as *unit cells* (non-primitive) of this structure. Each cell (height c, side a) contains *six* atoms.

The filling ratio for marbles in contact is clearly the same (0.74) as that of the fcc structure. The ratio c/a is $\frac{4}{3}(3/2)^{1/2} = 1.63$. However, since *atoms* are not solid marbles and since the constraint of Subsection 2.2.1 (to end with a cubic symmetry) does not apply, the ratio c/a is in practice rarely equal to 1.63. It may be *greater* (extended prism) or *less* (crushed prism) without altering the symmetry elements of the hexagonal structure (for some examples, see p. 108).

3 – Body-centred cubic structure

The unit cell of this structure is the centred cube (Figure 24). It contains two atoms.

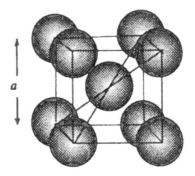

Figure 24   Body-centred cubic packing.

Exercise   This cell is *not primitive*. Determine the primitive cell then the cell of the reciprocal lattice (defined by (10)) and check that this reciprocal cell is the primitive cell of an fcc lattice. Thus, **the reciprocal lattice of a bcc lattice is an fcc lattice** (and conversely, see point iv) on p. 104). The cell of this reciprocal fcc lattice has edge $4\pi/a$.

If hard spheres are packed in a bcc structure, the packing is not compact. The filling ratio here is $2 \times \frac{4}{3}\pi(a\sqrt{3}/4)^3 a^{-3} = 0.68$.

This weak compactness lends itself well to the insertion of foreign atoms. The most favourable site for insertion is the centre of the faces of the cube (check that this is also the centre of the edges). This is the site occupied by carbon in the iron bcc structure (see **Problem 15**).

The most dense planes in decreasing order of density are: the {110} planes (equidistance $a\sqrt{2}/2$), the {100} planes (faces of the cube, equidistance $a/2$) and the {111} planes (equidistance $a\sqrt{3}/6$). The densest rows are the ⟨111⟩ rows.

N.B. 1 This bcc structure in which the object situated at the centre of the cube and those situated at the vertices are of the same type should not be confused with the structure of CsCl (see Figure 4, p. 7). Identify the Bravais lattice and the basis for the latter.

N.B. 2 We may now interpret the diffraction diagram of Figure 19(b) (p. 102). Sodium, initially ionised (Na^+), is accelerated in vacuum to an energy of 750 keV in an 'ion implanter' and impinges on the surface of a single crystal of magnesia MgO (NaCl structure, thus fcc). The ions penetrate this, are slowed and stop at a depth of around 700 nm in the magnesia. Treatment at 880° C enables the sodium atoms to migrate and cluster in globules (colloids), as shown in Figure 19(a).

Sodium is bcc, and its reciprocal lattice is fcc. Magnesia is fcc and its reciprocal lattice is bcc. The cells of sodium ($a = 0.429$ nm) and magnesia ($a = 0.422$ nm) are practically equal as are the cells of their reciprocal lattices. The diffraction diagram, which represents a plane section through these two reciprocal lattices (see p. 101) shows that these two lattices *coincide*. We observe (see Figure 19(c)) the square lattice of the {100} planes of the reciprocal lattice (bcc) of magnesia and the 'centred square' lattice of the {100} planes of the reciprocal lattice (fcc) of sodium. This coincidence shows that the *sodium colloids all have the same orientation* which is imposed by the crystalline matrix of magnesia. This orientation relation may be summarised as follows:

$$\{001\} \text{ Na } // \{001\} \text{ MgO} \quad \text{and} \quad \langle 100 \rangle \text{ Na } // \langle 100 \rangle \text{ MgO}.$$

This is a very simple example of an *epitaxy* relation, in which, thanks to the quasi-equality of the cubic cells, the centred cubes of sodium are arranged parallel to the face-centred cubes of magnesia.

Orientation relations between two phases in contact are often much more complicated.

4 – Diamond structure

This is a **fcc structure** of atoms X in which the *four* tetrahedral sites (out of *eight*, see p. 105) with coordinates $a\left(\frac{1}{4}\frac{1}{4}\frac{1}{4}\right)$, $a\left(\frac{3}{4}\frac{3}{4}\frac{1}{4}\right)$, $a\left(\frac{3}{4}\frac{1}{4}\frac{3}{4}\right)$ and $a\left(\frac{1}{4}\frac{3}{4}\frac{3}{4}\right)$ are occupied by atoms of the same nature X. The dense {111} planes of this structure are shown horizontally on Figure 25. This is, in particular, the structure of diamond (X = C) and also of silicon (X = Si).

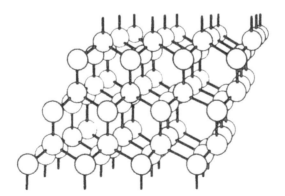

Figure 25 Diamond structure. Each atom is situated at the centre of a regular tetrahedron. The dense planes of the fcc structure are the 'horizontal' planes of the figure. As an exercise, find the fcc Bravais lattice on which the diamond structure is built.

In terms of equal, packed marbles, each marble of radius $a\sqrt{3}/8$ has *four* neighbouring balls at distance $a\sqrt{3}/4$ which form a *regular tetrahedron*. The packing *is not compact*. The filling ratio, equal to $8 \times \frac{4}{3}\pi(a\sqrt{3}/8)^3 a^{-3} = \pi\sqrt{3}/16 = 0.34$, is very low.

It is easy to check that the *diamond* lattice consists of two fcc lattices displaced from each other by a vector $a\left(\frac{1}{4}\frac{1}{4}\frac{1}{4}\right)$ and that, with the notation of Subsection 2.2.1, the packing is of type $AaBbCcAa \ldots$, where the (dense) planes a, b and c are vertically above A, B and C, respectively.

Exercise Determine the Bravais lattice and the basis of this structure.

5 – Allotropy

For a given material, for T and P fixed, there exists a crystalline structure which minimises its free enthalpy. This structure is the one which the material in principle adopts. For a given pressure (for example, standard) there may be several successive *allotropic phases* from 0 K to the melting point. Thus, titanium which has an hcp structure $(c/a = 1.60)$ at low temperatures takes the bcc structure at 882° C. The corresponding phase changes are usually accompanied by appreciable variations of physical properties and notably of the specific volume.

Sometimes, these changes may be *retarded*, which is often due to the presence of impurities (or foreign atoms); it is even possible for a high temperature phase to be maintained at low temperatures in very long-lasting (*years* or *centuries*) *metastable states*.

Table 1 Crystallographic structures of some elements in solid phase.

	Structure	a (Å)	c (Å)	c/a	Transformation
Aluminium	fcc	4.04			
Cadmium	hcp	2.98	5.63	1.89	
Chrome	bcc	2.88			
Copper	fcc	3.61			
Germanium	Diamond	5.65			
Iron	bcc	2.86	....		910° C → γ iron (fcc) 1400° C → δ iron (bcc)
Magnesium	hcp	3.20	5.20	1.62	
Nickel	fcc	3.52			
Silicon	Diamond	5.43			
Titanium	hcp	2.95	4.73	1.60	882° C → bcc
Helium 4 (2 K)	hcp	3.57	5.82	1.63	pressure > 25 atm.
Argon (20 K)	fcc	5.43			

A classical case is that of tin which has two crystal structures: *grey* tin (diamond structure) and *white* tin (tetragonal), where the former is stable at low temperatures. The transformation

$$Sn_g \rightarrow Sn_w \qquad \text{if } T \uparrow \tag{14}$$

occurs in principle at 13°C (286 K) with a transformation enthalpy ΔH_t equal to 0.5 kcal mole^{-1}. At 13° C, the reversible transformation is accompanied by an increase in entropy $\Delta S_t = \Delta H_t/T = 500/286 = 1.75$ cal mole^{-1} K^{-1} whereas the enthalpy (in fact, essentially the internal energy) of Sn_g at 25° C is lower than that of Sn_w.

Why is the entropy of Sn_w larger than that of Sn_g? The lower energy of Sn_g (in comparison with Sn_w) implies larger average interatomic force constants, in other words, larger return forces for any atom removed from its equilibrium site and higher frequencies of atomic vibration (see **Problem 18**). This diminution of the characteristic frequencies (symbolically: $\nu_w < \nu_g$) corresponds to a *variation of entropy*/atom $\Delta S = 3k \ln(\nu_g/\nu_w)$, which is *positive*.

Table 1 shows the structure at room temperature (unless otherwise indicated) together with the allotropic transformations of some elements with a simple structure. See Sections 2.2.1 and 2.2.2 for the meaning of a and c.

3. NEITHER GLASSES NOR CRYSTALS

Nature provides numerous examples of structures intermediate between the two large families discussed above. Let us consider some of them.

3.1 Polytypes

We have seen that in stacking compact hexagonal planes of atoms, there is a choice at the *third* plane between the fcc stacking $(\ldots ABCABCAB\ldots)$ and the hcp stacking $(\ldots ABABAB\ldots)$. This choice is determined by the relative energies $\mathcal{E}$ of the ABC structures on the one hand and the ABA structures on the other. Once this choice has been made, it is in principle repetitive, since naturally $\mathcal{E}(BCA) = \mathcal{E}(ABC)$ and $\mathcal{E}(BAB) = \mathcal{E}(ABA)$.

Sometimes, however, the energy differences are so small that the above choice becomes semi-random. This leads to the formation of packings in which the order of the letters is ill defined, or even random, for example, $\ldots CABABCBACAB\ldots$. Local hcp strips $(ABAB)$ and fcc strips $(CBAC)$ may be observed, but with no regularity. Thus, these structures, called **polytypes** are 1D non-crystalline stackings of 2D crystalline objects (the hexagonal planes).

3.2 Liquid crystals

There exist liquids, in the ordinary sense of the word (substances with low viscosity, which in a short time of observation take on the shape of the receptacle containing them) which exhibit a certain *degree of order*. These are the **liquid crystals**.

Without going into details of the many possible forms of organisation, we shall only list some of them. In all the cases below, these substances consist of elongated, quasi-rectilinear molecules.

The latter may be arranged parallel (or almost) to each other with the centres of the molecules having an arbitrary position: these are **nematic** crystals. They may be grouped in layers of equal thickness, with the molecules normal to the layers (**smectic** A crystals) or at an angle to the latter (**smectic** C crystals). They may be arranged in such a way that there exists a plane P to which they remain parallel, their local direction Δ turning regularly around a direction perpendicular to P: these are **cholesteric** crystals. It is clear that, in all the above cases, the physical properties (elastic properties, optical properties, molecular mobility, etc) are strongly anisotropic.

However, there are many other possibilities. Thus, Figure 26 shows packings observed for **water–soap** mixtures. The soap molecule is linear and ends with a 'hydrophilic head', represented here by a small circle. In arranging this head towards the water, according to the temperature and the concentration of the mixture, the molecules may form superposed planar sheets reminiscent of the smectic A structure (*a*), tubes packed hexagonally (*b*), spheroids (or **micellae**) packed in body-centred cubes (*c*), etc. Whether or not one may be said to be dealing with crystals depends on whether one is considering the packing of these objects (sheets, tubes, micellae) or that of the molecules.

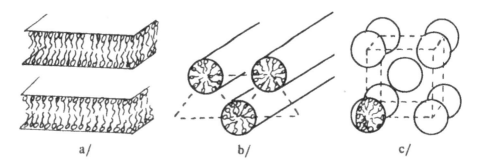

a/   b/   c/

Figure 26 Molecular packings in water–soap mixtures. Amphiphile soap molecules are supple filaments with a hydrophobic tail and a hydrophilic head. The water molecules not shown form the 'cement' between the planar (*a*), tubular (*b*) or spherical (*c*) aggregates. The stability of any given phase depends on the temperature and on the soap/water concentration.

3.3 Incommensurable phases

Two crystalline entities with incommensurable primitive cells may coexist within a single solid.

This is the case for the oxide FeO or *wüstite*. Ideally, this oxide has the *NaCl* structure, with the Fe^{2+} ions occupying the octahedral sites in an fcc packing of O^{2-} ions. But, wüstite is always under-stoichiometric ($Fe_{1-x} O$; $x \simeq 0.1$) and it can be shown that its true composition (which involves Fe^{3+} ions) may be written as

$$Fe^{2+}_{1-3x}(\text{oct}) \qquad Fe^{3+}_{2x-y}(\text{oct}) \qquad Fe^{3+}_{y}(\text{tet}) \qquad \boxed{O}_{x+y}(\text{fcc}) \qquad O^{-2}(\text{fcc})$$

where (oct) and (tet) denote the octahedral and tetrahedral positions (see p. 105) and $\boxed{O}$ denotes the oxygen vacancy. For sterical reasons, the Fe^{3+}(tet) ions, which are very compressed are surrounded by $\boxed{O}$ vacancies forming *clusters* of defects characterised by the ratio ρ of the number of oxygen vacancies to the number of Fe^{3+}(tet) ions ($\rho = x + y/y$). These clusters (for which ρ is measured to be 2.5) are distributed periodically in the host crystal (FeO). Thus, there are two phases here and two periodicities, but the latter are *incommensurable* (in each of the three spatial dimensions).

Under these conditions, the structure of wüstite must be described as that of a crystal *in a 6-dimensional space*[17]. The section of this crystal through 3D physical space reconstitutes the modulated tri-incommensurable solid which is wüstite.

3.4 Quasi-periodic crystals

If we return to Figure 15, it is clear that the parallelograms $(\overline{a}_1, \overline{a}_2)$ fill the plane and satisfy the following two properties:

i)   they form a complete *tiling* in the sense that they may be juxtaposed to cover the plane without gaps or superposition;

ii)   this tiling is *periodic* in the sense of equation (9).

Figure 27 gives examples of **periodic tiling** of the plane. We note that the *primitive tile*, which naturally has the same surface area (volume in 3D) as the primitive cell of the

[17] D Weigel, R Veysseyre, Cl Carel, *C.R. Acad. Sci.* **305**, 349 (1987); J R Galvari, Cl Carel, D Weigel *C.R. Acad. Sci.* **307**, 705 (1988). See another case of incommensurable waves in a superconductor in L Pierre, J Schneck, J C Tolédano, C Daguer *Phys. Rev.* **41** 766 (1990).

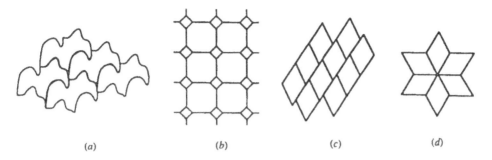

Figure 27 Various ways of beginning periodic tiling of the plane. (*a*), (*c*) and (*d*): only one kind of tile. (*b*): two kinds of tile. (*c*) and (*d*): tiles are rhombuses with vertex angle 60°.

Bravais lattice, may be formed from a single piece ((*a*) and (*c*)) or from a finite number of partial pieces. Thus, in (*b*), the two tiles are an octagonal piece and a square piece. We also note that a single tile (60°-rhombus shape in (*c*) and (*d*)) may lead to different periodic structures, according to the initial arrangement of tiles.

Exercise It is clear that a 2D- (or 3D-) object such as a circle (or sphere) is not generally a primitive tile in the above sense. Given the basis vectors of a Bravais lattice $\bar{a}_1, \bar{a}_2$ (and $\bar{a}_3$), determine the condition on the object which makes it the primitive tile of a periodic tiling.

Exercise Determine the shape of the primitive tile in (*b*). Check that it satisfies the above condition.

Exercise Continue the periodic tiling begun in (*d*) and trace the corresponding Bravais lattice. Is this lattice identical to the Bravais lattice of (*c*)? Draw the primitive tile in at least two different ways.

For a long time it was thought (and taught) that a complete tiling realised as in (*b*) using a *finite number* of pieces was necessarily crystalline. It is now known that this is not true. In 1979, Penrose conceived (completely away from the crystalline context) a non-periodic tiling of the plane using *two* rhombuses with angles multiples of $\pi/5$. Figure 28 shows a variant of Penrose's tiling: here, the plane is tiled in a *non-periodic* manner using two rhombuses with equal sides and acute angles $2\pi/5$ and $\pi/5$, respectively.

Exercise Use these rhombuses to begin a periodic tiling. Then begin another tiling by arranging five rhombuses ($2\pi/5$) in a star (imitating Figure 27(*d*)). Show that it is impossible to construct a periodic tiling based on this (refer to the possible rotational symmetries of Bravais lattices, p. 98 and to Figure 3). N.B. Figure 28 contains several of these star shapes formed by five rhombuses.

One important property of the Penrose tiling is that any finite fragment of this tiling is reproduced infinitely many times at other places in the tiled plane; this enables us to talk of **quasi-periodicity**.

This concept was made precise by Duneau and Katz, who generalised Penrose's tiling of the plane to an arbitrary number of dimensions and showed that *a quasi-periodic tiling is a projection of a periodic tiling* from a higher-dimensional space[18].

[18] **A Katz, M Duneau** *J. Physique* **47**, 181 (1986).

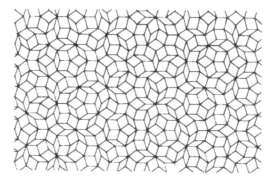

Figure 28 **Penrose tiling** based on two rhombuses with angles $2\pi/5$ and $\pi/5$. **M Duneau** and **A Katz**, *Phys. Rev. Letters* **54**, 2688 (1985).

Suppose, for example, that in the plane we have the square tiling constructed from the shaded square in Figure 29. We slide the square along any line Δ passing through one of its vertices A, thereby generating a *ribbon* between the parallel lines Δ and Δ'. This ribbon distinguishes a set of points of the lattice (in bold on the figure) which forms a broken line, the projection of which onto Δ' consists of two primitive *tiles* $\underline{a}$ and $\underline{b}$. The line Δ' is tiled by $\underline{a}$ and $\underline{b}$.

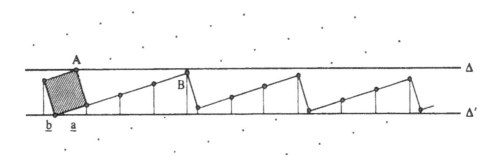

Figure 29 **A quasi-periodic tiling** of the straight line Δ' may be obtained by projecting the points of a square lattice between Δ and Δ', where Δ only intersects this lattice at the point A.

If the angle θ between Δ and the axis $\langle 01 \rangle$ of the lattice is such that $\tan\theta$ is rational, then the tiling is periodic. Thus, for $\tan\theta = \frac{1}{3}$, the sequence is *$\underline{aaab}$ $\underline{aaab}$, ...* On the other hand, if $\tan\theta$ is *irrational*, the tiling is not periodic. However, it does have quasi-periodic properties, which are apparent in Figure 29, where $\tan\theta$ is slightly less than $\frac{1}{3}$ and the sequence (*taken* for example, *from B*) *$\underline{baaa}$* is reproduced several times in a row before a change (non-periodic) to *$\underline{baaaa}$*[19].

---

[19] These quasi-periodic sequences were studied in the 12th Century by the mathematician Fibonacci (also called Leonardo of Pisa) for the example of the genealogy of rabbits. This problem was formulated as follows: rabbits are babies for one year (*l*) and adults after that; each adult gives birth to one baby per year; rabbits do not die. These conditions give rise to the following successive annual situations: $L \rightarrow L + l$ (denoted by Ll) $\rightarrow LlL \rightarrow LlLLl \rightarrow LlLLlLlL \rightarrow LlLLlLlLLlLl \rightarrow ....$. The sequence obtained in infinite time involves only doublets Ll and triplets LLl with quasi-periodic properties similar to those described above.

Similarly, we may obtain a non-periodic tiling of the space $\mathbf{R}^3$ using a suitable projection of a cubic lattice drawn in $\mathbf{R}^6$. The tiling shown in Figure 28 is in fact a plane section of such a tiling of $\mathbf{R}^3$.

One remarkable characteristic of these quasi-periodic lattices is that their Fourier transform consists of a lattice of points (not periodic) in the reciprocal space. If a quasi-periodic tiling consists of atoms, a diffraction experiment (for example, with electrons) gives rise to *well-defined diffracted beams*. The corresponding diffraction diagram exhibits a symmetry of order five, which is impossible to justify for *crystals*. It was this approach that led to the discovery of 'quasi-periodic crystals' or **quasi-crystals** in the metallic alloys Al-Mn[20], then in other alloys. Contrary to long-standing beliefs, the periodicity of the atomic arrangement is certainly a *sufficient* condition for observing diffraction, but it is *not necessary*. We note here that determination of a quasi-crystal structure does not provide detailed knowledge of the position of the atoms in this structure. This is a difficult point in the study of quasi-crystals[21].

Another unexpected characteristic is the existence of rows or planes as in periodic packings. Such arrangements may be seen by looking at Figure 28 at a low angle in different directions; they are observed experimentally in channelling experiments (see p. 97).

[20] D Shechtman, I Blech, D Gratias, J W Cahn, *Phys. Rev. Letters* 53, 1951 (1984). The date of this discovery (1984) should be compared with that of Penrose's first tilings (1979). This is another illustration of cross fertilisation in which initially distinct and diverse areas of research are suddenly drawn together to become one. Metallurgists, mathematicians, physicists, crystallographers, research engineers, etc were then able to exchange theoretical tools, methods, experiences and hypotheses and uncover new fundamental concepts and possible industrial applications in a topic which had become common and autonomous.

[21] For more details, see C Janot *Quasi-Crystals, A Primer*, Oxford Sc. Pub. (1992).

CHAPTER VI

BLOCH AND BRILLOUIN OR THE ELECTRONS IN A CRYSTAL

Jeux de Vagues
Claude Debussy
The Marriages between Zones Three, Four and Five
Doris Lessing

Our simple description of metallic bonding and cohesion (Chapters III and IV) has given us a first insight into a good number of properties of this class of materials. However, it has a number of shortcomings. To note but one, it does not enable us to understand why *silicon*, a material with valency four which should be a very good conductor (high value of N in formula (48), p. 56) is a factor of $\simeq 10^8$ *less* conductive than *silver* (valency one) at room temperature.

It is time to return to our most radical approximation, according to which the ions of the solid had no influence other than to erect a potential barrier to the departure of electrons. In the crystal, the potential was considered to be constant. This is clearly not the case. Each ion i exerts an attractive potential on the electrons $v_i(\bar{r})$. In what follows, we shall suppose that the potential 'seen' by each electron is simply

$$V(\bar{r}) = \sum v_i(\bar{r})$$

(this amounts to neglecting electron–electron interactions, as in most of Chapter III), whence that it has the *periodicity of the crystal* (Figure 1).

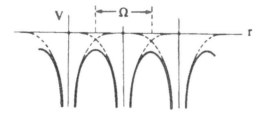

Figure 1 Crystalline potential. As the sum (continuous curve) of *atomic potentials* (dashes), it is little different from the latter around the centre of Ω.

This periodicity of the potential has many important consequences. Our approach to these will be at first intuitive (Section VI.1) and then more precise (Section VI.2 ...).

1. THE POTENTIAL TAKEN AS A PERTURBATION

1.1 Case of a 1D linear chain

Let us first consider a linear chain of atoms with period a. The periodic **potential** $V(x)$ (on Figure 1, replace r by x and Ω by a) will simply be taken to be **sinusoidal** (see Figure 2) and considered as a small **perturbation**:

$$v_p(x) = -V_l(e^{ilx} + e^{-ilx}) \qquad V_l > 0, \qquad l = \frac{2\pi}{a}, \tag{1}$$

where the origin of the abscissa x is taken on an atom. Note that l is the first vector of the **reciprocal lattice**, which is itself a linear lattice of equidistant points (separation $2\pi/a$).

N.B. Later, we shall generalise this to the case where l is *any* vector of the reciprocal lattice (see p. 126).

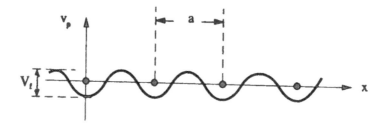

Figure 2 Sinusoidal potential of a linear chain as an approximation to the potential of Figure 1.

1.2 Perturbation of states

The *unperturbed* states are the plane waves (5) of p. 34, with spatial dependence:

$$\psi(x) = L^{-1/2}e^{ikx} \tag{2}$$

and energy:

$$E_n = \frac{\hbar^2}{2m}k^2. \tag{3}$$

To the first order, the perturbation energy may be written as (see Q.M.):

$$\Delta E = \int \psi^*(x)v_p(x)\psi(x)\,dx = L^{-1}\int v_p(x)\,dx = 0.$$

To this order, the energy (see (7), p. 34) is not modified by the presence of the potential v_p. If we denote by $|k\rangle$ the unperturbed state (2), its perturbation is of the form

$$|\varphi\rangle = \sum_{j \neq k} \frac{\langle j|v_p|k\rangle}{E_k - E_j}|j\rangle \tag{4}$$

where the $|j\rangle$ are the unperturbed eigenstates (2) and the $E_{j,k}$ are the unperturbed energies[1]. Since k, j and l belong to the k lattice (Figure 4, p. 35) the matrix elements $\langle j|v_p|k\rangle$ are **all zero** (since they are equal to

$$-L^{-1}V_l \int_L e^{i(k+l-j)x}\,dx - L^{-1}V_l \int_L e^{i(k-l-j)x}\,dx,$$

where the integrands are periodic on L) **except for two**, namely those corresponding to $j = k - l$ and $j = k + l$ (each equal to $-L^{-1}V_l$).

We deduce that *to the first order*:

i)   The energies are not modified from those of the free electron;
ii) After perturbation (*not taking into account* a final normalisation), the functions become:

$$\psi_k(x) \simeq L^{-1/2}\left(e^{ikx} - \frac{V_l}{E_k - E_{k-l}}e^{i(k-l)x} - \frac{V_l}{E_k - E_{k+l}}e^{i(k+l)x}\right). \tag{5}$$

In this function (5), the two terms in V_l (small) generally play a minor role **except** if the denominators may become small. Well, they may: for this, it is sufficient to have $k - l \simeq k$ or $k + l \simeq k$ (look again at the form of the $E_{j,k}$ in (3)), or, at the limit

$$\boxed{k = \pm l/2.} \tag{6}$$

Thus, as $k \to +l/2$ (*resp.* $-l/2$), it is the first (*resp.* second) term in V_l in (5) which becomes large (*large*, but not infinite, if one continues the calculation to next order of approximation), while the second (*resp.* first) term remains small. Then the function (5) turns out to be the superposition of **two plane waves**, both with notable weights, namely $\exp(ikx)$, i.e. $\exp(ilx/2)$, and $\exp(i(k - l)x)$, i.e. $\exp(-ilx/2)$:

$$\psi_{k=l/2} = L^{-1/2}(\alpha e^{i(l/2)x} + \beta e^{-i(l/2)x}) \qquad (\alpha^2 + \beta^2 = 1).$$

For this particular value $l/2$ of k, the wave k gives rise to a wave $-k$ $(= -l/2)$ which propagates in the opposite direction. If we now take account of the symmetry $(l/2, -l/2)$ (in other words, of the fact that we could have started with $k \to -l/2$), we may suppose (the proof can be found in (21'), p. 128) that, in fact, these waves have equal weights, $\alpha^2 = \beta^2$:

$$\boxed{\psi_{k=l/2}(x) = (2L)^{-1/2}(e^{i(l/2)x} \pm e^{-i(l/2)x}).} \tag{7}$$

N.B. 1   The occurrence of the wave $-k$ for $k \to l/2$ actually corresponds to the diffraction condition (13) of p. 101, since we have $\Delta k = k - (-k) = l$ and since l is a vector of the reciprocal lattice. Thus, the transition from the progressive, unperturbed plane wave (2) with velocity $v = \hbar k/m$ to the **stationary wave** (6) involves **diffraction** of the incident wave k.

---

[1] The reader may verify, as an **exercise**, that it is possible to use formula (4) although the problem is degenerate for the states $|k\rangle$ and $|-k\rangle$.

N.B. 2   The approximation which we use to replace the expression $\psi + \varphi$ ((2) and (4)) by the two terms of (6) is called *two-wave approximation*.

If, in condition (6), even the small potential v_p can modify the initial plane wave in such a radical manner, what then becomes of the conclusion **i)** (above) relating to the energies?

1.3 Perturbation of the energy

The stationary wave (7) is not random: it is a sinusoidal wave of period $2a$ (since $l = 2\pi/a$). More precisely, it varies as $\cos \frac{l}{2}x$ if k reaches $l/2$ 'from the left' ($k = l/2 - \varepsilon$, $\varepsilon > 0$) and as $\sin \frac{l}{2}x$ if k reaches $l/2$ 'from the right' ($k = l/2 + \varepsilon$). Indeed, the term $E_k - E_{k-l}$ of (5) is negative in the first case (since $k - l > k$), although it is positive in the second case, which changes the relative signs of α and β in the above.

The electron density $\psi^*\psi$, which thus varies as $\cos^2 \frac{l}{2}x$ or as $\sin^2 \frac{l}{2}x$, has the periodicity a of the chain in both cases. However, it 'is attached' in the first case *to* the atoms (thus to the minimum of the potential) and in the second case *between* the atoms (thus, to the maximum of the potential). Evidently, this gives rise to a potential energy term, absent of course in (3), which is *negative* in the first case and *positive* in the second case (see Figure 3).

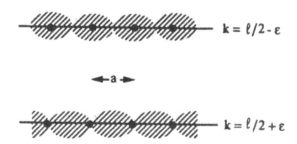

Figure 3   Electron density on the chain of Figure 2, for $k = l/2 = \pi/a$.

This term is equal to $\int_L \psi^*\psi v_p \, \mathrm{d}x$ or

$$\frac{L}{a} \int_a \psi^* \psi v_p \, \mathrm{d}x = -4 \frac{V_l}{a} \int_0^a \cos^2 \pi \frac{x}{a} \cos 2\pi \frac{x}{a} \, \mathrm{d}x,$$

or, finally:

$$\left.\begin{array}{ll} -V_l & \text{for } k = \frac{l}{2} - \varepsilon \\ +V_l & \text{for } k = \frac{l}{2} + \varepsilon. \end{array}\right\} \quad \varepsilon > 0 \qquad (8)$$

Similarly, for $k \to -l/2$, we find the energy variations

$$\left.\begin{array}{ll} -V_l & \text{for } k = -\frac{l}{2} + \varepsilon \\ +V_l & \text{for } k = -\frac{l}{2} - \varepsilon \end{array}\right\} \quad \varepsilon > 0. \qquad (8')$$

1.4 Forbidden energies: first consequences

All the above may now be summarised as follows. The linear chain perturbed by a periodic sinusoidal potential may now be treated, for most values of $\bar{k}$, in the free-electron framework (Chapter III). In particular, the energy is generally not sensitive to the perturbation and remains equal to $\hbar^2 k^2 / 2m$. However, when $k = \pm l/2$ (or $k = \pm\pi/a$), propagation becomes impossible because of interferences associated with diffraction. Thus, there is a decrease or an increase in the energy, depending on the position (right or left) of k in relation to $l/2$. This is summarised in Figure 4, which shows, by virtue of (8) and (8') a band of forbidden energies or **forbidden band** of width

$$\boxed{\Delta E_l = 2V_l.}$$

The domain of k between $\pm\pi/a$ (in which the energy deviates little (except at the edges) from the parabolic variation (3)) is called the **first Brillouin zone**.

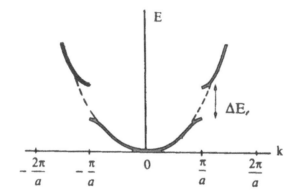

Figure 4 Forbidden energies. On the chain of Figure 2, discontinuities of the energy occur for values of the wavenumber $k = \pm\pi/a$. The dashes show the free-electron parabola.

N.B. Note that the exact form of the variation of $E(k)$ in the neighbourhood of the points $\pm\pi/a$ is not determined in this first approach.

Assuming that the chain of atoms has *one* free electron per atom, its Fermi energy is equal to $(\hbar^2/8m)\pi^2(1/a^2)$ (see (23), p. 40) with corresponding k_F of modulus $\pi/2a$. In this case, all the states are well within the *first* Brillouin zone. Thus, the forbidden band does not play a role for most properties of the chain and, in particular, for its conduction: the chain is in principle **metallic**.

If the chain has *two* electrons per atom, then k_F equals π/a. All the electrons are strictly within the first Brillouin zone. The electron excitations at the Fermi level then become either difficult or impossible, depending on the value of ΔE_l. The chain is **semi-conducting** in the first case and **insulating** in the second.

Let us return to the case of *one electron per atom*. In principle, as previously mentioned, the chain is metallic, with $k_F = \pi/2a$. Imagine a periodic *distortion* of the chain which, affecting one atom in two, *doubles the period* (in other words, the size of the primitive cell) of the chain (see Figure 5). The first Brillouin zone of this distorted chain is now $-(\pi/2a) + (\pi/2a)$. All the

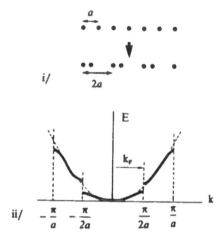

Figure 5 Peierls transition. A transition which changes the period of a chain from a to $2a$ (i) introduces an energy discontinuity for $k = \pi/2a$. (ii) If all the states below $k = \pi/2a$ are occupied, the energy of the electrons close to the Fermi level is decreased by the transition.

electrons of the chain are strictly within it, since $k_F = \pi/2a$. For the electrons near the Fermi level, the opening of a forbidden band for $k = \pi/2a$ is accompanied by a *decrease* in energy (Figure 5). This is often sufficient (despite the elastic distortion energy) for the free energy of the distorted chain (D) to be lower than that of the normal chain (N) at low temperature (see **Problem 2**).

The transformation

$$(N) \rightarrow (D)$$

is called *Peierls transition*. For the reasons described above, it gives rise to a *metal* $\rightarrow$ *insulator* (or *semiconductor*) transition. An example of a transition of this type is given in Figure 6, which shows how the electrical resistance of linear chains of an *organic conductor* (see p. 31) varies strongly at the transition.

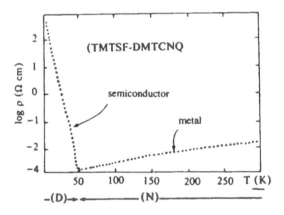

Figure 6 Electrical resistance of a quasi-one-dimensional organic conductor. The resistivity ρ of the organic conductor TMTSF-DMTCNQ, in the form of orthorhombic crystals, is measured along linear chains, as a function of the temperature. A transition occurs at $T \sim 45$ K.
(TMTSF-DMTCNQ = tetramethyl-tetraselenofulvalene-dimethyltetracyanoquinodimethane!).
L Forro and L Zuppiroli, *J. Physique* **43**, 977, (1982).

As examples of such organic chains turned into insulators by a Peierls transition, we mention *carotenoid pigments*. These molecules, derived from isoprene
$$CH_2 = \underset{\underset{CH_3}{|}}{C} - CH = CH_2$$
include, in particular, *lycopene* $C_{40}H_{56}$ (tomato pigment) and all the *carotenes*. They have a forbidden band of width ΔE, situated in the visible spectrum; for example, for lycopene, $\Delta E \simeq 2.4$ eV, in the green. The corresponding light is absorbed, inducing electron transitions across the forbidden band. Whence the red of tomatoes and carrots, the yellow of maize and eggs ....

1.5 Case of a 3D crystal

Everything that has been said in Subsections 1.1, 1.2 and 1.3 generalises to a 3D solid with a small periodic potential $v_p(\vec{r})$. Let us consider, for example, a simple cubic crystal (cube edge: a) and extend the sinusoidal potential (1) in 3D:

$$v_p(\vec{r}) = - V_1 \left[\left(e^{i\vec{l}_1 \cdot \vec{r}} + e^{-i\vec{l}_1 \cdot \vec{r}} \right) + \left(e^{i\vec{l}_2 \cdot \vec{r}} + e^{-i\vec{l}_2 \cdot \vec{r}} \right) + \left(e^{i\vec{l}_3 \cdot \vec{r}} + e^{-i\vec{l}_3 \cdot \vec{r}} \right) \right] \quad V_1 > 0$$

$$= - V_1 \left(\cos \frac{2\pi}{a} x + \cos \frac{2\pi}{a} y + \cos \frac{2\pi}{a} z \right)$$

where $\pm \vec{l}_1$, $\pm \vec{l}_2$ and $\pm \vec{l}_3$ are the six points of the reciprocal lattice (simple cubic) which are near neighbours of the origin ($l_1 = l_2 = l_3 = 2\pi/a$).

We again use (4) and we write the perturbed function in the form

$$\psi_k(\vec{r}) = L^{-3/2} \left(e^{i\vec{k} \cdot \vec{r}} - V_1 \sum_l \frac{1}{E_{\vec{k}} - E_{\vec{k}+l}} e^{i(\vec{k}+l) \cdot \vec{r}} \right)$$

where the sum $\sum$ which represents the perturbation, is over the six vectors $\pm \vec{l}_1$, $\pm \vec{l}_2$ and $\pm \vec{l}_3$. The six terms of this sum are in general small because V_1 is small. However, one of these may be important if $\vec{k}$ is such that $E_{\vec{k}} - E_{\vec{k}+\vec{\lambda}}$ is small, or zero (here, $\vec{\lambda}$ is one of the six vectors $\pm \vec{l}_1 \dots$). We again end up with a solution involving two waves:

$$\psi_k(\vec{r}) \simeq l^{-3/2} \left(\alpha e^{i\vec{k} \cdot \vec{r}} + \beta e^{i(\vec{k}+\vec{\lambda}) \cdot \vec{r}} \right).$$

In the limit, the condition on $\vec{k}$, namely $k^2 - (\vec{k} + \vec{\lambda})^2 = 0$ generalises (6) to

$$2\vec{k} \cdot \vec{l} + \lambda^2 = 0. \tag{8''}$$

If this condition, called the *Brillouin condition*, is satisfied, that is, if the vector $\vec{k}$ ends up on the median plane of $\vec{\lambda}$ (called the *Brillouin plane*), then we have a new solution involving two waves with identical weights, which can be written, like in (7), as:

$$\psi(\vec{r}) = 2L^{-3/2} \left(e^{i\vec{k} \cdot \vec{r}} \pm e^{i(\vec{k}+\vec{\lambda}) \cdot \vec{r}} \right)$$

$$= 2L^{-3/2} e^{i\vec{\kappa} \cdot \vec{r}} \left(\exp \left(i\frac{\vec{\lambda}}{2} \cdot \vec{r} \right) \pm \exp \left(-i\frac{\vec{\lambda}}{2} \cdot \vec{r} \right) \right) \tag{8'''}$$

where $\vec{\kappa}$ is the vector in the Brillouin plane in question such that $\vec{k} = \frac{\vec{\lambda}}{2} + \vec{\kappa}$.

If one sets out from (8''') to calculate the energy as in (8), one again finds in the Brillouin plane the same removal of degeneracy as in (8), namely $2V_1$, since in the expression for $\psi^* \psi$, the term $\exp(i\vec{\kappa} \cdot \vec{r})$ of (8'') disappears: thus, the calculation ending in (8) is rediscovered in an identical form here. We therefore deduce that *on crossing any one of the six Brillouin planes* (medians of the six vectors $\pm l_1 \dots$ of the reciprocal lattice) *considered here, the electron energy has a discontinuity of* $2V_1$. We shall come back to these results more precisely in the next paragraph.

2. BLOCH FUNCTIONS

We have just seen how, for a periodic potential perturbing the wavefunctions relative to a constant potential, energy discontinuities occur for particular values of the wavevector $\bar{k}$. We shall now consider this question more generally using wavefunctions corresponding to the existence of a periodic potential. These wavefunctions are the **Bloch functions**.

Bloch's Theorem (see **Annex 2**, p. 259) enables us to state that the eigenfunctions of the Schrödinger equation:

$$\left[-\frac{\hbar^2}{2m} \Delta + V(\bar{r}) \right] \psi(\bar{r}) = E\psi(\bar{r}), \tag{9}$$

where $V(\bar{r})$ is a *periodic function*, have the form

$$\boxed{\psi(\bar{r}) = U_{\bar{k}}(\bar{r}) e^{i\bar{k}\cdot\bar{r}}} \tag{10}$$

where

i) **the vectors $\bar{k}$ are the vectors of the '$\bar{k}$ lattice'** (see (30), p. 44);
ii) $U_{\bar{k}}(\bar{r})$ **is periodic with the same period as** $V(\bar{r})$.

Bloch's Theorem is a (three-dimensional) generalisation of Floquet's Theorem, according to which the solutions of differential equations of type

$$y'' + f(x)y = 0,$$

where $f(x)$ is periodic, have the form

$$y = g(x)e^{ikx}$$

where $g(x)$ is periodic with the same period as $f(x)$.

Let us now expand $V(\bar{r})$ and $U(\bar{r})$ in Fourier series:

$$\boxed{\begin{aligned} V(\bar{r}) &= \sum_l V_l e^{i\bar{l}\cdot\bar{r}} \quad \text{where} \quad V_{-l} = V_l^* {}^2 \\ U(\bar{r}) &= \sum_l U_l e^{i\bar{l}\cdot\bar{r}} \quad \text{or} \quad \psi(\bar{r}) = \sum_l U_l e^{i(\bar{k}+\bar{l})\cdot\bar{r}}. \end{aligned}} \tag{11}$$

The vectors $\bar{l}$ are the translation vectors of the **reciprocal lattice** (defined by (11), p. 100).

For example, if the periodicity of the crystal, and thus also that of V and U, is simple cubic (cube edge: a), the vectors $\bar{l}$ are the vectors of a simple cubic lattice (cube edge: $2\pi/a$). For an fcc crystal, the vectors $\bar{l}$ are the vectors of a bcc lattice etc. We note

---

[2] In fact, $V_{-l} = V_l^*$, since the Fourier coefficients of the function $V(\bar{r})$ are defined by $V_l = (1/\omega) \int_{\Omega} V(\bar{r}) \exp(-i\bar{l}\cdot\bar{r}) \, \mathrm{d}^3\bar{r}$ (ω is the volume of the crystalline cell Ω).

here, for what follows, that **the function** $\exp(i\bar{l} \cdot \bar{r})$ **has the periodicity of the crystal.** In fact, if $\bar{r}_T$ is a translation vector defined by (9) (p. 96), we may write

$$\bar{l} \cdot \bar{r}_T = \left(\sum_i n_i \vec{a_i} \right) \cdot \left(\sum_j m_j \bar{a}_j \right) = 2\pi \times \text{integer}$$

(see (11), p. 100). Thus, we have

$$
\boxed{
\begin{aligned}
&e^{i\bar{l}\cdot\bar{r}} = e^{i\bar{l}\cdot(\bar{r}+\bar{r}_T)} && \text{and} \\
&\int_\Omega e^{i\bar{l}\cdot\bar{r}}\, d^3\bar{r} = 0 && \text{except if } \bar{l} = 0
\end{aligned}
}
\tag{12}
$$

(Ω: primitive cell).

The reader should verify that it is possible to expand a periodic function $V(\bar{r})$ in terms of the vectors of the reciprocal lattice (RL) by calculating the coefficients V_l. Let $\overline{G}$ be one of the vectors $\bar{l}$ of the RL. Multiply both sides of the expansion of $V(\bar{r})$ (11) by $\exp(-i\overline{G} \cdot \bar{r})$ and integrate over a primitive cell Ω of the direct lattice. All terms of type $\int_\Omega V_l \exp(i(\bar{l} - \overline{G}) \cdot \bar{r})\, d^3\bar{r}$ are zero (by virtue of (12)), since $\bar{l} - \overline{G}$ is a vector of the RL. Thus, this leaves $V_G = \int_\Omega V(\bar{r}) \exp(-i\overline{G} \cdot \bar{r})\, d^3\bar{r}$, which establishes the existence of the coefficients V_l.

We retain the assumption of a *small potential* $V(\bar{r})$; in other words of small Fourier coefficients V_l. In the limit as $V_l \to 0$ (zero potential, free-electron model, Chapter III), all the U_l are zero, except for U_0 (corresponding to $\bar{l} = 0$). The Bloch function is then just the plane wave $\exp(i\bar{k} \cdot \bar{r})$. If V is small and only intervenes up to the first order, we can calculate the coefficients U_l and show that they too are small, provided that $\bar{k}$ is not equal to certain particular values, which we shall determine.

For this, we substitute the expansions (11) in equation (9) and retain the assumption that the U_l (except U_0) and the V_l are small. We obtain:

$$\left[U_0 \left(\frac{\hbar^2}{2m} |\bar{k}|^2 + V(\bar{r}) - E \right) + \sum_{l\neq 0} U_l V_{-l} \right] + \sum_{l\neq 0} U_l \left(\frac{\hbar^2}{2m} |\bar{k} + \bar{l}|^2 - E \right) e^{i\bar{l}\cdot\bar{r}}$$

$$+ \sum_{l,l'\neq 0} U_l V_{l'} e^{i(\bar{l}+\bar{l'})\cdot\bar{r}} = 0. \tag{13}$$

We multiply throughout by $d^3\bar{r}$ and integrate over a cell Ω of the crystal. Since $\exp(i\bar{l} \cdot \bar{r})$ is periodic (see (12)), all the terms of $\sum_l$ (outside brackets) and of $\sum_{l,l'}$ are zero and thus we have [ ] = 0. If we now choose the origin of the potentials so that $\int_\Omega V(\bar{r})d^3\bar{r} = 0$ and neglect the terms $U_l V_{-l}$ (considered to be second-order terms), we obtain

$$E = \frac{\hbar^2}{2m} |\bar{k}|^2 \tag{14}$$

and we have rediscovered (31) of p. 45: thus, *to the first order, the energy is the same as when the potential is constant*. This result was previously stressed in i) on p. 117.

In order to determine the coefficient U_l, we simplify (13) in the previous approximation to the form

$$U_0 \sum V_l e^{i\bar{l}\cdot\bar{r}} + \sum U_l \frac{\hbar^2}{2m} (2\bar{k} \cdot \bar{l} + |\bar{l}|^2) e^{i\bar{l}\cdot\bar{r}} = 0.$$

Whence, multiplying by $\exp(-i\bar{l} \cdot \bar{r})$ and integrating over Ω:

$$U_l = -\frac{2m}{\hbar^2} \frac{U_0}{2\bar{k} \cdot \bar{l} + l^2} V_l \qquad (15)$$

an expression which, in the general case, completely determines (to the first order) the Bloch function in terms of its Fourier coefficients U_l. The latter are proportional to the V_l: in particular, *the smaller are the V_l, the smaller are these coefficients.*

N.B.   An approximation to the Bloch functions of a crystal may be obtained by another method, that of *linear combination of atomic orbitals* (**LCAO** or the **tight binding method**). This method is described in **Annex 4** (p. 267) and used in **Problems 2, 3, 4, 5** and **6**.

It establishes a link between the *Bloch functions*, delocalised on the whole crystal (as are the plane waves) and the local *atomic functions* of the valence electrons (for example, $2s$ for lithium) of the constituent atoms. This link is shown in Figure 1, p. 268. This shows that the passage from atomic functions to Bloch functions at the time of construction of the crystal, is accompanied by a decrease in the mean *curvature* of the wavefunctions and thus also in the *kinetic energy* (see (Q.M.)). This is a major reason for the stability of crystalline material (see the beginning of Section V.2, p. 95).

3. BRILLOUIN PLANES AND ZONES

There are a number of special cases of the above, in which the U_l can no longer be determined, namely those for which the denominator of the expression (15) is zero; in other words,

$$2\bar{k} \cdot \bar{l} + l^2 = 0. \qquad (16)$$

This condition on the vectors $\bar{k}$, called the **Brillouin condition**, generalises (6) and (8″). It may be very simply stated in the reciprocal space. The vectors $\bar{l}$ are the translation vectors of the reciprocal lattice. The vectors $\bar{k}_0$ which satisfy (16) are the vectors which join the origin of this lattice to the points of the *median planes* of the vectors $\bar{l}$. These planes are called **Brillouin planes**. Figure 7 shows three points (including the origin O) of the reciprocal lattice and the two corresponding Brillouin planes.

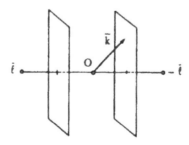

Figure 7   Two Brillouin planes corresponding to the vectors $\bar{l}$ and $-\bar{l}$ of the reciprocal lattice. The vector k_0 satisfies (16).

Exercise   Prove that condition (16) and the diffraction condition for a plane wave with wavevector $\bar{k}$ (formula (13), p. 101, or **Annex 1**, p. 255) hold.

Deduce, albeit in a semi-qualitative way, the link between **diffraction** and **propagation** for electrons which satisfy (or almost satisfy) condition (16). The two-wave form (19) of the corresponding Bloch function will be useful for this.

The Brillouin planes define successive regions of the reciprocal space, called *first, second* ... **Brillouin zones**, in the order that they are reached, beginning at the origin and passing through *zero, one* etc Brillouin planes. Figure 8 illustrates this definition for a square-plane crystal for which the reciprocal lattice (in the figure) is also square plane. The figure shows all the Brillouin planes (straight lines here) up to the 10th Brillouin zone.

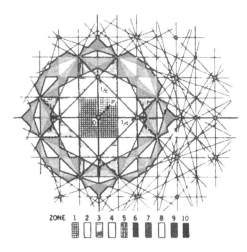

Figure 8   **Brillouin planes (lines here) for a square reciprocal lattice.** The origin of the square reciprocal lattice is 0 and $1/d$ indicates the edge of the square. p is the $\bar{k}$ vector of this book.
After **L Brillouin**, *Wave Propagation in Periodic Structures*, Dover Publ. (1953).

Figure 9 shows the *first Brillouin zone* for an fcc crystal and a bcc crystal.

Thus, in the case of the fcc crystal, this zone is formed by median planes of a bcc lattice (see **N.B. 2, iv**), p. 104). It is not difficult to show that in this case two families of planes are involved, namely: the median planes of the six vectors $\langle 100 \rangle$ (which form a cube) and the median planes of the eight vectors $\frac{1}{2}\langle 111 \rangle$ (octahedral planes which truncate the above cube).

Important Remarks

i)   We note that the first zone of an fcc crystal (*resp.* bcc) is the Wigner–Seitz cell of a bcc lattice (*resp.* fcc). Both cases involve volumes which, when stacked (see p. 72) occupy the space without gaps and without superposition. It follows that if a is the edge of the crystalline cubic cell, the volume of the first Brillouin zone of the fcc (*resp.* bcc) crystal shown in Figure 9(a) (*resp.* 9(b)) is equal to $(4\pi a)^3/2 = 32\pi^3/a^3$ (*resp.* $16\pi^3/a^3$) and that there are $(32\pi^3/a^3)(L^3/4\pi^3) = 2N$ (*resp.* $2N$) corresponding possible electronic states (see (32'), p. 45), where N is the number of *primitive cells* in the given volume L^3.

ii)   It can be shown that the *various Brillouin zones all have the same volume*, and, in
       particular, the same volume as the first one.

       Note here that the *n*th Brillouin zone is the set of points having the origin node of the
       reciprocal lattice as their *n*th neighbour node. Each point of the *n*th zone (unless it is on a
       zone boundary) has only *one* *n*th neighbour node in the lattice.
       The *n*th zone consists of portions which, when translated to the interior of the first
       zone by appropriate vectors $\bar{l}$ of the reciprocal lattice, reconstitute the latter exactly, without
       gaps and without superposition. For example, for the fourth zone of a plane square lattice
       (Figure 8), these translations are ⟨11⟩ and ⟨10⟩.

iii)   The following important conclusion follows from the above remarks:

       | each zone may accommodate two states per primitive cell | (16′)

       (or, for example, *two states per atom*, if there is one atom per primitive cell).

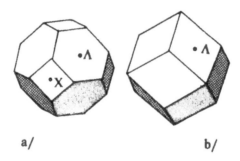

Figure 9   First Brillouin zone .... (*a*) for an fcc crystal; (*b*) for a bcc crystal. The points Λ are the points
of the Brillouin planes closest to the centre.

4. FORBIDDEN BANDS (continued)

At this level, a rapid return to Section 1 is desirable. First, as already mentioned, condition
(16) is a generalisation of (6) and (8″). On the other hand, by way of approximation,
we introduced a sinusoidal potential (1) with intensity V_l, relative to the 'first' vector of
the reciprocal lattice, $\bar{l}$. It is clear that if $\bar{l}$ had been not this particular vector but the
current vector of the reciprocal lattice, then the ideas developed and the results obtained
in Section 1 would have been identical: the sinusoidal potential (1) would have been
periodic with period a submultiple of *a* and thus also periodic on *a*, and we would also
have observed the creation of a forbidden band of amplitude $2V_l$ (see (8)). In fact,
expansion of the potential $V(\bar{r})$ in a Fourier series (see (11)) amounts to writing it as the
sum of sinusoidal potentials, each of which is associated with a particular vector $\bar{l}$ of the
reciprocal lattice (as was v_p in (1)) and has *weight* (Fourier coefficient) V_l.
       Thus, we generalise Section 1 with the following important conclusions.

i)   To each vector $\bar{l}$ of the reciprocal lattice there corresponds first a Brillouin plane
       (median of $\bar{l}$) and second a Fourier coefficient V_l of the potential $V(\bar{r})$ of the crystal.
ii)   Each direction $\bar{k}$ crosses infinitely many Brillouin planes. On each crossing of
       a plane of index *l* (*first* plane on Figures 4 and 11), the energy is subject to a
       discontinuity equal to $2|V_l|$, creating a **forbidden band**.

iii) These forbidden bands are surrounded by **bands of allowed energies**, in which the free-electron approximation is reasonably good for states of the reciprocal space located *far away from the Brillouin planes*, in other words, from the *boundaries of the zones*.

iv) As one approaches a zone boundary 'from the interior' (*resp.* 'from the exterior'), the **energy of the electrons decreases** (*resp.* **increases**) in comparison with the energy of the free electrons (look again at (7) and (7')).

Exercise Rediscover result **iv)** directly via (13) and (15) introducing the terms of type $U_l V_{-l}$ neglected earlier (see (14)).

v) For a given Brillouin plane defined by the vector $\bar{l}$, the discontinuity $2|V_l|$ is the same throughout the plane[3]. On the other hand, the average level E_0 varies from one point of the plane to another, as does $|\bar{k}_0|$ (see Figure 7). It is a minimum at the intersection of the Brillouin plane and the corresponding vector $\bar{l}$ of the reciprocal lattice. For example, in Figure 9, it is a minimum at the centres of the facets of the first two Brillouin zones shown (points X and Λ). Thus, in the plane relative to vector $\bar{l}$, we may write

$$\boxed{E = E_0 \pm |V_l|} \tag{17}$$

where E_0 (variable throughout the plane) is the unperturbed energy of the free electrons, given by (14).

We are now in a position to justify some of the previous conclusions and at the same time to make them more precise.

Let us consider a Bloch function $\psi_{\bar{k}_0}(\bar{r})$, for which the vector $\bar{k}_0$ is practically in condition (16) for a given vector $\overline{G}$ of the reciprocal lattice ($\overline{G} \ni \bar{l}$):

$$2\bar{k}_0 \cdot \overline{G} + G^2 \simeq 0. \tag{18}$$

The expansion (11) of $\psi_{\bar{k}_0}(\bar{r})$ is then dominated by two terms:

$$\psi_{\bar{k}_0}(\bar{r}) \simeq U_0 e^{i\bar{k}_0 \cdot \bar{r}} + U_G e^{i(\bar{k}_0 + \overline{G}) \cdot \bar{r}}. \tag{19}$$

We now substitute this approximation to $\psi_{\bar{k}_0}(\bar{r})$ into equation (9), taking account of (11) and (18). We obtain:

$$U_0 \left(E_0 - E + \sum_l V_l e^{i\bar{l} \cdot \bar{r}} \right) + U_G \left(E_0 - E + \sum_l V_l e^{i\bar{l} \cdot \bar{r}} \right) e^{i\overline{G} \cdot \bar{r}} = 0$$

where we have set $E_0 = (\hbar^2/2m)|k_0|^2$ and simplified the whole by dividing by $\exp(i\bar{k}_0 \cdot \bar{r})$. Multiplying this expression successively, first by $d^3\bar{r}$, then by $\exp(-i\overline{G} \cdot \bar{r})d^3\bar{r}$ and integrating each time over the volume of a primitive cell, recalling (12), we find:

$$(E_0 - E)U_0 + V_{-G}U_G = 0 \qquad V_G U_0 + (E_0 - E)U_G = 0, \tag{20}$$

which equations are compatible with the condition that

$$(E_0 - E)^2 = V_G V_{-G}.$$

---

[3] Except near the intersection of two planes.

It follows (see footnote to p. 122) that

$$E = E_0 \pm |V_G| \tag{21}$$

(which equation was previously given in (17)), and that:

$$U_0 = \pm U_G. \tag{21'}$$

which again incorporates (and justifies) the previous choice of $\alpha^2 = \beta^2$ (see (7)).

Exercise   Use (18) and (21') to rewrite the wavefunction (19) in a two-wave form (**N.B. 2**, p. 118) symmetric in $+\overline{G}/2$ and $-\overline{G}/2$.

The behaviour of E may be studied for values of $\overline{k}$ close to $\overline{k}_0$ using the previous approach for vectors $\overline{k}$ chosen, for example, collinear with $\overline{k}_0$ *just before* $(\overline{k}^- = \overline{k}_0(1-\varepsilon); \varepsilon > 0)$ or *just after* $(\overline{k}^+ = \overline{k}_0(1+\varepsilon))$ the Brillouin plane in question (Figure 10). This simple, but tedious calculation, for example, for $\psi_{\overline{k}^+}$, leads us (to the second order) to supplement equation (17) (G replacing l) by:

$$E^+ = E_0 \pm |V_G| + \frac{1}{2}(4E_0 - \xi)\varepsilon + \left(E_0 \pm \frac{\xi^2}{8|V_G|}\right)\varepsilon^2$$

where we have set $\xi = (\hbar^2/2m)G^2$.

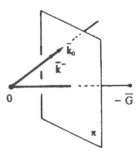

Figure 10   Vector $\overline{k}_0$ **satisfying equation (18).** π is the Brillouin plane for the vector $-\overline{G}$ of the reciprocal lattice. $\overline{k}^-$ is collinear with $\overline{k}_0$.

The choice of the sign of $|V_G|$ (whence also of $\xi^2/8|V_G|$) is dictated by the need for E to tend towards its free-electron value (formula (14)) when one moves away from the critical case $\overline{k}_0$. On the other hand, it is clear that $4E_0 - \xi = (\hbar^2/2m)(4k_0^2 - G^2) = P$ is positive (or zero when $\overline{k}_0 = -\overline{G}/2$), since we have $k_0 \geq G/2$ (see Figure 10). Thus, we have

$$E^+ = \begin{cases} E_0 + |V_G| + \frac{1}{2}P\varepsilon + \left(E_0 + \xi^2/(8|V_G|)\right)\varepsilon^2 & \text{for } \overline{k}_0 \neq -\overline{G}/2 \\ E_0 + |V_G| + E_0(1 + 2E_0/|V_G|)\varepsilon^2 & \text{for } \overline{k}_0 = -\overline{G}/2. \end{cases} \tag{22}$$

After calculating E^- (analogously), one may draw the curve $E(|\overline{k}|)$ for the direction $\overline{k}$ taken along $\overline{k}_0$ (see Figure 11).

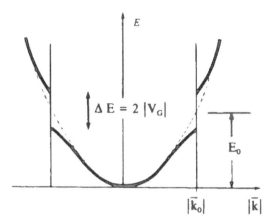

Figure 11 Variations of $E(k)$. In a crystal, the electron energy is subject to a discontinuity ΔE at E_0 for a value $|\bar{k}_0|$ of $|\bar{k}|$ satisfying (18). In the neighbourhood of $|\bar{k}_0|$, the electron energy has the parabolic form given in (22). Dashes: free-electron parabola. N.B. The figure corresponds to the case $\bar{k}_0 \neq -\overline{G}/2$ of (22).

5. ENERGY GAP

5.1 Surfaces of equal energy

The *surfaces of equal energy* remain spheres (see p. 45) far from the Brillouin planes and, in particular, in the interior of the first zone as long as the curve $E(\bar{k})$ does not move away from the free-electron parabola. For a small number of electrons, for example *one* per atom, it is then possible that the surface of maximum energy, the **Fermi surface**, may remain a sphere. This is what happens in Figure 12(*a*). In this case, the energy is purely kinetic, and we would expect the free-electron model (Chapter III) to apply.

1   Thus, let us consider a **monovalent metal** of **body-centred cubic** structure, with lattice parameter a. Its reciprocal lattice is fcc with parameter $4\pi/a$ (see p. 106). The twelve points of the surface of the first Brillouin zone nearest to the origin (Λ, Figure 9(*b*)) are thus at distance $k_\Lambda = 4\pi\sqrt{2}/4a = 4.44/a$ from the latter. In the free-electron approximation (see (33'), p. 46), we have $k_F = (3\pi^2 \times 2/a^3)^{1/3} = 3.90/a$. Clearly, in this case (which is that of *sodium*), k_F is distinctly less than k_Λ. Thus, we would expect the *Fermi surface* to be *completely contained within the first zone* (case of Figure 12(*a*)) and to be practically a sphere.

2   Let us now consider a **monovalent metal** of **face-centred cubic** structure with parameter a. Its reciprocal lattice is bcc with parameter $4\pi/a$. The points of the surface of the first Brillouin zone nearest to the origin (Λ, Figure 9(*a*)) are at distance $k_\Lambda = \pi\sqrt{3}/a = 5.44/a$, although the radius of the free-electron Fermi sphere is $k_F = (3\pi^2 \times 4/a^3) = 4.91/a$. Thus, there exist occupied states relatively close to points Λ. For these states, point **iv**) (above, p. 127) comes in to play, and this, increasingly as $|V_{111}|$ increases. In this region close to Λ, the surfaces of equal energy are thus deformed in comparison with the sphere, by protuberances towards the points Λ. If $|V_{111}|$ is sufficiently large, these protuberances touch the points Λ. When all the electrons are 'lodged' in the reciprocal space, the ultimate surface, called the **Fermi surface** must have a shape similar to that shown in Figure 12(*b*): at the same time, it must remain

almost spherical in the directions neighbouring $\langle 100 \rangle$ (k_x, k_y or k_z in the figure); thus it appears to be 'sucked up' by the Brillouin planes in the eight directions $\langle 111 \rangle$. Since it is expensive to cross these (energy cost: $2|V_{111}|$), we would expect it not to cross these planes: it remains attached to them, adopting a *neck* shape (eight necks in Figure 12(*b*)).

We shall see later how the shape of the Fermi surfaces may be determined experimentally. Suffice to note that the Fermi surface obtained for *copper* looks like that of Figure 12(*b*).

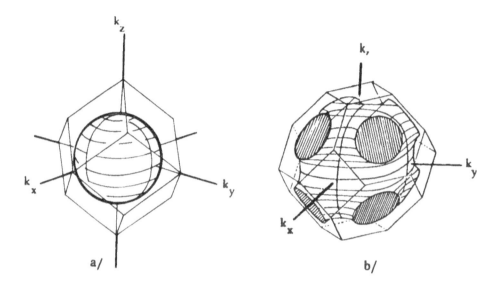

Figure 12   **Fermi surface.** (*a*) for a monovalent metal with a bcc structure (Na); (*b*) for a monovalent metal with an fcc structure (Cu). In both cases, only the first Brillouin zone is shown.

Let us evaluate the radius ρ_1 of the neck of this Fermi surface. For this, we consider the vector $\bar{k}_0$ corresponding to point Λ of the reciprocal lattice ($\bar{k}_0 = 1/4\langle 111 \rangle$) and an adjacent vector $\bar{k}$ situated *in* the first zone. We set

$$\bar{k} = \bar{k}_0 + \bar{\omega}$$

and project $\bar{\omega}$ parallel ($\omega_{\parallel}$) and perpendicularly ($\omega_{\perp}$) to $\bar{k}_0$ (Figure 13).

$E(|\bar{k}|)$ is calculated as in Section 4. Retaining the notation $\bar{G}$ ($= -1/2\langle 111 \rangle$), it is easy to show that:

$$E = E_0 + \frac{\hbar^2}{2m}|\omega|^2 \pm \left(4E_0 \frac{\hbar^2}{2m} \omega_{\parallel}^2 + |V_G|^2 \right)^{1/2} \qquad (23)$$

where, as before, $E_0 = (\hbar^2/2m)|\bar{k}_0|^2$. The Fermi energy E_F is measured from the bottom of the forbidden band, $E_0 - |V_G|$, by the quantity β:

$$E_F = E_0 - |V_G| + \beta.$$

It is clear that if $\beta < 0$, the Fermi surface Σ remains on this side of point Λ and zone boundary; we are in a situation similar to that of Figure 12(*a*). On the other hand, if the Fermi

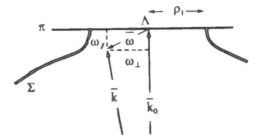

Figure 13 Neck of a Fermi surface Σ (see **Figure 12(b)**). State $\bar{k}$ near to a point Λ (see Figure 9) of the surface of the first Brillouin zone.

energy lies *in the forbidden band* $(0 < \beta < 2|V_G|)$, the surface Σ is in contact with the Brillouin plane π. The contact takes place on a circle of radius ρ_1, such that, for $\omega_{\parallel} = 0$ and $|\bar{\omega}| = \omega_\perp = \rho_1$, we have $E = E_F$. Using (23), we obtain

$$\rho_1 = \sqrt{2m\beta}/\hbar.$$

Exercise Using (23), prove that a surface of equal energy (and, in particular, the Fermi surface) rejoins the Brillouin plane π *perpendicularly*.

3 **If the valency increases**, a number of cases are possible, depending mainly on the value of the coefficients $|V_G|$. In the next two subsections, we shall examine the extreme cases of **small** coefficients (≤ 1 eV) and **large** coefficients (≥ 3 eV).

5.2 Case of small coefficients $|V_G|$

If the V_G (and, in particular, $|V_{111}|$ for an fcc crystal) are weak, the electrons may occupy states of the second Brillouin zone, then of the third ....

In the case examined above (Figure 13), states first appear around point Λ beyond plane π, forming a protuberance with a circular section on plane π (radius ρ_2; Figure 14). This case corresponds to a Fermi level *above the forbidden band* relative to the $\langle 111 \rangle$ direction of the reciprocal lattice, in other words, to the case $\beta > 2|V_G|$. The radius ρ_2 is calculated as above. It is equal to

$$\rho_2 = \sqrt{2m(\beta - 2|V_G|)}/\hbar$$

so that the difference between the areas of the two circles (ρ_1) and (ρ_2)

$$\pi(\rho_1^2 - \rho_2^2) = 4m\pi|V_G|/\hbar^2$$

is proportional to $|V_G|$. Measurement of these areas (see Section 6.2) gives access to coefficients $|V_G|$.

In general, the Brillouin planes here affect the Fermi surface only slightly.

This is, in particular, the case for *aluminium*, a metal for which the first zone is full and the second, third and fourth zones are partially occupied, with practically no alteration to the spherical shape of the Fermi surface (see p. 76).

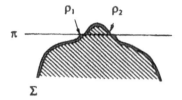

Figure 14   Radii of the neck (ρ_1) and of the protuberance (ρ_2) of a Fermi surface Σ situated on both sides of a Brillouin plane π.

In the case of aluminium, since all the faces of the first zone are covered by states occupied in the second, the feature of small $|V_G|$ comes into play only on the faces of the second and third zones.

The forbidden band relative to the Brillouin plane [100] (facet containing X on Figure 9(a)) may be validated by reflectance measurements. In fact, around X in this plane, the dispersion curves relative to the vectors $\overline{\omega}$ of the plane ($\omega_1 = 0$; see Figure 13) are given by (23)

$$E_{int} = E_0 - |V_{100}| + \frac{\hbar^2}{2m}|\overline{\omega}|^2 \text{ on the internal face of the plane, and}$$

$$E_{ext} = E_0 + |V_{100}| + \frac{\hbar^2}{2m}|\overline{\omega}|^2 \text{ on its external face.}$$

Thus, on the whole facet [100], the forbidden band is $E_{ext} - E_{int} = 2|V_{100}|$. A transition from the first to the second zone in this region of the reciprocal space thus absorbs an energy (greater than or) equal to $2|V_{100}|$. In an experiment involving the reflection of light by aluminium a *drop in the reflection coefficient* is observed, which corresponds to this absorption, for a photon energy of 1.5 eV (Figure 15). Thus, here, we have $|V_{100}| = 0.75$ eV.

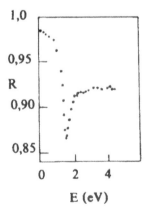

Figure 15   Coefficient of reflection R for aluminium as a function of the photon energy.
H E Bennet et al. *J. Opt. Soc. Am.* **53**, 1098 (1963).

5.3 Case of large coefficients $|V_G|$

Imagine a *divalent* crystal which is assumed to have simple structure (one atom per primitive cell). Following (16′) (p. 126), the first Brillouin zone of this crystal contains a number of states exactly equal to the number of valence electrons (two per atom). Suppose also that the coefficients $|V_G|$ relative to the planes bordering the first zone (here, the coefficients V_{100}) are 'large' in comparison with the thermal energies (for example, several eV).

On crossing the planes of the first zone, the energy jumps from $E_0 - |V_G|$ on the internal face (see (17)) to $E_0 + |V_G|$ on the external face, where E_0 varies along the planes, in other words, as a function of the direction of $\bar{k}$. For $|V_G|$ sufficiently large, the largest value of $E_0 - |V_G|$ (in a direction $\bar{k}_2$) may be lower than the least value of $E_0 + |V_G|$ (in direction $\bar{k}_1$). Figure 16 shows an example of this situation. Then, there exists a domain of forbidden energies *for all directions* $\bar{k}$ (E_g on Figure 16), in other words, *for the solid* (and no longer simply for a single direction $\bar{k}$ as in Figure 11), which is generally called an **energy gap**. Of course, we have $E_g < 2|V_G|$.

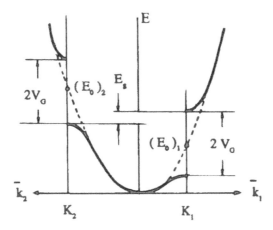

Figure 16 Curves $E(k)$ in the two directions $\bar{k}_1$ (and $\bar{k}_2$) for which the unperturbed energy E_0 is the least (the greatest). K_1 and K_2 are the two corresponding limits of the first zone.

In the case of our divalent metal, the situation is the following. The first Brillouin zone is exactly *full* (its surface merging with the Fermi surface) and the second is *empty*. For high values of E_g (for example, $E_g \simeq 5$ eV), since thermal excitations are impossible, the *electron specific heat* of such a solid would be *zero*. Since each electron ($\bar{k}$) is compensated by an electron ($-\bar{k}$), the *electrical conductivity* would be *zero* and the solid would be an **insulator**. For smaller coefficients $|V_G|$ (for example, down to $E_g \simeq 0.5$ eV), the conductivity would remain zero at low temperature, but would increase with temperature, by thermal promotion of electrons into the second zone: the solid would be a **semiconductor**.

In fact, this case *does not arise* for divalent solids. Their coefficients $|V_G|$ are actually such that some of the electrons occupy states in the second zone, and thus, divalent materials (beryllium, magnesium, zinc ...) are *metals*.

However, it *does arise* for certain **tetravalent materials**. For example, for tetravalent materials of diamond structure (two atoms – whence eight electrons – per primitive cell), the *first four* Brillouin zones are exactly full at 0 K (see (16') p. 126). Thus, these **tetravalent** materials are **insulators** (*diamond*) or **semiconductors** (*silicon, germanium*), for the above reason. This case is discussed in depth in the course on semiconductor physics (S.C.Ph.)[4].

5.4 Consequences for the density of states

The density of states for free electrons is of the form $n(E) = AE^{1/2}$ (see p. 45).

Around the centre of the first Brillouin zone, this form may be retained as a good approximation. However, when the surfaces S of equal energy *in the interior* of the first zone begin to be deformed by the proximity of a zone boundary π (region I on Figure 17), the density $n_1(E)$ of the states of the first zone *increases more rapidly* than for free electrons, first because the curvature of S is greater in I than in II (region distant from π) and second because the distance between $S(E)$ and $S(E + dE)$ is also greater: for example, at the apex, this distance is $dE/(dE/dk)$ where dE/dk is smaller than for free electrons (Figure 11) and tends to zero (see (21) for $\overline{k}_0 = \pm \overline{G}/2$). For energies above that corresponding to contact of S with π, the area of S continues to decrease, *as does* $n_1(E)$, which becomes zero when all the states of the first zone are occupied. This leads to a '*closed*' curve of density n_1 (see Figure 18).

The same arguments lead to curves of density $n_2, n_3 \ldots$, which are also closed (Figure 18).

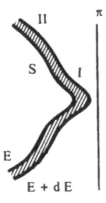

Figure 17 **Volume between two adjacent surfaces of equal energy (E and $E + dE$) in the first Brillouin zone.** Plane π: zone boundary.

Exercise Consider a simple cubic crystal (lattice parameter: a), where valence electrons are free (all $V_G = 0$). Show that the density of states in the first Brillouin zone, $n_1(E)$, equal to $A E^{1/2}$ for $k < \pi/a$, becomes equal to $(A/2\pi)^{2/3}(6\pi^2/a) - 2AE^{1/2}$ for $\pi/a < k < \pi\sqrt{2}/a$. Draw the graph of $n_1(E)$ where the similarity with n_1 of Figure 18 will be observed.

[4] B Sapoval and C Hermann, *Physics of semiconductors*, Springer-Verlag (1995).

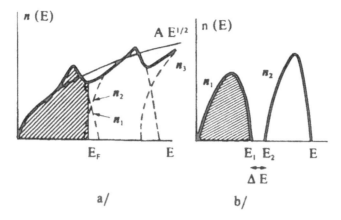

Figure 18   Density of states in the first zone (n_1), in the second (n_2).... (a) The bands n_1, n_2 ... intersect. Fine line: free-electron parabola. Dashes: curves n_1, n_2.... Bold line: total density $n(E) = n_1 + n_2$ .... The Fermi level corresponds to a divalent metal. (b) The bands n_1 and n_2 do not intersect. The band n_1 is assumed to be completely occupied (shaded) and the band n_2 completely empty.

Exercise   Show that the density $n(E)$ is of the form

$$n(E) = \frac{1}{8\pi^3} \int_S \frac{dS}{|\nabla_k(E)|} \tag{24}$$

for a unit volume, where S is a surface of equal energy E in the reciprocal space. Verify that (24) results in expression (33) (p. 45) in the case of free electrons.

5.5 Classification of solids

The two previous cases ($|V_G|$ *small* or *large*) then give rise to two extreme possibilities (Figures 18(a) and (b)).

i) **Either** ($|V_G|$ small; Figure 18(a)) the bottom of band n_2 has an energy lower than the top of band n_1 (case of Figure 14, where states are occupied both in the second and in the first zone). The total density is then $n(E) = n_1 + n_2 + \ldots$. It does not differ greatly from the parabola $AE^{1/2}$. The intersection of the bands means that there exist possible states for all energies (even if each direction $\bar{k}$ of the reciprocal space includes a forbidden band for each crossing of a Brillouin plane). The position of the Fermi level E_F depends on the valency z, via the condition $2 \int_0^{E_F} n(E) f(E) dE = zv$ (v is the number of atoms of the solid; see p. 48). Here, nothing opposes conduction in the sense of Chapter III (p. 55) and the corresponding solid is a **metal**.

ii) **Or** ($|V_G|$ large, with the occurrence of a gap E_g; Figure 18(b)) the two bands n_1 and n_2 do not intersect. In the case illustrated, the states of band n_1 are able to accommodate all the valence electrons of the solid exactly (see Section 5.3, p. 133). This band is called the **valence band**. At zero temperature it is full, while band n_2, called the **conduction band**, is empty. Taking into account the symmetry of the Brillouin zones about their centres, the conductivity is then zero at zero temperature.

N.B. Figure 18(b) illustrates the case of the fictional divalent crystal with large $|V_G|$ of Section 5.3; band n_1 corresponds to the first Brillouin zone, n_2 to the second. It does not represent a tetravalent material of *diamond* structure unless n_1 denotes the set of states of the first *four* zones and n_2 the states of the following four zones.

As the temperature increases, some electrons are **excited** from the top of the full band n_1 to the bottom of the empty band n_2. This excitation creates an equal number of *occupied states* in the latter and *empty states* in the former. These empty states, called **holes**, permit conduction *by holes* (in the top of band n_1; we shall return to this point later, p. 146) to which is added a *conduction by electrons* (in the bottom of n_2).

More precisely, the number of electrons N_e excited in band n_2 and of holes N_h created in band n_1 per unit volume, at temperature T are given by:

$$N_e = \int_{E_2}^{\infty} n_2(E)\frac{1}{e^{(E-\mu)/kT}+1}\,dE$$

$$N_h = \int_{-\infty}^{E_1} n_1(E)\left(1 - \frac{1}{e^{(E-\mu)/kT}+1}\right)\,dE \tag{25}$$

where μ is the *chemical potential* (equal to the Fermi level for $T = 0$) and E_1 and E_2 are as defined in Figure 18(b). At sufficiently low temperature such that $kT \ll E_2 - \mu$ and $kT \ll \mu - E_1$, it is easy to verify that (25) may be simplified to:

$$N_e = e^{-(E_2-\mu)/kT} \int_{E_2}^{\infty} n_2(E)e^{-(E-E_2)/kT}\,dE$$

$$N_h = e^{-(\mu-E_1)/kT} \int_{-\infty}^{E_1} n_1(E)e^{-(E_1-E)/kT}\,dE. \tag{26}$$

The nature of these expressions shows that the variation of N_e and N_h with temperature is dominated by the exponential which precedes the integral. Denoting the product of these two integrals by $B(T)$, we obtain:

$$\boxed{N_e N_h = B(T)e^{-E_g/kT}} \tag{27}$$

where $E_g = E_2 - E_1$ (see Figure 18(b)) is the **width of the gap**. The expression (27) is nothing other than the *law of mass action* applied to the reaction

$$\text{electron } (n_1) \rightleftharpoons \text{electron } (n_2) + \text{hole } (n_1)$$

where (n_1) and (n_2) denote the valence band and the conduction band, respectively. In the present case of a *pure*, undoped (or *intrinsic*) *solid*, the numbers N_e and N_h are clearly equal (each electron excited from (n_1) to (n_2) corresponds to a hole in (n_1)) and we have:

$$N_e = N_h = \sqrt{B(T)}e^{-E_g/2kT}. \tag{28}$$

Thus, the **electrical conductivity** (see (48) p. 56 where N_e and N_h replace N) must vary with temperature approximately as $\exp(-E_g/2kT)$. Depending whether this conduction is negligible (ΔE too large; case of diamond) or not, we are dealing with an **insulator** or a **semiconductor**. Figure 19 shows the conductivity σ of silicon. This does vary like $\exp(-E_g/2kT)$, whence it follows that $E_g = 1.2$ eV. Table 1 shows the values of E_g for a number of solids which crystallise into the *fcc diamond* system.

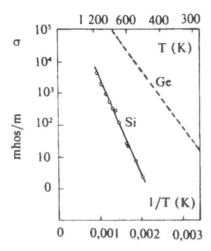

Figure 19 Logarithm of the conductivity σ of silicon (and of germanium) as a function of $1/T$.
F Morin et al. *Phys. Rev.* **96**, 28 (1954).

Table 1   Width E_g of the gap for a number of insulators and semiconductors.

Diamond	Ga As	Si	Ga Sb	Ge	Sn As	Sn (grey)
7.0	1.45	1.21	0.8	0.75	0.5	0.08

N.B. 1   Fermi level of an intrinsic semiconductor.

The position of the chemical potential (or *Fermi level*) was easy to define in the case of metals: it was at the energy of the electron with the largest energy at $T = 0$ K. What is it now?

Let us consider an *intrinsic semiconductor* (or insulator) (Figure 18(b)). At $T = 0$ K, all the states of the valence band n_1 are occupied and all the states of the conduction band n_2 are empty. At non-zero temperature T, N_e electrons have passed from n_1 to n_2 and N_h (= N_e) holes have appeared in n_1. By virtue of (26), the identity $N_e = N_h$ leads to:

$$E_1 + E_2 - 2\mu = kT \ln I_e/I_h$$

where I_e and I_h are the values of the integrals in (26). This equation fixes the position of the Fermi level μ. In particular, at *zero temperature*, we have $\mu = (E_1 + E_2)/2$ and **the Fermi level lies in the middle of the gap**. When I_e and I_h are not very different (frequently), μ varies only slightly in the gap as T increases.

In the case of doped (or extrinsic; see p. 148) *semiconductors*, the situation is very different (see S.C.Ph.).

N.B. 2   Can the gap of a semiconductor manifest itself in an immediately perceptible way?

Suppose that a solid with gap E_g (= $E_2 - E_1$) is lit by natural light. The energy of visible photons lies between $\simeq 1.6$ and 3.1 eV. Thus, electrons can only be *photo-excited* from n_1 to n_2 if $E_g < 3.1$ eV. Otherwise (since the excitation energies of the core electrons are always higher), the photons do not interact with the electrons of the solid, which is then *transparent*. This is the case for diamond ($E_g = 7$ eV).

Consider, however, two semiconductors with $E_g < 3.1$ eV. *Silicon* has a gap ($E_g = 1.2$ eV) such that all visible photons are absorbed. If it is sufficiently thick, it is *black*. *Sulphur*, a semiconductor with gap $E_g = 2.7$ eV, absorbs the photons from 2.7 to 3.1 eV, in other words, in the violet. Thus, it has a *yellow* colour, the complement of violet, and when one looks at a piece of sulphur, one literally 'sees' the gap in a semiconductor.

A reader with access to liquid nitrogen might cool a little sulphur in it. He/she will then see immediately that the latter becomes milky white. This change in colour is due to variations in the crystalline parameters of the sulphur which increase the width of the gap. Of course, the sulphur takes on its initial colour again when it is brought back to room temperature.

iii) **Between the above two extreme cases,** there may arise the case of a small intersection of the two bands n_1 and n_2 (E_g small and negative). Table 1 shows that *grey tin* comes close to this situation. The density of states is then as illustrated in Figure 20(a).

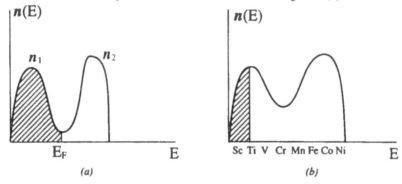

Figure 20 (a) **Semi-metals**. The density of states at the Fermi level $n(E_F)$ is low. (b) **Transition metals**. The Fermi level moves to higher values from Sc to Ti etc. The case of Ti is represented here.

The solids of valency five (As, Sb, Sb and Bi) have a rhomboidal primitive cell, with *two* atoms per cell. Thus, they have an even number (10) of valence electrons per primitive cell, which means that, as for Si and Ge, band n_1 is full and n_2 is empty. Under these conditions, the Fermi level is located in the region of intersection. Since this is small, the density of states at the Fermi level $n(E_F)$ is low. These solids exhibit a modest metallic conduction. They are the **semi-metals**.

If such a ($n_1 + n_2$) band is partially filled with $3d$ electrons of Sc, then Ti, then V, then Cr, then Mn ... the Fermi energy E_F moves to higher energies (Figure 20(b)). These are **transition metals** where a minimum of $n(E_F)$ is expected around the middle of the transition series. This explains the very low value of $n(E_F)$ measured in Cr as compared to V and Mn (see Table 2, p. 55).

Exercise Explain, from a qualitative argument, why the melting point has a maximum value at the middle of transition series (Cr for $3d$, Mo for $4d$ and W for $5d$): these are *refractory metals*.

6. DYNAMICS OF ELECTRONS

6.1 Remarks on the $\bar{k}$ vector of Bloch functions

1 The plane wavevector $\bar{k}$ used in Chapter III had a simple physical meaning. Since the plane waves $\exp(i\bar{k} \cdot \bar{r})$ are eigenfunctions of the momentum operator p:

$$p\psi = -i\hbar \overline{\nabla}\psi = \hbar \overline{k}\psi,$$

the free electrons are in *eigenstates of momentum*, and the latter thus takes well defined values proportional to $\overline{k}$:

$$\overline{p} = \hbar \overline{k} = m\overline{v}.$$

The same is no longer true for Bloch electrons. In fact, for a Bloch state $\psi_k(\overline{r}) = u_k \exp(i\overline{k} \cdot \overline{r})$ (see (10)), the operation

$$\begin{aligned}
p\psi_k &= -i\hbar \overline{\nabla} u_k e^{i\overline{k} \cdot \overline{r}} \\
&= \hbar \overline{k}\psi_k(\overline{r}) - i\hbar e^{i\overline{k} \cdot \overline{r}} \overline{\nabla} u_k(\overline{r})
\end{aligned}$$

does not create a state proportional to $\psi_k(\overline{r})$ except if u_k is a constant. This case corresponds in the reciprocal space to states far away from the Brillouin planes. Only then is the vector $\overline{k}$ an approximate but good representation of the *velocity*, up to the constant $\hbar/m$.

In the general case, an electron is in a state which is a *linear combination of Bloch states* with vectors $\overline{k}'$ close to a particular vector $\overline{k}$. This combination constitutes a **wave packet**:

$$\Psi(\overline{r}, t) = \int \psi_{\overline{k}'}(\overline{r}) \exp\left[-\frac{i}{\hbar}E(\overline{k}')t\right] f(\overline{k}') \, d^3\overline{k}' \qquad (29)$$

where $f(\overline{k}')$, which describes the 'weight' in the packet of the Bloch wave $\overline{k}'$, is non-zero only in a small volume μ of the reciprocal space situated around the extremity of the vector $\overline{k}$. The *group velocity* of the packet, in other words, of the electron, is given (see Q.M.) by

$$\boxed{\overline{v} = \overline{v}(\overline{k}) = \frac{1}{\hbar}\overline{\nabla}_k E(\overline{k}).} \qquad (30)$$

Thus, the velocity *is not* in general *collinear with* $\overline{k}$. Referring to Figure 11, one also sees that its projection v_p onto $\overline{k}$ decreases as the extremity of $\overline{k}$ approaches the zone boundaries, until it becomes zero at points such as X or Λ (Figure 9).

Exercise Prove the above proposition taking for $\overline{k}$ a vector $\overline{k}^+$ (defined on p. 128) the extremity of which is near to a Brillouin plane. Calculate $v_p = k^{-1}\overline{k} \cdot \overline{v}$ using (30) and (23) and then $\lim v_p$ ($\varepsilon \to 0$). Deduce that $\lim v_p$ is smaller than the value of v_p away from a Brillouin plane, and finally that $\lim v_p = 0$ when $\overline{k}_0$ (of (22)) $\simeq -\overline{G}/2$ (whence that $\overline{v}$ is perpendicular to $\overline{k}$) at a point Λ.

N.B. 1 Let us now describe more precisely the meaning and implication of the above condition relating to a 'small' volume μ (where f is appreciable). By small, we mean 'small in relation to the volume of a Brillouin zone'. Let us examine how $\psi(\overline{r})$ (29) varies around a given point $\overline{r}_0$, and, in particular, at a distance equal to the size a of a unit cell. Taking into account (10), we have:

$$\Psi(\overline{r}_0 + \overline{a}, t) = \int \psi_{\overline{k}'}(\overline{r}_0) \exp\left[i\left(\overline{k}' \cdot \overline{a} - \frac{E(\overline{k}')t}{\hbar}\right)\right] f(\overline{k}') \, d^3\overline{k}'.$$

Taken as a function of $\bar{a}$, for $\bar{r}_0$ fixed, this function is a superposition of plane waves with weight the function $F(\bar{k}') = \psi_{\bar{k}'}(\bar{r}_0) f(\bar{k}')$. If f (whence also F) has only non-trivial values in an interval Δk, then $\psi_{r_0}(\bar{a})$ extends to a region $\Delta a \simeq 1/\Delta k$ around $\bar{r}_0$. Since the volume μ ($\simeq (\Delta k)^3$) is very small in comparison with the volume of a zone ($\simeq (2\pi/a)^3$), it follows that Δa is large in comparison with a. The wave packet in question, with wavevector $\bar{k}$ rather well defined in its Brillouin zone, thus extends to a large number of unit cells.

N.B. 2   We note that in (30), we have rediscovered the fact that, for *spherical* surfaces of equal energy (in particular, Fermi surfaces), $\bar{v}$ is collinear with $\bar{k}$.

   Let us evaluate in this case, the flight time of a Fermi electron, assumed to be in the 'middle' of a zone ($k_F = \pi/2a$) from one ion to an adjacent ion (distance $\simeq a$). This time is $\simeq a/v_F$ or $\tau \simeq a/((\hbar/m)(\pi/2a)) = 2ma^2/\pi\hbar$, which, in a.u. (see p. 36) with $a = 4$, gives $\tau \simeq 10$ or $\tau \simeq 2 \times 10^{-16}$ s.

2   Let us consider a wavevector $\bar{k}$ with extremity situated in the nth Brillouin zone, together with the corresponding Bloch function $\psi_{\bar{k}} = u_{n\bar{k}}(\bar{r}) \exp(i\bar{k} \cdot \bar{r})$. There always exists a vector $\bar{k}_1$ **belonging to the first zone**, which may be derived from $\bar{k}$ by translation of a vector $\bar{l}$ of the reciprocal lattice:

$$\bar{k} = \bar{k}_1 + \bar{l}.$$

Then we have

$$\psi_k(\bar{r}) = u_{nk}(\bar{r})e^{i\bar{k}\cdot\bar{r}} = u_{nk}(\bar{r})e^{i\bar{k}_1\cdot\bar{r}}e^{i\bar{l}\cdot\bar{r}}.$$

The function $u_{nk}(\bar{r}) \exp(i\bar{l} \cdot \bar{r})$ has the periodicity of the crystal, as do both $u_k(\bar{r})$ and $\exp(i\bar{l} \cdot \bar{r})$ (see (12)). If we set $u_{nk_1}(\bar{r}) = u_{nk}(\bar{r}) \exp(i\bar{l} \cdot \bar{r})$, then the function

$$\psi_{\bar{k}_1}(\bar{r}) = u_{n\bar{k}_1}(\bar{r})e^{i\bar{k}_1\cdot\bar{r}}$$

is a Bloch function. It naturally has the same energy eigenvalue as $\psi_{\bar{k}}$ and at the same time has its wavevector in the *first Brillouin zone*. Thus, for a single wavevector ($\bar{k}_1$, for example) of the first zone, there exist a whole series of possible energies which are the eigenvalues of the Schrödinger equation:

$$\left[\frac{\hbar^2}{2m}(-i\bar{\nabla} + \bar{k}_1)^2 + V(\bar{r}) \right] u_{\bar{k}_1}(\bar{r}) = E_{\bar{k}_1} u_{\bar{k}_1}(\bar{r}).$$

These eigenvalues form a discrete sequence $E_{n\bar{k}_1}$ to which the eigenfunctions $u_{n\bar{k}_1}(\bar{r})$ correspond.

   These remarks enable us to rebuild the dispersion curves (see Figures 4 and 11) *in the interior of the first zone*. Thus, Figure 21 shows the dispersion curve of the linear chain (Figure 4, p. 119) localised inside the first zone. The branches (shown by dashes) corresponding to the second zone are translated respectively, by $+2\pi/a$ and $-2\pi/a$ in the first zone etc.

   This scheme by which all the branches of the dispersion curve are brought back into the first zone is called the **reduced zone scheme**.

   Here, the importance of remark ii) of p. 126 is easy to understand. All the states of the nth zone may in fact be brought back to the first zone, without omission or repetition.

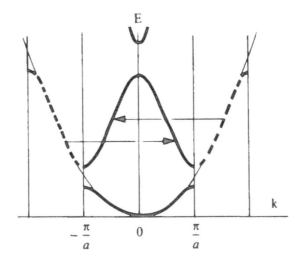

Figure 21 **Curve** $E(k)$ **of Figure 4 in the reduced zone scheme.** The parts (shown by dashes) situated in the second Brillouin zone have been brought back into the first zone by translations (see arrows) of $+2\pi/a$ and $-2\pi/a$.

6.2 Movement of an electron in the presence of an external force

1 – Electron in the presence of an electric field $\overline{\varepsilon}$

An electron, impelled with the velocity of its wave packet, namely $\overline{v}_g = \hbar^{-1}\overline{\nabla}_k E$, is subjected, in the presence of an electric field $\overline{\varepsilon}$, to the force

$$\overline{F}_{\text{ext}} = -e\overline{\varepsilon}.$$

The work of this force during time dt is

$$dE = -\frac{e}{\hbar}\overline{\nabla}_k E \cdot \overline{\varepsilon}\, dt.$$

By virtue of the fact that $dE = \overline{\nabla}_k E \cdot d\overline{k}$, we have[5]

$$\boxed{\frac{d\overline{k}}{dt} = \frac{1}{\hbar}\overline{F}_{\text{ext}}.} \tag{31}$$

  Rest assured: there is nothing in (31) which calls into question Newton's law, under the pretext that $\overline{F}_{\text{ext}}$ is proportional to d$\overline{k}$/dt rather than to d$\overline{v}_g$/dt. In fact, $\overline{F}_{\text{ext}}$ is the *external* force and not the resultant of the forces applied; it does not include the internal forces derived from the crystalline potential, which intervene in the form of $E(\overline{k})$. Of course, when the latter are negligible (case of Chapters II and III; constant crystalline potential), we have $m\overline{v}_g = \hbar\overline{k}$ and then (31) again gives Newton's law: $\overline{F} = \overline{F}_{\text{ext}} = m(d\overline{v}_g/dt)$.

[5] A more rigorous proof of (31) may be found in: **R Balian**, *J. Physique* **50**, 2629 (1989).

2 – Electron in the presence of magnetic induction $\overline{B}$

Here, we shall generalise the formula (31), without justification. In other words, we accept that the external force applied to the electrons by the induction $\overline{B}$ is simply the Laplace force $-e\overline{v}_s \wedge \overline{B}$ (in SI units). Thus, we write

$$\frac{d\overline{k}}{dt} = -\frac{e}{\hbar^2}\overline{\nabla}_{\overline{k}}E \wedge \overline{B}. \tag{32}$$

We shall describe the movement of an electron in the reciprocal space due to $\overline{B}$. For $\overline{B} = 0$, a point M_0 of this space, taken in the first Brillouin zone ($\overline{OM_0} = k_0$) defines an electron state. For M ($\overline{OM} = \overline{k}$), the application of B results in a movement described by (32): M moves like $\overline{k}$, in other words, perpendicularly to the gradient of the energy, in other words on a *surface of equal energy*. More precisely, the trajectory $\mathcal{L}$ of M is the intersection of this surface with the plane Q perpendicular to $\overline{B}$ passing through M_0. If, for the same reasons as described above (p. 56: the need to have vacant states in order to be able to observe transitions), we consider a *Fermi electron*, then $\mathcal{L}$ is the intersection of the Fermi surface and Q.

We can illustrate this rule for two special, extreme cases. To simplify the diagrams, let us consider a simple cubic crystal (with a simple cubic reciprocal lattice) and an induction $\overline{B}$ along the diagonal [111] of the cube. The first Brillouin zone is cubic.

i)   *Let us first suppose that there is one electron per atom* and that the Fermi surface Σ_F is spherical. The plane Q, perpendicular to $\overline{B}$, passing through M_0 (point of Σ_F) intersects Σ_F in a circle C_1 (Figure 22). The point M moves along C_1 in a direction on this circle given by equation (32) where $\overline{\nabla}_{\overline{k}}E$ points *towards the outside* of the sphere Σ_F. The velocity of the electron corresponding to M, collinear with $\overline{\nabla}_{\overline{k}}E$, remains normal to the sphere and thus describes a cone as M describes C_1. The 'classical trajectory' of the electron is a '*helix*' (shown in Figure 23 by a circular projection Γ_1). This is the classical **cyclotron trajectory** of electrons in the vacuum.

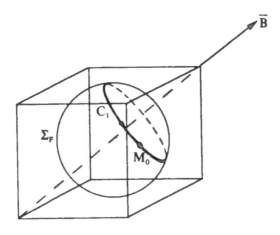

Figure 22 Trajectory of a point representing an electron state (initial state M_0) on a surface of equal energy (here, Fermi surface), assumed to be spherical, in the presence of an electromagnetic induction $\overline{B}$. Here, the first Brillouin zone is a cube. $\overline{B}$ has been drawn parallel to a diagonal of this cube.

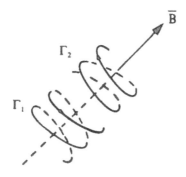

Figure 23 **Two possible types of electron trajectory in the presence of an induction** $\bar{B}$. For Γ_1, see Figure 22; for Γ_2, see Figure 24.

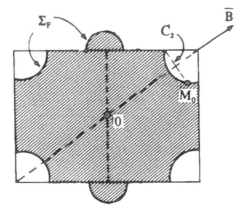

Figure 24 **Trajectory of point** M **(see Figure 22) on a concave part (pocket) of the Fermi surface.** Section cut by a plane {110} of a first Brillouin zone, assumed to be cubic.

ii) *Suppose now that there are two electrons per atom* and that the coefficient V_{100} is such (see Figure 14) that the surface Σ_F is partially located in the second zone and partially in the first (Figure 24). This situation creates empty 'pockets' in the eight corners of Z, which (again to simplify the diagrams) we shall assume to be spherical. For an initial point such as M_0 on Figure 24, we find that M again has a circular trajectory C_2, which, however, taking into account the direction of $\bar{\nabla}_{\bar{r}}E$, is described in a *direction opposite* to that of M on C_1. The electron trajectories are again helices, but this time described in the opposite sense to normal (Γ_2 on Figure 23).

This reverse movement of the electrons is that which would be exhibited in the vacuum by a *fictitious particle with a positive charge* $(+e)$. *This particle is simply the hole*, introduced above (p. 136) in the case of the thermal excitation of an electron from a valence band to a conduction band. The point representing the hole in the reciprocal space is an unoccupied point of the $\bar{k}$ lattice (whence the name *hole*; see p. 147).

N.B.   Here again, the laws of electrodynamics are not called into question. In fact, $-(e/\hbar)\bar{\nabla}_{\bar{r}}E \wedge \bar{B}$ is just the external force. The electrons are subjected to this force but also to the crystalline forces which become dominant at zone edges.

3 – Cyclotron resonance

Let us place a conducting single crystal in an induction field $\overline{B}$ parallel to one of its surfaces, assumed to be a plane. The electrons adopt *cyclotron trajectories*, or helices with axes parallel to $\overline{B}$, whence to the surface (Figure 25). We then subject the crystal to an electric field of microwave frequency $\overline{\varepsilon}$ (pulsation: ω), again parallel to the surface (so-called Azbel–Kaner geometry). A certain number of cyclotron helices penetrate the thin zone (*skin depth*) in which $\overline{\varepsilon}$ is not zero. If, in this case, the cyclotron period t_c is such that the electron penetrates the skin depth in phase with $\overline{\varepsilon}$, then it gains (or loses) energy. This resonance occurs when

$$t_c = n\frac{2\pi}{\omega} \quad (n = \text{integer}). \tag{33}$$

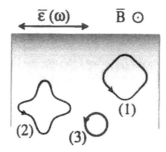

Figure 25 Cyclotron resonance. The induction $\overline{B}$ parallel to the surface gives the electron trajectories helical shapes, the projections of which are shown here. A microwave electric field penetrates the skin depth (in grey). Some helices cross this skin at regular intervals (1), others ((2) and (3)) do not.

The size and shape of the helices (see (2) and (3)) depend on the size and shape of the corresponding trajectory C (Figure 22) described on one of the surfaces of equal energy.

The cyclotron period can be evaluated from (32):

$$t_c = \int dt = \frac{1}{B}\frac{\hbar^2}{e}\int (\overline{\nabla}_k E)_\perp^{-1} dk = \frac{1}{B}\frac{\hbar^2}{e}J \tag{34}$$

where $(\overline{\nabla}_k E)_\perp$ is the projection of $\overline{\nabla}_k E$ onto a plane perpendicular to $\overline{B}$. This period varies as B^{-1}.

Measurements of *resonant absorption* of the microwave power, taken as a function of B for a fixed value of ω (for example, a few tens of gigacycles s^{-1}) in general clearly show the harmonics predicted by (33): the absorption has maxima for a sequence of values of B, $B_n = \text{constant}/n$. From these measurements, we are able to deduce the value of the integral J of (34). The absorption of energy implies electron transitions and thus, here again, only concerns *Fermi electrons*. The integration in (34) is taken over trajectories C on the *Fermi surface* Σ_F (see, for example, Figure 22); in other words, over sections of Σ_F with planes perpendicular to $\overline{B}$. Under these conditions, it is clear (except if Σ_F is a sphere) that the value of J depends on the orientation of $\overline{B}$ with respect to the crystal.

Without going into detail, and, in particular, without describing the circuits C which participate most actively in the resonance for a given direction $\overline{B}$, it is apparent that successive measurements of J for different directions $\overline{B}$ provide information about the shape of the Fermi surface.

Cyclotron resonance and annihilation of positrons (see p. 76) are two experimental methods which give information about the shape of Fermi surfaces.

6.3 – Concept of effective mass

It is now clear that the *acceleration* $d\overline{v}/dt$ of an electron in a solid (or that of a hole) is *not* generally *proportional* to the *external force* $\overline{F}_{ext}$. However, we may always write

$$\boxed{\overline{F}_{ext} = m^{*} \frac{d\overline{v}}{dt}} \qquad (35)$$

where m^{*} now no longer denotes a scalar (constant) but the *tensor* which transforms $\overline{F}_{ext}$ into $d\overline{v}/dt$. From (30), we deduce that

$$\frac{d\overline{v}}{dt} = \frac{1}{\hbar} \overline{\nabla}_k \frac{dE(\overline{k})}{dt}$$

and from the classical formula $dE/dt = \overline{F} \cdot \overline{v}$ (which remains valid in quantum mechanics for average values), we derive:

$$\frac{d\overline{v}}{dt} = \frac{1}{\hbar} \overline{\nabla}_k(\overline{F}_{ext} \cdot \overline{v}) = \frac{\overline{F}_{ext}}{\hbar^2} \left[\overline{\nabla}_k \cdot \overline{\nabla}_k E(\overline{k}) \right]$$

which is simply equation (35) where m^{*} is a tensor such that:

$$\frac{1}{m^{*}} = \frac{1}{\hbar^2} \begin{pmatrix} \partial^2 E/\partial k_x^2 & \partial^2 E/\partial k_y \partial k_x & \partial^2 E/\partial k_z \partial k_x \\ \partial^2 E/\partial k_x \partial k_y & \partial^2 E/\partial k_y^2 & \partial^2 E/\partial k_z \partial k_y \\ \partial^2 E/\partial k_x \partial k_z & \partial^2 E/\partial k_y \partial k_z & \partial^2 E/\partial k_z^2 \end{pmatrix}. \qquad (36)$$

The tensor m^{*} now plays the role played by the mass for a particle (electron or hole) moving in a medium with a constant potential. It is called the **effective mass**. It is easy to check that, in the case of the free electron ($E = (\hbar^2/2m)k^2$), the tensor m^{*} degenerates to a *scalar* m, which is simply the electron mass.

Let us consider the special case of an electron moving in the direction of a principal axis of the tensor (36) and choose this axis to be the x axis. The effective mass is then:

$$\boxed{m^{*} = \hbar^2/[\partial^2 E(\overline{k})/\partial k_x^2].} \qquad (36')$$

Thus, for an electron moving in the direction of the abscissa of Figure 11, the effective mass has the same sign as the curvature of $E(k)$. It is *positive* (and equal to m) if $\overline{k}$ is situated in the neighbourhood of the origin and *negative if $\overline{k}$ is near the edge of a zone* (or 'top' of a band). In the latter case, such an electron has an acceleration in the opposite direction to the *external* force which is applied to it.

This last point is not surprising. In fact, an external force parallel to x tends, according to (31), to bring the extremity of the vector $\overline{k}$ closer to the corresponding Brillouin plane, thereby achieving diffraction conditions which tend to transform a progressive wave into a stationary wave (see p. 117) and thus decelerate the former.

N.B.   The importance of this concept of *effective mass* stems from the fact that for Bloch electrons, the formula for electrical conductivity ((48), p. 56) remains valid if the scalar m is replaced by the tensor m^*. This property is proved in **Annex 3**, p. 265.

6.4 – Properties of holes

We now return to the concept of the **hole** as introduced above (p. 136 and p. 143). For this, we imagine a band which is **full** (for example, in the first Brillouin zone Z) except for **one unoccupied state** (Figure 26). This band, with a negative charge removed, is associated with a positive charge $+e$. This is the entity (*almost full band which has positive charge*) which is known as a *hole*.

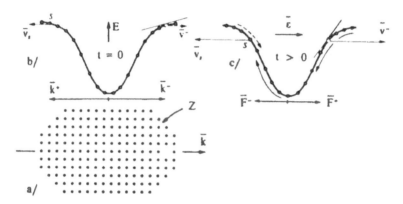

Figure 26   Hole in a quasi-full band. (*a*) Absence of an electron in the state $\bar{k}^-$; (*b*) a current is carried by the uncompensated electron (s) with velocity $\bar{v}_s$, or by the hole with velocity $\bar{v}^-$ (velocity here is the slope of the tangent of curve $E(k)$); (*c*) under the action of the electric field $\bar{\varepsilon}$, the states slide to the left (continuous arrows). Thus, the state s slides to the right (dashed arrow).

Thus, the first property of the hole is that it may be associated with a *positive charge* $+e$. Can it also be associated with a wavevector, a velocity, an effective mass, an energy ... ?

We attribute the index $(-)$ to the *missing electron*. Taking into account the central symmetry (index: s) of Z, to the unoccupied state $\bar{k}^{(-)}$ there corresponds an occupied state $\bar{k}_s$, which is thus uncompensated. Similarly, the whole of the almost full band (whence, the hole) is associated with the *wavevector* $\bar{k}^+$ equal to $\bar{k}_s$, namely

$$\boxed{\bar{k}^+ = -\bar{k}^-\,.}\tag{37}$$

The electron (s) has velocity $\bar{v}_s = \hbar^{-1}\bar{\nabla}_k E(\bar{k}_s) = -\bar{v}^-$ (on Figure 26, this velocity is negative). Uncompensated, this electron, carrying a charge of $-e$ gives rise to a current. The latter may be called the current of the positive hole $+e$, provided only that the hole is assigned the velocity $\bar{v}^+ = -\bar{v}_s$. Thus, *the velocity of the hole* is equal to the velocity which the missing electron would have had:

$$\boxed{\bar{v}^+ = \bar{v}^-\,.}\tag{38}$$

If an electric field $\overline{\varepsilon}$ is applied, it results in a force $\overline{F} = -e\overline{\varepsilon}$ on the electrons of the band, which drives the vectors $\overline{k}$ and, in particular, the vector $\overline{k}^-$, according to (31). This results in a variation in the velocity $\overline{v}^+$, whence an acceleration of the hole. Since the latter is subjected to the force $\overline{F}^+ = +e\overline{\varepsilon} = -\overline{F}^-$, application of (38) and (35) shows that the *effective mass of the hole* may be defined to be:

$$m^{*+} = -m^{*-}.$$

(39)

Thus, in Figure 26, the vectors $\overline{k}$ move (according to the continuous arrows) *to the left*; the velocity of the hole, which is positive like $\overline{\nabla}_k E$, increases under the effect of the positive force $\overline{F}^+$ applied to the hole; thus, the mass m^{*+} of the hole is positive, although in this region of the band, the mass m^{*-} of the electrons is negative (as is the curvature of $E(k)$, see (36')). We note here that, under the effect of the force applied, the uncompensated state (s) moves *to the right* (according to the dashed arrows) since it is symmetric with the state $\overline{k}^-$ of the missing electron.

Because the energy of the hole increases as the latter moves deeper into the band (since then the electron energy increases), the *energy of the holes* must be counted as positive in the direction opposite to that of the electrons (in other words, 'downwards' in Figure 26).

Thus, overall, the hole behaves like a **particle with a positive charge $+e$, velocity $\overline{v}^+ = \overline{v}^-$, wavevector $\overline{k}^+ = -\overline{k}^-$ and effective mass $m^{*+} = -m^{*-}$**, where $\overline{v}^-$, $\overline{k}^-$ and m^{*-} are the velocity, wavevector and effective mass of the missing electron. In the presence of an electric field $\overline{\varepsilon}$ and an induction $\overline{B}$, the vector $\overline{k}^+$ varies as implied by the equation

$$\frac{d\overline{k}^+}{dt} = \frac{e}{\hbar}(\overline{\varepsilon} + \overline{v}^+ \wedge \overline{B}).$$

(40)

7. BLOCH, ANDERSON, MOTT *et al.*

7.1 Real materials or established disorder

Everything in this chapter rests (via the Bloch states) on the ideal concept of perfect crystallinity, where the Bloch states are the eigenstates of translation operators (see **Annex 2**).

The Bloch functions describe non-local states, *extending* over all the crystal, without attenuation or diffusion terms; the Bloch electrons are not scattered by the ions of the perfect crystal. In fact, they are only scattered by *deviations* from the perfect order, in particular, by those due to *thermal agitation* (described in **Problem 18**) and by those constituted by *crystalline defects*. As far as the latter are concerned, we distinguish two extreme cases.

1 – Their concentration is small. This will be the case for crystal defects described in the following chapters. Their weak concentration in no way implies that their influence on the properties of materials is modest; quite the reverse. However, it does enable us to state that, to the first order, *the crystallinity remains well defined*, at least over distances which are large in comparison with interatomic distances.

One important example of this case is that of atoms of B dissolved in minute concentrations (for example, $c \simeq 10^{-6}$ at.) in a crystal of A. Let us consider the case in which A is a solid of valency *four*, such as *silicon*, with valence band (n_1) and conduction band (n_2) (see Figure 18(b)), and where B is an atom of valency *five* (for example, *phosphorus*), locally replacing a silicon atom by substitution. This case is discussed in detail in S.C.Ph. Suffice it to note here that four of the five electrons of B participate 'normally' in (n_1), the fifth being *bound* to the additional nuclear charge $+e$ of B, in the same way that the electron is bound to the proton in a hydrogen atom (Figure 27(a)). Thus, in this case, there exists a *bound* hydrogenoid state, **localised** around B. The corresponding ionisation energy δE is weak (0.04 eV), in particular, because of the dielectric constant of crystal A. The only available states which permit this ionisation are those at the bottom of the conduction band (n_2). Thus, the energy of the bound state created in this way around B lies *in the gap* at δE *below the bottom of the band* (n_2). This is referred to as a **donor level**, since the corresponding electron may be easily 'donated' to the conduction band. In the same way, localised states of *holes*, with energies in the gap at $\delta E'$ above the top of the conduction band (n_1) (**acceptor levels**; see Figure 27(c)), are created by dissolving atoms of B' of valency *three* (for example, *boron*). A semiconductor *doped* in this way is said to be *extrinsic*, as opposed to the *intrinsic* semiconductor (p. 137).

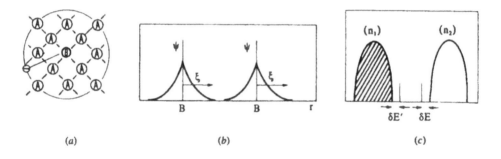

(a)   (b)   (c)

Figure 27 Localised states in a semiconductor. (a) Hydrogenoid orbital of an electron bound to impurity B (pentavalent) in a crystal A of valency four. (b) Hydrogenoid functions $\psi(r)$ localised on two donor centres (pentavalent atoms of B in a tetravalent semiconductor) and with extension ξ. (c) Energies of localised states in the gap of a semiconductor (or insulator). (n_1): valence band; (n_2): conduction band.

The wavefunctions of these localised hydrogenoid states are clearly not Bloch functions. They decay exponentially from their centre, and in their ground state, look like 1s functions in $\exp(-r/\xi)$. The radius ξ is the *extension* of the localised state (see Figure 27(a) and (b)).

These states cannot participate in the conduction *on their own*; however, the excitation of the localised electrons (*resp.* holes) towards Bloch states of the conduction band (*resp.* valence band) – on the condition that the energy δE (*resp.* $\delta E'$) be supplied – gives rise to the presence of quasi-free Bloch electrons in the conduction band n_1 (*resp.* of holes in the valence band n_2). These contribute – together with those excited through the gap (see p. 136) – as **free carriers** to the electrical conduction. If N_e (*resp.* N_h) is the total (temperature-dependent) number of free electrons in n_1 (*resp.* free holes in n_2), then the conductivity (see (48), p. 56) is:

$$\sigma = \frac{N_e e^2 \tau_e}{m_e^*} + \frac{N_h e^2 \tau_h}{m_h^*},$$

τ_e, τ_h, m_e^*, m_h^* being the relaxation times and effective masses of electrons and holes.

2 – Their concentration is large. This imprecise definition covers a considerable number of different situations involving materials of increasing technological importance and characteristic of so-called **ill-organised materials**: A–B *alloys* of the above type with a high concentration of B; *polyphase alloys*; *composite materials* in which fibres (ceramic, amorphous materials) are inserted into a matrix of a different type (metal, cement); *intercalated materials* in which atoms (for example, metallic) or molecules are introduced between the planes of a layered crystal such as graphite, where the intercalated entities may have repetition period commensurable or *incommensurable* with the host crystal, or even no period at all; *ill-crystallised ceramic materials*, formed by aggregation of minute crystallites (several tens of nm); and, going further, non-crystalline materials (see Chapter V): *amorphous glasses, quasi-crystals, polymers, gels* ....

Evidently, it is no longer possible to talk about Bloch states here. Must we thus conclude that these materials have no extended states, no gaps, no density of states ...? Certainly not.

By way of reassurance on this point, let us give three clues.

i) The handling of free electrons in Chapter III was not linked to any notion of crystallinity. Thus, it should be applicable, in certain cases, to ill-organised material. This is in fact the case, at the first order, for certain *liquid metals* (for example, sodium) the transport properties of which are not greatly affected when solidification occurs.

ii) A good number of *amorphous alloys* also have a metallic conductivity.

iii) A crystalline insulator of gap E_g (Figure 18) does not permit electron excitations of energy $< E_g$. In particular, it does not permit the absorption of photons with energy $< E_g$ to which it is therefore transparent. However, *glass*, a non-crystalline material, is transparent to visible light ($\simeq 1.6$–3.1 eV), which indicates the presence of a gap of width > 3 eV.

The study of electron states and of conduction in ill-organised materials constitutes a subject area as vast as the infinite variety of these materials (itself the daughter of the infinite inventiveness of solid-state chemists!). This subject area is rich in open questions and is now being developed rapidly; however, an elementary exposition such as this is not really the place in which to discuss it. We shall very briefly note just two of the most important points.

7.2 Anderson localisation

Anderson (1958) modelled ill-organised material by treating the problem of an electron in a 'random potential'. The latter consists of a set of square wells, centred on the geometrical points of a crystal, but not identical. The depth of these wells V varies randomly around a mean value V_0 with a distribution of width ΔV (see Figure 28). The

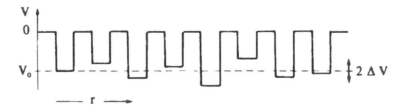

Figure 28   Anderson's random potential.

general nature of this treatment is a result of the fact that the nature of the disorder is not specified in detail (it may be *chemical* or *geometrical*). The 'intensity' of the latter is described by the phenomenological parameter ΔV.

The two most important results from Anderson's theory are the following.

(i)   Whenever ΔV is non-zero, *localised states* in the sense of **1** (p. 147) occur. If ΔV remains small, these localised states coexist with states which, whilst not Bloch states, are *extended states*. This so-called **Anderson localisation** clearly tends to decrease the electrical conductivity at low temperature by decreasing the number of carriers. An example of the formation of such localised states is given in **Problem 4**.

(ii)   When ΔV becomes larger than or equal to a critical value ΔV_c, *all the states become localised*. The corresponding transformation (metal $\rightarrow$ insulator) is called *Anderson transition*. The critical value of ΔV_c is of the order of W, the width of the valence band, which this same crystal would have if there were no disorder, i.e. if $\Delta V = 0$ were zero.

N.B.   Conversely, one should beware of identifying any metal–insulator transition as an Anderson transition associated with disorder.

Indeed, taking into account repulsions between electrons, it can be shown (see **Problem 5**) that a *crystallised* monovalent element (hydrogen, sodium...) must be either a conductor or an insulator, according to whether the number of atoms N per unit volume is greater than or less than a number N_c given by

$$N_c^{1/3} a_0 \simeq 0.2$$

where a_0 is the Bohr radius (= 1 a.u.). The corresponding metal–insulator transition is called **Mott transition**. It owes nothing to disorder, unlike Anderson transition.

Thus, a *hydrogen* crystal, which is naturally an insulator at normal pressure, should become a conductor at sufficiently high pressure. This is what appears to happen *in* the planet Jupiter, where there are good reasons to think that hydrogen (under several hundred GPa: $N > N_c$) is metallic.

7.3 Carrier hopping

We have just met **two cases of localisation**, namely a very simple case created by a foreign atom in a semiconductor or an insulator (see Section 7.1) and a less trivial case associated with a random disorder (see Section 7.2). In both cases, it is sensible to wonder about the effect of localisation on the **conduction**. In particular, we mentioned that Anderson localisation gives rise to a metal–insulator transition. This is true at low temperature, however, these insulators with strong localisation generally have a conduction which, zero at $T = 0$, increases with T. Under these conditions, what is the mechanism for **jump**, or **hopping**, of an electron (or of a hole) from one localised state to another? It is understandable that this question has become an obligatory point of passage if one wishes to comprehend the conduction in ill-organised material.

i) Let us now return to **the first case** and imagine again *donor* levels of phosphorus (in volume concentration c_P) in silicon crystal.

If c_P is sufficiently low, the localised functions *do not overlap* (which implies that $c_P \ll \xi^{-3}$, see Figure 27(*b*)). The localised electrons may then be thermally excited (energy δE, see Figure 27(*c*)) in the conduction band and diffuse *via* a Bloch state, and then, when the opportunity presents itself, re-localise at another donor level. This is the main source of conduction of doped semiconductors at low temperature (see S.C.Ph.).

If c_P is such that *the localised functions do overlap* ($c_P \geq \xi^{-3}$) then the previously degenerate donor levels (all with the same energy) create a band[6] which may give rise to metallic conduction. Here, as c_P increases, we have an insulator–metal transition which is *partially* Mott (by bringing together atoms of phosphorus) and *partially* Anderson (by the disorder of these atoms). It occurs in Si-P when c_P reaches $\simeq 4 \times 10^{18}$ cm^{-3} (see Figure 1 of **Problem 5**).

ii) Let us now consider the **second case**, that of a disorder creating localised states of random energies and spatial distributions. Many mechanisms are invoked to explain the *jump* of an electron (or a hole) from one site of localisation to another. We shall cite only one, proposed by Mott, in a formulation here simplified to the extreme.

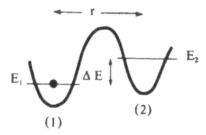

Figure 29   Jump (hopping) of a carrier from a localised state (1) to a localised state (2).

Let us imagine two sites of localisation (1) and (2) at distance r apart, separated by a potential barrier and characterised by two associated states and two levels E_1 and E_2, ΔE apart (Figure 29)[7]. There is initially one electron at (1). It may jump to (2) by *tunnelling* with a probability proportional to

$$s = \exp(-\Delta E/kT)\exp(-r/R). \tag{41}$$

---

[6] Or more exactly here, two bands (Hubbard bands) which join; see **Problem 5**.

[7] In fact, in the general case, the heights of these levels E depend on whether or not there is an electron on the site. Rearrangements of atoms may take place around the latter (for example, displacements of ions adjacent to the electron towards the former), decreasing the energy of the overall configuration via the *Jahn–Teller effect* (see p. 168 and **Problem 11**). This stabilised configuration (displaced atoms + electron) is called a **polaron**. Thus, it is possible (see Figure 29) that, for an initially polaronic site (1), site (2) may become polaronic after the jump, in such a way that $E_2' \simeq E_1$ (where $E_2' < E_2$ and $E_1' > E_1$), where the E' are the localised energies *after* the jump. In this case, the polaronic configuration jumps, in the material, from site to site with the electron.

The *first term* is the probability that the thermal agitation modifies the atomic environment of (1) and or (2) until the two levels are equalised. This is actually a necessary condition for the electron to be able to tunnel from (1) to (2). The *second term* is the probability that the tunnel jump occurs, where the parameter R depends on the shape (and, in particular, on the height) of the potential to be crossed (see Q.M.).

At *high temperatures* the first term is high and still permits jumps between first neighbours, whatever the value of ΔE. Thus, the conductivity, which is proportional to s (41), varies with the temperature as $\exp(-\Delta E/kT)$.

At *low temperatures*, the electron may 'seek' a more distant arrival site, if the corresponding difference ΔE is sufficiently low. On average, this difference ΔE varies as the inverse of the volume explored[8]: $\Delta E = \alpha r^{-3}$. The expression for s (41) then becomes

$$s = \exp(-\alpha r^{-3}/kT)\,\exp(-r/R)$$

which has a maximum ($ds/dr = 0$) for $r = (3\alpha R/kT)^{1/4}$. Thus, the conductivity, which is proportional to s, increases with T as

$$\boxed{\sigma = \sigma_0 \exp\left[-(T_0/T)^{1/4}\right]}$$

where σ_0 and T_0 are constants. This expression, which is said to be 'in $T^{-1/4}$' (**Mott's law**) is frequently observed in ill-organised materials (see Figure 30).

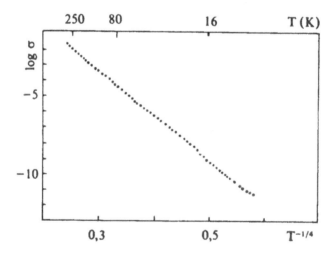

Figure 30   **Conductivity σ of boron carbide as a function of temperature T.** The graph of $\log \sigma$ as a function of $T^{-1/4}$ brings to the fore a conduction by hopping, well described by Mott's law. Following R Kormann and L Zuppiroli.

[8]   Indeed, the number of Fermi states between E and $E + \Delta E$ in the sphere of radius r around the site (1) is

$$n(r, \Delta E) = \tfrac{4}{3}\pi r^3 n(E_F)\Delta E$$

where $n(E)$ is the density of the localised states. To be certain that an arrival site is available, with energy difference ΔE, we must choose r such that $n(r, \Delta E) = 1$. From this, we deduce that

$$\Delta E = \left(\tfrac{4}{3}\pi n(E_F)\right)^{-1} r^{-3} = \alpha r^{-3}.$$

N.B. Throughout the above, we have assumed that the *carriers* were always electrons (or holes). Cases exist in which the **conduction occurs by diffusion** (see Chapter VIII) **of ions**. This is generally the case for *glass*. If the temperature is sufficiently high ($\gtrsim 700^{\circ}$ C), the Na^+ ions (and to a lesser extent K^+ and Ca^{++} ions) of glass become mobile and may conduct a current and give rise to Joule heating. This is the basis for the present research on electrical melting of glass on an industrial scale, which is more homogeneous and potentially more economical than melting by burners. Alternating current is used in order to avoid a migration of the ions.

CHAPTER VII

SURFACES AND POINT DEFECTS

Ruptures
Jules Supervielle
Holy Disorders
Edmund Crispin

Until now, except from p. 147, whenever we have required it, the crystalline periodicity has always been assumed to be perfect. This simplification has so far not been too restrictive. However, it is no longer possible the moment we begin to study phenomena such as **atomic mobility** (or *diffusion*) or **plastic deformation** in materials, phenomena which owe their very existence to breakings of periodicity which we shall call **crystalline defects**.

Make no mistake about it, in this case, crystalline defects, despite their generally weak concentration, do not constitute a sort of second-order refinement which is only useful for a more detailed understanding of the properties of materials. In fact, they play a 'leading role' in the phenomena referred to above, which control, amongst other things, *corrosion* of materials, *high-* or *low-temperature deformation*, *crystal growth*, the *photographic process* etc.

1. FREE SURFACES

These are the most inevitable defects.

In some cases, the properties of crystals are so anisotropic that they are visibly expressed in the appearance and the nature of the surfaces. Thus, for certain minerals (*calcite, feldspar, a fortiori mica*) the free surfaces created either at the time of crystallisation or by fracture or cleavage, involve a finite number of families of crystalline planes (*a single* family for mica). In other cases, this anisotropy is not very apparent. The same is also true for metals, a case described below.

1.1 Surface energy

Let us return to the 'good metal' described in Chapter III.

Earlier, we discussed the free electrons of the metal in plane waves $\exp(i\bar{k} \cdot \bar{r})$ (see p. 44). The choice of these solutions implied that the crystal was infinite, whence *without* free surfaces. We create these free surfaces by cutting this infinite metal along two parallel planes a distance L apart, perpendicular to the direction Oz. Let us consider the N electrons contained in a cube with edge L contained in this metallic 'sheet' (Figure 1).

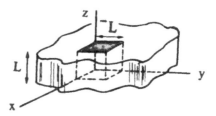

Figure 1 Cube with edge L in a sheet of thickness L. The free surface of the cube is $2L^2$.

One solution of the Schrödinger equation $-(\hbar^2/2m)\Delta\psi = E\psi$ which may be zero on the two free surfaces ($z = 0$ and $z = L$) of this cube is:

$$\psi = \begin{cases} (2L^{-3})^{1/2}e^{i(k_x x + k_y y)}\sin k_z z & \text{for } 0 < z < L \\ 0 & \text{for } z < 0 \text{ and } z > L. \end{cases} \tag{1}$$

The states with $k_z = 0$ were accessible for the electrons in an infinite crystal. In the k-space they formed the set of states in the plane $k_z = 0$ (Figure 13, p. 46), numbering $n_0 = \pi k_F^2/(2\pi L^{-1})^2$ in all. These states can no longer be considered, since they annihilate ψ everywhere. Now the corresponding electrons ($2n_0$ such) can only occupy free states **at the Fermi level**. Thus, the creation of the free surface has *increased the electron energy* by

$$\Delta E = 2n_0(E_F - \overline{E})$$

where E_F is the Fermi energy and $\overline{E}$ is the mean energy of these $2n_0$ electrons. The density of states of the latter is calculated as in Section III.2.2 (p. 45): the number of states between k and $k + dk$ is

$$2\pi k dk/(2\pi L^{-1})^2 \quad \text{where} \quad E = \frac{\hbar^2}{2m}k^2$$

and is proportional to dE. Thus, this density of states is constant and we have

$$\overline{E} = E_F/2;$$

whence,

$$\Delta E = \frac{k_F^2}{2\pi}\frac{E_F}{2}L^2 = \frac{m}{2\pi\hbar^2}E_F^2 L^2.$$

This energy is proportional to the free surface, namely $2L^2$. The constant of proportionality, called the **surface energy** is given (in atomic units) by:

$$\boxed{\Gamma = \frac{1}{4\pi}E_F^2.} \tag{2}$$

For a monovalent metal such as Na, we have $E_F \simeq 0.1$ (see p. 47), or $\Gamma \simeq \frac{1}{1200}$ or

$$\Gamma \simeq \frac{1}{1200} \times 27.2 \times 1.6 \times 10^{-16} \times (0.53 \times 10^{-10})^{-2} \simeq 1300 \text{ mJ m}^{-2}$$

which is close to observed experimental values (see later, Table 1). Note that (2) overestimates Γ in the case of trivalent metals such as Al.

Two points are immediately apparent:

(i) The surface energy is *positive*: a metal decreases its energy if it decreases the area of its free surface.

(ii) The above calculation does not exhibit any *anisotropy*. Whichever plane ($k_z = 0$ or another) is used to cut the Fermi sphere, the result is the same as far as Γ is concerned.

The atomic movements needed for a change in shape are not in general sufficiently rapid for point (i) to have consequences for crystals of macroscopic dimensions. On the other hand, if the free surfaces created are very small, the spherical (or pseudo spherical) shapes implied by points (i) and (ii) are clear to see. Figure 2 shows small cavities created by the accumulation of vacancies (see later, Section 2.1) in aluminium irradiated by He ions. These cavities (which form the free surface), which contain helium and can be called 'bubbles', are actually practically spherical. There is no marked anisotropy of the surface energy in this case (see **Problem 16**).

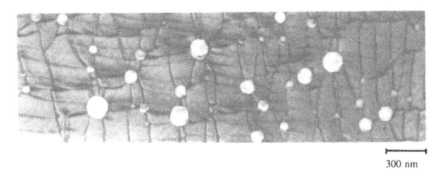

300 nm

Figure 2   Cavities in aluminium observed by electron microscopy.
Aluminium was bombarded by helium nuclei (α particles). The atoms displaced by impacts have created vacancies which have clustered together into cavities in which the helium has accumulated. The dark lines show a network of *dislocations* (see Chapter IX). After M **Caput**.

Point (ii) is no longer true for metals with a non-spherical Fermi surface. In this case, the term Γ depends on the *direction*, since now the area cut out in the Fermi volume depends on the orientation of the plane ($k_z = 0$ or another) chosen. The surface energy is then **anisotropic**. Thus, the irradiated cavities in metals such as Cu, Ni etc. have a more polyhedral or even cubic shape.

The value of Γ is usually *measured* by the **zero creep method**. A long sample such as a vertical rectilinear wire with length l and radius R is heated to a high temperature (to accelerate the atomic mobility) in a very good vacuum (to avoid pollution of the surface). If the weight of the wire is small in comparison with the vertical force derived from the surface energy $2\pi \Gamma l R$, the latter gives rise to a *decrease in the length* of the wire, which may be observed by optical measurement. By subjecting the wire to a weight P, it is possible to determine the critical weight P_c which balances the above force, and thus, consequently, to determine Γ. For $P > P_c$, the wire *lengthens* (by *creep*, see later, Chapter X). For $P < P_c$ it *becomes shorter*. A situation of 'zero creep' exists for $P = P_c$.

Table 1 shows values measured in this way (at temperature $T°$ C) for *copper, silver* and *gold*, together with the theoretical value evaluated using (2).

Table 1 Surface energy Γ measured at $T°$ C and calculated (at 0 K) for three monovalent metals (in mJ m^{-2}) (see N.B. on p. 159).

	T (°C)	Γ (measured)	Γ (calculated at 0 K)
Cu	850	1640 ± 100	3040
Ag	750	1310 ± 100	1790
Au	850	1480 ± 100	1790

1.2 Electron density

Let us return to the form (1) for the wavefunctions. This enables us to calculate the *electron density* $\rho(z)$ **near the surface** as a function of the depth z.

$$\rho(z) = \sum_{k<k_F} \psi_k^* \psi_k = \frac{2}{L^3} \int_{k<k_F} \sin^2 k_z z \frac{L^3}{2\pi^3} \, d^3\bar{k}.$$

We note here that, since k_z is positive, we are working on a Fermi *half* sphere and thus the density (32) of p. 45 should be doubled. Actually, the step of the k-lattice is still $2\pi/L$ in the directions k_x and k_y, but is reduced to π/L in the direction k_z. Thus, we have

$$\rho(z) = \frac{1}{\pi^3} \int_0^{k_F} \sin^2 k_z z \, dk_z \int_A dk_x \, dk_y$$

where the integral $\int_A$ is taken over the surface

$$A(k_z) = \pi(k_F^2 - k_z^2)$$

(see Figure 3). Consequently,

$$\rho(z) = \frac{1}{\pi^2} \int_0^{k_F} (k_F^2 - k_z^2) \sin^2 k_z z \, dk_z.$$

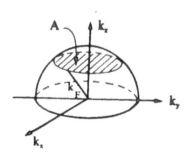

Figure 3 Determination of the surface $A(k_z)$. The bold line shows the Fermi half sphere.

Integration gives

$$\rho(z) = \rho_0 - 3\rho_0 \frac{\sin u - u \cos u}{u^3} \tag{3}$$

where

$$\rho_0 = k_F^3/3\pi^2 \quad \text{and} \quad u = 2k_F z.$$

After an initial behaviour in $(k_F z)^2$, as $r \to \infty$, the asymptotic form of (3) may be written as

$$\boxed{\rho(z) \simeq \rho_0 + 3\rho_0 \frac{\cos(2k_F z)}{(2k_F z)^2}.} \tag{4}$$

The main characteristic of this density is that it exhibits **oscillations** near the surface (see Figure 4). These oscillations have a wavelength equal to πk_F^{-1} and a z^{-2} attenuation. Thus, in a case such as Al (trivalent fcc, k_F large) almost all the oscillations take place in the first layer of atoms. In a case such as Na, the second atomic plane corresponds to a maximum of the electron density ($k_F(Na) < k_F(Al)$, see p. 47).

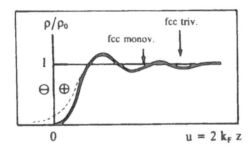

Figure 4 **Electron density of a good metal in the neighbourhood of its surface.** The diagram shows the second atomic plane (the first being $z = 0$) for monovalent or trivalent fcc metals.

The dashed line shows the case of a more realistic potential (not infinite on the surface). z denotes the depth from the surface.

N.B. In reality, the potential is *not infinite* on the surface and the wavefunctions do not vanish there. While the electron density retains the general appearance given by (4), it is thus not zero for $u = 0$. It is attenuated exponentially for $u < 0$ (in other words, in the vacuum), creating a surface dipole (− in the vacuum, + 'under' the surface).

We also note that the free surface has the effect of compressing the electron gas slightly. This compression gives rise to the increase in kinetic energy evaluated in (2). The functions (1) plainly lead to an overestimate of this compression; thus, it should come as no surprise that, in Table 1, the calculated values of Γ are systematically *overestimated* in comparison with the measured values.

1.3 Microscopic structure

This structure clearly depends on how the surface in question is prepared.

i)   If the surface is prepared in such a way that it is possible to approach **thermodynamic equilibrium**, then we obtain a structure which minimises the free energy F. The most common surfaces will then be those with a minimum surface *energy*; in addition, they will also involve a configurational disorder, which contributes to a decrease in F, by virtue of its positive *entropy*. This disorder consists largely of *atomic* **steps**. These have a stable configuration which is again a compromise between a minimum energy (they sit along *dense atomic rows*) and a maximum disorder (consisting of **kinks**, or atomic-size 'corners'), see Figure 5(*a*). The steps, and, more particularly, the re-entrant (or concave) kinks are very favourable sites for the **adsorption** of atoms falling onto the surface, especially at the times of crystal growth in vapour or liquid phases; see **Problem 10**.

Figure 5(*b*) shows the state of a surface of NaCl. In an ionic crystal such as NaCl, the surfaces of minimum surface energy are those which are dense and neutral. The most dense neutral surfaces are {100} surfaces (faces of cubes). The steps of minimum energy follow the most dense neutral rows, namely the (100) rows (edges of cubes). The steps shown in Figure 5(*b*) possess this geometry, the rounded corners globally indicating the presence of numerous kinks.

N.B. The technique used to obtain the surface micrography of Figure 5(*b*) is only one of many. One of the most recent and most promising techniques is *tunnel-effect micrography*: a very fine head is brought close (using piezoelectric crystals) to such a metallic surface so that a tunnel current can pass between the head and the metal. By keeping this current constant and moving the head along the surface one can obtain images of the latter at the atomic scale (see an example on p. 380).

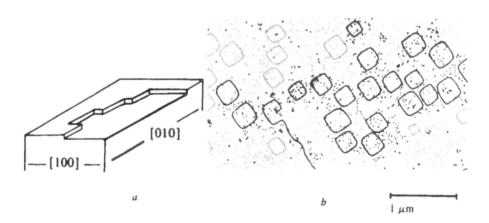

a b

1 μm

Figure 5   Atomic steps on free surfaces (see also pp. 214, 380). (*a*) Example of a step on a {100} surface of a cubic crystal. The elementary directions of the step are (100). The 're-entrant' kinks (there are four) are favourable positions for adsorption.

(*b*) Micrography of a surface of NaCl prepared by evaporation at 335° C. A small amount of gold is evaporated on the surface. This accumulates on the steps. The whole is covered with a continuous deposit of carbon, which is removed by dissolving the salt. This deposit (*replica*) is then examined by electron microscopy. The gold atoms attached to the replica, which are more absorbing for the electrons than the carbon atoms, give a black contrast which indicates the position of the steps.

The pseudo squares have sides parallel to (100). They delimit 'basins' with depth equal to *twice* the interplanar distance, i.e. a (a is the edge of the cell, Figure 22, p. 105). The rounded corners denote the presence of numerous kinks (see **Problem 10**).

After **H Bethge** and **K W Keller** *J. Crystal Growth* **23**, 105 (1974).

The microscopic structure has a major influence on most of the properties associated with the surface: adsorption, oxidation, corrosion, catalysis etc. This is particularly true for the kinks (and also for the steps) which generally have a large electric dipole moment.

Figure 6 gives an example of the catalytic action of steps and kinks. The transformation of cyclohexane (a cyclic molecule, C_6H_{12}) into n-hexane (a linear molecule, C_6H_{14}), which is a reaction necessitating the breaking of the C_6 cycle (in other words, breaking the bond C=C) is catalysed by platinum. Clearly, *both* the number of steps *and* the number of kinks have a large effect on its yield.

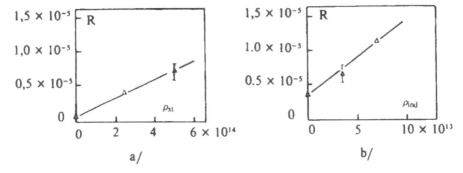

a/ b/

Figure 6 Example of the effect of the surface structure of the catalyser (platinum) on catalysis.
    The reaction cyclohexane → n-hexane is catalysed by platinum. The figure shows the catalytic yield R (in terms of the number of molecules produced per atom of platinum on the surface, per second) as a function of:
(*a*) the density of the *steps* ρ_{st} (step atoms cm^{-2});
(*b*) the density of the *kinks* ρ_{ind} (kinks cm^{-2}), for a constant density of steps $\rho_{st} = 2.5 \times 10^{14}$ cm^{-2}.
After **D W Blakely** and **G A Somorjai**, *J. Catal.* **42**, 181 (1979).

ii)   If the surface S is prepared **without giving equilibrium a chance** (for example, by machining), there is no reason for it to coincide with a surface P of minimum surface energy Γ. The surface then takes on a *terraced* structure, where the planes of the terraces are planes P delimited by a series of steps. In the particular case in which S is a plane *parallel* to a step of favourable direction, forming an angle φ with P, the height h of the steps and the width l of the terraces satisfy $\tan\varphi = h/l$. In this case, the steps form parallel equidistant straight lines.

Figure 11 (p. 214) shows an example of circular terraces corresponding to the case in which S is a cone.

2. POINT DEFECTS

A point defect is defined as a perturbation of the crystalline periodicity, with a small volume of the order of magnitude of the atomic volume Ω. Foreign atoms or **impurities** (atoms of B in a crystal of A), **vacancies** (absence of atoms in positions which are normally occupied) and **interstitials** (surplus atoms occupying positions which are normally empty) are common examples of point defects. We shall also include very small **aggregates** of these primitive defects, such as the association of two vacancies (called a *divacancy*), or of an impurity and a vacancy, etc.

The introduction of a point defect in a crystal which is assumed to be perfect increases the energy by a quantity E_f called the **formation energy** of the defect. However, it also increases the **configurational entropy**, since the object may be placed in different locations, sometimes in different ways. Thus, in the simplest, most common case, in which this object (for example, impurity or vacancy) occupies an atomic position and possesses the symmetry of the atom which it replaces, this entropy, calculated for n identical objects at N atomic positions is:

$$S = -Nk(c \ln c + (1 - c) \ln(1 - c)) \tag{5}$$

(see (Stat. Ph.)) where we have set $c = n/N$ (c is the **atomic concentration** of the defect). This term is always positive since we have $c < 1$. Moreover, it varies most violently for c small ($dS/dc \to \infty$ as $c \to 0$). The energy itself simply varies like NcE_F (see Figure 7). It follows that **the free energy always decreases**, whatever the value of E_f and whatever the temperature ($\neq 0$) **whenever one starts to introduce defects** (here, point defects) in an ideally perfect solid.

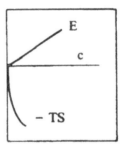

Figure 7 For a weak concentration of point defects c, the free energy $E - TS$ always decreases initially owing to the form of $S(c)$ (see (5)).

This explains why it is impossible to purify solids beyond a certain limit. The obstacle is fundamental rather than technological. In the same way, this explains why, as the performance of analyses increases, all the elements of the periodic table are discovered in a reputedly 'ultra pure' solid! For example, uranium, an element which may be analysed in minute quantities, thanks to its large fission cross section in the presence of slow neutrons, is detected in all solids in which it has been sought, in concentrations as low as $c \sim 10^{-14}$ atm.

The free energy of the crystal containing Nc defects (for example *vacancies*)

$$F = NcE_f + NkT(c \ln c + (1 - c) \ln(1 - c))$$

is a minimum when c satisfies the equation

$$\frac{c}{1 - c} = \exp(-E_f/kT)$$

or, when c is small, in other words, for E_f sufficiently large in comparison with kT:

$$c \simeq \exp(-E_f/kT). \tag{6}$$

N.B.   Strictly speaking, the free enthalpy $G_f = H_f - Ts_f$ should appear in formula (6) instead of E_f. This leads to a term $\exp(s_f/k)$ in front of the exponential, which is independent of the temperature, although in (6) we have implicitly taken $s_f = 0$; the entropy s_f of the defect, due largely to variations of the modes of atomic vibrations, is generally small. The enthalpy H_f is equal to $E_f + p\delta v$ where δv is the volume change in the crystal due to the introduction of the defect (p = pressure). Since defects of atomic 'size' are involved here, the atomic volume is a good approximation to δv. Thus, taking $\delta v = 20 \text{ Å}^3$, we obtain $p\delta v \simeq 10^{-5}$ eV for $p = 1$ atm. Therefore, it is reasonable to liken H_f to E_f. This would not be reasonable for materials subjected to high pressures (such as rocks in the Earth's mantle).

2.1 Vacancies

In what follows, we define a **vacancy** to be the defect obtained by *removing an atom* from its position and *replacing it on the surface*, so that we are always working with a constant number of atoms N. The energy brought into play in this operation is denoted by E_f. Although this is not the same as *sublimation* (in which atoms are *removed from the surface*), we see that it involves identical mechanisms (essentially the breaking of bonds). Initially, we deduce that E_f in a given solid is not very different from the sublimation energy E_s, being of the order of an electron volt.

Formula (6) shows both that *at any given temperature there exists a concentration of vacancies in thermodynamic equilibrium* in a crystal and that *this concentration is weak*: for $E_f = 1$ eV, we have $c \simeq \exp(-1/(1/40)) = 10^{-40/2.3}$ or $\simeq 10^{-17}$ at room temperature, or $\simeq 10^{-4}$ at $T = 1300$ K. In fact, the concentrations of vacancies rarely exceed 10^{-4} at the melting point of solids[1].

1 – Vacancies in metals

In metals with dense packing, the structure of the vacancy is not in doubt: it is that of the vacancy v on Figure 8(a). It is surrounded by 12 first neighbours. The structure (b) (two half vacancies separated by an atom) may be viewed as an excited form v^* of the vacancy v. In a body-centred cubic metal, the vacancy v has 8 first neighbours.

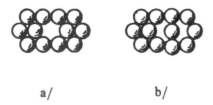

a/   b/

Figure 8   Vacancy in a dense metal (fcc or cph). In (a): vacancy v. This has 12 first neighbours, six in the plane of the figure, three above and three below. In (b): excited vacancy v^*.

[1] Here, we are talking about *pure solids*. The case of *compounds* of type $A_m B_n$ may be different, in particular when one moves away from the stoichiometric composition. Thus, the crystal may find interest as, in order to retain its structure up to some critical concentration, it accumulates vacancies on the sublattice of the deficient element. For example, carbide TiC, which crystallises into the NaCl structure, retains this structure until the composition $TiC_{0.7}$ at the price of 30% vacancies on the carbon (fcc) sublattice.

This first representation of the vacancy shows it as a small spherical void carrying a charge $-ze$ at its centre (where z = valency of the metal) corresponding to the absence of the ion $+ze$. This *negative charge* creates a repulsive potential $V(r)$ for the electrons of the metal (r: distance to the centre of the vacancy). This potential, which would be Coulombic in the vacuum, must be annihilated here at short distance, so that the electric field remains zero in the metal outside the vacancy.

How does this attenuation take place? The answer to this question is strongly hinted at by the results of Section 1.2 (p. 158) and, in particular, by Figure 4, which summarises these results. In the presence of a free surface, the gas of conduction electrons reacted with a substantial *local decrease in the density* and with *oscillations* of wavelength varying as k_F^{-1}. We anticipate a comparable behaviour of the electrons (on the 'free surface' of the vacancy, viewed as a small spherical void) with a **local decrease**, and **oscillations, of the electron density**. Calculations show that the change in symmetry alters the exponent in (4) from 2 to 3, whence, asymptotically ($r \to \infty$), we have

$$\rho(r) \simeq \rho_0 + 3\rho_0 \frac{\cos(2k_F r)}{(2k_F r)^3}. \tag{7}$$

These charge oscillations (known as *Friedel oscillations*) imply oscillations of the potential energy, whence also of the *local electric field*.

Two remarks are then in order:

i)   This decrease in the density of conduction electrons near the vacancy compensates the negative charge of the vacancy (and shields it beyond the volume of the vacancy).

ii)   Depending on the position of the first-neighbour atoms of the vacancy with respect to the peaks of the oscillating electric field, which are themselves linked to the distance k_F^{-1} and thus to the valency (see Figure 4), the force exercised on these atoms may be directed *towards the centre* of the vacancy or *outwards*. For example, in gold (fcc), the 12 first-neighbour atoms move *inwards* (by approximately 10%), while in aluminium, they seem to move slightly outwards.

It is clear that the *energy* of the vacancy E_f is directly associated with the electronic perturbations, as was the surface energy. Thus, using (2) and writing $E_f = 4\pi R^2 \Gamma$ (where R = radius of the vacancy), we obtain a first approximation to E_f. If for R, we take half the distance between the near-neighbour atoms, we obtain energies of the order of 2–5 eV, which are too high in comparison with experimental values. This is no surprise, since the electrons may tunnel through the vacancy, the 'surface' of the vacancy thus being less reflecting than a free surface with an infinite radius of curvature.

Let us return to Figure 8. The vacancy v^* may be viewed as being obtained from v by moving an atom adjacent to the vacancy by $r_0/2$ (r_0 = interatomic distance). For an fcc crystal the reader should check that (see Figure 21, p. 104) in space, the central atom of v^* is at the centre of a rectangle (length: $r_0\sqrt{2}$; width r_0) of *four* atoms and thus that it is compressed. The energy of the configuration v^* is *greater* than that of v by a quantity E_m, which is the energy needed for an atom adjacent to the vacancy (A on Figure 9) to jump into the vacancy v. This jump of A into v is also a jump of v onto the position of A. The (activation) energy E_m is called the **migration energy** of the vacancy. From the position A, the vacancy may ultimately return to v or, more probably, (probability $(Z-1)/Z$, where Z is the number of first neighbours) jump to the position of another of its Z neighbours. Generally, if no force is exerted on it, the vacancy describes a random walk. An atom adjacent to v (for example, A) vibrates with average frequency v, evaluated in **Problem 18**, which, for numerical applications, we may take to be equal to $\simeq 10^{13}$ s^{-1}. The number of jumps of the vacancy during time t is thus

$$n = (\text{'attack' frequency}) \times (\text{number of possibilities}) \times (\text{probability of jump}) \times t$$

$$\boxed{n = \left[vZ \exp(-E_{\mathrm{m}}/kT) \right] t.} \tag{8}$$

This random migration generally ends with the *disappearance* of the vacancy, which is absorbed by the **sinks** formed by the free surfaces or other defects, such as *dislocations* (see Chapter VIII). Thus, in parallel, the existence of an equilibrium concentration (6) also implies the presence of **sources** of vacancies (again the free surfaces, dislocations, boundaries etc.).

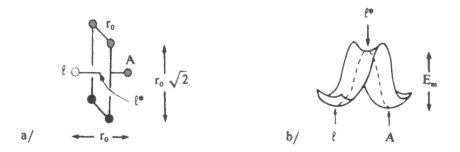

Figure 9   Jump of a vacancy in an fcc crystal. (*a*) The atom A must pass through the centre of a rectangular window of atoms in order to jump into the vacancy v. At the centre of the window, it constitutes the vacancy v^* (Figure 8(*b*)). The four atoms of the window are all near neighbours (distance r_0) of *both* v *and* A (see Figure 21, p. 104). (*b*) On the path from A to v, the energy reaches a maximum for the state v^* which is a saddle point.

Experimental studies of vacancies in metals currently involve *quenching*. In this technique, a sample is raised to a high temperature T so that the concentration c is sufficiently large (see (6)). The sample is then quenched, in other words, cooled very rapidly to T_0, so that the number of jumps (8) during the quench is small and the initial concentration $c(T)$ is frozen at T_0. If T_0 is sufficiently low (for example, 4 K), these artificially *supersaturated* vacancies are at rest. They are sufficiently numerous that they may be observed, for example, by the increase in the volume which they cause $\Delta V/V \simeq n\Omega/N\Omega \simeq c$ (or by the increase in length for a wire: $\Delta l/l \simeq c/3$) (Ω is the atomic volume). By measuring c as a function of T it is possible to determine E_{f} (see (6)).

In the same way, by *eliminating* the quenched-in vacancies by thermal treatments at moderate temperatures T_i, it is possible to measure the migration kinetics and so determine E_{m}.

Table 2 gives values found in this way for a number of metals.

Table 2   Measured values (in eV) of the energies of formation E_{f} and migration E_{m} of vacancies in a number of metals.

	Na	Cu	Ag	Au	Mg	Al	Pt	W
E_{f}	0.42	1.28	1.06	0.95	0.79	0.69	1.50	2.5
E_{m}		0.72	0.86	0.85	0.48	0.64	1.40	1.8

2 – Vacancies in ionic crystals

Here, as in Chapter I, we pick out alkali halides, which are among the best studied ionic solids.

We now have to consider *two types of vacancy: the cationic vacancy* v_C and the *anionic* vacancy v_A. Here, v_C is the defect resulting when a cation is extracted from a volume and placed on the surface. The vacancies occur in almost equal numbers in the volume, so that the electric neutrality of the crystal is conserved. Thus, each vacancy v_C corresponds to a vacancy v_A. The pair (v_A, v_C) is called a *Schottky pair*.

Let us consider this point in more detail. If there were no local rearrangement, the formation energies E_{fC} of these two vacancies would be equal, with common value $E_f \simeq \frac{1}{2} E_b$ (see p. 4). In fact, removing an anion (or a cation) from the volume and placing it on the surface amounts to changing its Madelung constant (p. 2) from M to $M/2$, and thus changing its energy from $-E_b$ to $-\frac{1}{2} E_b$. However, the effects of the *charge* of the vacancy ($+e$ or $-e$, respectively for v_A and v_C), i.e. *polarisation* of the ions and ionic *relaxations* (see Figure 10), amounting to a total of E_r, must be subtracted from this energy E_f. There is no reason this term should be identical for v_C and v_A. Thus, the energies $E_{fC} = E_f - E_{rC}$ and $E_{fA} = E_f - E_{rA}$ are different and formula (6) applied to each of the lattices C and A, leads to different numbers n of vacancies v_C and v_A.

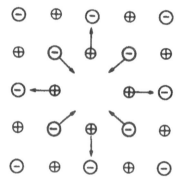

Figure 10   Anionic vacancy v_A in an alkali halide. The positive charge carried by this vacancy repels cations and attracts anions.

Note that the circles of this figure represent *ions*, while those of Figure 11 represent *ionic vacancies*.

Suppose, for example, that $n_C > n_A$.

Since the vacancies are charged, there is a risk of loss of electric neutrality inside the crystal. The latter reacts by arranging the surplus charged vacancies – $(n_C - n_A)$ v_C vacancies in the above case – on the surface in such a way as to create a dipolar layer called the *Debye layer* or *cloud*. This layer restores the electric neutrality in the volume and considerably decreases the range of the electric field created by the surface (Figure 11).

Let us calculate the *bulk* concentrations $c_C = n_C/N$ and $c_A = n_A/N$ (N is the total number of molecules) in these conditions, where we impose the constraint $c_C = c_A$. The introduction of the n_C and n_A vacancies with energies E_{fC} and E_{fA} modifies the free energy by the quantity

$$\Delta F = N c_C E_{fC} + N c_A E_{fA} - T \Delta S$$

where

$$-\Delta S = kN(c_C \ln c_C + (1 - c_C) \ln(1 - c_C)) + kN(c_A \ln c_A + (1 - c_A) \ln(1 - c_A)).$$

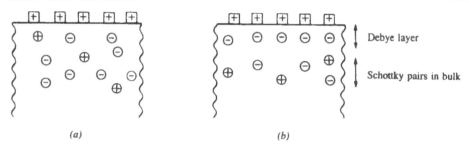

Figure 11 **Formation of a Debye layer**. (*a*) Cationic ($\ominus$) and anionic ($\oplus$) vacancies are distributed uniformly, in unequal numbers, in the crystal (here $n_C > n_A$). The surplus corresponding to cations + is on the surface. The electric fields are considerable. (*b*) A preferential movement of cationic vacancies towards the surface re-establishes the neutrality of the volume and creates a dipolar layer on the surface.

Minimisation of ΔF subject to the constraint $dc_C = dc_A$ immediately gives

$$\boxed{c_C c_A = \exp(-E_S/kT)} \tag{9}$$

where $E_S = E_{fC} + E_{fA}$; or, in the present case,

$$c_C = c_A = \exp(-E_S/2kT).$$

If the crystal contains charged impurities (for example, Ca^{++} in NaCl), additional vacancies must be created in order to neutralise these charges (for example, in the above example, as many cationic vacancies as there are Ca^{++} cations). Thus, we no longer have $c_C = c_A$, although the constraint $dc_C = dc_A$ remains. Therefore, formula (9) (in which the reader will have recognised the mass action law) remains correct.

The positive charge of the vacancy v_A gives the latter the important property of being able to *capture an electron*. This electron neutralises v_A and forms a hydrogenoid object giving rise, like a hydrogen atom, to characteristic energy levels and light-absorption or emission lines. This selective absorption, if it take places in the visible band, gives the normally transparent crystal a colour, whence the name **F centre**[2] given to the defect (v_A + electron).

For simplicity, the F centre may be viewed as a cubic cage containing the electron, where the edge of the cage αa is little different from the edge a of the crystalline cell. If the potential is taken to be zero in the cage and infinite outside it, the eigenvalues of the energy of the electron are (see Q.M.):

$$E = \frac{\hbar^2}{2m} \frac{\pi^2}{\alpha^2 a^2} (n_x^2 + n_y^2 + n_z^2) \tag{10}$$

where the n_i are non-zero integers.

Thus, for a given transition, the excitation energy varies in a^{-2}. For the transition from the ground state ($n_x = n_y = n_z = 1$) to the first excited state (one $n_i = 2$, the two others $= 1$), this energy is

$$\Delta E = \frac{\hbar^2}{2m} \frac{\pi^2}{\alpha^2 a^2} (6 - 3) = 3\frac{\hbar^2}{2m} \pi^2 (\alpha a)^{-2}. \tag{11}$$

---

[2] From the German 'Farbezentrum'.

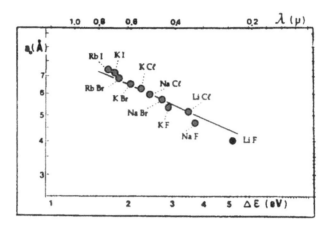

Figure 12 Absorption of light by alkali halides. The absorption lines of halides containing vacancies are measured. The energy ΔE of the first line is recorded here in logarithmic coordinates as a function of the edge a of the cubic cell. The variation in the power -2 (full line) is called the *Mollwo–Ivey law*.

This variation in a^{-2} is actually observed for all alkali halides, as Figure 12 shows (see also **Problem 11**). Moreover, combining the experimental results and formula (11), we obtain $\alpha = 1.13$. Thus, the effective size of the vacancy is a little larger than the size of the cell, which is normal, when one takes into account the six cation neighbours of the F centre which draw the electron slightly towards the outside.

N.B. 1 *Coloration.* Viewed in white light, the crystal containing the F centres absorbs a certain wavelength, whence a certain colour of the spectrum. Therefore, it has the *complementary colour* to the colour absorbed. For example, for NaCl (see Figure 12), $\lambda_{abs} = 0.47$ μm (or *violet*) and the coloured crystal is *yellow*. Let us suppose, in particular, that a crystal of NaCl is heated in a sodium (or potassium) vapour. An excess of cations is created in comparison with the stoichiometric composition together with a correlated excess of anionic vacancies which the electrons originating from the ionisation of the sodium transform into F centres. Very generally, *deviations from the stoichiometry* or the *presence of impurities* tend to colour ionic crystals, and minerals in particular.

The coloration of a non-crystalline mineral by *centres* or *impurities* is associated with the same causes. This applies to **glass**, which generally owes its slight tint to iron ions: the Fe^{2+} ions (*resp.* Fe^{3+}) absorb in the red (*resp.* in the violet) and thus give glass a blue (*resp.* yellow) colour. The usual green colour of glass is due to the simultaneous presence of these two ions. It was necessary to avoid any coloration in the thick (20 mm) glass of the *Louvre pyramid* and the company Saint-Gobain had to carry out very special research for this[3].

N.B. 2 *Jahn–Teller effect.* The cubic F centre states are degenerate (see (10)), except for the ground state. In particular, the first excited level exhibits a triple degeneracy: (2,1,1), (1,2,1) and (1,1,2). We shall show that a distortion which decreases the symmetry of the F centre partially removes this degeneracy and above all, decreases the energy of the defect.

For this, we deform the cube with edge $\alpha a = a_0$ into a parallelepiped with a square base (height b, side of the base c) preserving the volume ($a_0^3 = bc^2$). The deformation is defined with the help of the parameter $\eta = b/c$: for $b/c > 1$, the F centre is elongated, for $b/c < 1$, it is

[3] See **J Barton**, *La recherche*, **182**, 1428 (1986).

flattened. The axes Ox, Oy, Oz coincide with three edges of the parallelepiped, where the axis Oz is along the 'height'. Calculation of the ground state $(1,1,1)$ and the excited states $(2,1,1)$, $(1,2,1)$ and $(1,1,2)$ *via* the expression (10) (see **Problem 11**) enables us to plot the *energy–configuration* diagram of Figure 13. The diagram clearly shows that the deformation removes the degeneracy between $(2,1,1)$ (or $(1,2,1)$) and $(1,1,2)$ and that *the elongation* of the F centre stabilises the latter. This spontaneous deformation of the F centre, which is also observed for other defects and other crystals (for example, semiconductors) is called the *Jahn–Teller* effect (see also p. 151).

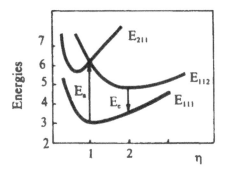

Figure 13   Jahn–Teller effect of an F centre. In its ground state, the F centre absorbs the energy E_a. After stabilisation of shape at $\eta = 2$ it re-emits the *fluorescent* energy E_c which is considerably less than E_a. The difference $E_a - E_c$ is retained by the crystal in the form of ionic vibrations (thermal energy). This difference is the *Stokes shift*. The energies are in units of $(\hbar^2/2m)(\pi^2/a_0^2)$.

2.2 Interstitials

Interstitials are atoms or ions inserted in positions which do not belong to the crystal lattice. These may be foreign atoms (*carbon*, inserted into *iron*, positions itself at the centre of the edges or at the centre of the faces of the body-centred cubic cell of the iron, see **Problem 15**), or atoms (or ions) of the crystal itself (Si in Si, Na^+ or Cl^- in NaCl etc).

We shall limit ourselves here to a few elementary facts about self-interstitials (or more simply, in what follows, just 'interstitials'), in other words, atoms of A inserted in A or in AB. In general, such a defect constitutes a much more severe perturbation than a vacancy. The deformations produced are large, at least in compact structures where the *formation energies* may reach 4–5 eV. Under these conditions, formula (6) leads to minute equilibrium concentrations ($\simeq 10^{-20}$ at 1300 K for $E_f = 5$ eV!). The high values of E_f associated with a very compressed configuration around the interstitial i correspond, quite naturally, to *low values* of the *migration energy* E_m. The large distortions around i make it easy to find an excited configuration i^* with energy little different from that of i. For example, $E_m \simeq 0.05$ eV is measured in copper.

One should not conclude from the above that interstitials are unobservable or that they play a minor role in solids! Moreover, there exist situations in which the energy E_f is moderate (c.f. *silver chloride*, see p. 200) and cases in which large concentrations of interstitials are created *out of equilibrium*. For example, in a solid subjected to *irradiation* (see **Problem 17**) by particles (fast neutrons in a nuclear reactor, protons in the solar wind etc), atoms which receive a sufficient energy from a direct impact are ejected from their position, leaving a vacancy, and insert themselves in the crystal in an interstitial position. The pair of antimorphic defects created in this way (vacancy plus interstitial) form a **Frenkel pair**. Similarly, in an ionic solid subjected to illumination by photons with sufficient energy (u.v. or X), ionising an anion ($A^- \rightarrow A^\circ$) amounts to local storage of energy which may be restored to the crystal by the creation of an anionic Frenkel pair. Here, we note that, in an ionic crystal, the creation of a Frenkel pair (whether anionic or cationic) does not destroy the local neutrality.

This creation of interstitials by absorption of photons seems to occur via the following mechanism, due to Pooley. Locally, the ionisation $A^- \rightarrow A^\circ$ creates a state with a positive charge (a *hole*) which reduces its energy via the Jahn–Teller effect by sharing two neighbouring anions with the formation of an A_2^- molecule (for example, Cl_2^- in NaCl, in the (110) direction, Figure 14). This very electrophilic molecule captures an electron, which, together with the hole, forms a hydrogenoid state (called an **exciton**). The local deformation (here, contraction), stabilises this state, just as in Figure 13 a deformation (in that case an elongation) stabilised the excited state denoted by (112). The radiative de-excitation energy (E_c on Figure 13), decreased by the deformation, may even be reduced to zero[4]. In this case, the energy of this **non-radiative transition** is completely yielded to the ions. Taking into account the symmetry, the most probable event is the ejection of one of the two A^- ions in the direction of the molecule itself. This direction ($\langle 110 \rangle$, see Figure 14) is dense. Moreover, it consists of A^- ions all with the same mass. Following the first impact, the subsequent impacts propagate with little loss. Thus, if the initial energy is sufficient (~ 5 eV), the initial A^- ion then expels its neighbour, which itself expels its neighbour etc. After this series of collisions, there exist a *vacancy* in the initial position and an *interstitial* A^- a few interatomic distances further on (Figure 14(b)). The non-radiative transition energy ends up as formation energy for the two defects and as losses by thermal agitation during the series of collisions.

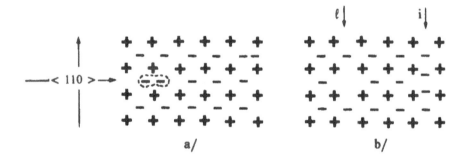

Figure 14 Frenkel pair induced by photons in an alkali halide. (a) A photon has ionised an anion: A^- $\rightarrow A^\circ$. Electron capture is followed by the formation of an *exciton* shared by two adjacent anions. (b) The anion on the left returns to its site by non-radiative de-excitation, while the anion on the right leaving the site v expels its right-hand neighbour, which expels etc. At the end of the process two anions share the site of a single one (i).

Of course, Frenkel pairs may also be created in monatomic solids such as metals. This is the case under irradiation (see **Problem 17**) by neutrons, electrons .... If the vacancy and the interstitial forming the Frenkel pair are close to each other, the corresponding 'close pair' is generally quite unstable and tends to *recombine*. Figure 15 shows an example of the disappearance (by direct recombination) of such defects in bismuth at temperatures as low as $\simeq 40$ K.

---

[4] For this, it is sufficient that the energy–deformation *configuration curves* (see Figure 13) of the ground and excited states should intersect.

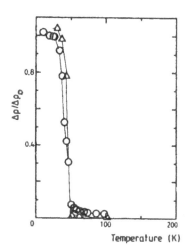

Figure 15 Two bismuth samples have been irradiated at $T = 4.2$ K by electrons of 1 MeV. These have displaced atoms and created close Frenkel pairs, which increase the electrical restivity by $\Delta\varrho_0$. When heated to temperatures T (in abscissae) as low as $\simeq 40$ K, the ratio of the remaining extra-restivity $\Delta\varrho$ to $\Delta\varrho_0$ decreases abruptly, indicating the disappearance of Frenkel pairs by mutual recombination.
After **P Bois** and **F Beuneu** *J. Phys. Cond. Mat.* **1**, 4535 (1989).

CHAPTER VIII

DIFFUSION AND PRECIPITATION

Wanderer Fantasie
Franz Schubert
The Pure and the Impure
Colette

One Friday afternoon in 1906, a German metallurgist, A Wilm, working on Al–4% Cu alloys, *quenched*[1] a sample. Leaving it on his bench, on his return on the Monday morning he was surprised to find it noticeably transformed. From being *ductile* just after quenching, it had in the meantime become *very hard*. The industry of *duralumins* (after Düren, the town where Wilm was working) had been born, with immediate and enormous consequences, particularly in aeronautics[2].

Later (Guinier, 1937 and Preston, 1937), it was shown that this violent variation of the mechanical properties was due to the agglomeration of copper atoms in small **clusters**, *Guinier–Preston* (or *GP*) '*zones*'. Initially (in other words, just after the quenching), dispersed in a solid solution on the normal sites of the aluminium crystal, the copper atoms had **migrated**, at room temperature, into the crystal to regroup several hours or days later.

This is just one example of the very general phenomenon of **atomic diffusion**. Another very important example is given by the *oxidation* of solids (see **Problem 14**). However, before we discuss the diffusion of atoms of B in a solid A (Cu in Al), we shall discuss the diffusion of atoms of A in A, in other words, self-diffusion.

1. SELF-DIFFUSION

Let us denote an atom of a pure solid A by A. It is possible to imagine several mechanisms by which A may diffuse in A. These include, for example, *insertion* of A into an interstitial position, followed by a *jump* between interstitial positions, or *interversion* A $\rightsquigarrow$ A' (where A' is an atom of A which is a near neighbour of A) repeated step by step. Even though such mechanisms may operate in certain particular cases they have never been observed in an indisputable way.

However, the **vacancy** mechanism has been studied in detail and observed without ambiguity in numerous cases. We have already met the elementary step of this, namely

---

[1] In other words, cooled it rapidly from a high temperature, $\simeq 650°$ C to room temperature.

[2] For all that, the weekend break is neither the most certain nor the most usual route to discoveries of this scale.

when a vacancy jumps by $\bar{r}_0$ (p. 165) and an atom of the crystal jumps by $-\bar{r}_0$. If this jump *of the atom* can be reproduced subsequently, the latter *diffuses* in the solid.

1.1 Random diffusion of vacancies

Let us first consider a vacancy which is able to make *random jumps*, such as those described in Section 2.1.1 (p. 163), in a crystal. Its movement is *Brownian* and Figure 9 (p. 90) provides an illustration for a cubic crystal. After n jumps, the vacancy is at a distance r_n from its point of departure, such that

$$\langle r_n^2 \rangle = n r_0^2 \tag{1}$$

(Einstein's equation, see also (1), p. 90). Here, r_0 is the length of an elementary jump. In fact, at the last jump, we have:

$$\bar{r}_n = \bar{r}_{n-1} + \bar{r}_0 \tag{2}$$

where $\bar{r}_0$ is one of Z possible jump vectors (for example, $Z = 12$ for an fcc crystal or 8 for a bcc crystal). We may then write:

$$\langle r_n^2 \rangle = \langle \bar{r}_n \cdot \bar{r}_n \rangle$$

or, by virtue of (2),

$$\langle r_n^2 \rangle = \langle r_{n-1}^2 \rangle + 2\langle \bar{r}_{n-1} \cdot \bar{r}_0 \rangle + r_0^2. \tag{3}$$

The term $\langle \bar{r}_{n-1} \cdot \bar{r}_0 \rangle$ is zero, since in the calculation of this average value, over the many possible cases for this last jump, the terms may be regrouped into equiprobable pairs ($\bar{r}_0$ and $-\bar{r}_0$). An obvious recursion leads from (3) to (1).

In the frequent case where n is proportional to time t (see, for example, the isothermal case, (8) on p. 165), it is clear that the *mean (square) distance covered by the vacancy is proportional to* $\sqrt{t}$.

The quantity

$$D_v = \frac{\langle r^2 \rangle}{6t} \tag{4}$$

or, by virtue of (8), p. 165,

$$D_v = \nu \frac{Z}{6} r_0^2 \exp(-E_m/kT) \tag{5}$$

is called the **diffusion coefficient** of the vacancy. The numerical factor 6 is included to make this definition coincide with another later definition (see (7)).

Let us now examine the case in which the crystal contains vacancies with a concentration $c(\bar{r})$ and consider *three* successive parallel atomic planes, P_{i-1}, P_i and P_{i+1}. To be specific, we shall suppose that the crystal is fcc ($Z = 12$) and that the planes P are dense planes {111} at distance λ apart. We shall also consider the case in which, in space, c is only a function of x, the direction perpendicular to P (see Figure 1): $c = c(x)$. Moreover, here c will denote the **volume concentration**, in other words, the number of vacancies per unit volume.

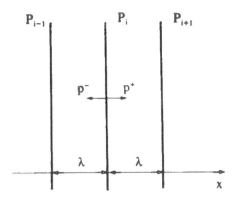

Figure 1 Successive atomic planes and probabilities of jumps p^+ and p^-.

At each jump, the vacancies have a probability p_0 ($= \frac{6}{12}$) of remaining in the plane in which they are currently located, a probability of p^+ ($= \frac{3}{12}$) of moving to the next plane 'on the right' and a probability p^- of moving to the next plane 'on the left'. Thus, in time dt, the concentration of vacancies in the plane P_i varies by:

$$dc_i = (p^+ c_{i-1} + p^- c_{i+1} - (p^+ + p^-)c_i)\nu Z \exp\left(-\frac{E_m}{kT}\right) dt$$

$$= 3\nu \exp\left(\frac{-E_m}{kT}\right)(c_{i+1} - c_i - (c_i - c_{i-1})) dt$$

$$\simeq 3\nu \exp\left(-\frac{E_m}{kT}\right)\lambda^2 \frac{d^2 c_i}{dx^2} dt.$$

Introducing the lattice parameter a (edge of the cubic cell), which is such that $\lambda = a/\sqrt{3}$ and $r_0 = a/\sqrt{2}$, and using (5), we obtain:

$$\boxed{\frac{\partial c}{\partial t} = D_v \frac{\partial^2 c}{\partial x^2}} \tag{6}$$

where

$$\boxed{D_v = \nu a^2 \exp(-E_m/kT)} \tag{7}$$

is the diffusion coefficient which we met earlier in (5).

N.B. It is easy to check that the above formulae also hold for a body-centred cubic (bcc) structure with $Z = 8$ instead of 12. Thus, for the planes P {100} (*resp.* {110}), we have $p^+ = p^- = 1/2$ (*resp.* 1/4) and $\lambda = a/2$ (*resp.* $a/\sqrt{2}$).

By counting the number of vacancies per unit time which cross a unit surface perpendicular to x (for example, a plane between P_i and P_{i+1}), in other words, by determining the **flux of vacancies** J_x, we obtain

$$\boxed{J_x = -D_v \frac{\partial c}{\partial x}.} \tag{8}$$

In the case, where c is a function of $\bar{r}$ and not just of x, all of the above may be immediately generalised to:

$$\boxed{\begin{aligned} \bar{J} &= -D_v\bar{\nabla}c \\ \frac{\partial c}{\partial t} &= D_v\Delta c \end{aligned}}$$

$$(9)$$

$$(10)$$

These equations are known as the first and second **laws of Fick**.

N.B. 1   In deriving (6), we assumed that the diffusion coefficient did not depend on the position. If this is not the case, then equations (6) and (10) become, respectively:

$$\frac{\partial c}{\partial t} = \frac{\partial}{\partial x}\left(D_v\frac{\partial c}{\partial x}\right) \qquad \text{and}$$

$$\frac{\partial c}{\partial t} = \text{div}\,(D_v\bar{\nabla}c).$$

N.B. 2   Care should be taken to use consistent units for J and c. Diffusion coefficients (see (4)) are expressed in terms of $m^2\,s^{-1}$ (or, more commonly in $cm^2\,s^{-1}$). If, as in the above, c is expressed in terms of (volume)$^{-1}$, then J is expressed in terms of (surface × time)$^{-1}$. If c is the *atomic concentration*, the reader will check that (9) should be replaced by

$$\bar{J} = -\frac{1}{\Omega}D_v\bar{\nabla}c$$

$$(9')$$

where Ω is the atomic volume. Then J is again expressed in terms of (surface × time)$^{-1}$.

N.B. 3   In the stationary state $(\partial c/\partial t = 0)$, the concentration of vacancies is given by equation

$$\Delta c = 0$$

which may be written in cylindrical coordinates as:

$$\frac{\partial^2 c}{\partial r^2} + \frac{1}{r}\frac{\partial c}{\partial r} = 0$$

$$(11)$$

and in spherical coordinates as:

$$\frac{\partial^2 c}{\partial r^2} + \frac{2}{r}\frac{\partial c}{\partial r} = 0.$$

$$(12)$$

N.B. 4   The reader will have recognised (10) as the *heat equation*, with D_v replacing the thermal diffusivity Λ and c replacing the temperature T.

N.B. 5   The minus sign in equations (8) and (9) is a reminder that, of course, vacancies diffuse globally in the direction opposite to the concentration gradient. Although the migration is random, everything takes place as though the vacancies were subjected to a force proportional to $-\bar{\nabla}c$.

1.2 Diffusion of vacancies subjected to a force

If the vacancies are subjected to a force $\bar{F}$, their migration is no longer random. The origin of this force may be *electrostatic* (if the vacancy is a charged defect), *elastic* (migration in a strain field) etc.

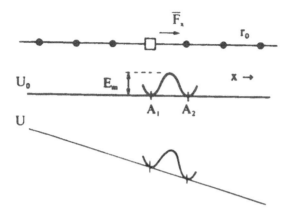

Figure 2 Energy profile for the jump of a vacancy from A_1 to A_2 in the absence (U_0) and then in the presence (U) of an external force $\overline{F}$ with projection F_x in the x direction.

Let us consider the movement of a vacancy in the x direction, where we suppose that the distance of the jump (projected) is r_0 and that the energy profile (see Figure 9(b), p. 165) is U_0 for $F = 0$ (Figure 2). The activation energy, which is normally independent of the direction of the jump is now, when $F \neq 0$, slightly different, depending on whether the jump is taken to the right or to the left. More precisely, returning to the geometry of Figure 1 and letting ν^+ denote the frequency of jumps in the direction of $\overline{F}$ and ν^- the frequency of jumps in the opposite direction, we have

$$\nu^+ = 3\nu \exp{-\left(E_m - F_x\frac{r_0}{2}\right)}\Big/ kT$$
$$\nu^- = 3\nu \exp{-\left(E_m + F_x\frac{r_0}{2}\right)}\Big/ kT$$

This leads to a velocity $\overline{v}_x$ of the vacancy in the direction of $\overline{F}_x$ equal to

$$v_x = 6\nu r_0 \sinh\frac{F_x r_0}{2kT} \exp\left(-\frac{E_m}{kT}\right).$$

In the limit of forces small in comparison with kT/r_0, we have

$$v_x \simeq 3\nu r_0^2 \frac{F_x}{kT} \exp\left(-\frac{E_m}{kT}\right)$$

or

$$v_x = \frac{D_v F_x}{kT} \rightarrow \boxed{\overline{v} = \frac{D_v \overline{F}}{kT}} \tag{13}$$

(Einstein's formula), where the force $\overline{F}$ gives the defect a **drift velocity** $\overline{v}$ parallel to itself and proportional to the diffusion coefficient.

N.B Despite the presence of kT in the denominator, this velocity *increases* with temperature, thanks to the exponential in the expression for D_v.

1.3 Vacancy-type self-diffusion

We shall now describe **self-diffusion** such as that, in a solid, which results from the diffusion of vacancies.

For an A atom of the solid $\mathcal{A}$ to jump in the crystal, it must be a near neighbour of a vacancy *and* an exchange (vacancy–A) must take place. The frequency of jumps of A is then the frequency of jumps of the vacancy *multiplied* by the probability that there exists a vacancy on a site adjacent to A. This probability is just the atomic equilibrium concentration c (see (6), p. 162). Thus, the A atom diffuses with a random motion defined by a diffusion coefficient D_A such that:

$$D_A = c D_v \tag{14}$$

$$\boxed{D_A = \nu a^2 \exp\left(-\frac{E_f + E_m}{kT}\right).} \tag{15}$$

Everything that was said about vacancies in Sections 1.1 and 1.2 may now be transposed to atoms. Previously, the object viewed as diffusing randomly was the *vacancy*; now, it is the A *atom*. Thus, in order to describe self-diffusion it is sufficient to *replace* D_v by D_A. In particular, Fick's equations (9) and (10) may now be written as:

$$\boxed{\overline{J}_A = -D_A \overline{\nabla} c_A \text{ and } \frac{\partial c_A}{\partial t} = D_A \Delta c_A} \tag{15'}$$

where c_A denotes the volume concentration of a discernible collection of atoms of A (for example, a certain isotope) in $\mathcal{A}$.

N.B. 1 Measurements of the self-diffusion coefficients (we shall see how these measurements are made later) as a function of the temperature generally show that the results obey the empirical law

$$\boxed{D_A = D_0 \exp\left(-\frac{E_A}{kT}\right).} \tag{16}$$

The energy E_A, which is determined by differentiating the values of $\ln D_A$ with respect to T^{-1} (*Arrhenius's diagram*):

$$E_A = -k \frac{\mathrm{d}\ln D_A}{\mathrm{d}T^{-1}}$$

is called the *self-diffusion energy*. Comparing (15) and (16), we obtain the identity

$$E_A = E_f + E_m. \tag{16'}$$

Clearly, one decisive test of the validity of the above model (of vacancy-type self-diffusion) involves a comparison of the experimental values of E_A with the sum of *independently* determined values of E_f and E_m (see later, p. 181 and Table 2 of p. 165).

N.B. 2 Formula (15), like (7) is valid for a cubic crystal. For other symmetries, D_A should be multiplied by a numerical factor, which does not alter the present description of the phenomenon.

1.4 Measurement of self-diffusion

The most direct measurements are those of the diffusion of (discernible) *radioactive* isotopes A^* of A in $\mathcal{A}$. An ideal experiment involves preparing two cylindrical bars of $\mathcal{A}$ with axis $x'x$, evaporating a thin uniform deposit of A^* on the (well-polished) surface $x = 0$ of one of these and resoldering the two bars along this surface so that the deposit of A^* is buried on the plane $x = 0$. The sample is then heated at a temperature T for a time t. After this, the concentration $c_A^*(x)$ of A^* along the axis $x'x$ (one-dimensional geometry) is determined[3] using a device to measure radioactivity.

If the initial thickness ε^* of the deposit is small in comparison with the distances involved, the solution of (15') which satisfies the condition of conservation of the isotope A^* (quantity m_0^*):

$$\int_{-\infty}^{\infty} c_A^*(x, t)\, dx = m_0^* \tag{17}$$

is

$$\boxed{c_A^*(x, t) = \frac{m_0^*}{2\sqrt{\pi D_A^* t}} \exp(-x^2/4D_A^* t).} \tag{18}$$

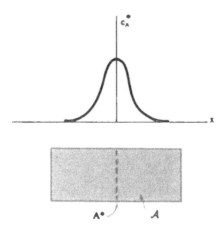

Figure 3 A self-diffusion experiment. The isotope A^* deposited initially on the plane $x = 0$ of the bar $\mathcal{A}$ diffuses at temperature T. After time t, its concentration c_A^* is Gaussian (see (18)).

N.B. 1   A total of $n_0 = \varepsilon^*/\Omega$ atoms of A^* were deposited per unit section of the bar, where Ω denotes the atomic volume of A (and of A^*). If c denotes the *atomic concentration* (c_A^* = number of atoms of A^*/total number of atoms in a given volume), then clearly, c_A^* is simply equal to Ωdn^*, where dn^* is the number of atoms of A contained in the section dx. The integral (17) is then $\Omega \int dn^* = \varepsilon^*$. The quantity m_0^* of (18) is then simply equal to the thickness of the deposit ε^*.

---

[3] In fact, it is normally sufficient to lay the deposit *on the surface* ($x = 0$) and study the diffusion in the bar ($x > 0$, semi-infinite geometry). The solution of the diffusion equation is then analogous to (18), without the factor 2 in the denominator.

N.B. 2   The major characteristics of the solution (17) are shown in Figure 4.

Figure 4(a) shows the Gaussian distribution of $c_A^* = c_A^*(x)$ for a given time t. This Gaussian distribution 'spreads' with time, although the area beneath it (see (17)) remains constant. The quantity $c_A^*(0)$ decreases like $1/\sqrt{t}$ while the half width of the distribution at the point of inflection of the Gaussian distribution ($x_i = \sqrt{2D_A^* t}$) increases like $\sqrt{t}$. Figure 4(b) shows the gradient dc_A^*/dx; this quantity is proportional to the *flow of matter* across a plane $x = $ constant (see (15')). This flux is zero for $x = 0$ and for very large $|x|$. Lastly, Figure 4(c) is a graph of the quantity $d^2c_A^*/dx^2$, which is proportional to the *rate of accumulation* of matter at a given point. The central region with a *negative curvature of $c_A^*(x)$* loses matter which is gained by the two regions with positive curvature.

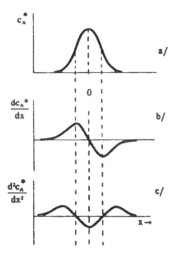

Figure 4   Self-diffusion concentration c_A^* and its first two derivatives: $(b) - (dc_A^*)/dx$ proportional to the flux of atoms; $(c) - (d^2c_A^*)/dx^2$ proportional to the rate of accumulation $\partial c_A^*/\partial t$ (see (15')).

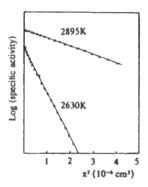

Figure 5   Self-diffusion of tungsten. Specific activity (proportional to c_A^*) in arbitrary units as a function of x^2 ($x = $ penetration distance) for two different temperatures (2895 and 2630 K) and identical treatment times ($t = 9.2 \times 10^4$ s).

The isotopes used were ^{187}W (γ emitter) for $T = 2895$ K and ^{185}W (β emitter) for $T = 2630$ K.
After **J N Mundy** *et al. Phys. Rev.* **18**, 6566 (1978).

Figure 5 illustrates the results obtained by this method for a tungsten crystal treated at high temperature. The graph of $\ln c_A^*$ (where c_A^* was obtained by measuring radioactivity) against x^2 is a straight line with gradient $-\frac{1}{4}D_A t$ (see (18)) and may be used to determine D_A. The variation of D_A with T is determined by a series of analogous measurements at various temperatures. We note that, very generally, D_A satisfies an Arrhenius equation (16), from which D_0 and E_A may be deduced.

Table 1 shows the self-diffusion parameters D_0 and E_A for a number of solids.

Table 1   **Self-diffusion parameters D_0 and E_A for a number of solids.**

	Na	Cu	Ag	Au	Al	Pt	Mo	Ge	Si
D_0, cm^2 s^{-1}	0.24	0.20	0.40	0.10	0.18	0.22	0.5	7.8	32
E_A, eV	0.46	2.05	1.92	1.81	1.30	2.89	4.22	3.0	4.26

If one refers to Table 2 of p. 165, which lists the measured values of E_f and E_m, it is clear that equation (16'), the test of the vacancy mechanism, is palpitably true.

N.B.   The diffusion coefficient D_A^* measured in this way relates to the diffusion of A* in $\mathcal{A}$. It is slightly different from the coefficient D_A of A in $\mathcal{A}$. There are two reasons for this.

First, if m and m^* are the atomic masses of the two isotopes A and A*, then the vibration frequencies (ν in (15)) vary as $m^{-1/2}$ (see, for example, **Problem 18**) and thus we have $D_A^* = (m/m^*)^{1/2} D_A$. The corresponding correction is generally minor.

More fundamentally, there is a difference between the self-diffusion of A in $\mathcal{A}$ and the self-diffusion of the tracer A*. The latter is *traceable* and retains its identity after a jump. This is not the case for an A atom, which is indistinguishable from other A atoms. The theory of random walks, which is rigorous for A is not quite rigorous for A*, since immediately after a jump the latter has a probability of an *inverse jump* greater than that of other jumps. For *this particular atom*, which we trace, the jumps are not totally random, but are in fact **correlated**. An approximation to the *correlation coefficient* f may be obtained from the observation that the correlation (or the inverse jump) has probability close to $1/Z$ (exactly $1/Z$ if one considers only the immediate inverse jump, without passing to the circlet of second neighbours) and results in the annihilation of the effect on the diffusion of *two* jumps of the atom. Thus, we have

$$f \simeq 1 - \frac{2}{Z}$$

or 0.83 for an fcc crystal ($Z = 12$). A complete calculation gives $f = 0.78$.

Therefore, overall, we pass from the measured coefficient D_A^* to the self-diffusion coefficient D_A via the equation

$$D_A = \frac{1}{f}\sqrt{\frac{m^*}{m}}\, D_A^*.$$

2. HETERODIFFUSION

By *heterodiffusion* we mean the diffusion of a B atom in a solid $\mathcal{A}$. This is the case of the example of the diffusion of Cu in Al given at the beginning of this chapter.

Limiting ourselves in what follows to the case of concentrations of B in A which are sufficiently weak that B–B encounters may be neglected (*dilute solid solutions*, B is often called an **impurity**), we shall highlight two different mechanisms.

2.1 The vacancy mechanism

This is based on the same idea as the self-diffusion mechanism of Section 1. If B is *substituted* in the crystal A and it is a near neighbour of a vacancy, it may interchange with the latter and thus *diffuse*.

However, unlike in the case of self-diffusion, we must now take into account the possible **interaction of the vacancy with B**. We recall (p. 164) that, in the case of a metal, a vacancy is surrounded by oscillations of electron charge which may, depending on their position (i.e. on the Fermi wavevector k_F of A), *attract* or *repel* a B atom with valency different from that of A. We also recall (p. 167) that, in an *ionic crystal*, a cation with charge greater than that of the cations of A (for example, Ca^{++} in NaCl) tends to attach itself to a cationic vacancy which restores the neutrality. In these two examples, the immediate neighbourhood of the impurity B contains either *more* or *less* vacancies than the equilibrium concentration c of vacancies in A; thus, we can no longer follow the reasoning of the beginning of Section 1.3. Suppose that v is a vacancy adjacent to B and that A denotes the atoms of A (for example, fcc structure) adjacent to v (Figure 6). We must replace the single jump frequency $w_A = v \exp -(E_m/kT)$ used for self-diffusion by the frequencies w_1 (jump which displaces B), w_2 (jump which causes the vacancy to turn around B), w_3 (which dissociates the pair l–B) and, of course, w_A (migration of the vacancy away from B). These frequencies are not equal to one another and many cases are possible.

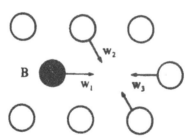

Figure 6   Jump frequencies of a vacancy adjacent to an atom of an impurity (shaded).

Only one is ideally simple.   We denote the energy of a v–B interaction by E_i ($E_i < 0$ means that the vacancy is *bonded* to B). The formation energy of a vacancy *on the Z sites adjacent to B* is $E_f + E_i$ and the concentration of vacancies there is

$$c_i = \exp\left[-(E_f + E_i)/kT\right].$$

Suppose also that B perturbs the lattice of A to a sufficiently small extent that the *average vibration frequency* v and the *migration energy* (E_m) are unchanged both for the A atoms and for the B atoms: $w_1 = w_2 = w_3 = w_A$. Then the diffusion of B in A is defined by a diffusion coefficient D_B obtained by replacing c by c_i (or E_f by $E_f + E_i$) in (14):

$$D_B = va^2 \exp\left[-(E_A + E_i)/kT\right]. \tag{19}$$

In this case, the *diffusion energy* is the *self-diffusion energy* E_A plus the *interaction energy*. In the frequent cases in which the impurity B and the vacancy *attract one another* ($E_i < 0$), we have $D_B > D_A$ and B *diffuses in A more rapidly than* A.

N.B. In Sections 1.3 and 2.1, we assumed explicitly that the vacancies in a crystal were in thermodynamic equilibrium. This is not always the case. Earlier (p. 165), we saw that large supersaturations of vacancies may exist in a crystal, in particular, under **irradiation** or after **quenching**. In the first case, *steady-state* vacancies are created and enhanced diffusion and self-diffusion take place. Thus, in the expression for D_A (14), the concentration c may be a factor of 10^4 or 10^5 larger than the equilibrium concentration at the temperature in question (case of materials subject to irradiation by neutrons in *nuclear reactors*, see **Problem 17**). In the second case, immediately after the quenching, at the low temperature T_0, the crystal contains a stock of vacancies of *supersaturated* concentration $c(T)$ (T is the temperature before quenching). These vacancies accelerate the diffusion at temperature T_0, however, this acceleration is attenuated as they disappear in the available *sinks*.

This is why, in the example at the beginning of the chapter, the copper atoms succeed in diffusing at room temperature and regrouping in GP clusters in *quenched* aluminium. In fact a vacancy in a metal is surrounded by an oscillation of charge (whence also of potential). In *aluminium* (trivalent), k_F is sufficiently large (see point ii) of p. 164) that on the first-neighbour site of a vacancy, this oscillation *inverts* the sign of the potential created by the latter, making the normally *negative* potential produced by the vacancy *positive*. A copper atom (valency one) has a *negative* effective charge in aluminium. In a near-neighbour position for a vacancy, this atom is thus bonded to the vacancy by a **negative** *interaction energy* E_i (J Friedel and A Blandin). Consequently, the quenched-in vacancies together with the copper atoms form *stable* v–Cu complexes (or pseudo molecules), which diffuse easily (see Figure 7) at room temperature T_0, resulting in the diffusion of copper with diffusion coefficient D_{Cu} approximately *a factor of* 10^7 *larger* than the normal coefficient $D_{Cu}(T_0)$ (19) (in other words, that measured in *non-quenched* aluminium).

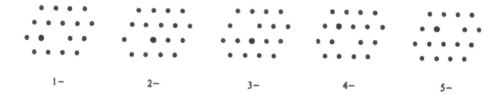

1-   2-   3-   4-   5-

Figure 7 Diffusion of a vacancy–copper 'molecule' in aluminium. Small circles: aluminium atoms (plane {111}). Large circles: copper atom. The jumps $1 \rightarrow 2$ and $3 \rightarrow 4$ are v–Cu exchanges (w_1 on Figure 6). The jumps $2 \rightarrow 3$ and $4 \rightarrow 5$ are exchanges v–Al (w_2 on Figure 6). Overall, the v–Cu molecule progresses by *one* interatomic distance between 1 and 5.

2.2 The interstitial mechanism

If the B atom is normally in an interstitial position in the crystal A, it diffuses by jumps from its interstitial site to other interstitial sites, *without the intervention of crystalline defects*. This is true in *iron* for *carbon*[4] which occupies the octahedral positions (centres of faces and centres of edges, both crystallographically equivalent) of the bcc cell of

---

[4] This example is particularly important in practice since, not taking into account various extra additive elements, Fe–C solid solutions (in atomic concentrations of the order of 10^{-2}) constitute the dominant phase of steels.

this metal. The distance between two near-neighbour sites (z) and (x) (see Figure 8) is $r_0 = a/2$ (a = lattice parameter). During the movement from (z) to (x), the energy passes through a saddle point of height E_{mi}, the *migration energy of the interstitial* (see **Problem 15**).

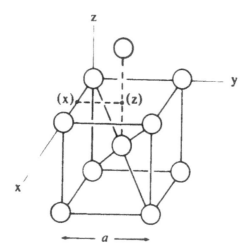

Figure 8 Octahedral positions (x) and (z) of an interstitial carbon atom in a bcc crystal (for example, iron). The interstitial atom diffuses by jumps of type $(z) \rightarrow (x)$.

The site (z) gets its name from the fact that two nearest iron atoms are found in the z direction on either side of this site at distance $a/2$; thus, locally, the carbon atom produces a tetragonal distortion of the main axis z of the iron crystal. When the carbon jumps to the site (x), the distortion switches to the x direction.

Denoting the average frequency of vibration of the interstitial atom (for which there are four possible jumps) by ν_i, the diffusion coefficient for carbon in iron may be written (using (4)) in the form:

$$D_i = \frac{1}{6}\nu_i a^2 \exp(-E_{mi}/kT). \qquad (20)$$

By way of example, the energy E_{mi} measured for the diffusion of carbon in iron is 0.83 eV, where the term before the exponential is equal to 1.5×10^{-2} cm^2 s^{-1}.

2.3 Diffusion equations

Earlier, in (9) and (10), we established the two diffusion (or Fick's) equations for the case where the migrating object was a vacancy. When the migrating object in which we are interested is the B atom *highly diluted* in the solid A, exactly identical reasoning leads to two equations identical to (9) and (10):

$$\boxed{\vec{J} = -D\vec{\nabla}c} \qquad (21)$$

$$\boxed{\frac{\partial c}{\partial t} = D\Delta c} \qquad (22)$$

which may be used to calculate the flux $\overline{J}$ of B as a function of the *gradient* of the volume concentration c (of B) and the variation of c with time (*accumulation* or *loss*) *as a function of the Laplacian of c. D* is the *diffusion coefficient* of B in $\mathcal{A}$, which is assumed (in (22)) not to vary with c or r. If this is not the case, (22) may be generalised as in **N.B. 1** of p. 176.

N.B. 1 Similarly, the elementary results about random motion described in Section 1.1 may be directly transposed here, and, in particular, the fact that *the average distance travelled during time t, at a given temperature, by a B atom varies as* $\sqrt{Dt}$.

N.B. 2 The equations (9) and (10), whence also (21) and (22), were obtained for the special case in which the elementary jumps involve *identical steps* in *discrete directions* (edges of crystalline cells). Part I of **Problem 13** contains a generalisation to the case of objects (for example, *electrons* in a semiconductor, *neutrons* in a nuclear reactor etc) with arbitrary directions of movement. If the *velocity* of these objects is *constant* (v) and they are subject to *collisions* (relaxation time: τ), equations (21) and (22) remain valid with the value for D:

$$D = \frac{v^2\tau}{3}. \tag{22'}$$

If, in addition to the 'Brownian' random motion, the objects are subjected to a **force** $\overline{F}$ which gives them a *drift velocity* $\overline{V}$, equations (21) and (22) become

$$\overline{J} = -D\overline{\nabla}c + \overline{V}c$$
$$\frac{\partial c}{\partial t} = D\Delta c - \overline{V} \cdot \overline{\nabla}c$$

where $\overline{V}$ and $\overline{F}$ are related (as in (13)) by $\overline{V} = D\overline{F}/kT$. Examples of diffusion in the presence of a force are given in **Problem 13** (force of elastic origin) and **Problem 14** (force of electrostatic origin).

N.B. 3 The mechanisms for diffusion by **thermal activation** discussed in earlier paragraphs are not the only ones known. **Problem 12** gives an example of (surface) diffusion by the **tunnel effect**.

N.B. 4 If the assumption of *large dilution*, whence of low gradients $\overline{\nabla}c$, is not true and one is studying the diffusion of a concentrated alloy A–B, equations (21) and (22) must be modified. In fact, the fluxes J generally 'descend' the gradient of the *chemical potentials* μ and not necessarily that of the *concentrations* c. The return to equilibrium, of which diffusion is a manifestation, gives rise to fluxes, which, at first order and in one dimension, may be written in the form:

$$J = -M\frac{\partial}{\partial x}\mu \qquad M > 0.$$

Exercise Show that the mobility coefficient M and the diffusion coefficient D of (21) are related by $D = M \, \partial^2 \mathcal{G}/\partial c^2$ where $\mathcal{G}(c)$ is the free enthalpy of the alloy.

One interesting instance of this situation is that of a concentrated solid solution such that $\partial^2 \mathcal{G}/\partial c^2$ is negative (see below, Figure 9(d)). The reader should convince himself that, in this case, if there exists a gradient of c, the diffusion tends to *accentuate* it (and not *soften* it as equation (21) would lead one to imagine). This **transition** is said to be **spinodal**.

3. PRECIPITATION

It is plain that atomic diffusion, in particular that of a species B in a matrix $\mathcal{A}$, permits, or at least favours, *structural modifications* which give a decrease in the free enthalpy. Thus, in a material, diffusion favours the **approach to equilibrium.**

In what follows, we describe some of the mechanisms which may, in particular, give rise to *phase precipitation*. We shall consider only the case of solids with two constituents A and B. First, we have to determine the **nature of the equilibrium phases** in such materials.

3.1 Equilibrium of an alloy with two constituents

Let us consider the very simple case of a **solid solution** A–B comprising N atoms including Nc atoms of B and $N(1-c)$ atoms of A, which is assumed to be a *substitutional solution* in which every site of a *unique crystalline lattice*[5] with coordination number Z ($Z = 12$ for an fcc structure) is occupied **at random** *either* by an A atom *or* by a B atom. We shall determine the free enthalpy $G = H - TS$ of this alloy as a function of the *temperature* T and the *concentration* c.

1 – Calculation of H

Let us denote the enthalpies of the bonds A–A, A–B and B–B at 0 K by H_{AA}, H_{AB} and H_{BB}, respectively, and restrict the expansion of the crystal enthalpy to these near-neighbour terms. If the solid is completely randomised it includes

$$
\begin{aligned}
P_{AA} &= \tfrac{1}{2}N(1-c)Z(1-c) \\
       &= \tfrac{1}{2}NZ(1-c^2) \qquad \text{pairs A–A} \\
P_{AB} &= NcZ(1-c) \\
       &= NZc(1-c) \qquad \text{pairs A–B} \\
P_{BB} &= \tfrac{1}{2}NcZc \\
       &= \tfrac{1}{2}NZc^2 \qquad \text{pairs B–B.}
\end{aligned}
$$

The enthalpy H_0 at 0 K is then

$$
H_0 = P_{AA}H_{AA} + P_{AB}H_{AB} + P_{BB}H_{BB}
$$

or

$$
\boxed{H_0 = \tfrac{1}{2}NZ((1-c)H_{AA} + cH_{BB} + 2c(1-c)\varepsilon)} \tag{23}
$$

where we have set

$$
\boxed{\varepsilon = H_{AB} - \tfrac{1}{2}(H_{AA} + H_{BB}).} \tag{24}
$$

The expression (23) shows that the enthalpy H_0 is the sum of the enthalpy H_A of a crystal $\mathcal{A}$ containing the $(1-c)N$ A atoms, the enthalpy H_B of a crystal $\mathcal{B}$ containing the cN B atoms, and an *extra enthalpy* **with the same sign as** ε. If this sign is positive (*resp.* negative), in other words, if the attraction between dissimilar atoms is less (*resp.* larger) than that between similar atoms, then the enthalpy of the random solid solution is larger (*resp.* less) than that of two separate pure crystals $\mathcal{A}$ and $\mathcal{B}$.

---

[5] This implies that the pure solids $\mathcal{A}$ and $\mathcal{B}$ are isomorphic, in other words, they have the same crystal structure.

If $\varepsilon = 0$, the **solid solution** is said to be **ideal**. Its entropy varies linearly from the enthalpy of pure $\mathcal{A}$ to that of pure $\mathcal{B}$.

Thus, at temperature 0 K, at which the enthalpy H and the free enthalpy G are equal, the problem is solved, although the solution is trivial. Completely predictably, *total miscibility* occurs if the formation of unlike pairs is favourable and *non-miscibility* (two distinct pure phases $\mathcal{A}$ and $\mathcal{B}$) otherwise.

When T is non-zero, the enthalpy is

$$H = H_0 + \int_0^T C_p \, dT$$

where C_p is the specific heat of the solid at constant pressure.

2 – Calculation of S

To the expression for the configurational entropy:

$$\boxed{S_m = -kN(c \ln c + (1 - c) \ln(1 - c))} \qquad (25)$$

we must add the term

$$S_{th} = \int_0^T \frac{C_p}{T} \, dT.$$

We shall neglect all other entropy terms.

3 – Variation of the free enthalpy with the concentration

The free enthalpy of the random solid solution is:

$$G = H_0 - TS_m + \int_0^T C_p \, dT - T \int_0^T \frac{C_p}{T} \, dT.$$

Insofar as, for a given T, we are interested in the *variation of G with c*, it is reasonable to neglect the variations of C_p (in other words, principally those due to thermal agitation) with c, since they are slow, particularly if the atomic masses of A and B are not too different. The free enthalpy then varies like the function:

$$\boxed{\mathcal{G} = H_0 - TS_m} \qquad (25')$$

where H_0 and S_m are given by (23) and (25).

Recalling the form of the function S_m (and, in particular, its rapid variation for c near to 0 or 1; see Figure 7, p. 162) and observing that the variation of H_0 is *parabolic*, it is easy to illustrate the different possible forms of $\mathcal{G}(c)$ graphically.

In Figure 9, the origin of the energies was chosen arbitrarily to correspond to a pure $\mathcal{A}$ crystal ($H_{AA} = 0$) and it is assumed that $H_{BB} > 0$. The free enthalpy $\mathcal{G}'$ of a mixture of the two phases $\mathcal{A}$ and B (assumed pure), which is equal to H_{AA} for $c = 0$ and H_{BB} for $c = 1$, varies linearly with the concentration[6]. Its variation is represented by the straight line ab (Figure 9(a)). The interpretation of the cases (a) ($\varepsilon < 0$) and (b) ($\varepsilon = 0$) is then immediate: $\mathcal{G}$ is always less than $\mathcal{G}'$ and the random solid solution is always more stable

---

[6] At least if one neglects the energy of the $\mathcal{A}$–B interface.

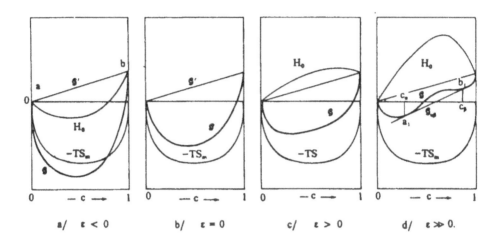

Figure 9 Variation of the function $\mathcal{G}$ of (25′) with the concentration in an alloy A–B in random solid solution, for four different values of the term ε. H_0 is the parabolic function (23). S_m is given by (25).

than a mixture of the two phases $\mathcal{A}$ and $\mathcal{B}$. This is known as **total miscibility** in the solid state.

In case (*c*), where the term ε is just positive, the enthalpy H_0 is slightly larger than $\mathcal{G}'$. At $T = 0$ K, desorption into two phases $\mathcal{A}$ and $\mathcal{B}$ would be favourable, but at temperature T, the gain in entropy is sufficiently great to impose stability of the solid solution throughout the domain of concentration. We again have *total miscibility*.

As ε grows and becomes highly positive ($\varepsilon \gg 0$, Figure 9(*d*)), the A–B bonds become more and more expensive in comparison with A–A or B–B bonds. $\mathcal{G}$ continues to decrease for *small* (but increasing) concentrations c of A, or $(1-c)$ of B. However, for *medium concentrations, $\mathcal{G}$ exhibits a negative curvature*. What is the most stable structure for the alloy in this case?

The answer to this question may be found by drawing the common tangent $a_1 b_1$ to the two portions of $\mathcal{G}$ with a positive curvature. Let c_α and c_β be the abscissae of the two points a_1 and b_1. It is immediately apparent from Figure 9(*d*) that:

i)   For $0 < c < c_\alpha$ the stable phase is a (random) *solid solution of* B *in* $\mathcal{A}$. This isomorphic phase of $\mathcal{A}$ will be called, for example, the α *phase*.

ii)   Similarly, for $c_\beta < c < 1$, the stable phase is the β *phase*: a (random) solid solution of A in $\mathcal{B}$.

iii) On the other hand, for $c_\alpha < c < c_\beta$, it is advantageous to mix the α and β phases (with concentrations c_α and c_β, respectively). In fact, the free enthalpy $\mathcal{G}_{\alpha\beta}$ of this **mixture of two phases** α and β varies linearly from $\mathcal{G}(c_\alpha)$, the free enthalpy of the α *phase saturated in* B, to $\mathcal{G}(c_\beta)$, the free enthalpy of the β *phase saturated in* A. The variation of $\mathcal{G}_{\alpha\beta}$ is represented by the straight line $a_1 b_1$ which lies *below* the curve $\mathcal{G}(c)$: the mixture of the two phases is more stable than the random solution. Figure 10 shows an example of such a two-phase alloy.

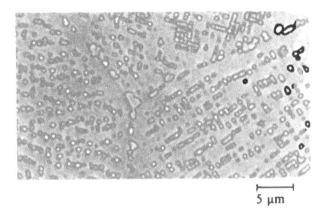

5 µm

Figure 10 Example of a two-phase alloy. Optical micrography of the polished surface of an alloy Ni–7%Si. Two phases occur. The first, dominant, grey phase is a solid solution of silicon in nickel. The other, white, phase consists of precipitates of Ni₃Si composition.

The alloy was treated for 64 hours at 850° C. There are three clearly visible grains of the alloy, resulting in three different orientations of the precipitates and their epitaxial relation to the nickel matrix. A number of large precipitates mark the grain boundaries, together with the triple point.

After **J Manenc**.

Let us now justify the linear variation of $\mathcal{G}_{\alpha\beta}$ with c. Let c ($c_\alpha < c < c_\beta$) be the global concentration of an alloy formed from the mixture in the proportions x and $y = 1 - x$ of the α and β phases, respectively. The alloy consists of Nc B atoms and $N(1 - c)$ A atoms, distributed amongst the α and β phases. The α phase has xNc_α B atoms and the β phase has yNc_β. Thus, we may write

$$Nc = \text{(atoms of B in } \alpha) + \text{(atoms of B in } \beta)$$
$$= xNc_\alpha + yNc_\beta;$$

whence, it follows that

$$x = \frac{c_\beta - c}{c_\beta - c_\alpha} \quad \text{and} \quad y = \frac{c_\alpha - c}{c_\beta - c_\alpha}. \tag{26}$$

Of course, the fact that x is linearly dependent on c implies that $\mathcal{G}_{\alpha\beta}$ varies linearly with c.

The expression (26), which for obvious reasons is called the **rule of the lever arm** may be used to determine the quantities (or volumes) of the α and β phases at equilibrium in the alloy of concentration c.

To summarise this discussion (corresponding to the case $\varepsilon \gg 0$), we see that, at temperature T, when one progressively adds atoms of B to an initially pure matrix $\mathcal{A}$, one *always* begins by forming a solid solution of B in $\mathcal{A}$ (called the α phase here). Above a certain limiting concentration c_α (**solubility limit**) the A–B pairs become too numerous and a second phase, the β phase, consisting of A atoms in solution in $\mathcal{B}$ must *precipitate*, where the proportion of these two phases is given by the rule of the lever arm (26) (Figure 10).

4 – Free energy of the liquid phase

Here, we simply note that a close analysis of the above may be carried out for the *liquid* A–B alloy, and that, since the term $-T S_m$ is dominant, we have an instance of *total miscibility* (the most common case) which gives the curve $\mathcal{G}_l$ (term $\mathcal{G}$ for the free enthalpy of the liquid phase) a form similar to that of $\mathcal{G}$ on Figures 9(a), (b) or (c).

5 – Equilibrium diagrams

It is now easy to determine the phases in equilibrium at a given pressure for a point of the space (c, T), in other words, to construct the **equilibrium diagram** of the A–B alloy.

We shall do this, by way of example, for the case of *partial miscibility in the solid state*, where $\varepsilon \gg 0$ (Figure 9(d)), by examining the evolution of the curves $\mathcal{G}_l(c)$ and $\mathcal{G}(c)$ *with respect to one another* as a function of the temperature. Suppose we have a temperature T greater than that of the melting point T_{mB} of pure B. At the point with abscissa $c = 1$ (pure B), the curve $\mathcal{G}_l(1)$ lies *below* the curve $\mathcal{G}(1)$. More precisely, we have:

$$\mathcal{G}(1) - \mathcal{G}_l(1) = -\Delta G_{mB}$$

where

$$\Delta G_{mB} = \Delta H_{mB} - T \Delta S_{mB}.$$

At the melting point $T = T_{mB}$, we have:

$$\Delta G_{mB} = \Delta H_{mB} - T_{mB} \Delta S_{mB} = 0.$$

If we neglect the (slight) dependence of ΔH_m and ΔS_m on temperature, we may write:

$$\mathcal{G}(1) - \mathcal{G}_l(1) = \Delta H_{mB} \left(\frac{T}{T_{mB}} - 1 \right). \tag{27}$$

This expression, or the equivalent expression for $\mathcal{G}(0) - \mathcal{G}_l(0)$, shows that a decrease in T from a high temperature at which any alloy is liquid to a low temperature at which any alloy is solid corresponds to a relative shift in the curves $\mathcal{G}_l(c)$ and $\mathcal{G}(c)$ moving the first upwards with respect to the second. This shift appears on each of the five sequences of Figure 11, which each (in other words for each temperature $T_1, \ldots, T_5$ considered) enable us to determine the **phases in equilibrium**. Thus, for $T = T_3$, Figure 11(c) shows the succession of equilibria as c increases from 0 to 1: first a *solid* α phase (solid solution of B in $\mathcal{A}$); then a mixture of the α phase (with composition c_α) and liquid (with composition c_a; then *liquid*); then a *mixture* of the β phase (with composition c_β) and liquid (with composition c_b); and finally a *solid* β phase (solid solution of A in B). At temperature T_4 *three* phases are in equilibrium, namely the α phase (with composition c'_α), the β phase (with composition c'_β) and the liquid with composition c_E.

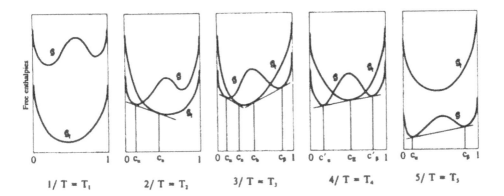

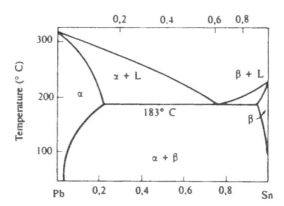

Figure 11   Relative positions of the free enthalpy curves of the solid ($\mathcal{G}$) and the liquid ($\mathcal{G}_l$) for a binary alloy with total miscibility in the liquid phase and partial miscibility in the solid phase.

The temperatures $T_1, \ldots, T_5$ are in decreasing order. They correspond, for example, for the alloy Pb–Sn (Figure 12) to the following sequence: $T_1 \simeq 400°$ C, $T_2 \simeq 300°$ C, $T_3 \simeq 200°$ C, $T_4 = 183°$ C (*eutectic temperature*) and $T_5 \simeq 100°$ C.

These five cases are observed in the lead–tin equilibrium diagram in Figure 12. This is a **eutectic diagram**, the temperature T_4 being the *eutectic temperature*. The point (c_E, T_4) is the eutectic point. The eutectic concentration c_E is the *only* concentration which permits direct melting of the solid without coexistence of liquid and solid.

Figure 12   **Equilibrium diagram of the lead–tin alloys.** The lower axis shows the atomic concentrations of tin. The upper axis shows the concentration by weight.

Exercise Construct a sequence $T_1, \ldots, T_5$ analogous to that of Figure 11, but where the shapes of $\mathcal{G}_l$ and $\dot{\mathcal{G}}$ are such that, at the temperature T_4, the point of common tangency on $\mathcal{G}_l$ is *outside* the segment $c'_\alpha c'_\beta$ (rather than inside it as in Figure 11(d)). Draw the corresponding equilibrium diagram, and recognise it as a **peritectic diagram**.

Construct an analogous sequence for the simple case in which the curvature of G is always positive (total miscibility in the solid state). This gives a **diagram of total miscibility**, similar to the example of germanium–silicon alloys given in Figure 13.

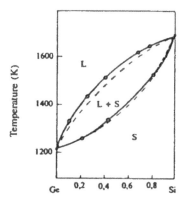

Figure 13 **Equilibrium diagram of the germanium–silicon alloys.** Continuous line: experimental diagram. Dashed line: calculated diagram.

The diagrams which we have just formed while attempting to minimise the free enthalpy describe the *equilibrium* state. This equilibrium state is by no means always attained. Consider, for example, a Pb–Sn alloy (Figure 12) with composition (of Sn) $c = 0.1$ (atom.). Suppose this is heated to 200° C. The stable state is clearly an α phase, random solid solution of Sn in Pb. Let us now cool this alloy rapidly (by quenching in water) to 0° C. During this fast cooling, it is highly likely that the alloy will stay in the initial state (α phase), although the equilibrium state is now a mixture of the α phase (dominant, rule of lever arm) and the β phase. At 0° C, the alloy is in a **metastable** state and will tend to *decompose* with time by **precipitation** of β phase islets. This precipitation, which represents a local regrouping of B atoms (here, of tin) necessitates a diffusion of these atoms.

In what follows, we describe one of the most common processes of precipitation, based on diffusion, namely the **nucleation and growth** mechanism.

N.B. The case discussed in the next two sections is the particularly important case of the precipitation of a β phase in an *alloy*. However, the phenomenon of precipitation is more general and may involve the occurrence of a *crystallographically* (as opposed to *chemically*) new phase.

It is easy to convince oneself of this by looking at glasses (for example, in a canteen) which have become 'milky' after a certain period of use. Glass, which is an *amorphous* material, initially contains *crystalline* nuclei which are too small to scatter light. The high temperature of dishwashers leads to a progressive increase in this crystalline phase which is slightly more stable than the amorphous phase. When these crystalline 'granules' reach a sufficient size they scatter light and make the glass opaque.

Thus, here, the precipitation involves local rearrangements of atoms in a material which remains chemically homogeneous throughout the treatment.

3.2 Nucleation

In the above discussion, which involves only the static nature of equilibrium, we used the term *phase* to refer to the set of homogeneous zones of the solid with identical macroscopic parameters (composition, temperature and pressure). In fact, atomic diffusion leads to **fluctuations** in the composition and/or in the arrangement of the atoms. These open up the possibility of transformations of a phase which has become out of equilibrium, for example by a change in the temperature.

In what follows, we shall consider very localised fluctuations of this type, which affect regions of a few atomic volumes (so-called heterophase fluctuations). Inside a homogeneous α phase, which we shall call the **mother phase**, these may create **embryos** of another phase, β. By absorbing or emitting atoms, these embryos may be born, disappear, grow or shrink. We shall denote the free enthalpy of an embryo of n atoms by ΔG_n and the difference in the average free enthalpy per atom between the β phase and the mother phase by Δg. Finally, we shall liken the region of transition between the core of the nucleus β and the matrix α to a clear-cut surface, called the α/β **interface** with surface energy γ and area A_n. Thus, we have

$$\Delta G_n = n\Delta g + \gamma A_n$$

or

$$\boxed{\Delta G_n = n\Delta g + K\gamma n^{2/3}} \tag{28}$$

where K is a shape factor, for example equal to $(4\pi)^{1/3}(3\Omega)^{2/3}$ (where Ω is the mean atomic volume in the β phase) if the embryo is spherical.

Let us consider an A–B alloy of the type illustrated in Figures 11 and 12. At a point (c, T_1), where the unique α phase is stable, we have $\Delta g > 0$ and ΔG_n is positive (since γ is positive). Some embryos (n) exist in the mother phase α, but the number of these decreases rapidly as n (whence also ΔG_n) increases (apply (6) of p. 162 and the subsequent N.B. to the embryos). Let us now cool the alloy rapidly to a point (c, T_0) ($T_0 < T_s$, see Figure 14) where the equilibrium state is the coexistence of the two phases α and β. This is a *quench* ($T_0 < T_1$, see Figure 14) similar to the example given at the very beginning of this chapter. At the point (c, T_0), we have $\Delta g < 0$ (see the positions of the line a_1b_1 and the curve $\mathcal{G}$ on Figure 9(d)). Under these conditions, ΔG_n increases initially with n, passes through a maximum ΔG_{n_c} for a critical embryo size n_c and then decreases and becomes negative (see Figure 15). It is easy to check that, by virtue of (28), we have

$$n_c = -\left(\frac{2K\gamma}{3\Delta g}\right)^3 \quad \text{and} \quad G_c = \Delta G(n_c) = \frac{4K^3\gamma^3}{27\Delta g^2}. \tag{29}$$

These equations clearly imply that the variations $\Delta G(n)$ evolve with the temperature as shown in Figure 15, since $|\Delta g|$ first increases as T decreases (see Figure 9(d)). The reader should also check that each critical embryo n_c taken in isolation is in unstable equilibrium since we have $\partial^2 \Delta G_n/\partial n^2 < 0$ for $n = n_c$; thus, embryos smaller than n_c tend to disappear, while, on the other hand, those larger than n_c tend to grow. Above the critical size, the latter are generally called **nuclei**.

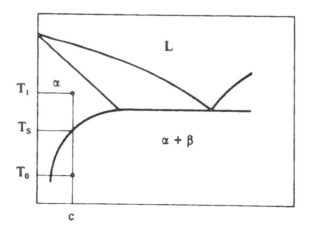

Figure 14 Quenching of an alloy (concentration *c*) from temperature T_1 to temperature T_0.

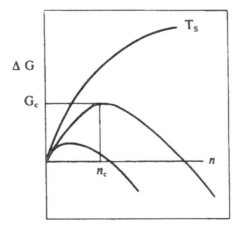

Figure 15 Free enthalpy of a β phase embryo in an A–B alloy containing *n* atoms (see (28)) at various temperatures. Upper curve: temperature T_s (see Figure 14); middle curve: $T_0 < T_s$; lower curve: $T_0' < T_0$.

At the transition temperature T_s, where we have $\Delta g = 0$ (see Figure 14), the enthalpy G_c (the formation enthalpy of a stable nucleus) is infinite; thus, the transformation $\alpha \rightarrow \beta$ cannot occur at this transition temperature. This remark explains the experimental fact that some undercooling below T_s is generally necessary before one can observe the phase change. Similarly, overheating is necessary in order to observe the inverse transformation. This double effect produces a **transformation-related hysteresis**.

To summarise, in a non-equilibrium phase (here, the mother phase α) there exist embryos of the stable phase (here, β), characterised by their size *n*. Above a critical size n_c which corresponds to a free formation enthalpy G_c, these embryos grow and become the nuclei of the future domains of the β phase.

We shall not discuss the important problem of the *nucleation rate* (in other words, that of the number of embryos which exceed the critical size per unit time) here. This rate certainly has a decisive influence on the *precipitation kinetics* of the β phase. We mention only that it is usual to distinguish between:

i) **homogeneous nucleation** where the probability of observing an embryo of a given size is the same at all points of the mother phase;

ii) **heterogeneous nucleation** where local modifications of the values of the parameters of (29) (in particular, that of γ), as a result of the structural inhomogeneities of the mother phase (in particular, crystal defects), enable the nucleation to develop more easily at certain points than at others.

This is generally what happens in the presence of *dislocations* or *grain boundaries* (see Chapter IX). Thus, the energy of an embryo forming astride a boundary is decreased by the energy of that portion of the boundary which it has suppressed. Figure 2 (p. 157) shows a typical example of heterogeneous nucleation in which the embryos, then the nuclei, formed originally at the triple points of a pre-existing dislocation network, thereby decreasing the line energy of these dislocations (see also **Problem 16**).

3.3 Growth

The experiment in which we are interested here is the same as the previous one (quenching of the A–B alloy from point (c, T_1) to point (c, T_0), see Figure 14), however, the question now is to determine *how the β phase nuclei grow once they have been formed*.

There is no single answer to this question. The elements which may influence the growth rate of nuclei (which will later become **precipitates**) include, for example:

i) the exact mechanism governing the propagation of the *interface* between the precipitate and the mother phase;

ii) the relationship between the *crystalline orientations* of the two phases;

iii) the nature and concentration of the *crystal defects* present;

iv) the rate of *atomic diffusion* of the two species (A and B) in the mother phase and in the precipitate.

This last point is among the most important, since atoms are always *redistributed* during the precipitation. We shall single this out in what follows, giving it the dominant role in the kinetics of growth.

More precisely, we assume we are in a situation in which the *rate of absorption* of the atoms on the interface (for example, atoms of B arriving to constitute the β phase) is *very much larger* than the *rate of arrival* of B atoms by diffusion near the β nucleus. Under these conditions, on either side of the interface, this absorption creates regions α and β which are in *equilibrium* with their concentrations c_α and c_β. For simplicity, we shall consider the case of a **spherical nucleus** (radius ρ) at the centre of a spherical region of radius R ($R \gg \rho$). The concentration of B at a very large distance[7] from the nucleus is $c(R) = c_R$, which is the initial concentration of the mother phase and is greater than c_α (see Figure 16).

When the interface progresses by $d\rho$, the nucleus captures $(1/\Omega)c_\beta \, d\rho$ atoms of B per *unit area* of the interface (Ω is the atomic volume). Of these atoms, $(1/\Omega)c_\alpha \, d\rho$ were

---

[7] In other words, at a very large distance in comparison with $\sqrt{Dt}$, where D is the diffusion coefficient of B in the α phase.

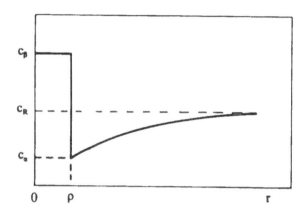

Figure 16 Concentration of the element B around a spherical nucleus of the β phase of radius ρ in an A–B alloy. The flux of B atoms towards the nucleus is a result of the gradient $\partial c/\partial r$ outside the nucleus.

present; the remainder $(1/\Omega)(c_\beta - c_\alpha)$ must be provided by diffusion. In principle, we can handle this diffusion problem and, in particular, determine how the mean concentration c decreases in the mother phase or how the radius ρ varies as a function of time, by solving the diffusion equations (21) and (22). However, even under the very simple conditions which we have assumed, these equations can only be solved numerically. Thus, we shall introduce another approximation, that of the **stationary state**. If $\bar{c}$ denotes the mean concentration, we suppose that the *supersaturation* $\bar{c} - c_\alpha$ (or $c_R - c_\alpha$) is sufficiently weak that $\partial c/\partial t$ is very small and that, in the limit, it is zero. Fick's second law (22) may then be written as:

$$\frac{\partial^2 c}{\partial r^2} + \frac{2}{r}\frac{\partial c}{\partial r} = 0.$$

Attaching the boundary conditions (see above)

$$c = \begin{cases} c_\alpha & \text{for } r = \rho \\ c_R & \text{for } r = R \end{cases} \tag{30}$$

this has the solution

$$c(r) \simeq -\frac{\rho(c_R - c_\alpha)}{r} + c_R. \tag{31}$$

The flux of B atoms on the sphere $r = \rho$ is then (see (21) and (9′))

$$J(\rho) = -\frac{D}{\Omega}\left(\frac{\partial c}{\partial r}\right)_{r=\rho} = -\frac{D}{\Omega}\frac{c_R - c_\alpha}{\rho(t)} \tag{32}$$

where D is the diffusion coefficient of B in the mother phase. This flux supplies the β nucleus to the detriment of the mean concentration $\bar{c}$ of the mother phase and we have:

$$J(\rho)\,4\pi\rho^2 = \frac{4}{3}\pi\frac{R^3}{\Omega}\frac{d\bar{c}}{dt}$$

or, by virtue of (32):

$$\frac{d\bar{c}}{dt} = -\frac{3D}{R^3}(c_R - c_\alpha)\rho(t). \tag{33}$$

The conservation of the species B, located *either* in the nucleus (ρ) *or* in the sphere (R) enables us to write:

$$\rho^3 c_\beta = R^3 (c_R - \bar{c}) \tag{34}$$

and thus to transform (33) into:

$$\frac{d\bar{c}}{dt} = -\frac{3D}{R^2}\frac{c_R - c_\alpha}{c_\beta^{1/3}}(c_R - \bar{c})^{1/3}.$$

The impoverishment of the mother phase to the profit of the nucleus thus takes place, at least in the initial period of growth of the latter, according to the law:

$$\boxed{\bar{c} = c_R - \left(\frac{2Mt}{3}\right)^{3/2}} \quad \text{where} \quad M = \frac{3D}{R^2}\frac{c_R - c_\alpha}{c_\beta^{1/3}} \tag{35}$$

(where we have set $\bar{c} = c_R$ for $t = 0$). From (34) and (35) we deduce the law for the growth of the nucleus:

$$\boxed{\rho = \left(2\frac{c_R - c_\alpha}{c_\beta}Dt\right)^{1/2}.} \tag{36}$$

This law, in $t^{1/2}$ for the radius (or in $t^{3/2}$ for the volume) of the nuclei then of the precipitates, is very frequently observed during precipitation experiments, at least, in the case of moderate supersaturations considered here.

N.B. 1 It is clear that in a real case, there is more than one nucleus. There are in fact N nuclei per unit volume. If these nuclei are distributed at random, the problem is usually reduced to that of *cells* each containing a nucleus at its centre. Each cell is taken as a sphere of radius $(3/4\pi N)^{1/3} = R$ on the surface of which the concentration gradients are assumed to be zero: $\overline{\nabla}c(r = R) = 0$. Each nucleus is then fed within 'its' sphere. This problem is evidently similar to that discussed above.

N.B. 2 Equation (35) is linked to the simple fact that the radius of the region drained by the nucleus varies as $(Dt)^{1/2}$ (see **N.B. 1** of p. 185). Spherical nuclei draining a *sphere*, result in an impoverishment $c_R - \bar{c}$ varying like $(Dt)^{3/2}$. Similarly, it is also apparent here that a precipitate which is not spherical but *linear*, draining in a *cylinder*, would give a law analogous to (35), but with a variation in $(t^{1/2})^2$, whence with exponent 1 instead of $\frac{3}{2}$. In the same way, for *planar* precipitates or precipitates in strip, such as are sometimes found on grain boundaries, the exponent is equal to $\frac{1}{2}$. Conversely, a precise experimental study of the kinetics of precipitation and a determination of the exponent n of the law $c_R - \bar{c} = \text{constant} \times t^n$ give a valuable indication of the shape of the nuclei.

N.B. 3 Among the numerous terms we have neglected in the above discussion, we would mention the possible *anisotropy* of the interface energy γ, the *elastic energy* created when the specific volume of the β phase is different from that of the mother phase and the *stability of the excrescences* which appear on the nuclei (fluctuations of shape). According to these effects, various *instabilities* may appear in the growth of these excrescences, with a number of possible final shapes for the precipitate: **spheroids**, but also **platelets** (like the Guinier–Preston zones mentioned on p. 173), **rods, dendrites** (having the shape of ferns) etc.

3.4 The photographic process

Here we give an example of a particularly important phenomenon of microprecipitation which forms the basis for the photographic process. The corresponding nucleation mechanism is *as heterogeneous as possible* (see p. 195).

1 – Experimental generalities[8]

We recall that a photographic film consists of a suspension of small silver halide crystals (AgBr or AgCl) in a film of gelatine. The crystals (or *grains*) are usually single crystals of tabular shape with two dimensions of the order of 1 μm and the third being around 0.1 μm. The two 'larger' surfaces often have a {100} orientation and sometimes a {111} orientation (fcc structure).

After exposure of the film to light, very small *clusters* (several atoms) of *metallic silver* appear on the *surface* of certain grains. During *development*, the film is dipped in an organic reducing solution and each cluster[9] catalyses the reduction of the silver for the *whole of the corresponding grain*. These reduced grains will then appear in *black*. Grains without clusters are not reduced, and after the *fixing* treatment they remain transparent. The final image is constituted from the set of these black points.

A given emulsion is characterised by its H–D (Hurter and Driffield) graph, which shows the *optical density* δ of the developed film as a function of the logarithm of the *exposure to light* ϵ. Figure 17 shows a typical H–D graph. We shall retain *one* important point, namely the existence of an *incubation* period, such that the film 'reacts' to the light only after a critical exposure ϵ_c. The value of ϵ_c defines the *speed* of the emulsion, the slope of the linear part defines its *contrast*.

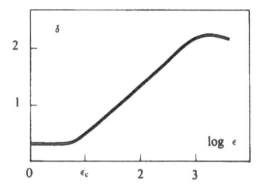

Figure 17   H–D graph of a photographic film.

$$\delta = \frac{\text{luminous intensity transmitted by the developed film}}{\text{incident luminous intensity}}$$

ϵ = luminous intensity of exposure.

Detailed experimental studies of the domain adjacent to ϵ_c have led to the establishment of the following three remarkable facts:

(i)   The probability that a grain may be reduced (in other words, the catalytic efficiency of the clusters of n atoms of silver precipitated on the surface) increases sharply as n **changes from 3 to 5. Thus, the critical cluster** appears to contain $n = 4$ **silver atoms.**

---

[8] See L Slifkin in *Solid State Dosimetry*, Gordon and Breach (1969); *Sci. Prog. Oxf.*, **60**, 151 (1972).

[9] Provided only that its *redox potential* (which depends on its size) is sufficiently large (see *Pour la Science*, p. 17, July (1990)).

(ii) The **number of photons** absorbed by a grain, for an exposure ϵ_c is also **of the order of 4**.

(iii) For a given number of absorbed photons, there exists a **critical luminous intensity** below which there is no further development, or at least, no further reduction of grains. More precisely, if the above four photons are absorbed at time intervals greater than ~ 1 s (at room temperature), the grain is not reduced during development.

2 – Mott and Gurney's model

Long before the above experimental data were accumulated, Mott and Gurney (1938) had laid the basis for a description of the photographic process, the broad outlines of which are not contradicted by the above data.

In alkali halides, the valence electrons of the cations are found in a *full valence band* (see n_1 on Figure 18(*b*), p. 135), separated from an *empty conduction band* n_2. A photon with sufficient energy may create an electron–hole pair, in which a *photo-electron* is promoted into n_2, leaving *a hole* in n_1. If they do not recombine, they diffuse independently in the crystal and finally end up on the surface. There, together with an anion, the hole forms a halide *atom* which reacts with the gelatine and plays no further role in what follows. The electron settles in a **surface trap** (step? kink? foreign atom?) where it attracts a mobile (and thus, almost certainly interstitial) cation (here Ag^+) from the grain. This cation is reduced by the photo-electron into an *atom* (Ag) on the surface next to the trap. The sequence may then be reproduced with a *second* photon. If, after migration, the second photo-electron settles *on the same surface trap*, it may attract another cation Ag^+, which, together with the previous cation, will form an Ag^2 molecule. This continues until the critical cluster Ag^n is achieved.

3 – The miracle of photography

It now remains to compare this model with the above results (i), (ii) and (iii) which provides an opportunity to marvel at the astonishing yield of this process and even at the very fact of its existence.

Comparison with points (i) and (ii) shows immediately that:

1) *each photo-electron survives* until it meets an interstitial Ag^+;

2) the crystal *contains sufficiently many interstitials*;

3) the *meeting place* of photo-electrons and cations *is on the surface* (if it was inside the volume, development would be impossible) and, most importantly, *is unique*. We shall call it S.

The result (iii) is linked to the fact that an Ag *atom* on the surface is unstable as far as re-ionisation in the halide is concerned. Its lifetime τ is of the order of a second (at $T \sim 300$ K). If the rate of precipitation of the Ag^+ ions at the point S is lower than $\sim \tau^{-1}$, a cluster cannot form; this explains the result (iii).

Review of point 1) If the photo-electron migrates (diffusion coefficient D, see **N.B. 1**, p. 185) for a time t_0 before reaching S, the distance covered is $\simeq \sqrt{Dt_0}$. From experience of migration under electric fields, we know how to measure D and t_0. In AgCl at 300 K, we have $D \simeq 1$ cm^2 s^{-1} and $t_0 \simeq 10$ μs. Thus, $\sqrt{Dt_0} \simeq 3 \times 10^{-3}$ cm, which distance is approximately 100 times greater than the typical dimensions of the grains; whence, the photo-electron collides many times with the surfaces of the grain before reaching the point S. The result (ii) proves that during this long migration the photo-electron is never trapped for any considerable time in the volume and thus is never 'lost' to the photographic process. This happy circumstance is linked to the high value of the *dielectric constant* of these crystals (12.5 for AgBr) and to the relatively low *effective mass* at the bottom of the conduction band ($0.29m$ in AgBr). Thus, for example, a hydrogenoid ionisation energy of 13.6 eV is reduced there to $13.6 \times 0.29/(12.5)^2 = 0.03$ eV and traps are of little effect within the volume. As far as the numerous 'rebounds' made by the electron on the surface before it reaches S are concerned, these are due to the presence of a *surface charge* comparable with the Debye charges (see p. 166), but here associated with the balance between cation interstitials and vacancies (which are in the majority in comparison with anionic defects). At room temperature, this charge is negative and repels the electron towards the interior until the latter succeeds in finding in site S a defect in the armour which enables it to escape and reach the surface.

Review of point 2)

Table 2   Enthalpies (in eV) corresponding to cationic point defects in *thick* samples of AgCl and AgBr.

	AgCl	AgBr
H_f	1.47	1.06
H_{mv}	0.05	0.06
H_{mi}	0.29	0.34

Table 2 lists a number of values of the enthalpies (in eV) corresponding to cationic point defects in *thick* samples of AgCl and AgBr. H_f is the formation enthalpy of a *Frenkel pair*, H_{mv} is the migration enthalpy of the cationic vacancy and H_{mi} is the migration enthalpy of the Ag^+ interstitial.

The enthalpy H_f may be used to calculate the concentrations of Ag^+ interstitials at room temperature (see p. 169). The latter are such that there is approximately *only one* interstitial in each normal-size photographic grain! Here again, the surface charge saves the situation, in that it creates a variation of the electrostatic potential δV which increases the concentration of Ag^+ interstitials by a factor $\exp(e\delta V/kT)$. These potential differences have been measured. For δV (measured) $= 0.15$ volts, this factor is approximately 400. Thus, the halide grains contain (and, what is more, near their surfaces) sufficient quantities of interstitials which migrate sufficiently rapidly (see H_{mi} in Table 2) that one of them may reach the site S which is negatively charged by the electron.

Review of point 3)   Let us again marvel at the 'miracle' of the photographic process. The fact that the halides AgBr and AgCl absorb *in the visible band* (which we have not justified and which is associated with a very special band structure) and the fact that the photo-electrons (at the rate of *one* per absorbed photon) *all* reach the surface and *all* capture an interstitial Ag^+ are already remarkable. The fact that, in addition, this precipitation of four silver atoms takes place on the same one of the two faces of the grain and *at a unique point* S, amounts to an exceptional piece of good fortune as far as photography is concerned.

The possible nature of this trap S is not currently known. It may correspond to the point of emergence on the surface of a unique dislocation of the crystal, which would both drain the photo-electron at S and, eventually, constitute a line of easy diffusion for the Ag^+ interstitial. However, this role of dislocations has not yet been established beyond doubt.

A number of laboratories, universities and companies are working on this enigma and also, more specifically, on the possibility of finding a photographic material other than AgBr or AgCl (very expensive). Despite considerable efforts on the second point, no other solid has yet been found which combines the admirably adapted optical, electronic and crystalline performances of silver halides.

CHAPTER IX

DISLOCATIONS AND BOUNDARIES

Italian Concerto   and
English Suites
Johann Sebastian Bach
The Turn of the Screw
Henry James

1. GENESIS OF A CONCEPT

Our relatively recent understanding of the nature and role of dislocations in solids is based on the meeting of two distinct approaches.

i) The first approach, which is both *theoretical* and *fundamental* (although the rubber manufacturer Pirelli showed interest in it at a very early stage) involved studying the lines of *elastic* singularity in continuous media from a mathematical point of view, by solving Navier's equation for various boundary conditions (Italian school: Volterra, 1907).

ii) The second *experimental* and *intuitive* approach aimed to discover the mechanisms for the deformation of solids. Despite considerable work[1], few useful data were available until around the beginning of this century. It was at least known that, in many cases, the deformation of crystals involved **glides** along planes, much as the sheets in a ream of paper glide on top of one another. This led to the idea (Yamaguchi, 1929) that these planes might well be nothing other than the support of elementary *linear distortions* propagating[2] from one edge of the sample to the other during the deformation. These distortions were called **dislocations**. The idea was refined in the 1930s (Taylor, Orowan, Polanyi).

It was at the end of the 1930s (Burgers, 1939) that the link between these two approaches was established. The theory was then deepened rapidly although experimental data were only accumulated in the 1950s with the development of *electron microscopy* (Cambridge School, Hirsch, 1956).

---

[1] By a pleiad of mechanicians, physicists, chemists etc, and important ones at that, including Mariotte, Galileo, Leibniz, Réaumur, Coulomb etc. For some details of this historical aspect, see B **Jouffrey** in *Dislocations et déformation plastique*, Yravals, Éditions de Physique, Paris (1980).
[2] Like the fold which one creates at the end of a carpet which is too heavy to be slid *en bloc*, which one propagates right to the other end along the ground, this process being repeated as many times as necessary in order to achieve the desired displacement.

The first linear defects observed and analysed in detail were in **liquid crystals** (G Friedel, 1910), or, to be more precise, in *smectic crystals* (stacks of molecular layers similar to those of Figure 26(a) (p. 110)). It was the observation of the form of these lines (*focal conics*), which are visible by optical microscopy, which led G Friedel to conjecture that the smectic crystals were based on a stacking of parallel equidistant layers. In fact, if these are deformed, the property of equidistance is retained and they adopt the shape of *Dupin cyclids*, or surfaces with singularities along focal conics[3].

For the present, we shall adopt the approach ii) (above). What then are the experimental data which led to postulation of the existence of these linear distortions?

They are of two types (Sections 1.1 and 1.2).

1.1 The phenomenon of glide

When a single crystal (Al, NaCl etc) is deformed, for example, in a **tensile test**, a series of parallel lines appears on the surfaces as soon as the deformation begins. It is easy to reconstitute these lines in three dimensions as being the intersections of the *external surface* of the sample with parallel *planes*, and to see that these planes are, in

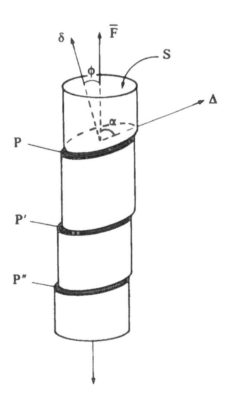

Figure 1 Glide geometry. The force $\overline{F}$ is applied along the axis of the cylinder. The latter is deformed by glides of portions of the cylinder σ in parallel planes P, P′, P″ ..., in the directions Δ. Direction δ is perpendicular to P.

---

[3] See, for example: **M Kléman**, *Points, lignes, parois*, Éditions de Physique, Paris (1977).

fact, the dense crystallographic planes P of the crystal (for example, {111} planes for aluminium (fcc)). Moreover, the planes P glide in a fixed direction Δ in P (for example, ⟨110⟩ directions of the {111} planes in fcc crystals). The *glide system* (P,Δ) may vary in a given solid, depending on the external conditions and, in particular, on the temperature T.

The **glide** only occurs on these planes if the applied stress, σ, exceeds a critical value σ_c. For a given crystal and for fixed T, this value depends on the orientation of the effort applied with respect to the system (P, Δ); however, it is observed that *it is fixed if it is measured in this system*. For example, in the classical geometry of the *tensile experiment* (Figure 1), in which one pulls (force $\overline{F}$) on a bar (cross section S), and in which the glide system is referenced with respect to $\overline{F}$ via the angles ϕ and α (Figure 1), *it is the stress*

$$\sigma_c = \frac{F_c}{S} \cos\phi \cos\alpha$$

which is constant for the various possible orientations (ϕ, α), where F_c is the critical force needed to *initiate* the glide (law of Schmid and Boas, 1924). The stress σ_c, which is called the *resolved* **yield stress**, is just the shear stress exerted on the plane P in the direction Δ.

Actually, $F_c \cos\alpha$ is the component of the force in the direction of glide Δ and $S/\cos\phi$ is the cross section of the sample along the glide plane.

1.2 Very low measured value of the resolved shears

Let us imagine the glide (P, Δ) as being that of two adjacent atomic planes P_i and P_{i+1} (distance a apart) of the family P, gliding globally, one above the other ('coherent glide'). Let x be the instantaneous displacement of the atoms of P_{i+1} with respect to P_i (Figure 2). Under the present assumptions, all the displacements are in phase. How does the applied (shear) stress σ vary as a function of x? σ is clearly a periodic function of x, with period b (atomic *distance* in the direction of glide). Moreover, the proportionality between σ and x in the elastic domain (called *Hooke's law*), is written as:

$$\sigma = \mu \frac{x}{a}$$

where μ is, by definition, the *shear modulus*, and x is small or close to b. The simplest function $\sigma(x)$ which satisfies these conditions is

$$\sigma = \frac{\mu b}{2\pi a} \sin 2\pi \frac{x}{b} \tag{1}$$

and the stress necessary for the glide to take place is

$$\sigma_0 = \mu b / 2\pi a \sim \mu / 2\pi. \tag{1'}$$

This value is **a factor of** 10^3 **or** 10^4 **greater** *than the measured yield stresses* (current values of σ_c in fcc metals are $\sigma_c \simeq 10^{-3}$–$10^{-4} \mu$).

Thus, for Al, a yield stress of the order of $\mu/2\pi$ (where $\mu = 2.7 \times 10^{10}$ N m^{-2} or $\sigma_0 \simeq 0.4 \times 10^{10}$ N m^{-2}), would imply that it is possible to suspend approximately 400 kg on the end of an aluminium wire with cross section 1 mm^2 without the latter undergoing any permanent deformation! In fact, σ_c measured in a single crystal of pure aluminium is of the order of 50 g mm^{-2} or $\simeq 0.5 \times 10^6$ N mm^{-2}, or $10^{-4}\sigma_0$!

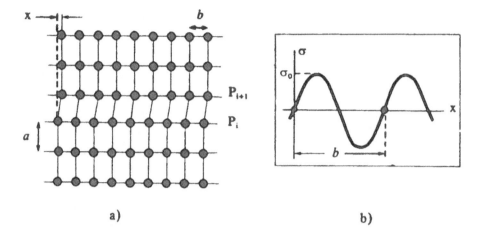

a)   b)

Figure 2 Coherent glide and shear stress. (a) The upper half of the crystal is subject to a shear towards the right. The crystal is fcc. The glide planes P are planes {111}. The direction of glide (in the plane of the figure) is (110). Thus, we have (see Figure 21, p. 104) $a = 2b\sqrt{2}/\sqrt{3} = a_0\sqrt{3}/3$ (where a_0 is the edge of the unit cell). (b) Corresponding variation of the shear stress as a function of the displacement x.

The enormous difference between the calculated σ_0 and the measured σ_c frees us from all scruples regarding the simplification made in (1); it is not the precise choice of the function $\sigma(x)$ which is in question, but the mechanism of coherent glide, shown in Figure 2(a).

In order to get out of this impasse, the existence of an object (actually, a *defect*) which would prevent the glide from taking place in phase for all the atoms was postulated. For this, this defect should concentrate the deformation in a reduced region of P and progress in P. It should be *linear* (so as to sweep the whole of P), advance in the direction Δ and, above all, even before the deformation begins, store enough elastic energy that the glide is *locally* easy (in the same way as on p. 169 the high energy of the interstitial gave it a high mobility).

The defect illustrated in Figure 3(a) (amongst others) corresponds well to the profile outlined above. This was created from an initially perfect crystal (assumed to be cubic here) by *removing* a half plane of atoms bounded by the line L and letting the atoms close up the void created in this way. This configuration may also be obtained by *inserting* an extra half plane of atoms (n) bounded by L, and ensuring that after insertion, the inserted atoms are in a good crystalline position, except, of course, near L. This configuration (l) created along the line L (rectilinear here) is called a **dislocation** (see Figure 3(a) or (c)).

With minor atomic rearrangements, (l) may be transformed into (l^*) (an excited form of l) which may be retransformed into l either in the initial position or in an immediately adjacent position. In the latter case, the reaction may be written as:

$$(l_n) \rightarrow (l^*) \rightarrow (l_{n+1}) \qquad (2)$$

This represents a jump of the line of dislocation L as a whole by *one* interatomic distance in the plane P. The **dislocation** is said to **glide** along the plane P.

The reader will easily verify from Figure 3 that this jump has moved not the atoms of the half plane n but the configuration *'extra half plane'* from position n to position

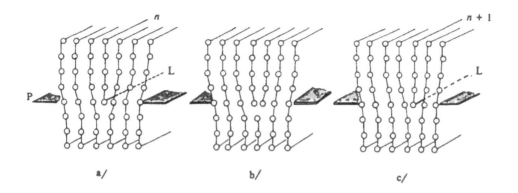

Figure 3 A dislocation. (a) A half plane of atoms has been inserted at the position n of a cubic crystal, up to the plane P, respecting the interatomic distances. The line (a straight line here) which delimits this half plane defines a region of *bad crystal* called dislocation L, with configuration l. (b) Some atomic rearrangements have modified the structure of L, which now appears to border two half planes above and one below (configuration l*). (c) The unstable structure l* has restabilised to l, but at position n + 1. If the process has taken place along the whole of the dislocation L, *the latter has moved to the right in the plane P by one interatomic distance.*

N.B. This particular dislocation is called 'edge' dislocation.

n + 1, and that, with successive jumps in the same direction, this configuration may be transported from one face of the sample to the other. Thus, when the dislocation L has passed from the left surface to the right surface, we will have shifted the upper half of the sample (above plane P) by *one* interatomic distance with respect to the lower half. Since the efforts exerted on the atoms in different parts of P are no longer in phase, we expect a massive reduction in the resolved shear stress needed to initiate the glide, as compared to the previous hypothetical case of coherent glide.

Thus, the movement of dislocations as shown in Figure 3 is a plausible mechanism which may account for the glide. But, do dislocations exist? We shall see that this is in fact the case; however, before we do so, we have to describe this defect in more detail.

2. ELASTIC MODEL OF DISLOCATIONS

2.1 Volterra's procedure

The dislocation of Figure 3 is only a special case of a more general defect (Volterra dislocation) which we create as follows. We apply the following steps to an *elastic, continuous, homogeneous* medium, which we shall also assume to be *isotropic* and initially *unstressed*.

i)   We make a cut along a surface S bounded by the line L.

ii) Using an external stress, we move the lips S_1 and S_2 of S with respect to one another.

iii) We fill the newly created void, or remove the excess matter;

iv) Finally, we 'reglue' the lips and suppress the external stress applied in stage ii).

A system of *internal stresses* is then present in the medium; these stresses have a discontinuity on L and are primarily concentrated along this line. In what follows, we shall consider only the case in which the displacement (point ii), above) is *uniform* and equal to $\bar{b}$. The linear defect located along L is then called a *translation dislocation* (we shall call it a *dislocation* in what follows), $\bar{b}$ is its **Burgers vector** (we shall often abbreviate this to BV).

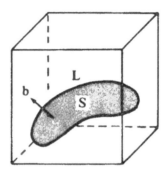

Figure 4 Creation of a dislocation along line L limiting surface S.

In what follows, we shall examine in more detail the case in which the surface S is a plane Σ bounded by a straight line L which is the axis of a cylindrical sample. In this case, the dislocation is *rectilinear* with the two extreme possibilities being $\bar{b} \perp L$ and $\bar{b} \parallel L$.

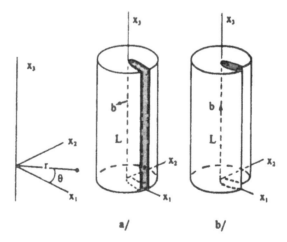

a/   b/

Figure 5 Dislocations in an elastic cylinder. (*a*) **Edge dislocation**. In grey: matter added after cutting and separation of the lips. $\bar{b} \parallel x_2$. (*b*) **Screw dislocation**. In grey: plane of the cut and of the gliding of the lips. $\bar{b} \parallel x_3$.

The first case (Figure 5(*a*), insertion or removal of matter) corresponds exactly to that described for a crystal in Figure 3. Thus, in the latter, the Burgers vector is a vector perpendicular to L, situated in P with length a (the interplanar distance). The corresponding dislocation is called an **edge dislocation**.

The second case (Figure 5(*b*)) clearly creates a *helical symmetry*. A surface which is initially a plane in the sample and is perpendicular to the L axis is transformed into a helix. The corresponding dislocation is called a **screw dislocation**.

An *arbitrary* rectilinear dislocation of BV $\bar{b}$ may be viewed as a superposition of an edge dislocation of BV $\bar{b}_\perp$ and a screw dislocation of BV $\bar{b}_1$, where $\bar{b}_\perp$ and $\bar{b}_1$ are, respectively, the projections of $\bar{b}$ onto a plane perpendicular to L and onto L.

N.B. The previous procedure may be retained while modifying stage ii). Having made (for example) the planar cut delimited by a straight line D, the lips are displaced by *rotation* through an angle ϕ around D (instead of by *translation*). After stages iii) and iv), a linear defect, known as a **disinclination** appears along D. Figure 6 shows an example, in cross section. Beginning with a planar hexagonal lattice, a cut in the plane was made (perpendicular to the figure), the two lips were opened by a rotation through $\phi = 60°$ and matter was introduced into the opening. The disinclination is a line in the centre and perpendicular to the figure.

Figure 6 **Example of a disinclination.** The reader may also wish to refer to N.B. 2 of p. 85.

Disinclinations play an important role in liquid crystals. We shall not refer to them otherwise in the remainder of the book.

2.2 Stresses and strains

Let us consider an elastic cylindrical solid (radius R, unit height) with axis L and create a *screw dislocation* (BV: $\bar{b}$) along the line L as in Figure 5(*b*). We first evaluate the *displacement vector* $\bar{u}$ associated with the dislocation. $\bar{u}$ must be a solution of the Navier equation, written without external forces[4]:

$$(\lambda + 2\mu)\overline{\text{grad}}\,(\text{div}\,\bar{u}) - \mu\overline{\text{curl}}\,(\overline{\text{curl}}\,\bar{u}) = 0,$$

which is equivalent, at least in orthonormal cartesian coordinates, to

$$\boxed{(\lambda + \mu)\overline{\text{grad}}\,(\text{div}\,\bar{u}) + \mu(\overline{\Delta})\bar{u} = 0} \tag{3}$$

---

[4] See: **J Salençon**, *Mécanique des milieux continus*, Ellipses (1988).

(λ and μ are the Lamé *elasticity moduli*), and must also be such that $\bar{u} = 0$ and $\bar{u} = \bar{b}$, respectively, on the two lips S_1 and S_2 of the surface of the cut $S_1(\theta = 0)$ and $S_2(\theta = -2\pi)$. The reader should check that, for a cylinder of infinite length, the solution

$$u_1 = u_2 = 0 \qquad u_3 = \frac{b\theta}{2\pi} = \frac{b}{2\pi} \arctan \frac{x_2}{x_1} \tag{4}$$

satisfies these requirements.

For this, we note that div $\bar{u}$ is zero, as are Δu_1, Δu_2 and Δu_3.

The non-zero components of the **strain tensor** $e_{ij} = \frac{1}{2}(\partial u_i/\partial x_j + \partial u_j/\partial x_i)$ may now be deduced immediately, namely:

$$e_{31} = e_{13} = -\frac{1}{2}\frac{b}{2\pi}\frac{x_2}{x_1^2 + x_2^2} = -\frac{b}{4\pi}\frac{\sin\theta}{r} \qquad e_{32} = e_{23} = \frac{1}{2}\frac{b}{2\pi}\frac{x_1}{x_1^2 + x_2^2} = \frac{b}{4\pi}\frac{\cos\theta}{r}. \tag{5}$$

Hooke's law (proportionality of stresses and strains) then gives the components of the **strain tensor**:

$$\sigma_{ij} = \lambda\delta_{ij}\sum_n e_{nn} + 2\mu e_{ij} \tag{5'}$$

or, here:

$$\sigma_{31} = \sigma_{13} = 2\mu e_{13} = -\frac{\mu b}{2\pi}\frac{\sin\theta}{r} \qquad \sigma_{32} = \sigma_{23} = 2\mu e_{23} = \frac{\mu b}{2\pi}\frac{\cos\theta}{r} \tag{6}$$

or again

$$\boxed{\sigma_{\theta 3} = \sigma_{3\theta} = -\sin\theta\sigma_{13} + \cos\theta\sigma_{23} = \frac{b\mu}{2\pi r}} \tag{7}$$

as the only non-zero stresses. These correspond to strains

$$\boxed{e_{\theta 3} = e_{3\theta} = \frac{b}{4\pi r}.} \tag{8}$$

The reader will have noted that the volume expansion (or div $\bar{u} = \sum_n e_{nn}$) is zero.

We now recall the meaning of σ_{ij}. Suppose that, in a solid, a surface element ds normal to x_i is situated at the point (x_1, x_2, x_3) and belongs to a surface S. The latter delimits two regions of the solid. The element ds in one region exerts a force σds on the other, where the component of this force on x_j is $\sigma_{ij}ds$. Thus, $\sigma_{\theta 3}$ ($\sigma_{\theta 1}$ and $\sigma_{\theta 2}$ being zero) is the force per unit surface area exerted on a surface normal to θ. Figure 7 illustrates the (very simple here) system of stresses (7) produced by a rectilinear screw dislocation. Since, in both cases, the force lies in the surface of application, both stresses are pure *shears*.

The solution of equation (3) for a *straight edge dislocation* is more complicated than (4). We just give here the expressions for the stresses in cartesian and cylindrical coordinates:

$$\sigma_{11} = -D\frac{\sin\theta(2 + \cos 2\theta)}{r} \qquad \sigma_{rr} = \sigma_{\theta\theta} = -D\frac{\sin\theta}{r}$$

$$\sigma_{22} = D\frac{\sin\theta\cos 2\theta}{r} \qquad \sigma_{r\theta} = D\frac{\cos\theta}{r} \tag{9}$$

$$\sigma_{33} = \nu(\sigma_{11} + \sigma_{22})$$

$$\sigma_{12} = \sigma_{21} = D\frac{\cos\theta\cos 2\theta}{r} \qquad \sigma_{33} = -2D\nu\frac{\sin\theta}{r}$$

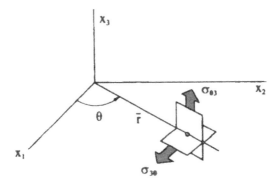

Figure 7 Shear stresses, exerted by a rectilinear screw dislocation along x_3.

where $D = \mu b/[2\pi(1 - v)]$. v is the *Poisson ratio* of the solid ($v = \lambda/2(\lambda + \mu)$). We shall retain the fact that, like for screw dislocations, these stresses decrease as r^{-1}.

2.3 Elastic energy

The work of the stresses introduced by the dislocations produces an **elastic energy** stored in the solid, which we shall calculate, again for the case of the rectilinear screw dislocation. This energy amounts to

$$w = \frac{1}{2}\sum_{ij}\sigma_{ij}e_{ij} \tag{10}$$

per unit volume; or, using (7) and (8):

$$w(r) = \frac{\mu b^2}{8\pi^2}\frac{1}{r^2}. \tag{11}$$

Thus, the elastic energy stored between r and $r + dr$ amounts to

$$dw = \frac{\mu b^2}{4\pi}\frac{dr}{r}$$

per unit length. Summed from $r = 0$ to $r \to \infty$, the energy W would diverge. In fact, letting r tend to zero, takes us into a domain in which, since the stresses and the deformations increase in r^{-1}, use of the linear elasticity is no longer admissible. Limiting the use of the latter to regions in which the deformations are small (in practice, less than $\simeq 0.05$) we introduce a limiting radius r_0, which is essentially equal to b (for $e = 0.05$, see (5)). The cylinder of radius r_0 is called the **core of the dislocation**. This is a region which, to a first approximation, may be treated as *incompressible*. Thus, its energy is generally negligible. For a cylinder of *finite* radius R, the energy of the screw dislocation is then

$$W = \frac{\mu b^2}{4\pi}\ln(R/r_0) \tag{12}$$

per unit length.

Let us return for a moment to the energy of the core, and compare the latter to an incompressible cylinder in which the deformation remains equal to that obtained at the limit of validity of the linear elasticity, in other words, at r_0. The energy density (10) may be written as $\sum_{ij} \mu(e_{ij})^2$ (see (5')) or, here (see (8)): $\mu b^2 / 8\pi^2 r_0^2$. For a cylinder of radius $r_0 = b$, the energy of the core is thus $\mu b^2 / 8\pi$ per unit length, which is negligible in comparison with (12).

Formula (12) shows that the *energy would be infinite in an infinite solid*. In a cylinder of radius $R = 10^5 r_0$ (for example, $r_0 = b = 2$ Å and $R = 20$ μm) the energy, which varies slowly with R is essentially

$$\boxed{E \simeq \mu b^2} \tag{13}$$

per unit length, or, if the interatomic distance along the dislocation is b:

$$E \simeq \mu b^3$$

per atom. In many solids, μb^3 *is of the order of 5 eV*. Thus, the energy stored by a 'dislocation atom' is considerable.

N.B. It can be shown that, in the case of an *edge* dislocation with BV $\bar{b}$, formula (12) remains valid, provided only that the 4π in the denominator is replaced by $4\pi(1 - \nu)$ as in (9). This factor alters the previous results only numerically; for simplicity, we shall continue our treatment for the case of the *screw dislocation*.

Formula (12) may be used in a solid of any shape (not cylindrical), where R is the mean distance of the dislocation from the surfaces or, more generally, a characteristic dimension of the 'domain' of the dislocation. Thus, in the general case, in which several dislocations with random orientations coexist, R may be taken as the average half distance between dislocations: $R \simeq \rho^{-1/2}$, where the quantity ρ (which is a *number per unit surface area*) is the **dislocation density**; it is the number of dislocations cutting a unit surface area, or the total length of dislocations per unit volume.

In a crystal, the configurational entropy increases when a dislocation is created, but the above energies are far too large for a measurable equilibrium concentration to be created, contrary to the case of vacancies (p. 162). However, all solids contain dislocations in an appreciable density, for example, $\rho \simeq 10^8$ m^{-2} for a 'good' single crystal of silicon, 10^{10} m^{-2} for a well annealed metal and up to 10^{16} m^{-2} for highly deformed materials (*rolled* or *wiredrawn* metals, rocks of a *fold* region etc). Thus, **these dislocations are always out of equilibrium**. A solid with a high dislocation density always decreases its free enthalpy if it is able to decrease this density. This is the driving force behind the phenomenon of **recrystallisation** (see later, p. 235).

Let us evaluate the energy stored in a metal with a dislocation density $\rho = 10^{16}$ m^{-2}. We shall take $b = 2.5$ Å, $\mu = 7 \times 10^{11}$ dynes cm^{-2} and $r_0 = b$. Per unit volume, the dislocation length is ρ and the energy is $\rho(\mu b^2 / 4\pi) \ln(\rho^{-1/2}/b)$. We find $\simeq 0.2$ J cm^{-3}.

The existence of an *energy per unit length* suggests that a sinuous dislocation must tend to become rectilinear, like an elastic rope. Thus, we define a **line tension**, which is the ratio of the variation of the elastic energy to the variation of the length $dW = \tau dl$. The line tension depends on the curvature of the dislocation, but taking it equal to the *line energy* (per unit length) gives a good approximation. Thus, using the simple formula (13), we shall write

$$\tau \simeq \mu b^2. \tag{14}$$

For example, a dislocation in the form of an arc o⌐ ⌐ circle (radius of curvature R_c) of length δl is subjected to the force τ at each of its extremities A and B (see Figure 8) along the tangent to the arc. The resultant of these two forces is a force δf directed towards the centre of the circle:

$$\delta f = 2\tau \sin \delta\theta \simeq 2\tau\delta\theta = \tau\delta l/R_c.$$

Thus, the curved dislocation is **subjected to the force**

$$F \simeq \frac{\mu b^2}{R_c} \qquad \text{per unit length.} \tag{15}$$

Therefore, the curvature can only be maintained if an external force (here, δf) is applied to the dislocation.

Figure 8   A curved dislocation δl (radius R_c) is subjected to the line tension τ.

2.4 Forces exerted on dislocations by elastic stresses

The concept of dislocation has emerged via a hypothetical movement (glide) of this defect when it is subjected to an external effort. Does this mean that there is a force exerted on a dislocation by an applied stress?

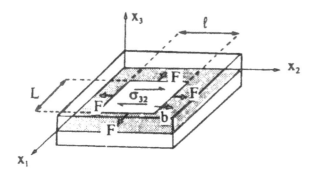

Figure 9   Force on a dislocation. A rectangular loop of a dislocation (two sides ∥ x_1: edge; two sides ∥ x_2: screw) is subjected, in the presence of the shear σ_{32}, to a force which stretches it. Lengths of the sides of the loop: L and l.
    Figure 2 of Annex 5 (p. 272) shows a section through this loop in a crystal.

Suppose that a solid, with axes x_1, x_2 and x_3 as shown in Figure 9, is subject to a shear stress σ_{32} which pulls the upper part ($x_3 > 0$) to the right. Let us now introduce a rectangular **dislocation loop** in the plane $x_3 = 0$ (sides L and l parallel to x_1 and x_2) by the operation of Figure 4: we cut the solid along the rectangle and move the *upper* lip of the cut to the right by a vector $\overline{b}$ parallel to x_2, so that two sides of the rectangle are *edges* and two are *screws*. The *energy* of the solid has decreased, in the operation, by the work of the stress during the displacement, namely: (force) × (movement) = $(-\sigma_{32}Ll) \times (b) = \Delta E$. Thus, the *force* exerted on the loop, here, tends to *stretch* it. This force acts on the four segments of the rectangle and its intensity (*absolute value*) is given by:

$$\partial\Delta E / \partial l = \sigma_{32}Lb \qquad \text{for the sides} \parallel x_1 \text{ with length } L$$
$$\partial\Delta E / \partial l = \sigma_{32}lb \qquad \text{for the sides} \parallel x_2 \text{ with length } l.$$

In other words, for each of the four sides of the loop:

$$\boxed{f = \sigma_{32}b} \qquad (16)$$

per unit length.

N.B. 1   As one might expect, this force is parallel to x_2 for the edge segments. However, it is parallel to x_1 for the screw segments; thus, the glide in the plane $x_3 = 0$ and in the direction x_2 involves a movement of edge dislocations in the direction of the glide (x_2) and a *movement of screw dislocations in the perpendicular direction*.

N.B. 2   Formula (16) is only a special case of the general expression for the force exerted by a system of stresses defined by the tensor $\overline{\overline{\sigma}}$ on a dislocation segment $\overline{dl}$ with Burgers vector $\overline{b}$. This force is:

$$\overline{df} = (\overline{b} \cdot \overline{\overline{\sigma}}) \wedge \overline{dl} \qquad (16')$$

(law of Peach and Koehler) where $\overline{b} \cdot \overline{\overline{\sigma}}$ is the vector with components $p_i = \sum_k \sigma_{ik}b_k$.

In the above, the stress σ was an *applied stress* (efforts exerted on a sample). What was said remains valid for **internal stresses** *associated with the presence of other defects*. In particular, two dislocations exert a force on one another, which may be attractive or repulsive, depending on the signs of the two Burgers vectors (see (16')). Similarly, a point defect, a foreign atom, or a small cluster ('precipitate', see above, p. 190) generally exert a force on a dislocation. By way of example, referring to Figure 3, a larger (*resp.* smaller) atom than the atoms of the matrix reduces the elastic energy of the solid by positioning itself in the region of tension immediately beneath the dislocation L (*resp.* in the region of compression immediately above L). *In both cases*, the dislocation and the foreign atom are linked by an *attractive force*. An example of the interaction between a dislocation and a precipitate is given on p. 239 (Figure 13). See also **Problem 13**.

Exercise   Calculate the force F exerted between two parallel rectilinear screw dislocations D_1 and D_2 with the same Burgers vector $\overline{b}$ at distance r_0 apart. **Hint**: use formula (16'); or create D_2 by the Volterra procedure in the stress field of D_1, calculate the energy involved $E(r_0)$ and deduce the force.
Result $F = \mu b^2/2\pi r_0$ *per unit length*. This force is *repulsive* if the two Burgers vectors are in the same direction and *attractive* if they have opposite directions.

3. DISLOCATIONS IN CRYSTALS

For the most part, the model described in the last section did not involve the crystallinity. The medium was simply elastic. We shall now describe the dislocations actually observed in crystals.

3.1 Structure of dislocations

The discontinuous nature of crystalline matter makes an arbitrary displacement $\bar{b}$ in the procedure of Section 2.1 impossible for fear of the creation of prohibitive energies. The vectors $\bar{b}$ are, *in principle*, translation vectors of the crystal, so that the two lips S_1 and S_2 (Figure 4) may be glued together again perfectly. The corresponding dislocations are said to be **perfect**. On the other hand, since the energy varies as b^2 (see (12)), given a displacement nb of the lips, it is more advantageous to create n dislocations with BV $\bar{b}$ (energy $\simeq nb^2$) rather than *a single* dislocation with BV $n\bar{b}$ (energy $\simeq n^2b^2$). Thus, in general, the Burgers vectors are *the vectors of the primitive cell*. For example, for an fcc crystal, the most common Burgers vectors are $\frac{1}{2}\langle 110 \rangle$ and sometimes $\langle 100 \rangle$. In bcc crystals, they are $\frac{1}{2}\langle 111 \rangle$.

N.B. Until now, we have viewed the Burgers vector only as a *displacement*. A more precise definition (particularly as far as the *direction* of this vector is concerned) is given in **Annex 5** (p. 271).

Figure 10 shows an edge dislocation viewed 'end on' (as in Figure 3) by electron microscopy in germanium. Such a dislocation does not modify the surface on which it emerges. On the other hand, a screw dislocation (see Figure 5(*b*), p. 206, or Figure 1(*c*), p. 271) which ends on a free surface creates an *atomic step* on the latter (see p. 160). The height of this step is equal to the Burgers vector. Thus, by permitting external atoms to attach themselves to the surface, making the steps turn and advance in spirals (Figure 11), screw dislocations play a crucial role in *crystal growth* (see **Problem 10**).

Figure 10 Frontal view of an edge dislocation (perpendicular to the figure) in germanium. High-resolution electron micrograph of a germanium crystal. The atomic rows $\langle 110 \rangle$ are perpendicular to the plane of the photo: the bright spots are the projections onto this plane. Since the Burgers vector $\bar{b}$ $(= \frac{1}{2}\langle 110 \rangle)$ of this dislocation is perpendicular to these rows, they remain rectilinear.

Taking into account $\bar{b}$, this dislocation may be observed by inclining the photo, *either* as two extra (low-density) planes with a 'northerly' direction *or* as one extra plane (dense) in two inclined directions (to the 'North East' or to the 'North West'). Distance between dense planes: 0.32 nm.

After **A Bourret**, (1981).

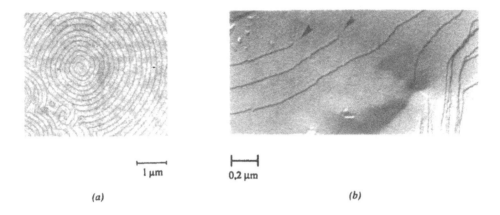

1 μm   0,2 μm

(a)   (b)

Figure 11 Two examples of screw dislocations. (*a*) A screw dislocation emerging on the surface of an NaCl crystal creates a step which winds in a *spiral*, and favours crystal growth from an aqueous solution. See **Problem 10**.

Height of the step: 0.28 nm (distance between the planes {100}). Same technique as for Figure 5, p. 160. After **H Bethge** and **K W Keller** (1974).

(*b*) A *soap* with a *lamellar* structure (in other words, consisting of regularly repeated lamellae, which themselves consist of a layer of soap and a layer of water, see Figure 26, p. 110), has been quenched to $\simeq 80$ K, then solidified, then broken. The observation shows the fractured surface and it is possible to distinguish (on the right) steps (edges of lamellae) either forming terraces or interrupted (arrows). Each interruption represents the point of emergence of a screw dislocation on the surface (see Figure 5(*b*), p. 206, or Figure 1(*c*), p. 271).

Thickness of the lamella: 4.5 nm.
After **M Allain** (1986).

However, there is one important mechanism, namely the **dissociation of dislocations**, leading to the existence of dislocations with BVs which are *not* vectors of the primitive cell. The corresponding dislocations are said to be **imperfect**.

Let us return to the Volterra procedure, and see that it leaves us free to write the displacement $\bar{b}$ as the sum of two displacements $\bar{b}_1$ and $\bar{b}_2$, such that $\bar{b}_1 + \bar{b}_2 = \bar{b}$. If $\bar{b}_1$ and $\bar{b}_2$ satisfy the inequality:

$$|\bar{b}_1|^2 + |\bar{b}_2|^2 < |\bar{b}|^2 \tag{17}$$

then the overall energy of the two dislocations L_1 and L_2 with BVs $\bar{b}_1$ and $\bar{b}_2$ is less than that of the dislocation $L(\bar{b})$. If $\bar{b}$ is a vector of the primitive cell, then $\bar{b}_1$ and $\bar{b}_2$ are not. Thus, the price one has to pay for this gain in energy is imperfect resewing of the lips S_1 and S_2 between L_1 and L_2, which is not necessarily prohibitive. The reaction $L \rightarrow L_1 + L_2$ is called dissociation of the perfect dislocation L. L_1 and L_2 are said to be imperfect or *partial* dislocations.

One classical case of dissociation involves perfect dislocations $\frac{1}{2}\langle 110 \rangle$ of fcc crystals. In a dense (111) plane (Figure 12), we may write $\bar{b} = \bar{b}_1 + \bar{b}_2$ in the form

$$\frac{1}{2}[10\bar{1}] = \frac{1}{6}[2\bar{1}\bar{1}] + \frac{1}{6}[11\bar{2}]. \tag{18}$$

The vectors $\bar{b}_1$ and $\bar{b}_2$ satisfy (17). Thus, the reaction $\bar{b} \rightarrow \bar{b}_1 + \bar{b}_2$ is favourable, although $L_1(\bar{b}_1)$ violates the ABCABCA...fcc stacking (see p. 105), transforming it into ABCBCABC...(the bold letters denote the two planar lips S_1 and S_2). Thus, the dissociation (18) creates two partial

Figure 12   Dissociation of a perfect dislocation. In a compact hexagonal plane of an fcc stacking, the displacement $\bar{b}$ is equivalent to the sequence of displacements $\bar{b}_1$ then $\bar{b}_2$.

dislocations L_1 and L_2 with an imperfection in the fcc stacking between them; this imperfection is called a **stacking fault**. Locally, there appears a sequence BCBC, which is that of an hcp stacking (p. 106), on either side of the defect.

3.2 Movement of dislocations

We have already discussed movement by **glide**. This relates to the movement of a dislocation in a glide plane *containing* the Burgers vector. This movement involves rotations of atomic bonds, *without transport of matter.*

A movement *perpendicular* to the glide plane of an edge dislocation is called a **climb**. Any movement may be analysed in terms of a glide component and a climb component. The climb requires transport of matter. Let us return to Figure 3(*a*). It is clear that an arrival of atoms (*resp.* vacancies) along L on the extra half plane causes the dislocation to 'move down' (*resp.* 'up') in a direction perpendicular to the glide plane P.

This ability of an edge dislocation to capture (or emit) atoms is facilitated by the presence of **jogs** (see Figure 13), or steps of atomic scale, on the edge of the extra half plane.

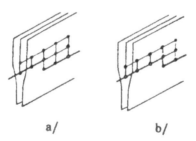

a/   b/

Figure 13   Jog on an edge dislocation. From (*a*) to (*b*), the jog migrates along the dislocation by an interatomic distance by *absorption* of a vacancy (or emission of an atom). From (*b*) to (*a*) involves *emission* of a vacancy.

Only a few atoms of the extra half plane are shown.

These jogs exist *in thermodynamic equilibrium* along dislocations, with a concentration

$$n_j = n_0 \exp(-E_j/kT) \tag{19}$$

where n_0 is the number of atomic sites per unit dislocation length and E_j is the formation energy of the jogs (often of the order of 0.1 eV).

N.B.   The expression (19) may clearly be derived in the same way as the expression (6) (p. 162) for point defects.

Jogs also exist *out of equilibrium*, either in order to give curvature to a pure edge dislocation (see Figure 14) or because they are created when two dislocations *cross*. For example (Figure 15), when two edge dislocations with parallel BVs cross one another, they create two jogs, each consisting of small dislocation elements of screw type.

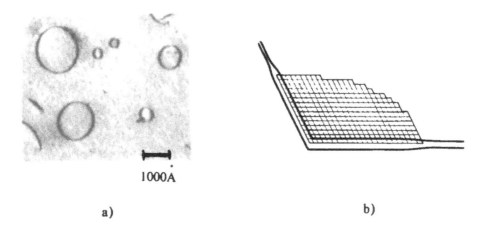

a)   b)

Figure 14   Dislocation loops. Tungsten has been irradiated by neutrons. The interstitials thus created have grouped into discs a single-atom thick, forming edge dislocations along the rim of the discs.

(*a*) Electron microscopy. The dark lines are due to local distortions induced by the dislocations. The scale bar represents 100 nm.

After **L Zuppiroli**, 1977.

(*b*) Schematic section of such a loop. The curvature of the loop implies the presence of a large number of jogs on the loop.

After **L Zuppiroli**, 1977.

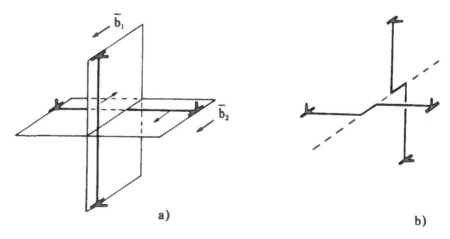

a)   b)

Figure 15   Formation of jogs. (*a*) Two edge dislocations glide towards one another, each in its own glide plane. (*b*) After intersection, each has acquired a jog. Each of these jogs is parallel to the Burgers vector of the other dislocation. Thus, they are both of screw type here.

N.B. The symbol ⊥ usually denotes an edge dislocation, in terms of its extra half plane | and its glide plane ＿, where the Burgers vector is parallel to the latter.

4. SUB-BOUNDARIES AND BOUNDARIES

Only very rarely does a crystalline solid occur in the form a **single crystal**. It generally consists of an aggregate of small crystals or **grains** linked together by connecting surfaces, or **grain boundaries**. The material is then said to be **polycrystalline** (see Figure 16).

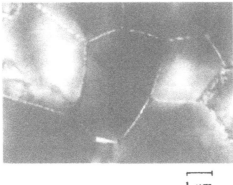

1 μm

Figure 16 Grains of boron carbide. This carbide, with composition B_4C, is a ceramic obtained by hot pressing of powders. It has been irradiated by He^+ ions. The helium has collected in the grain boundaries, which are actually accentuated by helium bubbles (see **Problem 16**). Transmission electron microscopy.

After **T Stoto** and **L Zuppiroli** (1987).

N.B. 1 A reasonable representation of the granular structure may be obtained by shaking a (transparent) bottle containing a little shampoo. The polyhedral soap bubbles which fill the space have such a structure. Boundaries, triple lines and quadruple points may be seen.

N.B. 2 The grains are generally formed at the actual time of the **elaboration**. For example, in a metal ingot cooling *from the liquid phase*, solid nuclei appear in the bulk, grow and come together, in general without there being any relationship between their orientations. As another example, in the case of *sintering* where chemically prepared powders are subjected to hot pressing (in the solid phase; see Figure 16), the grains often correspond to the initial solid particles.

In the first case, the nuclei grow more quickly the more dislocations they contain (see Section 3.1), which explains the often high dislocation densities (out of equilibrium) observed in 'as-received' materials.

N.B. 3 In a given polycrystalline sample, the crystalline directions relative to each grain are not in general random. They group around fixed directions, themselves associated with the elaboration process (for example, direction of rolling etc). The study of the statistical distribution of certain axes (for example, ⟨100⟩, ⟨110⟩ etc) in three dimensions, allows one to define the **texture** of the polycrystal. Thus, we may talk of the *rolling texture, the drawing texture*, the *recrystallisation texture* etc.

4.1 Sub-boundaries

If the disorientation between two adjacent grains is low ($\lesssim 10°$), the grain boundaries have a simple structure. They are called **sub-boundaries**. Figure 17 illustrates a very simple example of a sub-boundary. This consists of a planar *wall* of regularly spaced (distance: l) edge dislocations (BV: b) which produce a **tilt** with angle $\theta \simeq b/l$. In the same way, **twist** sub-boundaries joining two grains slightly disoriented by a rotation about an axis perpendicular to the boundary consist of a *network* of screw dislocations.

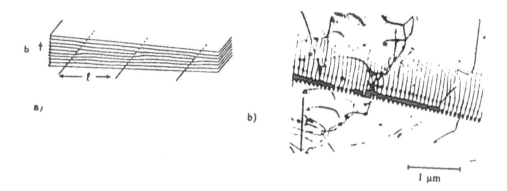

Figure 17 **Tilt sub-boundaries**.

  (a) The parallel, regularly spaced (distance: l), edge dislocations with Burgers vector $\bar{b}$, create a *tilt* between the upper and lower parts of the crystal (represented by its atomic planes) with angle $\theta \simeq b/l$.

  (b) Electron micrography of a tilt sub-boundary in an austenitic stainless steel (fcc). Transmission of electrons in a thin film reveals various dislocations, and, in particular, those of a sub-boundary cut by the two surfaces of the film, with $l \simeq 800$ Å

  After G **Dupouy** and F **Perrier**.

In a sub-boundary, the stress field of a dislocation does not extend beyond the distance l because of the presence of its neighbours. Thus, the surface energy of the tilt sub-boundary of Figure 17 is (see (12) and (9)):

$$\gamma \simeq \frac{\mu b^2}{4\pi(1-\nu)l} \ln \frac{l}{b} \simeq \frac{\mu b}{4\pi(1-\nu)} \theta |\ln \theta| \tag{20}$$

which increases very rapidly with θ for small θ and is in excellent agreement with experimental results (see later, Figure 19).

4.2 Grain boundaries

Above a certain disorientation ($\simeq 10°$) the above analysis, in terms of individual dislocations belonging to one or other of the two grains, is no longer valid as the distance between dislocations becomes too small. Here, we shall define the *grain boundary* only as the region in which two grains with arbitrary orientation are adjacent, and state the following small number of properties:

i)   The grain boundary is a *thin* region of 'bad crystal' (approximately one or two interatomic distances thick, see Figure 18).

ii) The fluctuations of atomic positions along a boundary provide numerous *favourable sites* (either under compression or under tension) in which atoms of impurities may be accommodated in the solid (see later, Section 4.4).

iii) Being a region of *positive* (surface) *energy*, any boundary tends to become planar when the temperature is such that equilibrium may be approached. This tendency can be seen on Figure 16. The energy of the boundaries is also the cause of their *reactivity* (see Section 4.5).

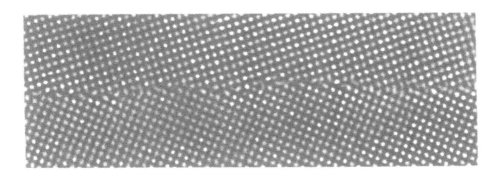

Figure 18   A grain boundary in rutile (TiO_2).
Rutile is a tetragonal crystal. Here, the boundary is viewed in projection (by high-resolution electron microscopy) in the very special case in which the two adjacent grains are derived from one another by a simple rotation around an axis (here [001]) perpendicular to the figure (*tilt boundary*).
By viewing along the boundary, its step-and-terrace structure is clearly visible.
After **V Dahmer, S Paciornik, I G Solorzano** and **J B Vandersande**, Interface Science, 2, 125 (1994).

4.3 Interfaces

It is common that a material should comprise several distinct **phases**[5]. The boundary between two phases constitutes an *interface*. Interfaces may be *coherent* or *incoherent*, depending on whether or not there is a simple relationship between the crystallographic orientations of the two phases. In the second case, the interface takes on the characteristics of grain boundaries. In the first case, the structure of the interface is generally that of a wall of dislocations (see Figure 19).

One important special case of coherence is that of **epitaxy**. Suppose we prepare (by polishing), a well-defined surface with crystallographic indices {$h\,k\,l$} of a sample A called a *substrate*. Next we grow (in vapour[6] or liquid phase) a solid A' of a different type. If the crystalline structure of

---

[5] This may be true in equilibrium (see p. 189) as well as out of equilibrium (for example, in the case of **composite materials**).

[6] In particular, using the technique of **molecular beam epitaxy**. Using ovens at a controlled temperature, atoms of A' *and/then* of A" (as desired) are emitted and collected on the substrate. This leads to the formation of alloys, or of multilayers, with a thickness controlled to within one atomic plane. The latter may be periodic (A' A" A' A" ...). This is the way in which *quantum wells* are obtained.

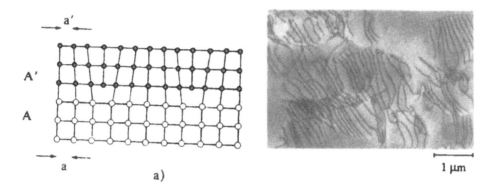

a)

Figure 19 Epitaxial or interface dislocations.

(*a*) If the interatomic distances a and a' of the two phases A and A' are close to each other, the sticking is done through a lattice of parallel dislocations, in this case, equidistant (distance: $a \cdot a'/\delta a$, where $\delta a = a' - a$) edge dislocations.

(*b*) Transmission electron micrography of a nickel-rich 'superalloy'. The interface dislocations between the matrix and precipitates of a second (so-called γ') phase can be seen.

After **M G Bunting** and **D W Hogan.**

A' contains a plane $\{h' k' l'\}$ in which the interatomic distances are similar to those of A in $\{h k l\}$, then, in general, A' will grow by a stacking of planes $\{h' k' l'\}$. The (small) difference between the interatomic distances of A and A' is 'corrected' by a wall of dislocations in the interface plane (see Figure 19). This is known as *epitaxial growth* (see also p. 107).

4.4 Segregation at the boundaries

Let us consider a boundary in a solid A containing *impurities* in the form of atoms of B. We have already stressed that, for reasons associated with elastic energy, the atoms of B are generally stabilised *in* the boundary. On the other hand, localising them there decreases the configuration entropy. Thus, there must exist an equilibrium concentration of B in the boundary. Let us denote the interaction energy between a B atom and the boundary by E_i (the energy change is E_i when B moves from the crystal to the boundary; it is negative if B is stabilised in the boundary); we also let C and J denote the numbers of crystalline sites *outside* and *inside* boundaries, respectively, and n and p denote the numbers of atoms of B in the crystal *outside* and *inside* boundaries. The presence of atoms of B in a boundary is accompanied by the variation of the free energy:

$$\Delta F = pE_i - T\Delta S$$

where

$$\Delta S = k \ln \frac{C!}{n!(C-n)!} \, \frac{J!}{p!(J-p)!}.$$

Expanding ΔS using Sterling's formula ($\ln N! \simeq N \ln N - N$), assuming low concentrations ($n/C \ll 1$) and minimising ΔF for a fixed number of B atoms ($dn + dp = 0$), we finally obtain:

$$j = \frac{c}{c + \exp(E_i/kT)} \tag{21}$$

where $j = p/J$ is the *concentration of B atoms in the boundary*, and $c = n/C$ is the *concentration* in the crystal *outside the boundary*. This formula shows that, for $E_i < 0$, a *supersaturation* (or **segregation**) of B atoms is produced in the boundary. This supersaturation (or ratio of the concentrations of B in the boundary and in the bulk: $j/c = (c + \exp(E_i/kT))^{-1}$) decreases at high temperature, where the entropy term tends to make the concentration of B in the solid uniform. It also decreases for large concentrations, when the boundary is *saturated*.

Example For $c = 10^{-2}$, $E_i = -0.1$ eV and $T = 473$ K (= $200°$ C), we find $j/c \simeq 10$ and the boundary has 10 times more impurities of B than the bulk of the sample.

4.5 Reactivity of boundaries

The energy of boundaries (or of interfaces) and the segregations which they drain generally make these the *places of least resistance*.

The *melting point* of the material is decreased, sometimes by several degrees on the grain boundaries.

The *sublimation energy* is reduced on the boundaries. If a polycrystalline solid is placed in a vacuum at a high temperature, the evaporation is faster on the boundaries than elsewhere and a furrow is left (thermal etching).

Grain boundaries are often *brittle* (tend to rupture under load) and serve as a privileged path for the propagation of the front of a fracture (this is referred to as *intergranular brittleness*).

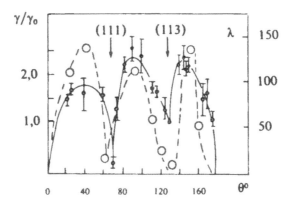

Figure 20 Energy of boundaries and corrosion. Two aluminium crystals are linked by a tilt boundary. The angle of disorientation around a common direction [011] is θ.

Black circles: the energy of the boundary divided by the value γ_0 measured when the plane of the boundary coincides with the plane (113).

White circles: the rate of corrosion of the boundary. The quantity λ is the depth (in μm) of penetration of pressurised water at $150°$ (over a period of 48 h).

After **J Y Boos** and **C Goux** (1971).

Figure 21 Grain boundary corrosion. Electron micrography (so-called scanning microscopy) of a surface of *aluminium* containing 50×10^{-6} atoms of copper and attacked by hydrochloric acid.

The figure shows a furrow corresponding to a preferential attack along a grain boundary. The boundary (invisible in the micrography) is situated at the bottom of and along the furrow. We also note the lateral attack on the two adjacent grains, along perpendicular {100} planes of the fcc structure. The scale bar represents 5 μm.

After O Iwao, *Thesis*, University of Toronto (1970).

Most of all, boundaries are susceptible to *chemical corrosion* (which is often dangerous). This susceptibility is closely correlated with the energy of the boundaries, as illustrated in Figure 20, which shows both the energy of tilt boundaries and the rate of intergranular corrosion as a function of the angle of tilt, measured for an aluminium bicrystal. The corrosion rate increases as the *energy* of the boundary increases, and is also larger the greater the *disorientation* of the two grains (see (20)). Figure 21 provides micrographic evidence of this *grain boundary corrosion*.

CHAPTER X

DEFORMATION AND FRACTURE

The Spirit of Laws
Montesquieu

The Deformed Transformed
Lord Byron

Forest Scenes
Robert Schumann

Fear of Fragmentation
Arnold Wesker

The mechanical behaviour of materials and, in particular the possibility that some may be **plastic** and undergo *permanent deformations*[1], while others may be **brittle**, depends on the nature of the solid in question, on its microstructure, on its 'history' (previous deformations), on external conditions (temperature, pressure etc) and, of course, on the manner in which the effort is applied.

Plastic deformation may be viewed as a *beneficial* event (this is clearly the case for a steel maker who rolls metal sheets or who forges steel; it is also the case for a seismologist who knows that earthquakes are generally associated with a resistance to deformation of a rock which when subjected to an intense effort, breaks instead of deforming plastically) or a *malefic* event (this is the case for an engineer who knows that the machine parts which he draws will certainly be deformed in time). In both cases, it is advantageous to know how deformation is produced on the **microscopic scale** and what thwarts it. Such microscopic descriptions complete, illuminate and, in certain cases, explain the **macroscopic laws** of deformation such as is established, and used, in mechanical engineering. We shall not describe these laws in detail here. We shall attempt to consider these in spirit rather than to the letter.

1. GENERALITIES

1.1 Tensile test

When an increasing effort is applied to a solid (for example, an increasing tensile stress applied to a *traction sample*), the latter is initially deformed in a reversible elastic fashion. When the stress exceeds a certain threshold called the **yield stress** (σ_E)[2], a permanent deformation occurs. The manner in which this deformation evolves, either as a function of the stress applied or as a function of time, is studied experimentally in **mechanical tests**. There are very many of these, each of which is adapted to particular conditions of use of the material; we mention in particular *traction* tests, *torsion* tests, *hardness* and *micro-hardness* tests, *creep* tests etc.

---

[1] During operations such as *rolling, wiredrawing, stretching, stamping, hammering, forging* etc.
[2] This was called σ_c on p. 203.

In practice, the most common test is the **tensile test,** in which a *traction machine* is used to pull on a sample with uniform cross section S, thus establishing a *uniaxial, uniform stress system.* The force applied F and the extension $\Delta l / l$ are then measured simultaneously. Next, a *stress–strain curve* $F = f(\Delta l / l)$ or, better still, a *rational* traction curve $\sigma = g(\varepsilon)$, is then constructed using the **true stress:**

$$\sigma = \frac{F}{S} \simeq \frac{F}{S_0} \left(1 + \frac{\Delta l}{l_0} \right) \tag{1}$$

(which is the initial stress F/S_0 (S_0 is the initial cross-section of the sample) corrected by the shape factor due to the continuous reduction of the cross-section (the volume is considered to be constant)) and the **rational deformation:**

$$\varepsilon = \ln \frac{l}{l_0} = \ln \frac{S_0}{S} = \ln \left(1 + \frac{\Delta l}{l_0} \right) \tag{2}$$

defined by $d\varepsilon = dl/l$.

Figure 1 shows a typical (at least for a metal) stress–strain curve. This begins with an elastic portion OE which is *reversible.* Above the yield stress ($\sigma = \sigma_E$), the solid undergoes a **permanent deformation.** If the applied stress is suppressed when an arbitrary point A is reached, the point on the figure follows the line AB parallel to the initial elasticity curve OE and passing through A. At this stage of the test, OB represents the permanent deformation of the sample. If the stress is re-applied to the sample, the new curve essentially describes the line BA which is the new elasticity curve of the deformed (or **cold-worked**) sample. Thus, the stress σ_A may be viewed as the yield stress of the material *after cold work.* It is actually called the *flow stress.* After A, the deformation continues until the sample breaks (point R).

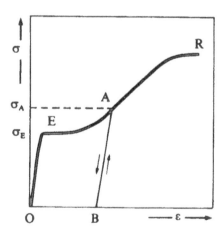

Figure 1 **Example of a tensile stress–strain curve of a solid.** σ is the applied stress or, better still, the true stress (see (1)). ε is the extension observed or, better still, the rational deformation (see (2)).

The gradient of the rational ($d\sigma/d\varepsilon$) curve is a measure of the difficulty of deformation; in other words, of the **hardness.** Often (but not always), the hardness increases with the deformation rate and then tends to decrease before the fracture (case of Figure 1).

1.2 Plastic flow

The deformation which occurs above the yield stress may be described in terms of flow. This is characterised by the system of *applied stresses* (for example, uniaxial uniform stress σ in the case of the tensile test) and by the *deformation rate* $\dot{\varepsilon}$ $(= d\varepsilon/dt)$. Only one of these quantities may be specified for the material during an experiment. Thus, the following two cases arise, working with constant T and P:

i) in a classical **tensile test** (or **compression** test) the rate $\dot{\varepsilon}$ is specified and the stress $\sigma(t)$ needed to obtain this rate is measured;

ii) in a **creep test**, on the other hand, a constant stress is applied and the deformation rate $\dot{\varepsilon}(t)$ is measured.

At the beginning of Chapter IX (p. 201), we mentioned the crucial role played by the movement of dislocations and, in particular, their glide during plastic deformation of solids. In this chapter, we shall study in more detail how this movement explains certain experimental data and certain macroscopic laws in different experimental contexts (in other words, for various temperature conditions, applied stresses and material microstructures).

Let us return to the case of the **glide of an edge dislocation**, illustrated in Figure 3 of p. 205. Let us suppose that such a dislocation (Burgers vector: $\bar{b}$) crosses a parallelepiped-shaped sample (sides A, B and C) from one surface to the other, as shown in Figures 2(c) and (d). This crossing leads to an elementary deformation (called a **shear** deformation here[3]), $\varepsilon' = b/A$. If there are N dislocations, all parallel to the first, uniformly distributed in the sample and all moving in the same direction with average velocity v, then the number crossing the sample during time t is $N(vt/B)$. Thus, during this time, the deformation is

$$\varepsilon(t) = \frac{b}{A} N \frac{vT}{B} \tag{3}$$

where the quantity N/AB is simply the **density of mobile deformations** (see p. 210) or ρ_{m}. The **deformation rate** $\dot{\varepsilon}$ is then given by

$$\boxed{\dot{\varepsilon} = \rho_{\mathrm{m}} v b} \quad \text{.(Orowan's formula).} \tag{4}$$

N.B. 1 If the dislocations **climb** rather than *glide* (see p. 215), the denominator AB of (3) is replaced by BA and Orowan's formula is still valid in the form (4).

N.B. 2 If, for the same dislocation movement as above, we are interested in a deformation other than the shear deformation (for example, the extension of a diagonal of the parallelepiped) the expression (4) for $\dot{\varepsilon}$ must be multiplied by a shape factor. For example, the reader should check that for the rod of Figure 1 this factor is equal to $\sin\alpha\cos\alpha$.

---

[3] The *shear* is the particular mode of deformation shown on Figure 2 (see also p. 204). A force $\sigma \times B \times C$ is applied to the upper face of the parallelepiped (here, to the right). The deformation is defined by the angle ε (Figure 2(b)). For $\sigma \le \sigma_{\mathrm{E}}$ (yield stress), ε is proportional to σ and the **shear modulus** μ of the solid is defined by $\varepsilon = \sigma\mu$ (Hooke's law, see p. 203). For $\sigma \ge \sigma_{\mathrm{E}}$, a permanent deformation ε takes place, which increases with σ as in a traction test. The elementary 'quantum' ε' of this permanent deformation ($\varepsilon' = b/A$) is shown in Figure 2(d).

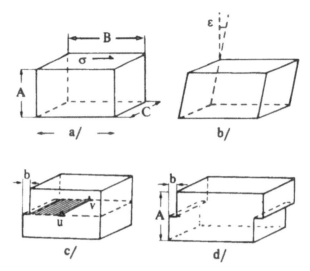

Figure 2 **Elementary shear deformation produced by the glide to the right of an edge dislocation with**
BV $\bar{b}$. The permanent deformation (*d*) from the undeformed crystal (*a*) is $\varepsilon' = b/A$. (*c*) shows the intermediate
situation in which the dislocation *uv*, which began on the left, has swept the area of the glide plane shaded in
grey. An elastic deformation ε is shown in (*b*). The symbol $\perp$ is as defined in the N.B. of p. 217.

In what follows, we shall describe a number of mechanisms for the deformation of
solids **at low** and **at high temperature**, where this distinction actually corresponds to
the two cases in which *the role played by atomic diffusion is insignificant* or *significant*,
respectively, during a time characteristic of the experiment.

2. DEFORMATION AT LOW TEMPERATURE

Here, we return in more detail than in Chapter IX to the process of **glide**, which is
actually the process which led us to introduce the concept of dislocation. Let us recall
the idea. If a dislocation with **Burgers vector** (BV) $\bar{b}$ glides in a glide plane (containing
$\bar{b}$) and terminates (in self-destruction) on a surface (for example, a free surface), it causes
an elementary deformation corresponding to the irreversible creation of a step $(\bar{b})$ (see
Figure 2).

Knowing that dislocations in the presence of a stress are subjected to a force (see
p. 211), it is easy to understand that they may be a potential cause of deformation.
However, for the deformation not to be transitory ('exhausting' the initially present
dislocations) and for it to become permanent, these dislocations disappearing on the
surfaces must be replaced by others. Thus, the crystal must contain **sources of**
dislocations. We shall show how this is possible by describing one of the most effective
sources.

2.1 Frank and Read's source

Let us consider a segment of a screw dislocation AB with BV $\bar{b}$ ($\bar{b} \parallel AB$), which can
glide freely in a plane P and suppose that the points A and B are fixed (Figure 3). This
is, in particular, the case when these points are **pinned down** by two dislocations MM'
and NN' outside the plane P (Figure 3).

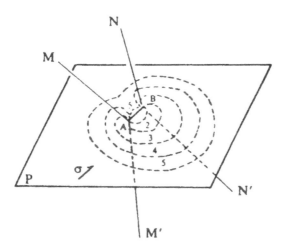

Figure 3 Frank and Read's source. The segment of screw dislocation AB, which is pinned down by the fixed dislocations MM' and NN', curves under the effect of the shear stress σ (stages $\underline{1}$ to $\underline{4}$). At stage $\underline{5}$ a dislocation loop is formed and the segment AB is regenerated.

In the presence of a stress σ (for example, a *shear stress* parallel to AB), the segment AB curves as a result of the force applied to the dislocation (see pp. 211–212). The curvature reaches a maximum (as does the force applied to the segment) when the dislocation has the shape of a semicircle with diameter AB. Beyond this position, which is achieved for a critical stress σ_c, the extension of the arc AB may continue freely in forwards, sideways and even backwards directions at the same time (here we recall that on Figure 9 of p. 211, the shear stress σ_{32} stretched the rectangular loop in *all* directions). Figure 3 shows various stages ($\underline{1}$, $\underline{2}$, $\underline{3}$ and $\underline{4}$) of this extension in the plane P. There comes a time (after stage $\underline{4}$ on Figure 3) when the two 'rear' portions of the loop, which are still extended under the effect of the applied stress, meet and *coalesce*, giving birth to the configuration $\underline{5}$, consisting of a large quasi-circular loop, which continues to stretch, *plus* an arc $\widehat{AB}$ which soon returns to the initial position of the segment AB.

Thus, during the operation, we have **created a new loop** and **restored the segment** AB ready to restart the cycle. Consequently, under the effect of the applied stress, this dislocation *factory* creates a series of quasi-concentric loops which develop in the solid (on the plane P) like a circular wave propagating on the surface of a liquid.

N.B. 1 Returning to the evaluation on p. 212 of the force which must be applied to an arc of a dislocation to ensure that it remains curved (see (15)) and to the relationship between the force and the shear stress (see (16), p. 212), we see that the shear stress σ curves the arc AB with radius of curvature R such that

$$\sigma \simeq \mu b / R \qquad (5)$$

where AB is a screw dislocation. For an edge dislocation, R must simply be multiplied by $(1 - v)$, in accordance with the remark following formula (13) on p. 210. This stress is a maximum when the arc is semicircular, in other words, for $R = l/2$ (l is the length of AB). Whence, the critical stress σ_c (see above) for which the arc AB has a semicircular shape is:

$$\sigma_c = 2\mu b / l.$$

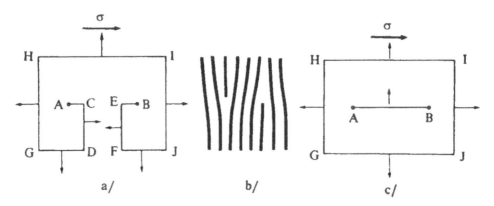

Figure 4   Coalescence of dislocations during Frank and Read's mechanism (Figure 3). (*a*) and (*c*): plane of the figure = plane P of Figure 3. (*b*): section of the segments CD and EF perpendicular to CD; the superposition of CD and EF restores the perfect crystal and thus erases the segments CD and EF of the dislocation. The small arrows indicate the directions of the forces applied to the various portions of the dislocation by the stress σ, thus, also, the movements of these portions.

Thus, for a source length $l = 1$ μm (trivial case), the critical stress needed for the source to operate is $\sigma_c \simeq 10^{-4}$ μ, which is of the order of magnitude of the reduced shears measured ordinarily (see p. 203).

N.B. 2   Let us come back for a moment, on the other hand, to the phenomenon of **coalescence** which occurs between stages $\underline{4}$ and $\underline{5}$ of the above. Let us suppose, to simplify the drawings, that the loop at stage $\underline{4}$ has a rectangular shape (see Figure 4(*a*)). The portions CD and EF of the loop, which face one another and which the stress (shear parallel to AB) causes to advance towards one another, are both *edges* (since they are perpendicular to AB and thus to the Burgers vector $\bar{b}$). They have the structure shown in Figure 4(*b*). It is immediately clear that when they meet, these two portions wipe each other out, leading to the configuration of Figure 4(*c*) (analogous to stage $\underline{5}$ of Figure 3) where the initial segment is regenerated, accompanied by a loop $GHIJ$, which continues to stretch under the effect of the stress.

Exercise   Show, using the *node equation* (see p. 273) that when they meet the segments CD and EF form a 'dislocation' with zero BV.

2.2. Glide of dislocations

We know (see p. 202) that one common mode of plastic deformation involves the glide of the (generally dense) crystalline planes P, one on top of the other, where this gliding is the consequence of the glide (same word, different phenomenon) of corresponding dislocations (in other words, with their BV *in* P) in the planes P. Having described in the above section how dislocations are created in a plane P, we now return to discuss their glide in more detail than on p. 204.

Figure 5(*a*) shows a section of an edge dislocation with BV $\bar{b}$. This figure should be compared with Figure 3 (p. 205). Only the positions of the atomic rows (black circles) immediately above (under compression) and immediately below (under tension) the glide plane are shown. We are interested in the displacements of the former with respect to the latter, which are assumed to be fixed. Figure 5(b) shows the displacements relative to the stationary dislocation. These may be used to define a **width** of the dislocation, w (for example, that corresponding to a displacement greater than $b/4$). When the dislocation glides *one* interatomic distance *to the right* (see arrow) its centre passes from A to B and each row within the width w glides a distance $\simeq b^2/w$ to the right.

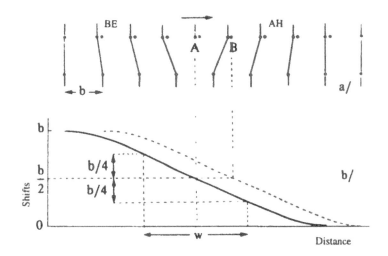

Figure 5 Displacements of atoms near an edge dislocation. (*a*) Positions of atoms when the dislocation is at A (black circles) and after glide to B (white circles). (*b*) Displacements of atoms of the upper rows with respect to the lower rows, as a function of the distance, for position A (continuous line) and position B (broken line). The displacement at a large distance from the dislocation is equal to zero on one side and b on the other.

It is immediately clear that a row AH 'ahead' of the dislocation resists this movement and thus exerts a backwards-pointing force on it, while a row BE 'behind' the dislocation exerts a forwards-pointing force. In the symmetrical positions A and B, these two opposing forces are equal for each pair AH–BE which is symmetrical with respect to the dislocation; thus, the latter is in equilibrium. During the movement of A to B, this is not totally exact, since 'symmetrical' rows such as AH and BE are not quite equidistant from the moving dislocation. However, it is apparent here that the differences between these almost equal forces remain very small; this is even more true the greater the width of the dislocation w. The corresponding stress needed to cross the energy barrier half-way between the two stable positions A and B is called the **Peierls–Nabarro stress**.

These stable positions A, B etc correspond in three dimensions to valleys in the potential (**Peierls valleys**) perpendicular to the plane of Figure 5. The glide of the dislocations occurs by jumps between adjacent valleys, where these jumps involve *a stress σ_P which is weaker the wider the dislocation is.*

The calculation of this stress, first carried out by Peierls and Nabarro, leads to the formula

$$\sigma_P \simeq \mu \exp(-2\pi w/b) \tag{6}$$

where μ is again the shear modulus (one of the two Lamé moduli, see p. 208) and w is the width of the dislocation[4].

What does the width of a dislocation depend on?

This width is the result of a compromise between two opposing tendencies: if w is increased, the elastic energy distributed across a large number of rows decreases; but at the same time the number of atomic bonds deformed increases. Thus, we expect **wide dislocations in 'good' metals** in which the interatomic bonds are sufficiently isotropic, and **narrow dislocations in solids such as insulators or covalent semiconductors** in which bonds are highly directional.

Calculation of this width gives:

$$w \simeq \frac{\mu b}{2\pi(1 - \nu)\sigma_0} \tag{7}$$

where σ_0 is the theoretical glide stress (see (1'), p. 203) and ν is the Poisson ratio (see p. 209). The presence of σ_0 (whence also that of b/a, where a is the interplanar distance, see (1'), p. 203) in the denominator of w reflects the fact that the directional nature of the bonds together with the crystalline structure and the choice of glide planes P are critical elements as far as the width w is concerned.

We shall summarise the above as follows:

i)   Glide of the dislocations occurs by jumps between adjacent potential valleys (successive parallel rectilinear valleys A, B...: *Peierls valleys*).
ii) These jumps require the application of a stress σ_P.
iii) This stress is weaker the greater the width w is, and also the weaker the *theoretical glide stress* σ_0 (see p. 203).

We shall now distinguish between two extreme cases of **hard** and **soft** materials, corresponding to **large** or **small values** of σ_P in comparison with the stresses generally applied.

1 – Intrinsically hard materials

Solids with *highly directional interatomic bonds* (high value of σ_0) have *deep Peierls valleys*. Thus, they are in principle *hard* at low temperature, where glide is difficult. This is generally the case for ionic crystals (in particular, for *minerals*), in which it is unfavourable to bring two ions of the same sign closer together. This is also the case for covalent crystals with directional bonds (*diamond, silicon* etc).

This is the reason for the current development (1990s) of numerous industrial **ceramics**[5] of a covalent nature. In addition to their main property of *hardness*, these may also have the following characteristics (depending on the case and on the application): resistance to *fracture* (see p. 250), to *friction*, to *thermal shocks* and to *corrosion* etc, or *transparence* to electromagnetic waves of a given wavelength etc.

We mention the following in particular:

*   *Alumina* Al_2O_3, which is used for taps (domestic or industrial) or in single crystalline platelets (Figure 6) to abrade or reinforce cutting tools.
*   *Silicon nitride* Si_3N_4, which is used for ball bearings working at very high speeds (60 000 revolutions per minute) without lubrication.

---

[4] See for example, A H Cottrell, *Theory of Crystal Dislocations*, Documents on modern physics, Gordon and Breach (1964).

[5] Here, we call compounds of carbon, nitrogen or oxygen with partly covalent, partly ionic bonds, 'ceramics'. They are insulators (Al_2O_3 etc) or conductors (TiC etc) of very high hardness and low density.

- *Silicon carbide* SiC, which is shaped by uniaxial pressing of powders in the manufacturing of water pump rings (cars) or fittings for pumps for the chemical industry.
- *Zirconia* ZrO_2, which, after sintering, constitutes the heads of prostheses (hips) with good tribological properties (low friction and wear of the polyethylene female part).
- *Alumina–zirconia–titanium carbide TiC alloys* which are used to machine cast iron.
- *Cordierite* Al_2O_3-SiO_2-MgO which is used to make filters for liquid metals before casting (to eliminate scum).

10 µm

Figure 6   **Single crystalline platelets of alumina used to reinforce cutting tools.** Micrography: Société Atochem.

A large b/a factor (see the previous page) may also give rise to hardness. This is the case for many intermetallic compounds (for example, $CuAl_2$) with complicated crystalline structures, which, despite their metallic nature, are often hard and brittle.

This intrinsic hardness[6] decreases very generally as the temperature increases. For a dislocation width w as small as $\simeq b$, the jump from valley to valley involves jumps of atoms similar to those in diffusion and, even at quite a low temperature, one would expect the hardness, or more precisely, the stress σ_P to be thermally activated: $\sigma_P = \sigma_P^0 \exp(-E/kT)$. This is actually what is observed in hard materials such as LiF or TiC.

2 – Intrinsically soft materials

Good metals (such as Al, see Chapter IV) with highly non-directional bonds are the prototypes of *soft* materials. Here, the dislocations, which are generally dissociated (see p. 214) are wide. The b/a factor is small for the dense planes (large a), which are precisely the glide planes. The Peierls–Nabarro stresses here are very weak (for pure crystals) and these solids remain plastic to very low temperatures.

[6] Intrinsic in the sense that it is not associated with the presence of foreign atoms and that it exists in the ideally pure solid.

Solids other than metals may be soft at low temperature, thanks to a favourable b/a factor. This is the case, for example, for *graphite* and, more generally, for solids consisting of rigid atomic planes weakly bonded together (large a), for example *talc*, *molybdenum bisulphide* (used as a lubricant), *soap flakes* etc which are solids of very weak Peierls–Nabarro stresses.

In the next paragraph, we draw up a list of **extrinsic** factors which oppose glide and increase the hardness of these **intrinsically** soft solids.

2.3 Obstacles to glide

After having seen that, in an intrinsically soft material an applied stress tends both to create dislocations (Section 2.1) and to cause them to glide (Section 2.2), thus provoking a plastic flow (see (4)), we must now examine a number of sources of **hardening** in these materials.

1 – Hardening by grain boundaries
In a solid consisting of *grains* (see p. 217) of average size ϕ, the dislocations emitted by a source under the effect of a stress σ and gliding in a plane P *come up against* a grain boundary without being able to cross it in the first instance (see Figure 7(a)). These dislocations, which are mutually repulsive (Figure 7(b)) are pushed forwards by

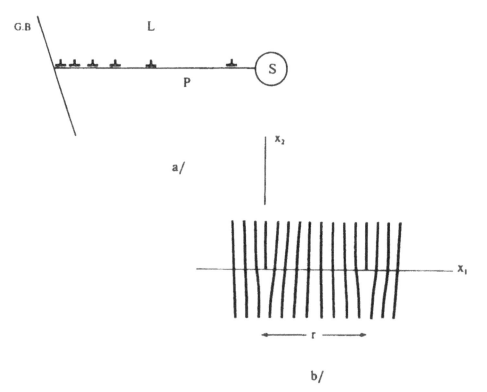

Figure 7 Pile-up of edge dislocations emitted by a source (S) in a plane P which come up against an obstacle (here, a grain boundary). (a) The pile-up as a whole. (b) Repulsion of two adjacent dislocations with the same Burgers vector. This figure should be compared with Figure 4(b), in which, unlike here, there is elastic attraction between the two dislocations.

the stress σ but blocked by the stress σ_{gb} from the barrier (grain boundary) which the leading dislocations come up against. They stop, forming a **pile-up** of length L containing n dislocations. Each of these is subjected to the force

$$F = \sigma b$$

per unit length (see (16), p. 212). If the pile-up as a whole moves by dx, the work provided is $nF\mathrm{d}x = n\sigma b\mathrm{d}x$, which is equal to the work executed by the leading dislocation, namely $\sigma^*b\mathrm{d}x$, where σ^* denotes the stress at the head of the pile-up. Thus, we have

$$\boxed{\sigma^* = n\sigma.} \qquad (8)$$

Equation (8) shows that the pile-up behaves as a **concentrator of stress**. It is this accumulated stress σ^* which finally enables the dislocations *either* to cross the boundary, when it becomes equal to the stress σ_{gb}, *or* to trigger remotely a source in the adjacent boundary, thereby propagating the deformation.

If we accept (for justification, see **N.B.** below) that the *length* of the pile-up is *proportional to the number of dislocations n* composing it and *inversely proportional to the stress* compressing it σ, then we have

$$L = \text{constant}\, n/\sigma. \qquad (8')$$

Thus, taking into account that this length is of the order of the size ϕ of the grain, we see (using (8) and (8')) that the stress to be applied to overcome the 'resistance' of the boundary is

$$\sigma = \frac{\sigma_{gb}}{n} \propto \frac{\sigma_{gb}}{L\sigma} \quad \text{whence} \quad \sigma \propto L^{-1/2} \propto \phi^{-1/2}. \qquad (8'')$$

If we add to this stress that needed to make the dislocations in a single crystal glide, namely σ^0 (equal to at least σ_P), the yield stress σ_E of a polycrystal takes the form

$$\boxed{\sigma_E = \sigma^0 + K\phi^{-1/2}} \qquad (9)$$

where K is a constant. This is the **law of Hall and Petch** which generally provides a good description of experimental results (an example is shown on Figure 8). This law is a clear statement of the technologically important fact that *a polycrystal with fine grains is stronger than a polycrystal with coarse grains* or, *a fortiori, than a single crystal.*

N.B. Let us now return to Figure 7(*b*) and justify what we have just said. With the axes defined on this figure, the dislocation (1) parallel to x_3 passing through the origin, with BV $\bar{b}_1$ ($\parallel x_1$), creates the system of stresses $\sigma_{r\theta} = \sigma_{12} = \sigma_{21} = \mu b_1/2\pi(1-v)r$ at the point ($x_1 = r$, $x_2 = x_3 = 0$) (see (9), p. 208). If a dislocation (2), also parallel to x_3, passing through this point, with BV $\bar{b}_2$ ($\parallel x_1$) is introduced, it is subjected by the first dislocation to a force which is repulsive if $\bar{b}_1 \cdot \bar{b}_2 > 0$ and which has intensity given by (16') (p. 212; see also the **Exercise** of p. 212):

$$F = \frac{\mu b_1 b_2}{2\pi(1-v)r}$$

per unit length. In fact, here, the vector $\bar{b} \cdot \bar{\sigma}$ of (16') has components $p_1 = 0$, $p_2 = \sigma_{21}b_2$ and $p_3 = 0$. If the two parallel dislocations are *identical* ($\bar{b}_1 = \bar{b}_2 = \bar{b}$, case of Figure 7(*b*)), this repulsive force is equal to

$$F = \frac{\mu b^2}{2\pi(1-v)r} = \frac{\mu b^2}{\lambda r} \qquad (10)$$

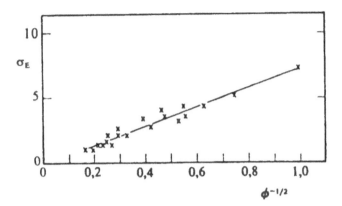

Figure 8   Yield stress σ_E (in kg mm^{-2}) of aluminium polycrystals as a function of the average grain size ϕ (in μm). We note that the law (9) is obeyed with $\sigma^0 \simeq 0$.
   After D McLean, *Mechanical Properties of Metals*, J Wiley, New York (1962).

per unit length, where $\lambda = 2\pi(1 - \nu)$.

The equilibrium of the pile-up of n dislocations results from the equilibrium between these repulsive forces and those due to the applied stress σ which push the n dislocations forwards. In particular, the distance r_1 between the leading dislocation and the second results from the equilibrium between the repulsion $\mu b^2/\lambda r_1$ and the force which pushes this second dislocation forwards, namely $(n - 1)\sigma b$ (see (8)). The third dislocation is pushed by the force $(n - 3)\sigma b$ and held back by the repulsion between the first two, which may itself be likened to the repulsion of a single dislocation with BV $2\overline{b}$. Thus, step by step, we obtain[7]:

$$r_1 = \frac{\mu b}{\lambda(n - 1)\sigma}, \quad r_2 = \frac{2\mu b}{\lambda(n - 2)\sigma} = \frac{2(n - 1)r_1}{n - 2}, \dots \quad r_p = \frac{p(n - 1)r_1}{n - p}.$$

Therefore, the total length L of the pile-up is:

$$L = r_{n-1} = (n - 1)^2 r_1 \simeq \frac{n\mu b}{\sigma}. \tag{11}$$

This is proportional to the number of piled dislocations and to the inverse of the applied stress, as we conjectured on the previous page.

Thus, the stress exerted at the head of the pile-up (see (8)) is:

$$\sigma^{\bullet} \simeq \frac{L\sigma^2}{\mu b}. \tag{12}$$

When this becomes equal to a critical stress σ_c which *either* permits the crossing of the boundary *or* activates a source in the adjacent boundary, the plastic deformation may evolve from grain to grain. By combining (10) and (11), noting that the length L can only be of the order of the grain size ϕ and adding a glide stress σ^0 to σ (where σ^0 is generally small in comparison with σ), we obtain:

$$\sigma \simeq \sigma^0 + \left(\frac{\sigma_c \mu b}{\phi}\right)^{1/2} \tag{13}$$

which agrees with the empirical law (9).

[7] **J** Friedel, *Les dislocations*, Gauthier-Villars (1956), p. 158; or *Dislocations*, Pergamon (1964).

2 – Forest hardening

Large numbers of dislocations are created during a deformation (see Section 2.1). The dislocation density ρ grows, increasing both the density of mobile dislocations ρ_m (see (4)) and that of dislocations which are stationary or almost stationary (*either* because they are pinned down, *or* because the force to which they are subjected ((16) and (16'), p. 212) is weak or zero). The latter generally impede the movement of the former and decrease the velocity v of formula (4). This is the origin of the *cold-work* hardening observable on Figure 1[8].

Let us consider the glide of a dislocation D with BV $\bar{b}$ in the plane P, where this plane is cut by a set (or **forest**) of parallel dislocations (called **trees**) (see Figure 9). The crossing of the forest by the mobile dislocation D creates a hardening, which has the following two principal origins.

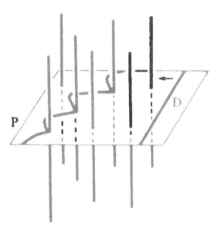

Figure 9 **Crossing of a forest of dislocations by a dislocation D.** P: glide plane. To the right: initial position of D; to the left: final position of D, curved by crossing the trees.

i) **Dislocation–tree interaction** We have already mentioned that there exists a force between two dislocations $(\bar{b}_1)$ and $(\bar{b}_2)$. Depending on the individual case, **this force may be attractive, zero or repulsive**.

This force may be calculated using (16') (p. 212). For two rectilinear dislocations $(\bar{b}_1)$ and $(\bar{b}_2)$, it is found to be proportional to $\bar{b}_1 \cdot \bar{b}_2$. In particular, it is repulsive for $\bar{b}_1 = \bar{b}_2$ (case of (10) and Figure 7(b)) and attractive for $\bar{b}_1 = -\bar{b}_2$ (see **Exercise**, p. 212).

Whether it is *attractive* or *repulsive*, this force creates a hardening which increases proportionally with the density of the forest. In fact, crossing the forest requires an additional effort beyond that of the simple glide. *In the first case*, D must be drawn away from each of its interactions with the trees and *in the second*

[8] Conversely, annealing of a *cold-worked* material at high temperature favours, by glide and climb, the mutual annihilation of the dislocations created at the time of the cold work. The resulting decrease in the energy of the crystal is often accompanied by a modification of the structure and the crystalline orientation of the grains. This is known as **recrystallisation** (see p. 210). The hardness then decreases sharply. This is why the plumber *reheats* his initially cold-worked copper pipe with his blowlamp in order to be able to bend it.

one, it must be forced to approach the trees. Under these conditions, a dislocation (D) which is initially rectilinear, rapidly takes on a zig-zag shape as it crosses the forest (Figure 9), which characterises the friction to which it is subjected.

In fact, *attractive junctions* are more effective for hardening material than *repulsive* ones. Let us imagine two dislocations D_1 $(\bar{b}_1)$ and D_2 $(\bar{b}_2)$ approaching and attracting one another $(\bar{b}_1 \cdot \bar{b}_2 < 0$, see above). The force of attraction, which varies as $1/r$ (see (10)), deforms the two dislocations in their zone of proximity so that the two portions $A_1 B_1$ and $A_2 B_2$ (Figure 10(a)) become parallel. These two portions then tend to coalesce into a single dislocation segment (AB on Figure 10(b)) with BV $\bar{b}$. The reader should check, by introducing two Volterra surfaces S_1 and S_2 passing through D_1 and D_2 (see also **Annex 5**), that $\bar{b} = \bar{b}_1 + \bar{b}_2$. Since $\bar{b}_1 \cdot \bar{b}_2 < 0$, whence $b^2 < b_1^2 + b_2^2$, it follows that the coalescence $A_1 B_1 + A_2 B_2 \rightarrow AB$ *decreases* the energy (see (13), p. 210) and thus is very likely. These *attractive junctions* are difficult to break. Thus, they contribute very effectively to hardening.

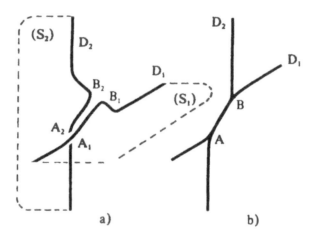

Figure 10   Attractive junction of two dislocations D_1 and D_2. (*a*) As they approach each other, the two dislocations deform one another by attraction. (*b*) Recombination of the segments $A_1 B_1$ and $A_2 B_2$ is followed by the formation of a junction AB. S_1 and S_2: Volterra surfaces of dislocations D_1 and D_2.

ii)   Formation of jogs We have already noted (see Figure 15, p. 216) that the junction of two dislocations is accompanied by the formation of *jogs*. Another example is given in Figure 11, which shows a screw dislocation D_1 $(\bar{b}_1)$ glide, meeting an edge *tree* D_2 $(\bar{b}_2)$, a jog of edge type (since it is perpendicular to $\bar{b}_1$) being created on D_1 after the crossing.

Often, jogs glide less easily than the dislocation which carries them. Thus, as far as the global movement of the latter is concerned, they constitute 'brakes', introducing a friction, whence also a hardening in this case.

This is the situation for the edge jog CC' of Figure 11(*b*). The movement which is required of it in order to accompany the dislocation D_1 in its glide must take place in a plane perpendicular to its Burgers vector. This is a *climb* movement, which requires *atomic diffusion* and is thus very slow at low temperature.

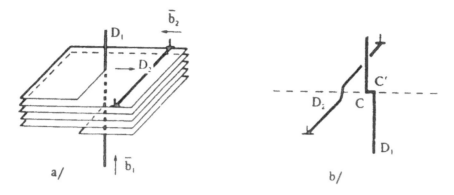

Figure 11 Formation of a jog CC' of edge type on a screw dislocation D_1 after crossing the edge dislocation D_2.

3 – Alloy hardening
Foreign atoms in solid solution, or in **precipitates**, interact very generally via their stress field with the dislocations, thus creating an additional source of hardening.

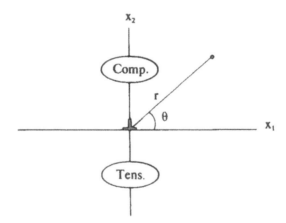

Figure 12 Regions under compression and under tension near an edge dislocation lying along x_3. See N.B. of p. 217, for an explanation of the symbol $\perp$.

It is easy to convince oneself of this by imagining an edge dislocation analogous to that of Figure 12 and examining the regions immediately *above* ($\theta = \pi/2$ in (9), p. 208) and immediately *below* ($\theta = -\pi/2$) the dislocation. The former is a region **under compression**, the latter a region **under tension**. If now an atom of the crystal is replaced by an atom of a different *size*, the theory of elasticity predicts a **decrease in energy** for *either* a **smaller** atom situated *above* *or* a **larger** atom situated *below* (see Part III of **Problem 13**).

It is actually a classical result of the theory of elasticity of isotropic media that the elastic energy of a sphere of radius $R(1 + \varepsilon)$ introduced into a spherical cavity of radius R, due to the presence of a hydrostatic-pressure stress p, is:

$$W = p\Delta V \quad \text{where} \quad \Delta V = 4\pi R^3 \varepsilon$$

and that the hydrostatic-pressure component of the stress field σ_{ij} is

$$p = \tfrac{1}{3}(\sigma_{11} + \sigma_{22} + \sigma_{33}). \tag{14}$$

Using the expression (9) (p. 208) for the stresses created by the edge dislocation (see Figure 12), it is easy to see that

$$W = \frac{4\mu b R^3 \varepsilon}{3r} \frac{1+\nu}{1-\nu} \sin\theta. \tag{15}$$

It follows immediately that 'above' the glide plane ($\theta > 0$) this energy is in fact negative for atoms smaller than the atoms of the crystal ($\varepsilon < 0$) and that 'below' ($\theta < 0$) it is negative for larger atoms ($\varepsilon > 0$).

N.B. An elasticity calculation for a *screw dislocation* leads to an interaction energy $p\Delta V$ which is *zero* to the first order, since in this case, we have $p = 0$ (see (14), above and (7), p. 208). In fact, a rectilinear screw dislocation L which passes near a foreign atom tends to bend so that locally it adopts a direction perpendicular to L, whence an edge nature near this atom; this *then* enables it to decrease its energy.

The important thing is that *in both cases* (large or small atoms) there exists a region which makes it possible to *decrease the energy* of the *foreign atom–dislocation* as a whole. Thus, in both cases, the foreign atoms are liable to **pin down** the edge dislocations, since the interaction energy must be provided to draw the dislocation away from this foreign atom. This pinning down explains the considerable hardening of solid solutions such as Au-Pt, Au-Ag etc in comparison with the pure metal (Au etc). This is properly called *alloy hardening* (see point i) of p. 11). We distinguish two cases here:

i) *The temperature is too low* for the foreign atoms to diffuse. It is then that the dislocations, gliding under the effect of a stress, meet these atoms and are pinned down.

ii) *The temperature permits diffusion.* The atoms diffuse under the effect of the stress field of the dislocations towards the latter and cluster around these in **Cottrell clouds**. This formation of clouds (described in Part III of **Problem 13**) tends to block the sources **and/or** inhibit the glide of dislocations and thus leads to hardening.

All we have said and proved above for the case of inclusions of atomic size can easily be extended to most *clusters* of foreign atoms and, in particular, to **precipitates** accompanied by a size effect ΔV. The general idea remains the same: since a given edge dislocation is mobile, in the presence of a precipitate surrounded by a stress field it may in general arrange itself so that the interaction energy is negative, which may imply that the dislocation is then pinned down or blocked by this precipitate (see Figure 13). This is, in particular, what happens with copper precipitates (see p. 173 and p. 183) in *Al-Cu alloys*. These precipitates, in the form of planar clusters parallel to the {100} planes of the aluminium, create long-distance distortions. It is the interaction of these deformations with the stresses of the dislocations which impedes the movement of the latter, producing the so-called **structural hardening** typical of this type of alloy[9].

[9] In order to decrease the density of these alloys even further, in the 1980s, the Société Péchiney developed an alloy of the same type as the duralumin, but in which the copper is replaced by *lithium*, a lighter atom. This means that a given aircraft could gain the weight of 10 to 20 passengers.

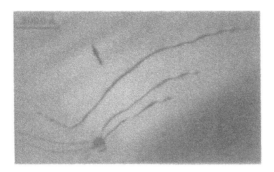

Figure 13 Interaction of gliding dislocations with a precipitate in a stainless steel (Fe 55% – Ni 20% – Cr 25%). Transmission electron micrography. The precipitate is the black spot seen at the bottom. The stress tends to displace the three dislocations in the direction of the arrow. The dislocations are interrupted by the surfaces above and below the sample. The dark fringes on the bottom right correspond to a *grain boundary*. After **L Boulanger**.

In a nuclear reactor, materials are subjected to irradiation by fast neutrons (see **Problem 17**). The latter produce small clusters of vacancies (*Seeger 'zones'*) which contribute, in the same way, to the *irradiation hardening*[10].

3. DEFORMATION AT HIGH TEMPERATURES

We explained earlier (p. 225) what we mean by high temperatures here. These temperatures at which atomic diffusion plays a significant role are generally those above approximately 0.4 or 0.5 T_m (where T_m is the melting point).

Diffusion may change (sometimes radically) the conditions of creation or mobility of the dislocations, whence also the modes of deformation. Blacksmiths were well aware of this when they *heated* pieces which were difficult to deform in their cold state. In the same way, the materials are **heated** when *rolling* metal or *forging* large items (wagon wheels, axles etc).

In what follows, we shall give a number of examples of deformation mechanisms in which the **atomic mobility** intervenes directly and which are connected with **creep**.

N.B. *Creep* refers to a deformation (generally slow) under stress, observed as a function of *time*, which makes a solid behave like a *fluid*; the term *flow* is also used. In practical experiments, a *creep test* involves studying the deformation $\varepsilon(t)$ of a solid subjected to a *constant stress*, at a given (usually high) temperature. The deformation curves obtained under these conditions are often *linear in t*, after an incubation period, which enables us to define a typical *flow rate* $\dot{\varepsilon}$ for the stress and the temperature of the test.

3.1 Newtonian creep

In this section, we shall consider an extreme case of the above, in which the mobility of point defects is the *sole* factor determining the creep.

[10] **A Seeger**, U.N. Conference on Atomic Energy, **6**, 250, (1958).

Let us consider a **single crystal** initially in the shape of a cube (edge: L) (or a **grain of a polycrystal**) subjected, at temperature T, to a system of stresses obtained by superposing a *tension* σ_1 along x_1 (see Figure 14) and a *compression* σ_2 along the perpendicular direction x_2. We set

$$\sigma = \sigma_2 = \sigma_1.$$

We assume that the crystal has no dislocations. The only contributions to the flow will come from point defects, essentially vacancies, for which the unique sources and sinks are the *free surfaces* (or the *grain boundaries* in the case of a polycrystal).

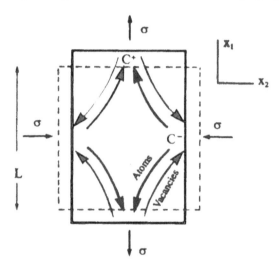

Figure 14 Principle of Newtonian creep by diffusion of vacancies. The dashes show the initial shape of the sample.

The bulk equilibrium concentration of vacancies (see (6), p. 162) is not exactly the same as that in the immediate vicinity of the surfaces *under tension* or those *under compression*. A vacancy is created by extracting an atom from *beneath* the surface and placing it *on* the surface. The stress exerted on the latter produces a work $\pm \sigma b^2 = \pm \sigma b^3$, depending on its sign (here, b is the interatomic distance and b^2 is the 'atomic surface area' in a crystal with a simple cubic structure). Thus, very close to the surface under *tension* (*resp.* under *compression*) the formation energy of a vacancy, which is equal to E_f in the unstressed solid, becomes $E_f - \sigma b^3$ (*resp.* $E_f + \sigma b^3$). The equilibrium concentrations are then:

$$c^+ = c_0 \exp(+\sigma b^3/kT) \qquad \text{near surfaces under } tension$$
$$c^- = c_0 \exp(-\sigma b^3/kT) \qquad \text{near surfaces under } compression \qquad (16)$$

where $c_0 = \exp(-E_f/kT)$.

This difference in the concentrations of vacancies at different points of the solid results in a **flux of vacancies** (Fick's equation, p. 176) from the surfaces under tension to the surfaces under compression. There is a corresponding equal and opposite **flux of**

atoms, which, leaving certain surfaces to settle on others, thus imply a deformation of the solid (Figure 14).

Without permanent knowledge of the detailed shape of the gradient of the concentration of vacancies, we may, to the first order, consider it to be constant (see (10), p. 176, with $\partial c/\partial t = 0$). If l is the mean path of the vacancies (very similar to L), we may then write:

$$|\overline{\mathrm{grad}\, c}| = \frac{c^+ - c^-}{l} \simeq \frac{c^+ - c^-}{L}. \tag{17}$$

The number of vacancies N transferred from one face (with cross sectional area L^2) to the other per unit time is then proportional to the flux J (see (9′), p. 176):

$$N = JL^2 = \frac{1}{\Omega} D_l L^2 |\overline{\nabla} c|. \tag{18}$$

where $\Omega = b^3$ is the atomic volume. The corresponding N atoms contribute to an elementary extension δL of the sample:

$$\delta L = N \frac{b^3}{L^2}$$

per unit time, whence to a **deformation rate**

$$\frac{\mathrm{d}}{\mathrm{d}t}\left(\frac{\delta L}{L}\right) = \dot{\varepsilon} = \frac{Nb^3}{L^3}. \tag{19}$$

Thus, combining (16), (17) and (18), we obtain:

$$N \simeq \frac{c_0 D_l L}{\Omega}\left(\exp(\sigma b^3/kT) - \exp(-\sigma b^3/kT)\right)$$
$$= \frac{2c_0 D_l L}{\Omega}\sinh \sigma b^3/kT.$$

For a stress (already large) of $10^{-3}\mu$ (see p. 203), or $\sigma b^3 = 10^{-3}\mu b^3 \simeq 5 \times 10^{-3}$ eV (see p. 210) and for $T = 1200$ K (or $kT \simeq 0.1$ eV), we have $\sigma b^3/kT \simeq 5 \times 10^{-2}$. For standard conditions of use of the material, we may therefore set $\sigma b^3 \ll kT$. Thus,

$$\dot{\varepsilon} = \beta c_0 D_l L \frac{\sigma}{k} \frac{b^3}{T} \frac{1}{L^3}$$

where β is a numerical coefficient ($\simeq 2$), or, finally:

$$\boxed{\dot{\varepsilon} = \frac{\beta D_{\mathrm{SD}}\Omega}{L^2 kT}\sigma} \tag{20}$$

where D_{SD} is the *self-diffusion coefficient* (called D_{A} in Chapter VIII; see (14), p. 178), and we have replaced b^3 by the *atomic volume* Ω. The quantity $\dot{\varepsilon}$ essentially describes the deformation of a cubic monocrystal with side L. For a polycrystal with average grain size L, this formula is also valid for the deformation of the solid. In fact, since (20) describes the behaviour of *each grain*, we shall assume here that the elementary deformations of each grain are *additive* and, in particular, that there is no decohesion between grains.

This process involving vacancies, which is generally known as **Nabarro–Herring creep**, depends on the temperature chiefly through the exponential of the self-diffusion coefficient: it is thermally activated, with activation energy the self-diffusion energy. The thermal activation included in D_{SD} and the variations in L^{-2} and in σ predicted by (20) are observed in certain experimental cases[11] (metals, ionic crystals etc), particularly for small L.

If we recall *Newton's law*

$$\dot{\varepsilon} = \alpha\sigma_c = \frac{\sigma_c}{\eta} \tag{21}$$

which gives the *flow rate* of a fluid with **viscosity** η ($\alpha = \eta^{-1}$ is the *fluidity*) as being *proportional to the stress*, we see that the Nabarro–Herring creep is a **Newtonian flow**. We note that in (21), σ_c is, by convention, a *shear stress*, although the stress σ in (20) is a *tension*. For the same value of η, Newton's law (21) written for the case of a *tension* σ, becomes:

$$\dot{\varepsilon} = \frac{\sigma}{3\eta}. \tag{22}$$

Thus, the viscosity implied by the Nabarro–Herring creep has the form

$$\boxed{\eta = \frac{L^2 kT}{3D_{SD}\Omega}.} \tag{23}$$

It is heavily dependent on the temperature (*via* D_{SD}) and on the dimension L (dimension of the solid, or average grain size for a polycrystalline material). Thus, for $L = 1$ cm, $T \sim 750$ K ($kT \sim 10^{-13}$ erg), $D = 10^{-10}$ cm^{2}s^{-1} and $\Omega = 10^{-23}$ cm^3, formula (23) gives $\eta = 2.5 \times 10^{19}$ poise; this is a considerable viscosity which enables us to neglect the flow. If, for the same parameters, we take $L = 0.01$ cm ($= 100$ μm, a common grain size for polycrystals), we have $\eta = 2.5 \times 10^{15}$ poise, which is a thoroughly measurable viscosity.

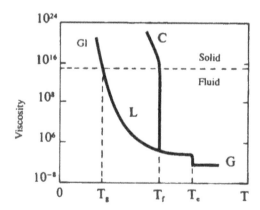

Figure 15 Evolution of the viscosity η (in poise) as a function of the temperature T. T_m: melting point. T_b: boiling point. T_g: temperature of vitreous transition (examples: $T_g = 1200°$ C for SiO$_2$, 85° C for polystyrene. 30° C for selenium, 180° C for alcohol). Gl = glass, L = liquid, G = gas, C = crystal (case described by (23)).

[11] See, for example, **R B Jones**, *Creep by transport mechanisms*, J. Sheffield Univ. Met. Soc. **12**, 34 (1973).

Figure 15 provides a schematic reminder of the evolution of the viscosity as a function of the temperature for different states of matter. By convention, the boundary between a solid and a fluid is usually fixed at 10^{15} poise.

It is tempting to extend formula (23) to the case of **liquids**. The viscous flow of a liquid may be compared with that of a solid, as described above, if we now confine the atomic movements to a volume of the order of the atomic volume Ω (instead of to a volume of the order of L^3 in the solid). We then have:

$$\eta \simeq \frac{\Omega^{2/3}kT}{3D_{SD}\Omega} \simeq \frac{kT}{4.8D_{SD}R} \tag{24}$$

where D_{SD} is the self-diffusion coefficient of the liquid and R is the radius of the sphere of volume Ω. We note the kinship between this formula and the *Stokes–Einstein* formula of hydrodynamics:

$$\eta_{S.E.} = \frac{kT}{6\pi Da} \tag{25}$$

which gives the viscosity of small particles of radius a, with diffusion coefficient D in suspension in a fluid.

N.B. A calculation of D for small particles can be found in **Problem 13**.

3.2 Non-Newtonian creep

In the previous section, we neglected completely the influence of dislocations. This undoubtedly explains why the Newtonian Nabarro–Herring creep (20) is only observed well for solids with very small grains, where the number of dislocations is small. The laws of creep observed experimentally are often *non-Newtonian*, and the deformation rates $\dot{\varepsilon}$ vary as σ^n (for $n = 3, 4, 5$ etc) rather than as σ.

In general, the presence of dislocations is not only non-negligible, but even critical to an understanding of the plastic properties at high temperature. In what follows, we shall describe a model due to Weertman in which the deformation ε is due to the *glide* of the dislocations, but in which the rate $\dot{\varepsilon}$ is governed by their **climb** (see p. 215).

Let us return to a crystal subjected to a stress (for example, shear σ), which activates a *source S* (see p. 227) and results in a *pile-up* of dislocations (see p. 232) (in a glide plane P_1) coming up against any conceivable obstacle (*precipitate, wall, boundary* etc). If the temperature is sufficient for the dislocations at the head of the stack to *climb* (in other words, leave the plane P_1) by supplying matter (or vacancies), then the dislocations may cross the obstacle one by one and restart their glide in a plane P_2 parallel to P_1 (see Figure 16). The *slower* of the two processes (glide and climb) is the climb, which thus imposes its kinetics on the deformation.

Figure 16 Creep by glide and climb. Dislocations piled up against an obstacle in the plane P_1 may restart their movement in another glide plane P_2 after having *climbed* out of the plane P_1.

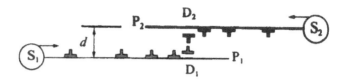

Figure 17 Pile-ups blocked by the interaction of their leading dislocations D_1 and D_2. The latter are in the process of *climbing* towards one another.

Let us consider the particular case in which the blockage is due simply to the fact that the dislocations D_1 of the plane P_1 are faced by the dislocations D_2 of *another pile-up* in a parallel plane P_2. The reader should check that the configuration of Figure 17 is implied by a single stress activating the sources S_1 and S_2. Dislocations (taken to be *screw dislocations*) of type D_1 and D_2 are mutually attractive. Since, at low temperature, each remains in its glide plane (P_1 and P_2, respectively) blockages occur *both* for D_1 *and* for D_2, followed by the formation of pile-ups 'behind' D_1 and D_2. If the temperature is such that the climb is possible (with velocity v_{cl}) and if the mean distance between planes P_1 and P_2 is d, then D_1 and D_2 climb towards and annihilate one another in an average time:

$$t_{cl} \simeq \frac{d}{v_{cl}}. \tag{26}$$

This annihilation releases a dislocation in each pile-up and enables the sources S_1 and S_2 to emit a new dislocation each. Since the climb time t_{cl} is long in comparison with the glide time, the former controls the frequency of operation of the sources. This leads to the establishment of a permanent regime in which the sources regularly produce dislocations. It can be shown that this results in a deformation with rate $\dot{\varepsilon}$

$$\boxed{\dot{\varepsilon} = \frac{D_{SD}b^3\sigma^{4.5}}{\alpha\mu kT}.} \tag{27}$$

Let us indeed assume that there are several sources, the number of which (M per unit volume, assumed fixed) we shall suppose to be independent of the stress applied. Each source produces its dislocations on a surface area of the order of $2L^2$ (L is the length of the pile-ups, see p. 233). The distance d separating the opposing dislocations which climb towards one another should be proportional to σ^{-1}, assuming that the configuration of the dislocations is in equilibrium with the applied stress and that the stress field produced by a dislocation varies as r^{-1}. Thus, we suppose that

$$d = \lambda\sigma^{-1} \tag{28}$$

with, moreover

$$2MdL^2 = 1. \tag{29}$$

Using Orowan's formula (see (4), p. 225), and assuming that the only mobile dislocations are those of pile-ups ($\rho_{cl} = nd/L$, where n is the number of dislocations piled on each side of a source), we have

$$\dot{\varepsilon} = \frac{nb}{dL}v. \tag{30}$$

The rate of glide v of the piled dislocations is determined by the climb time t_{cl} (and thus also the annihilation time) of the leading dislocations:

$$v = \frac{L}{n t_{cl}}. \tag{31}$$

If such a dislocation is subjected to a stress σ^*, the order of magnitude of the force which is exerted on it is

$$\sigma^* b \text{ per unit length,} \qquad \text{or}$$

$$\simeq \frac{\sigma^* b}{1/b} = \sigma^* b^2 \text{ per atomic distance.}$$

The climb velocity (v_{cl}) corresponding to this force is:

$$v_{cl} = \frac{D_{SD}\sigma^* b^2}{kT}$$

(see (13), p. 177), where, by virtue of (26),

$$t_{cl} = \frac{dkT}{D_{SD}\sigma^* b^2}. \tag{32}$$

The stress σ^* under which the leading dislocations evolve was calculated earlier (see (12)):

$$\sigma^* = \frac{L\sigma^2}{\mu b}$$

where σ is the stress applied. Combining (28)–(32), we obtain:

$$\dot{\varepsilon} = \frac{D_{SD}\Omega\sigma^{4.5}}{(2M)^{0.5} b\lambda^{2.5} \mu kT}$$

which is formula (27) where the term $\alpha = (2M)^{0.5} \lambda^{2.5}$ is a constant which depends on the material and the sample studied.

Although an exponent $\simeq 4.5$ for σ is very often observed experimentally, there is no question of giving it a universal value here. Certain of the arguments of the above calculation cannot be carried over to all cases and certain high temperature creep experiments cannot be analysed directly via (27). On the other hand, as a whole, these experiments are generally well described by *Weertman's phenomenological equation*:

$$\boxed{\dot{\varepsilon} = \dot{\varepsilon}_0 \left(\frac{\sigma}{\mu}\right)^n \exp(-E_{SD}/kT)} \tag{33}$$

together with the following remarks:

i)   The exponent n almost always lies between 3 and 5. The case $n = 4.5$ corresponds to (27). The *climb-induced creep* is **not a Newtonian flow**, although it is controlled by a phenomenon (the diffusion of vacancies, which permits climbing) which is itself Newtonian. This is due to the fact that the number of dislocations in movement, the area which they sweep and the speed at which they climb are *all* dependent on σ.

ii) The energy E_{SD} (called E_A in Chapter VIII) is the **self-diffusion energy** if the material contains only one type of atom. If it has more than one type, then E_{SD} is the diffusion energy of the slowest type.

iii) $\dot{\varepsilon}_0$ is a constant, which is determined not only by the nature of the material but also by its microscopic structure (number and nature of precipitates, number of sources etc).

4. BRITTLE FRACTURE

The ultimate end of a *tension* test (see Figure 1), a *creep* test etc is the **fracture** of the sample. More dramatically, an *item* or a *structure* subjected to static or cyclic stresses (connecting rod, bridge, pane of glass, cable etc) may break, either after a certain plastic deformation or even before the yield stress has been reached.

Taking into account the features of fracture surfaces, we distinguish between *brittle fracture* and *ductile fracture*. In the first case, the two surfaces of the broken item may be replaced in contact with one another with a correspondence which is almost perfect down to the atomic scale. This is not so in the second case. Brittle fracture generally occurs without plastic deformation, in other words, the material gives way without warning[12], and may, in many respects, be viewed as particularly dangerous. In what follows, we shall only discuss this case of brittle fracture.

4.1 Theoretical cleavage stress

If brittle fracture takes place along a dense crystallographic plane (for example, a hexagonal plane of an hcp metal: classical case of *zinc*), **cleavage** is said to occur. Let us evaluate the stress needed to *cleave* a crystalline solid along a plane P.

Let P_0 be one of these planes, P_1 the adjacent plane at a distance d and $\sigma(x)$ the (tensile) stress needed to displace P_1 from its equilibrium position, where x is the displacement perpendicular to P. The stress $\sigma(x)$ should be of the form illustrated in Figure 18, which we shall represent by:

$$\sigma(x) = \sigma_{th} \sin(\pi x / \alpha) \tag{34}$$

where α is an unknown distance of the order of d. By the definition of *Young's modulus of elasticity Y*, we have (Hooke's law):

$$(d\sigma/dx)_0 = Y/d. \tag{35}$$

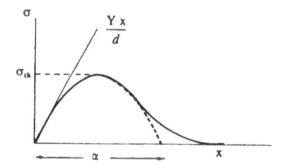

Figure 18 Cleavage stress.

---

[12] At least macroscopically. In reality, the fracture is preceded by an emission of sound or ultrasound which is audible using detectors (F Sébilleau).

The work needed to cleave the crystal, in other words, to separate the planes P_0 and P_1 may be evaluated *either* as being equal to W, the work of the stress $\sigma(x)$ during the displacement of the crystal, *or* as that needed to create two free surfaces on P_0 and P_1. Whence, if γ is the surface energy of the crystal, for a unit surface area we have:

$$W = \int_0^\alpha \sigma(x)\,dx = 2\gamma \tag{36}$$

from which it follows, *via* (34) and (35) that:

$$\sigma_{th} = \left(\frac{\gamma Y}{d}\right)^{1/2}. \tag{37}$$

N.B. Since the distance α is similar to d, we may use (36), taking $2\gamma = \int_0^d \sigma(x)\,dx$ to evaluate γ. Recalling the classical relation of isotropic elasticity

$$Y = 2(1 + v)\mu \simeq 3\mu \tag{38}$$

(v is the *Poisson coefficient* ($\simeq 0.5$) and μ is the *shear modulus*) and using (35), the following approximations are immediate:

$$\sigma_{th} = \mu \text{ and } \gamma = \frac{3\mu d}{\pi^2}; \text{ or } \gamma \simeq \frac{\mu d}{3}. \tag{39}$$

Here, σ_{th} is the **theoretical stress** needed to initiate the *cleavage. Its numerical value is approximately 100 times larger than the values of measured fracture stresses.*

In fact, the latter are of the order of 100 MPa (for example, 300 MPa in α iron at 200° C) while σ_{th} calculated *via* (37) is of the order of 10 GPa.

Exercise Calculate σ_{th} for common numerical values of γ and μ (see p. 210). Replace Y in (37) by 3μ (see (38)).

This is a paradox similar to that of the yield stresses (p. 203): solids *resist brittle fracture far less well than they should* (just as they resisted deformation less well). This law is general and holds for a material such as glass, even though, in the absence of crystalline *planes*, the fracture in this case is not, strictly speaking, a *cleavage*.

4.2 Griffith's model and criterion

This paradox was resolved by Griffith, who postulated the existence in solids of **microcracks** which contain the potentiality of the fracture (like dislocations are potentialities of deformation).

Let us consider an elliptical microcrack of length $2l$, with minimum radius ρ, in a thin plate subjected to the tension σ (Figure 19). A classical elasticity calculation shows that *at the end of the crack*, the stress is *amplified* and reaches the value

$$\sigma_m = 2\sigma\left(\frac{l}{\rho}\right)^{1/2}$$

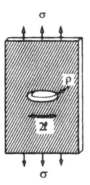

Figure 19 Elliptical microcrack.

The applied stress which satisfies $\sigma_m = \sigma_{th}$ (see (37)) is:

$$\sigma = \frac{1}{2}\left(\frac{\gamma Y \rho}{dl}\right)^{1/2} \tag{40}$$

where it is clear that the smaller ρ is (thin cracks) the less resistant the material is to fracture. In fact, even for the extreme case $\rho = 0$, the material retains some resistance because of the surface energy of the surfaces which must be created to extend the crack. To evaluate this resistance, Griffith writes that the extension (or *propagation*) of the crack must decrease (or, in the limit, conserve) the energy. The latter, call it E, for a plate of *unit thickness* involves:

i)   A *surface term* equal to $4\gamma l$.
ii) An *elastic term*. It is clear that in the neighbourhood of the two lips of the crack, the stresses are zero. The crack is said to *relax* the applied stress. The stress, which is zero on the lip and equal to σ far from the latter, may be viewed, to the first order, as being relaxed in a region of radius l around the crack. The gain in elastic energy, of density $\frac{1}{2}\sum \sigma_{ij}e_{ij}$, or in this case, $\frac{1}{2}\sigma \times \sigma/Y$ (see (35)) $= \sigma^2/2Y$, is of the order of $(\sigma^2/2Y)\pi l^2$. A rigorous calculation, taking into account the exact analytical distribution of σ around the crack leads to a *doubling* of this result.

Griffith's criterion, which is obtained by taking the energy to be stationary, is then

$$\frac{dE}{dl} = \frac{d}{dl}\left(4\gamma l - \frac{\pi\sigma^2 l^2}{Y}\right) = 0$$

or

$$\sigma_G = \left(\frac{2\gamma Y}{\pi l}\right)^{1/2} \quad \text{or} \quad \boxed{\sigma_G \simeq \left(\frac{\gamma Y}{l}\right)^{1/2}}. \tag{41}$$

This gives the value of the stress (**Griffith's stress**) needed to *propagate a crack* of dimension $2l$. Comparison of (40) and (41) then shows that an effective limit of the fineness of a crack is $\rho \simeq 4d$. Finer microcracks (in the limit, those with ρ *zero*) do propagate despite equation (37) under Griffith's constraint, which thus measures the *resistance* of a material *to fracture*.

4.3 Initiation of cracks

Given the experimental fracture stress σ_F, and writing $\sigma_F = \sigma_G$, formula (41) may be used to calculate the size l of corresponding microcracks. Three examples of this are given below:

Table 1 Size l of Griffith's cracks. σ_F: fracture stress (in GPa). γ: surface energy (in J m^{-2}). Y: Young's modulus (in GPa).

	σ_F	γ	Y	l
Glass	0.2	0.2	60	0.3 μm
Zinc (polycrystal)	0.2×10^{-2}	0.8	35	7 mm
NaCl	0.2×10^{-2}	0.15	50	2 mm

These values of l merit examination. While it is plausible (and has been verified) that cracks of several tenths of a micron may exist in glass, it is inconceivable that well crystallised zinc or sodium chloride (which are both subject to brittle fracture should initially contain millimetric cracks. Thus, here, one must hypothesise **either** that other objects play the role of the cracks of Griffith's model **or** that the cracks are created in the material during the deformation, prior to the fracture.

An example of the **first case** is provided by certain *brittle cast irons* which contain *lenticular* precipitates of **graphite**. The graphite is only weakly bonded to the metallic matrix and the cavities in which it is found take on the role played in the above by cracks. Conversely, the same cast irons, thermally treated so as to ensure that the graphite precipitates are *nodular* (spheroidal) are plastic and non-brittle.

The **second case**, involves the *glide* of *dislocations* (Zener, Mott, Friedel, Cottrell etc), in particular, when this leads to the formation of *pile-ups* (see p. 233). It appears that one way for a crystal to release the stresses σ^* existing at the *head of a pile-up* involves the creation of a microcrack. The necessary condition for this is

$$\sigma^* \geq \sigma_{th} \qquad (\simeq \mu \text{ see (39)}). \tag{42}$$

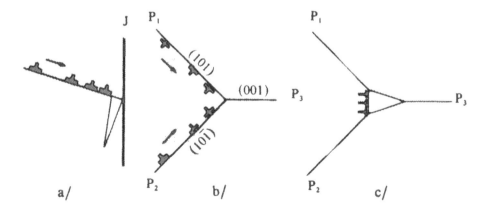

Figure 20 Microcracks produced by glide and pile-up. (*a*) a single pile-up coming up against an obstacle J. (*b*) Double pile-up in a bcc crystal (Cottrell's mechanism). (*c*) Microcrack (of three dislocations) opened by the pile-ups of (*b*).

Once created, this microcrack will propagate, provided a source upstream continues to supply the pile-up. Two examples of this mechanism are shown in Figure 20.

Let us consider, following Cottrell, two perpendicular glide planes {110} of a bcc crystal, namely $P_1 \equiv (101)$ and $P_2 \equiv (10\bar{1})$ and the dislocations D_1 of P_1 with Burgers vector $\bar{b}_1 = \frac{1}{2}[\bar{1}1\bar{1}]$ and D_2 of P_2 with BV $\bar{b}_2 = [111]$ (see Figure 20(b)). The following are left to the reader to check as an **exercise**:

i)   That $\bar{b}_i$ is contained in P_i and that P_i is thus a glide plane for D_i ($\bar{b}_i$).
ii)   That, the *coalescence* reaction $D_1 + D_2 \rightarrow D_3$ is favourable as far as energy is concerned (see p. 212).
iii)   That the resulting dislocation D_3 has BV $\bar{b}_3 = [001]$, that this vector is contained in the plane of Figure 20(b) and that D_3 at the intersection of P_1 and P_2 is a pure edge dislocation.

Under these conditions, if sources located on P_1 and P_2 feed the above reaction, dislocations D_3 for which the BV has a large modulus (equal to the edge of the body-centred cube, in other words, greater than the edge of the primitive cell $\frac{1}{2}\langle 111 \rangle$) may initiate a microcrack in the plane $P_3 \equiv (001)$ (Figure 20(c) shows one such microcrack formed from three dislocations D_3).

4.4 Fracture versus deformation

We may now attempt to understand, in broad outline, why and how such material has a preferential tendency either towards brittle fracture or towards deformation.

Let us apply an external stress σ (for example, a tension) to a material, and take up what was said earlier.

i)   If this material contains microcracks (like glass) or the 'germs' of fracture (like brittle cast irons) of size $2l$, *a priori*, even the application of a stress $\sigma_G(l)$ (see (41)) which is very much less than the cleavage stress σ_{th} ($\simeq \mu$, see (39)) is sufficient to propagate the cracks and provoke a brittle fracture.

ii)   If this material does not contain such cracks, the stress σ may or may not initiate them. This requires the creation of a high *local* stress equal to σ_{th}.

- If *glide is easy* (shallow Peierls valleys, numerous glide planes etc) and so too is the activation of the *sources*, then we expect a modest yield stress (for example, $\simeq 10^{-4}\mu$, see p. 203). When σ exceeds σ_E (see Figure 1, p. 224), the material deforms, so that the stresses can only be weak (of the order of σ) throughout the material, and cannot be equal to σ_{th} anywhere, except possibly after substantial cold work, in other words, after strong deformation. However, the fracture observed in this last case is not generally brittle. It owes little to cleavage.

  Classical examples of this case include fcc metals (or alloys) (Al, Cu, austenitic stainless steels etc) or bcc metals (or alloys) ('low-carbon' steels etc) capable of withstanding considerable deformations without breaking[13].

- *If glide is not easy* (hard materials, see p. 230) and/or the sources are difficult to activate, the material has a **high yield stress** ($\sigma_E = 10^{-2}\mu$ in bcc iron at low temperature). Thus, in this case, the applied stresses σ may also be high. If, in addition, they are susceptible to amplification (in particular, by the *pile-ups*), it is possible that stresses of the order of σ_{th} (or, for a pile-up, $\sigma^* = \sigma_{th}$) may be attained **locally**. Thus, the conditions for the birth of a microcrack are met.

---

[13] The *drink can*, obtained by deep stamping of a flat circular plate (either aluminium or steel) is a good example. Approximately 140 billion of these are produced annually (1995) with an insignificant reject rate for microcracks.

In this case, a brittle fracture may soon occur. In fact, the *propagation stress* σ_G is always less than the *cleavage stress* σ_{th} (compare (37) and (41), where, when a crack of size l is opened we have $l > d$).

Let us evaluate the yield stress needed for the second scenario (above) to be plausible. We showed earlier ((12), p. 234) that a pile-up (size: L) of dislocations (BV: $\bar{b}$) creates a stress $\sigma^* = L\sigma^2/\mu b$ at its head, where σ is the applied stress. This formula assumed that the source operates for all applied stresses σ. If, in fact, there is a certain *source stress* σ_s which must be exceeded (for example, due to a *cloud* of impurities, see **Problem 13**, which creates friction) before it operates, then σ of (11) must be replaced by $\sigma - \sigma_s$. Thus, instead of (12), we will have

$$\sigma^* = L\sigma(\sigma - \sigma_s)/\mu b. \tag{43}$$

If plastic deformation is triggered at a certain critical value σ_c of σ, then the *yield stress* σ_E is simply the solution of the equation

$$\sigma_c = L\sigma_E(\sigma_E - \sigma_s)/\mu b \tag{44}$$

and fracture rather than deformation (second scenario) occurs provided $\sigma_{th} \leq \sigma_c$. If this condition is realised with equality ($\sigma_c = \sigma_{th} \simeq \mu$; see (39)), solving (44), simplified with the assumption that $\sigma_s = 0$, we obtain:

$$\sigma_E = \mu\left(\frac{b}{L}\right)^{1/2};$$

in other words, a very large value of $\sigma_E = 10^{-2}\mu$ for $L = 10^4 b$. The introduction of a non-zero σ_s increases this value even more, as does decreasing L.

Thus, it is clear here that a high yield stress is a necessary condition for the occurrence of brittle fracture.

4.5 Influential parameters

Among the numerous *structural* (additive elements, grain size etc) or *external* (temperature etc) parameters which have a particular influence on the brittleness of a material, we shall consider only those listed below.

i)   **Initial presence of cracks or of germs of cracks.** We have already noted the case of *glass*, in which microcracks exist *a priori* and that of *alloys with precipitates* in which the latter are liable (depending on their shape, their nature etc) to *initiate* cracks.

On the other hand, the brittleness of a glass may be reduced considerably by annihilating those cracks which end on the surface. This may be carried out by **chemical quenching**, in which ions (for example, K^+) are introduced on to the surface of the glass. By a sterical effect, these ions create compression stresses on the surface, which 'reseal' the surface cracks.

ii)   **Grain size.** Condition (42) (above), relating to the initiation of cracks, implies a sufficient stress σ^*, whence a sufficient *pile-up size* L, and thus also a sufficient *grain size* ϕ.

From (8), and (8''), it follows that σ^* varies as $L\sigma^2$ (σ is the applied stress). From (42), we thus expect the stress needed for fracture ($\sigma = \sigma_F$) to vary linearly with $L^{-1/2}$, whence also with $\phi^{-1/2}$. In general, this (or at least a tendency towards this) is what is actually observed experimentally, and we note that *the resistance to fracture tends to increase as the grain size decreases.*

More precisely, if σ_s is sufficiently small in comparison with $\mu(b/L)^{1/2}$, then, from (42) and (43) the fracture stress is:

$$\sigma_F \simeq \frac{\sigma_s}{2} + \mu\left(\frac{b}{L}\right)^{1/2}$$

or, equating grain size and pile-up size:

$$\sigma_F \simeq \frac{\sigma_s}{2} + \mu\left(\frac{b}{\phi}\right)^{1/2}$$

iii) **Temperature**. The tendency towards brittle fracture is accompanied by high values of the yield stress σ_E. These high values are more common at low temperature, in particular, when *glide* or the operation of *sources* are inhibited by clouds of impurities (see **Problem 13**) or by deep Peierls valleys. Thus, often, a material may be brittle at low temperature and ductile when thermal activation permits glide and/or operation of sources. The corresponding **brittle–ductile transition** is generally clear-cut, as Figure 21 shows.

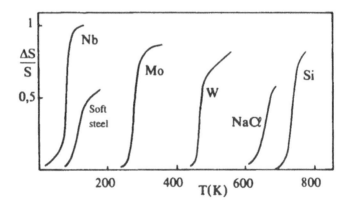

Figure 21   Brittle–ductile transitions of a number of materials. Tensile experiments were used. The reduction $\Delta S/S$ in the cross-sectional area of a rod at fracture is measured. $\Delta S/S \simeq 0$ corresponds to *fracture* of the rod without deformation.

In fact, except for fcc metals and alloys, all materials (metals, insulators, polymers etc) exhibit a brittle–ductile transition. The transition temperature T_{BD} is approximately $0.1T_m$ (T_m is the absolute melting point) for transition metals and of the order of $0.5T_m$ for a number of other materials. We note that T_{BD} depends to a large extent on the nature of the test carried out (tension, compression, impact etc).

The transition temperatures of a certain number of *ferritic steels* (with a bcc structure) lie between -100 and $0°$ C. Thus, it is very important to know the temperature at which they are to be used, if one is to avoid serious problems! For example, at the time of the retreat from Russia, cannons and swords of the Great Army, which had been cooled to temperatures lower than T_{BD}, broke under the effect of simple impacts. More recently, during the Second World War, *liberty ships* in convoys to the USSR via the Arctic Ocean broke in two for the same reason, the propagation of the crack being facilitated by the 'all-welded' technique which made the hull continuous.

iv) Others. Many other parameters often play an essential role as far as the conditions for the occurrence of fracture are concerned. Here is an incomplete list: the *nature of the test* (tension, compression, torsion, impact with or without gashes etc) together with the *deformation rate*, the *content of impurities*, the *crystallographic texture* of the material, and the presence of *grain boundaries* or *sub-boundaries*.

The last point is worth a final note.

Let us consider a sub-boundary of screw dislocations. If a cleavage crosses the plane of the sub-boundary, the two free surfaces form *steps* beyond the sub-boundary, one per dislocation. The latter modify (and increase) the term γ of formula (41) and the cleavage propagates more easily before than beyond the steps. Taken to its extreme, this *brake* may actually cause the *propagation to stop* during the crossing of a boundary. Thus, as a result, boundaries may inhibit a fracture process initiated in one grain and attempting to propagate itself to an adjacent grain.

DIFFRACTION BY A
CRYSTAL

Les rayons et les ombres
Victor Hugo

Let us consider a crystal, assumed to be monatomic, whose elementary lattice, defined by the three vectors $\bar{a}_1$, $\bar{a}_2$ and $\bar{a}_3$ comprises *one* atom. Let us suppose that the crystal is parallelepipedal and consists of N_i elementary lattices in the direction x_i, the support of $\bar{a}_i$. Let us take the origin O on one of the eight atoms which are vertices of the parallelepiped. An arbitrary atom α of the crystal (Figure 1(a)) is defined by

$$\bar{r}_\alpha = n_1\bar{a}_1 + n_2\bar{a}_2 + n_3\bar{a}_3 \tag{1}$$

(here, we neglect the *thermal vibrations*, which, in reality displace the atom α slightly from the *site* $\bar{r}_\alpha$.; see **Problem 18** on this subject).

Let ψ be the amplitude of a plane wave with wavevector $\bar{k}$ incident on the crystal

$$\psi = A\mathrm{e}^{\mathrm{i}(\bar{k}\cdot\bar{r}-\omega t)}.$$

We make no other assumptions about the nature of this wave other than the following: when the wave touches an atom α, it is scattered **spherically**, in other words isotropically, and **elastically**, that is, without modification to the amplitude of $\bar{k}$.

In the case of an electromagnetic wave (for example, X-rays), this scattering is due to the excitation of the electrons attached to the atom α which, accelerated by the wave, re-emit a wave of the same wavelength. For other waves (for example, those associated with a flux of neutrons), the scattering mechanism may be radically different (bringing into play the nuclei, in the case of neutrons), without the following development being modified.

Let us consider the amplitude scattered by all the atoms α at a point P away from the crystal, in other words, at a distance much greater than the crystal size. We set $\overline{OP} = \bar{R}$. This point will be reached by waves with vector $\bar{k}'$ such that $|\bar{k}'| = |\bar{k}|$ and $\bar{k}' \parallel \bar{R}$.

The wave reaching P from the scattering centre α has the general form

$$\psi_{\alpha P} = A'_\alpha \mathrm{e}^{\mathrm{i}(\bar{k}'\cdot\bar{R}-\omega t)}.$$

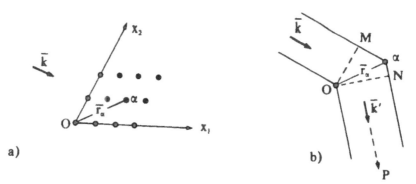

Figure 1   Incident wave ($\bar{k}$) and scattered wave ($\vec{k}'$).

A'_α is a constant associated with the scattering process, which is independent of t and inversely proportional to $|\overline{R} - \bar{r}_\alpha|$, and thus also, practically to $|\overline{R}|$. With respect to the wave reaching P and scattered by O, that scattered by α exhibits a path difference corresponding to the trajectory $M\alpha N$ (see Figure 1(b)). Taking into account the fact that

$$M\alpha = \frac{1}{|\bar{k}|}\bar{k} \cdot \bar{r}_\alpha \quad \text{and} \quad \alpha N = -\frac{1}{|\vec{k}'|}\vec{k}' \cdot \bar{r}_\alpha$$

the *phase difference* between the waves ψ_{OP} and $\psi_{\alpha P}$ is

$$-\overline{\Delta k} \cdot \bar{r}_\alpha \quad \text{where} \quad \overline{\Delta k} = \vec{k}' - \bar{k}.$$

Thus, we may write

$$\psi_{\alpha P} = A' e^{i(\vec{k}' \cdot \overline{R} - \omega t)} e^{-i\overline{\Delta k} \cdot \bar{r}_\alpha}.$$

The total amplitude of the wave scattered at the point P by the crystal is the sum of all these partial amplitudes: $\psi_P = \sum_\alpha \psi_{\alpha P}$. It is proportional to

$$\boxed{A = \sum_\alpha e^{-i\overline{\Delta k} \cdot \bar{r}_\alpha}.} \qquad (2)$$

Let us introduce the coordinates Δk_i of the vector $\overline{\Delta k}$ on the **reciprocal lattice** defined by (11) (p. 100). The product $\overline{\Delta k} \cdot \bar{r}_\alpha = (\Delta k_1 \bar{a}_1^* + \ldots) \cdot (n_1 \bar{a}_1 + \ldots)$ reduces to $2\pi(n_1 \Delta k_1 + n_2 \Delta k_2 + n_3 \Delta k_3)$. Thus, we have

$$A = \sum_{n_1 n_2 n_3} e^{-i2\pi(n_1 \Delta k_1 + \ldots)} = A_1 A_2 A_3$$

where

$$A_1 = \sum_1^{N_1} e^{-i\pi n_1 \Delta k_1} = \frac{1 - \exp[-i2\pi(N_1 + 1)\Delta k_1]}{1 - \exp[-i2\pi \Delta k_1]} \simeq \frac{1 - \exp[-i2\pi N_1 \Delta k_1]}{1 - \exp[-i2\pi \Delta k_1]}.$$

The signal received experimentally at P (on a screen or in a counter) is proportional to the **energy** transported by the wave, thus to the square $I = \psi_P^* \psi_P$, whence

$$I = A'^2 I_1 I_2 I_3$$

where $I_1 = A_1^* A_1$, whence also

$$I_1 = \frac{\sin^2 \pi N_1 \Delta k_1}{\sin^2 \pi \Delta k_1}.$$ (3)

Figure 2 shows the variations of the function $I_1(\Delta k_1)$, whose main characteristic is that it is practically zero almost everywhere, except for at **integer values** of Δk_1. Thus, the intensity $I(\overline{\Delta k})$ is zero in almost all spatial directions except in those such that $\overline{\Delta k}$ is a **vector of the reciprocal lattice**: since the vector $\overline{k}$ is fixed in space (incident beam) the diffracted beams emerge from the crystal in the directions $\overline{k}'$ such that the $\overline{k}' - \overline{k}$ are the vectors of the reciprocal lattice. Ewald's construction sets this condition in concrete form (Figure 18, p. 101).

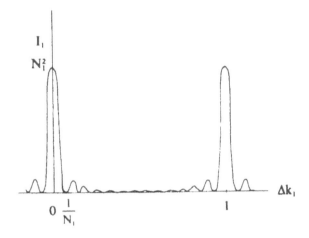

Figure 2   Variations of the function $I_1(\Delta k_1)$ defined by (3).

Remarks

i)   The diffracted beams have a certain angular width. For an ideally defined direction $\overline{k}$, this width is shown on Figure 2 by the extension of the function I_1 around the integer values of Δk_1, which is an extension of the order of $1/N_1$: the smaller the crystal is, the wider the diffraction beams are.

ii)   It is evident that a knowledge of the directions of diffraction is equivalent to a knowledge of the reciprocal lattice, whence to a knowledge of the direct lattice (see (12), p. 100). This is the basis of all X-ray crystallography.

iii)   If, more generally, one studies the scattering of X-rays by a sample of condensed matter, whether it be solid or liquid, crystallised or not, with electron density $\rho(\overline{r})$, considered to be continuous, the sum $\sum_\alpha$ of formula (2) becomes

$$\int \exp[i(\overline{k} - \overline{k}') \cdot \overline{r}]\rho(\overline{r})d_3\overline{r} :$$

the **amplitude scattered** in the direction $\overline{k}'$ is proportional to the **Fourier transform of the electron density** in the sample. This property (which includes remark ii)) is the starting point of X-ray crystallographic studies of liquids, glasses, organic matter, biological matter, ..., where the fundamental difficulty however arises from the fact that the quantity measured is the *intensity* $\mathcal{A}^*\mathcal{A}$ rather than the amplitude $\mathcal{A}$.

iv) In the above, the scattering centres α were assumed to be fixed on exact crystalline sites. However, the thermal vibrations displace them constantly from these sites. It can be shown that this only results in a *decrease in the diffracted intensity*. This decrease, which depends on the amplitude of the vibrations and on the vector $\overline{\Delta k}$ in question, is described by the **Debye–Waller factor**: see **Problem 18** and, in particular, formula (5).

BLOCH'S THEOREM

On the knowledge whiche maketh a wise man
Thomas Elyot

We wish to solve the Schrödinger equation

$$H(\bar{r})\psi(\bar{r}) = E\psi(\bar{r})$$

for an electron in a crystal, in other words, in a medium with **periodic potential**:

$$V(\bar{r}) = V\left(\bar{r} + \overline{T}_{(n)}\right)$$

where $\overline{T}_{(n)}$ is a translation vector for the crystal:

$$\overline{T}_{(n)} = n_1\bar{a}_1 + n_2\bar{a}_2 + n_3\bar{a}_3.$$

The n_i are integers, the $\bar{a}_i$ are the vectors of the Bravais lattice. The crystal contains N_i cells in the direction x_i, the support of $\bar{a}_i$. Finally, the electrons of the crystal are subject to Born–von Karman periodic conditions (see p. 34):

$$\psi(\bar{r}) \equiv \psi(\bar{r} + N_i\bar{a}_i). \tag{1}$$

Let $T_{(n)}$ be the translation operator which transforms $\bar{r}$ into $\bar{r}' = \bar{r} + \overline{T}_{(n)}$. Writing

$$T_{(n)}V(\bar{r}) = V(\bar{r}') = V(\bar{r})$$

and

$$T_{(n)}\frac{\partial^2}{\partial x^2} = \frac{\partial^2}{\partial x'^2} = \frac{\partial^2}{\partial x^2},$$

we see that we have

$$T_{(n)}H(\bar{r}) = H(\bar{r}).$$

Let χ and η be two functions related by

$$H(\bar{r})\chi(\bar{r}) = \eta(\bar{r}).$$

Operating on χ by $HT_{(n)}$ and $T_{(n)}H$, successively, we obtain:

$$H(\bar{r})T_{(n)}\chi(\bar{r}) = H(\bar{r}')\chi(\bar{r}') = \eta(\bar{r}')$$

$$T_{(n)}H(\bar{r})\chi(\bar{r}) = T_{(n)}\eta(\bar{r}) = \eta(\bar{r}'),$$

and we see that H and $T_{(n)}$ commute, as do $T_{(n)}$ and $T_{(m)}$: $[H, T_{(n)}] = [T_{(n)}, T_{(m)}] = 0$.

Thus, these operators possess a complete orthonormal system of common eigenfunctions. Let $\psi(\bar{r})$ be one of these. We write successively, for example along x_1:

$$H\psi(\bar{r}) = E\psi(\bar{r})$$

$$T_1\psi(\bar{r}) = \psi(\bar{r} + \bar{a}_1) = \lambda\psi(\bar{r})$$

$$T_2\psi(\bar{r}) = \psi(\bar{r} + 2\bar{a}_1) = \lambda^2\psi(\bar{r}) \dots$$

$$T_{N_1}\psi(\bar{r}) = \psi(\bar{r} + N_1\bar{a}_1) = \lambda^{N_1}\psi(\bar{r}).$$

From (1), it follows that λ is an N_1th root of unity.

$$\lambda = e^{i2\pi p_1/N_1} \qquad p_1 \text{ being an integer} < N_1.$$

Recalling that 2π can be defined by $2\pi = \bar{a}_j \cdot \bar{a}_j^*$ (see (11), p. 100), and setting $k_j = p_j/N_j$ $(j = 1, 2$ or $3)$, we write;

$$T_{(n)}\psi(\bar{r}) = \exp[i(\bar{a}_1\bar{a}_1^* n_1 k_1 + \bar{a}_2\bar{a}_2^* n_2 k_2 + \bar{a}_3\bar{a}_3^* n_3 k_3)]\psi(r)$$

$$= \exp[i(n_1\bar{a}_1 + \dots) \cdot (k_1\bar{a}_1^* + \dots)]\psi(\bar{r}),$$

whence

$$T_{(n)}\psi(\bar{r}) = e^{i\bar{k}\cdot\bar{T}_{(n)}}\psi(\bar{r}). \tag{2}$$

Here $\bar{k}$ is the vector having the components k_j in the reciprocal lattice. We note immediately that this vector $\bar{k}$ is quantised: the vectors $\bar{k}$ are the vectors of a lattice constructed from the three vectors $\bar{\beta}_j$ defined by $\bar{\beta}_j = \bar{a}_j^*/N_j$. It is not difficult to check that this lattice coincides with that of the vectors of plane waves $\bar{k}_n$ which were defined by (30) on p. 44 in the special case of a sample with a cubic form and a cubic lattice.

Exercise   Verify that this property remains true if the lattice is fcc or bcc.

Let us define a function $u_{\bar{k}}(\bar{r})$ by the relation

$$u_{\bar{k}}(\bar{r}) = \psi(r)e^{-i\bar{k}\cdot\bar{r}}.$$

We write directly:

$$T_{(n)}\psi(\bar{r}) = u_{\bar{k}}(\bar{r} + \bar{T}_{(n)})e^{i\bar{k}\cdot(\bar{r}+\bar{T}_{(n)})}$$

and also, by virtue of (2):

$$T_{(n)}\psi(\bar{r}) = u_{\bar{k}}(\bar{r})e^{i\bar{k}\cdot\bar{r}}e^{i\bar{k}\cdot\bar{T}_{(n)}}.$$

It follows that

$$u_{\bar{k}}(\bar{r}) = u_{\bar{k}}(\bar{r} + \bar{T}_{(n)})$$

and thus that the eigenfunctions we seek may be written in the form

$$\psi(\bar{r}) = u_{\bar{k}}(\bar{r})e^{i\bar{k}\cdot\bar{r}}$$

where the $u_{\bar{k}}(\bar{r})$ are **periodic functions** with the same period as the potential $V(\bar{r})$.

BOLTZMANN'S EQUATION. ELECTRICAL CONDUCTIVITY

Wake me when it's funny
Garry Marshall

1. GENERALITIES

Let us consider a gas of particles constrained to move in one dimension and its phase space with two dimensions x, $v = dx/dt$ (Figure 1).

Let $g(x, v, t)$ be the **distribution function** (or **distribution**, for simplicity) of this gas, which is such that at time t, $g\, dx\, dv$ is equal to the number of particles contained in the element $ds(x, x + dx; v, v + dv)$.

In the time dt, these representative points move continuously towards the element ds' (equal to ds to the first order: $ds' = ds$). However, in the time dt, particle–particle collisions which make v vary rapidly, imply that certain points are 'lost' and others 'gained' ((a) and (b), respectively in Figure 1).

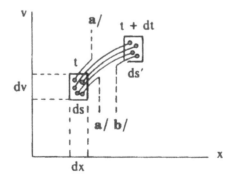

Figure 1 Phase space with two dimensions x and v. Evolution with time.

The result is:

$$g(x + v\,dt, v + \gamma\,dt, t + dt) = g(x, v, t) + \left(\frac{\partial g}{\partial t}\right)_{coll} dt$$

where γ denotes the acceleration; or, to the first order:

$$\frac{\partial g}{\partial t} + v\frac{\partial g}{\partial x} + \gamma\frac{\partial g}{\partial v} = \left(\frac{\partial g}{\partial t}\right)_{coll}.$$

This is **Boltzmann's equation** which is immediately generalised to the case of distributions $g(\overline{r}, \overline{v}, t)$ in three dimensions (or six for the phase space), giving:

$$\frac{\partial g}{\partial t} + \overline{v} \cdot \overline{\nabla}_r g + \overline{\gamma} \cdot \overline{\nabla}_v g = \left(\frac{\partial g}{\partial t}\right)_{coll}.$$

In a steady state, that is, when $\partial g/\partial t = 0$, Boltzmann's equation reduces to

$$\boxed{\overline{v} \cdot \overline{\nabla}_r g + \overline{\gamma} \cdot \overline{\nabla}_v g = \left(\frac{\partial g}{\partial t}\right)_{coll}.} \tag{1}$$

The right-hand term $(\partial g/\partial t)_{coll}$ describes the rate of the variations of the distribution g due to the collisions. This term is difficult to analyse in the general case. It groups together both collisions occurring at velocity $\overline{v}$ which lead to the loss of representative points and those occurring at velocities other than $\overline{v}$ which lead to gains. This problem is frequently simplified by means of the **relaxation-time hypothesis**, which involves writing

$$\boxed{\left(\frac{\partial g}{\partial t}\right)_{coll} = -\frac{g - g_0}{\tau(v)}} \tag{2}$$

in other words, postulating that if the distribution g moves away from its equilibrium value g_0, the return to equilibrium is brought about, by collisions, with a rate proportional to the separation $|g - g_0|$, where the coefficient of proportionality, τ^{-1} is assumed to depend only on $\overline{v}$. τ is called the *relaxation time*.

2. ELECTRICAL CONDUCTIVITY OF FREE ELECTRONS

First we transpose equation (1) (with the assumption (2)) to the case of free electrons studied in Chapter III. In the phase space, $\overline{k}$ will from now on play the role of $\overline{v}$ (through the intermediary of $\overline{v} = (\hbar/m)\overline{k}$, where m is the mass of the electron) and we write

$$g \equiv g(\overline{r}, \overline{k}, t).$$

We shall consider the case in which the distribution g is modified from its equilibrium value g_0 by the presence of a uniform *constant force* $\overline{F}$ acting on the electrons. Equation (1) becomes:

$$\frac{\hbar}{m}\overline{k} \cdot \overline{\nabla}_r g + \frac{\overline{F}}{\hbar} \cdot \overline{\nabla}_k g = -\frac{g - g_0}{\tau(\overline{k})}. \tag{3}$$

For $\overline{F} = 0$, the distribution is simply the product of the density in the $\overline{k}$ space (see (32'), p. 45) by the Fermi–Dirac function $f(E)$ (see (25), p. 42):

$$g_0(\overline{r}, \overline{k}, t) = \frac{1}{4\pi^3} \frac{1}{\exp[(E(\overline{k}) - \mu)/kT] + 1} \tag{4}$$

the expression given here being for a *unit volume*. In the general case, that of an inhomogeneous solid, the dependence $g(\overline{r})$ is introduced by the spatial dependence of $E(\overline{k})$ and of T: this applies to a metal subjected to a temperature gradient, whose thermal conductivity we study. In the case considered here, namely that of a **homogeneous** Sommerfeld **metal**, the term $\overline{\nabla}_r$ is zero. Finally, if $\overline{F}$ is the force $-e\overline{\varepsilon}$ produced by an electric field $\overline{\varepsilon}$, in the absence of a magnetic field, the Boltzmann equation reduces to

$$\boxed{\frac{-e\overline{\varepsilon}}{\hbar} \cdot \overline{\nabla}_k g = -\frac{g - g_0}{\tau(\overline{k})}} \tag{5}$$

In the absence of an applied force ($\overline{\varepsilon} = 0$) the distribution g_0 does not give rise to an electric current. On the other hand, the current density

$$\overline{j} = -e \int \overline{v} g \, d^3\overline{k} \tag{6}$$

corresponds to the distribution g. Thus, in a general fashion, the calculation of $\overline{j}$, i.e. that of the **electrical conductivity**, implies that of g via the Boltzmann equation. In the present case, this calculation is simple. In fact, to the first order, g may be replaced by g_0 in the left-hand member of (5). Whence:

$$g = g_0 + \frac{\tau(\overline{k})e\overline{\varepsilon}}{\hbar} \cdot \overline{\nabla}_k g_0. \tag{7}$$

Recalling that $\overline{v} = (\hbar/m)\overline{k}$ and $E(k) = \hbar^2 k^2 / 2m$ and not forgetting that the current $e \int \overline{v} g_0 \, d^3\overline{k}$ is zero, one sees that (6) becomes:

$$\overline{j} = -\frac{e^2}{m} \int \tau(\overline{k})\overline{k} \left[\overline{\varepsilon} \cdot \overline{\nabla}_k g_0 \right] d^3\overline{k} \tag{8}$$

$$= -\frac{e^2}{m} \int \tau(\overline{k})\overline{k} \left[\overline{\varepsilon} \cdot \overline{k} \right] \frac{\hbar^2}{m} \frac{\partial g_0}{\partial E} 4\pi k^2 \, dk$$

$$= -\frac{e^2}{\pi^2 m} \int \tau(\overline{k}) \left[\overline{\varepsilon} \cdot \overline{k} \right] k \frac{\partial f}{\partial E} \, dE \tag{9}$$

where f is the Fermi–Dirac function (see (4)). Now, we recall that, for moderate temperatures in comparison with the Fermi temperature T_F (for example, $T \lesssim 2000$ K), the function $\partial f / \partial E$ is very 'pointed' around $E = \mu \simeq E_F$ and may be compared with a Dirac delta:

$$\frac{\partial f}{\partial E} \simeq -\delta(E - E_F).$$

It follows that the integral of (9) reduces to the mean value $\langle \tau(\overline{k})\overline{k} \left[\overline{\varepsilon} \cdot \overline{k} \right] k \rangle$ for $k = k_F$, namely $\tau_F \langle \overline{k}[\overline{\varepsilon} \cdot \overline{k}]k \rangle$, which is equal to $\tau_F(\overline{\varepsilon}/3)k_F^3$ [1]. Taking into account the fact that

---

[1] Let θ be the angle between $\overline{k}$ and $\overline{\varepsilon}$. We write $\langle\ \rangle = \overline{\varepsilon}k_F^3 2 \int_0^{\pi/2} \sin\theta (\sin\theta/2) \cos\theta \, d\theta = (\overline{\varepsilon}/3)k_F^3$.

$k_F^3 = 3\pi^2 N$ for a unit volume (see (33'), p. 46), we end up with:

$$\bar{j} = \sigma\bar{\varepsilon} \quad \text{with} \quad \sigma = \frac{Ne^2\tau_F}{m} \tag{10}$$

where we again find the conductivity calculated earlier ((48), p. 56) with a relaxation time $\tau_F = \tau(k_F)$ explicitly taken at the *Fermi level* (or on the *Fermi sphere*) in accordance with the more intuitive discussion on p. 56.

3. ELECTRICAL CONDUCTIVITY OF ALMOST-FREE ELECTRONS

We shall generalise the above to the case of 'almost-free' electrons which are the subject of Chapter VI. For this, we return to the Boltzmann equation and the current in their forms (5) and (6); however, in what follows, we have to use the expression $\bar{v} = 1/\hbar\,\overline{\nabla}_k E$ for the velocity (see (30), p. 139). Then we have:

$$\bar{j} = -\frac{e^2}{4\pi^3}\int \tau(\bar{k})\bar{v}\,[\bar{\varepsilon}\cdot\bar{v}]\,\frac{\partial f}{\partial E}\,\mathrm{d}^3\bar{k}.$$

If we trace surfaces of constant energy in the $\bar{k}$ space, the volume element is written as:

$$\mathrm{d}^3\bar{k} = \frac{\mathrm{d}S\,\mathrm{d}E}{|\overline{\nabla}_k E|}.$$

Whence:

$$\bar{j} = \frac{e^2}{4\pi^3}\int_{\Sigma_F} \tau(\bar{k})\bar{v}\,[\bar{\varepsilon}\cdot\bar{v}]\,\frac{\mathrm{d}S}{|\overline{\nabla}_k E|} \qquad (\Sigma_F = \text{Fermi surface}). \tag{11}$$

The relationship between $\bar{j}$ and $\bar{\varepsilon}$ is no longer linear, but tensorial:

$$\bar{j} = \overline{\sigma}\bar{\varepsilon}$$

where $\overline{\sigma}$ is the **conductivity tensor** which links the two non-collinear vectors $\bar{j}$ and $\bar{\varepsilon}$ ($j_1 = \sigma_{11}\varepsilon_1 + \sigma_{12}\varepsilon_2 + \sigma_{13}\varepsilon_3\ldots$). Assuming that τ is constant on the Fermi surface (= τ_F), the following expression for $\overline{\sigma}$ may be deduced from (11):

$$\sigma_{ij} = \frac{e^2\tau_F}{4\pi^3}\int_{\Sigma_F} v_i v_j \frac{\mathrm{d}S}{|\overline{\nabla}_k E|}. \tag{12}$$

If we wish to find an expression for $\overline{\sigma}$ similar to that for the free electrons (10), we may write:

$$\boxed{\sigma_{ij} = \frac{Ne^2\tau_F}{\langle m_{ij}^*\rangle}} \tag{13}$$

subject to the condition

$$\frac{1}{\langle m_{ij}^*\rangle} = \frac{1}{4\pi^3 N}\int_{\Sigma_F} v_i v_j \frac{\mathrm{d}S}{|\overline{\nabla}_k E|} = \frac{1}{4\pi^3 N}\frac{1}{\hbar^2}\int_{\Sigma_F} \frac{\partial E}{\partial k_i}\frac{\partial E}{\partial k_j}\frac{\mathrm{d}S}{|\overline{\nabla}_k E|} \tag{14}$$

in which the integral $\int_{\Sigma_F}$ may be identified with

$$\int_{\mathcal{V}_F} \frac{\partial^2 E}{\partial k_i \partial k_j} \, d^3\overline{k}$$

where $\mathcal{V}_F$ is the volume within $\Sigma_F{}^2$. Since this volume is equal to $4\pi^3 N$, we see that the $1/\langle m_{ij}^* \rangle$ of (13) is simply the mean value in the interior of the Fermi surface of the inverse of m_{ij}^*, an element of the *effective mass* tensor defined in (36) (p. 145).

---

^{2} In fact, we have

$$I = \int_{\mathcal{V}_F} \frac{\partial^2 E}{\partial k_i \partial k_j} \, d^3\overline{k} = 2 \int \frac{1}{2} \left(\frac{\partial E}{\partial k_j} \right)_{\Sigma_F} dk_j \, dk_l$$

where the factor 2 is associated with the fact that the two values taken by $\partial E/\partial k_j$ on Σ_F, for $k_i > 0$ and $k_i < 0$, are equal and opposite (because of the symmetry of Σ_F with respect to the plane $k_j k_l$). Then we replace the integration in the plane $k_j k_l$ by an integration over Σ_F (whence the factor 1/2). Since the vector $\overline{\nabla}_k E$ is perpendicular to the surface element dS of Σ_F we have:

$$dk_j \, dk_l = dS \cos\theta \quad (\theta = \text{angle between } \overline{k}_i \text{ and } \overline{\nabla}_k E)$$
$$= dS \frac{\partial E/\partial k_i}{|\overline{\nabla}_k E|}.$$

Whence

$$I = \int_{\Sigma_F} \frac{\partial E}{\partial k_j} \frac{\partial E}{\partial k_i} \frac{dS}{|\overline{\nabla}_k E|}$$

which is the integral of (14).

TIGHT BINDING METHOD

Most likely to succeed
John Dos Passos

1. APPROXIMATE FORM OF THE BLOCH FUNCTIONS

The eigenfunctions of independent electrons in a periodic potential, which have the form (see p. 122):

$$\psi_{\bar{k}}(\bar{r}) = u_{\bar{k}}(\bar{r})e^{i\bar{k}\cdot\bar{r}},\qquad (1)$$

can be known only approximately. First we ask, what is the nature of these functions $\bar{u}_k$, above and beyond their property of being periodic?

In response, we substitute (1) in the Schrödinger equation

$$\left[-\frac{\hbar^2}{2m}\Delta + V(\bar{r})\right]\psi = E\psi.$$

Thus, we immediately obtain

$$\frac{\hbar^2}{2m}(k^2 - 2i\bar{k}\cdot\bar{\nabla} - \Delta)u_{\bar{k}} + V(\bar{r})u_{\bar{k}}(\bar{r}) = Eu_{\bar{k}}(\bar{r}).$$

If we assume a situation in which k is small, this equation reduces approximately to

$$-\frac{\hbar^2}{2m}\Delta u_{\bar{k}} + V(\bar{r})u_{\bar{k}} = Eu_{\bar{k}}.\qquad (2)$$

The potential $V(\bar{r})$ is taken to equal the sum of the **atomic potentials** $v_i(\bar{r})$ associated with each site i, which are all identical (see Figure 1, p. 115). Thus, in a period Ω of the crystal, $V(\bar{r})$ is little different from $v(\bar{r})$. Therefore, equation (2) may itself be approximated by

$$-\frac{\hbar^2}{2m}\Delta u_{\bar{k}} + v(\bar{r})u_{\bar{k}} = Eu_{\bar{k}}\qquad (3)$$

(this is only considered in Ω, since $u_{\bar{k}}(\bar{r})$ is periodic) which is just the *Schrödinger equation for atoms* relative to an arbitrary site of the crystal.

We deduce that in the limit $k \to 0$, the functions $u_{\bar{k}}(\bar{r})$ are close to the **atomic functions** $\varphi(\bar{r})$ of the element studied, for the electrons considered before being delocalised into Bloch states. For example, for *lithium*, the Bloch functions of the delocalised 2s electrons may be written approximately as

$$\psi_{\overline{k}}(\overline{r}) \simeq \phi_{2s}(\overline{r})e^{i\overline{k}\cdot\overline{r}} \quad (\text{as } k \to 0) \tag{4}$$

where ϕ_{2s} is obtained by extending the atomic function φ_{2s} of lithium periodically in space. Whence the approximate form of $\psi_{\overline{k}}$ (Figure 1).

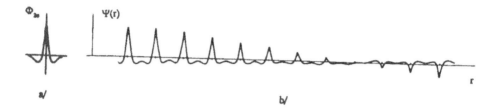

Figure 1 **Approximate form of a Bloch function (4). Real part.** The case illustrated is that of lithium (*a*). The axis shown passes through a series of atomic sites (*b*).

2. LINEAR COMBINATIONS OF ATOMIC ORBITALS (LCAO)

The above results suggest that as first approximations to the electronic states of a crystal, we should consider functions which are:

1. *Bloch functions*;
2. *linear combinations of atomic orbitals*.

   The first suggestion is obligatory since we are working on a crystal. The second suggestion is the generalisation of what chemists do, for molecules, when they construct molecular orbitals from atomic orbitals.

   Thus, following 2., we set:

$$\psi_{\overline{k}}(\overline{r}) = N^{-1/2} \sum_j A_{j\overline{k}}\varphi(\overline{r} - \overline{R}_j) \tag{5}$$

(N is the number of atoms in the crystal; $N^{-1/2}$ is the normalisation constant; the $\overline{R}_j$ are the coordinates of the crystalline sites).

   According to assumption 1. we write (5) in the Bloch form

$$\psi_{\overline{k}}(\overline{r}) = N^{-1/2}e^{i\overline{k}\cdot\overline{r}} \sum_j A_{j\overline{k}}e^{-i\overline{k}\cdot\overline{r}}\varphi(\overline{r} - \overline{R}_j)$$

where $\sum_j$ plays the role of $u_{\overline{k}}$ in (1). If we set

$$A_{j\overline{k}} = a_{j\overline{k}}e^{i\overline{k}\cdot\overline{R}_j},$$

the sum $\sum_j$ becomes

$$\sum_j = \sum_j a_{j\bar{k}} e^{-i\bar{k}\cdot(\bar{r}-\bar{R}_j)} \varphi(\bar{r} - \bar{R}_j). \tag{6}$$

The same sum taken at the point $\bar{r} + \bar{r}_T$ ($\bar{r}_T$ is a translation vector of the lattice (see (9), p. 96)) may be expressed as

$$\sum_j = \sum_j a_{j\bar{k}} e^{-i\bar{k}\cdot(\bar{r}+\bar{r}_T-\bar{R}_j)} \varphi(\bar{r} + \bar{r}_T - \bar{R}_j). \tag{7}$$

These two sums (6) and (7) are equal if and only if $a_{j\bar{k}}$ is constant, since $\bar{r}_T - \bar{R}_j$, like $\bar{R}_j$, denotes a crystalline site. We shall take $a_{j\bar{k}} = 1$.

Thus, the linear combination (5) is a Bloch function if it is of the form

$$\boxed{\psi_{\bar{k}}(\bar{r}) = N^{-1/2} \sum_j e^{i\bar{k}\cdot\bar{R}_j} \varphi(\bar{r} - \bar{R}_j).} \tag{8}$$

3. ENERGIES OF LCAO

Let us consider one of the LCAO defined by (8), namely $\psi_{\bar{k}}$. The corresponding energy is

$$E_{\bar{k}} = \langle \psi_{\bar{k}} | \hat{H} | \psi_{\bar{k}} \rangle \tag{9}$$

where

$$\hat{H} = -\frac{\hbar^2}{2m}\Delta + \sum_j v(\bar{r} - \bar{R}_j) = -\frac{\hbar^2}{2m}\Delta + \sum_j v_j.$$

For the sequence, we suppose, taking into account the rapid decrease of φ in an atomic cell Ω, that:

1. the φ are orthogonal (and normalised): $\langle \varphi(\bar{r} - \bar{R}_j) | \varphi(\bar{r} - \bar{R}_m) \rangle = \delta_{jm}$.
2. The matrix elements of type $\langle \varphi(\bar{r} - \bar{R}_j) | v_n | \varphi(\bar{r} - \bar{R}_m) \rangle$ are all zero except those relating to *two near-neighbour sites* j and l, which are set equal to $-\alpha$ and $-\gamma$ (α and $\gamma > 0$):

$$\langle \varphi(\bar{r} - \bar{R}_j) | v_l | \varphi(\bar{r} - \bar{R}_j) \rangle = -\alpha \qquad \langle \varphi(\bar{r} - \bar{R}_j) | v_j | \varphi(\bar{r} - \bar{R}_l) \rangle = -\gamma. \tag{10}$$

Then, we expand (9) in the form

$$E_{\bar{k}} = \frac{1}{N} \left\langle \sum_j e^{i\bar{k}\cdot(\bar{r}-\bar{R}_j)} \varphi(\bar{r} - \bar{R}_j) \left| -\frac{\hbar^2}{2m}\Delta + \sum_j v_j \right| \sum_m e^{i\bar{k}\cdot(\bar{r}-\bar{R}_m)} \varphi(\bar{r} - \bar{R}_m) \right\rangle.$$

Recalling that, by construction, we have

$$\left(-\frac{\hbar^2}{2m}\Delta + v_i \right) \varphi(\bar{r} - \bar{R}_i) = -\varepsilon_0 \varphi(\bar{r} - \bar{R}_i)$$

($-\varepsilon_0$ is the energy of the *atomic levels*), it follows immediately that

$$E_{\bar{k}} = -N\varepsilon_0 - N \sum_{j \text{ neighbour of } l} \alpha \quad - N\gamma \sum_{j \text{ neighbour of } l} e^{i\bar{k}\cdot(\bar{R}_j - \bar{R}_l)}.$$

In the case in which the crystal is simple cubic (lattice constant a), an atom has six near neighbours, and per atom we have

$$\varepsilon_{\bar{k}} = -\varepsilon_0 - 6\alpha - 2\gamma(\cos k_x a + \cos k_y a + \cos k_z a). \tag{11}$$

Figure 2 shows the variation of the energy $\varepsilon(k)$, for example along the diagonal axis ($k = k_x = k_y = k_z$) of the reciprocal lattice. The figure shows the shape of the first energy band (see Figure 4, p. 119 or Figure 11, p. 129) in the first Brillouin zone ($-\pi/a, \pi/a$). The other bands can be obtained using the LCAO of atomic functions $\varphi_{1s}, \varphi_{2p}, \varphi_{3s}, \varphi_{3p}, \ldots$ other than φ_{2s} (case of lithium, for example).

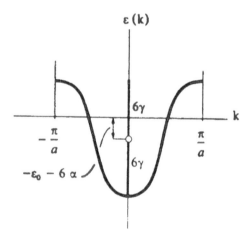

Figure 2 Energy band of an LCAO. In a simple cubic crystal, k is taken along a diagonal of the reciprocal lattice (111). The energies ε_0, α and γ are positive. Width of band: 12γ (see (10) and (11)).

N.B. Other examples of the use of the tight binding method will be found in **Problems 2, 3, 4, 5** and **6**.

BURGERS VECTOR

Ultima verba
Valéry
All's well, that ends well
Shakespeare

1. DEFINITIONS

The definition of the Burgers vector (denoted by BV) given in Chapter IX, based on 'displacements' of matter contained an ambiguity relating to the *direction* of the BV.

Let us consider the (rectilinear) edge dislocation of Figure 1(*a*). Let us trace *a closed circuit* around the dislocation (**Burgers circuit**) through S–1–2–3–F (S = Start, F = Finish). If we suppress the dislocation and reconstruct the circuit defined by the same displacements in the *perfect crystal* (Figure 1(*b*)), this circuit is no longer closed. The *non-closure* distance SF is, by definition, the *modulus* of the BV. To describe the exact *direction* of this vector, we must choose a **direction of movement** along the dislocation L (for example, here, the direction ⊗ from front to back) and adopt a convention. The most popular is called **FS/RH**. This involves defining the direction of rotation on the Burgers circuit by the corkscrew rule (Right Hand). Then, the **vector which completes the closure** $\overline{FS}$ is defined to be the Burgers vector (Figure 1(*b*)), where S denotes the start of the circuit and **F** its finish.

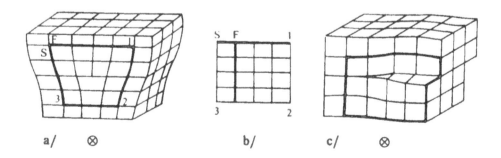

a/ ⊗ b/ c/ ⊗

Figure 1 Burgers vector and circuit. (*a*). Edge dislocation; direction of movement along dislocation ⊗. (*b*) Same circuit as in (*a*), but without dislocation. (*c*) Screw dislocation; direction of movement along dislocation ⊗.

Exercise 1   Apply this procedure to the case of the screw dislocation of Figure 1(c) to define the modulus and direction of its BV. The circuit shown in bold may be used.

Exercise 2   In Figure 9 on p. 211, we introduced a dislocation loop with a rectangular shape, obtained by a displacement **in the plane** of the loop.

   Verify that Figure 2, which shows the 'vertical' (i.e. perpendicular to the plane of the loop) atomic planes, illustrates the same situation. The bold curve shows the line of dislocation $\mathcal{L}$ and only *half a loop* is visible (a whole segment l of p. 211 and two segments $L/2$).

   Choose a direction of movement along $\mathcal{L}$ and define the BV of $\mathcal{L}$ at three places: on each of the two segments $L/2$ and on the segment l. It may be helpful to place atoms regularly on the atomic planes.

   Show that these three vectors are identical. Identify the **edge** and **screw** portions of the loop.

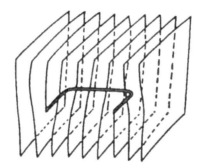

Figure 2   **Section of a dislocation loop created like that of Figure 9 (p. 211).** Note the arrangement of a family of atomic planes perpendicular to the BV, which is in the plane of the loop. The *bold line* denotes the line of dislocation.

2. NODES

Let us consider three dislocations L_1, L_2 and L_3 (BV: $\bar{b}_1, \bar{b}_2, \bar{b}_3$) linked by a node N (Figure 3) and let us trace the Burgers circuits C_1, C_2, C_3, C_4 and C_5. We choose the direction of movement on the dislocations to be that indicated by the arrows written *on* the dislocations. From the continuous passage from $C_2 + C_3$ to C_4 then to C_5 and from the fact that C_1 and C_5 are identical, we deduce that

$$\bar{b}_1 = \bar{b}_2 + \bar{b}_3.$$

If we now define the direction of movement on the dislocations **from the node** N (or converging towards N) as shown by the arrows written *beside* the dislocations, the BV $\bar{b}$ become $\bar{b}$ such that $\bar{b}_1 = -\bar{b}_1$, $\bar{b}_2 = \bar{b}_2$ and $\bar{b} = \bar{b}_3$. The preceding identity becomes

$$\bar{b}_1 + \bar{b}_2 + \bar{b}_3 = 0$$

which generalises in the case of a node of n dislocations with direction of movement

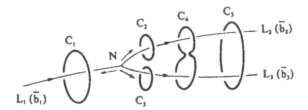

Figure 3 Law of nodes. The Burgers circuits C_1 and C_5 (i.e. also $C_2 + C_3$) define the modulus of a common Burgers vector.

defined *from* (or *to*) the node, to:

$$\sum_j \bar{b}_j = 0$$

which is analogous to Kirchoff's law, where the BV play the role of electrical intensities.

PROBLEMS ...

Great Expectations
Charles Dickens

Most Likely to Succeed
John Dos Passos

Happy Days
Samuel Beckett

Quiet Waters
Ivan Sergeevich Turgenev

The Examiner
Jonathan Swift

What Is To Be Done ?
Lenin

Hard Times
Charles Dickens

Tales of Mystery and Imagination
Edgar Allan Poe

Labyrinths
Jorge Luis Borges

Lawless Roads
Graham Greene

Darkness at Noon
Arthur Koestler

Valley of Fear
Conan Doyle

Lost Illusions
Balzac

Dramatic Gibberish
Voltaire

Despair
Vladimir Nabukov

The Fall
Albert Camus

Long Day's Journey into Night
Eugene O'Neill

As I Lay Dying
William Faulkner

De Profundis
Oscar Wilde

Apocalypse
St. John

... with SOLUTIONS

Much Ado About Nothing
Shakespeare

N.B. The questions to be answered explicitely in the following problems are indicated by *italics*.

LITHIUM PLATELETS

Ionel Solomon

This problem follows the results of the following two articles:

- *Précipitation de lithium dans les monocristaux de fluorure de lithium irradiés aux neutrons thermiques* by **M Lambert**, **C Mazières** and **A Guinier**, J. Phys. Chem. Solids **18** (1961) 129;
- *Spectres de résonance magnétique nucléaire et électronique de plaquettes de lithium dans le fluorure de lithium* by **C Taupin**, Comptes Rendus **262** (1966) 1617.

When one puts a crystal of LiF in a nuclear reactor (irradiation by slow neutrons) a part of the fluorine atoms is driven away and the excess lithium atoms regroup to form two-dimensional metallic platelets, or clusters. Up to 1966, little was known about these clusters: X-rays indicate that they undoubtedly consist of a square-cell lattice, with parameter a of the same order as that of the crystal of metallic lithium (i.e. a is of the order of 3 Å$= 3 \times 10^{-8}$ cm) and that they contain a number of lithium atoms N which is certainly less than 500.

We propose here a very simple model which permits the prediction of certain physical properties of these clusters: we shall rely on the fact that lithium in bulk is very well described by a model of almost-free electrons in a box, with one electron per atom.

We shall show that the small size of these clusters gives them properties which may be very different from the properties of ordinary metals.

1. *In this model, find the formula giving the electron energy levels in a planar platelet with dimensions L_1, L_2 and monatomic thickness.*

2. We shall not seek to classify these energy levels explicitly (difficult for high quantum numbers); instead, we shall work with the mean 'distance' between these levels (concept of 'density of states' introduced in Chapter VI).

Calculate this order of magnitude of the distance between two quantum levels (at the Fermi level) for a platelet of area 2000 Å^2 (approximately 200 atoms).

3. *Deduce that there exists a characteristic temperature θ such that for $T \ll \theta$ this system has special physical properties (determination of θ was one of the ways of measuring the order of magnitude of the size of platelets).*

4. *Which physical measurement might one use in an attempt to determine the parity of the number of electrons N?*

SOLUTION TO PROBLEM 1

Since the distances between atoms in the platelets are of the same order as in the metal, we propose using the Sommerfeld theory (Chapter III), but in a *two-dimensional* box.

1. We shall consider a platelet with dimensions $N_1 a = L_1$ and $N_2 a = L_2$, which thus contains $N = N_1 \times N_2$ atoms and the same number of free electrons N.

The wavefunction of one of these electrons may be written as:

$$\psi = A \sin k_1 x \sin k_2 y$$

(A is a normalisation factor). The dependence on z (perpendicular to the platelet) which corresponds to an atomic wavefunction is not involved in this model. The sines are chosen so that the wavefunction is zero for: $x = 0$ and $x = L_1$ and for $y = 0$ and $y = L_2$. Thus, the possible values for the wavenumbers k_1 and k_2 are

$$k_1 = \frac{n_1 \pi}{L_1} \quad \text{and} \quad k_2 = \frac{n_2 \pi}{L_2}$$

where n_1 and n_2 are positive integers. From this, we deduce the energy levels

$$E = \frac{\hbar^2 \pi^2}{2m} \left(\frac{n_1^2}{L_1^2} + \frac{n_2^2}{L_2^2} \right)$$

for non-negative integers n_1 and n_2.

We remark that in this model neither of the two quantum numbers n_1 and n_2 may be zero, since then we would have $\psi(xy) = 0$. On the other hand, for a model based on the Born-von Karman boundary conditions, n_1 and n_2 could be zero.

2. For arbitrary values of L_1 and L_2, classification of the energy levels is a complicated arithmetical problem. By way of illustration, for a square platelet ($L_1 = L_2 = L$), the figure below shows the energy levels in units of $\varepsilon_0 = \hbar^2 \pi^2 / 2mL^2$ together with their degeneracy.

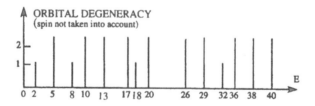

For larger quantum numbers n_1 and n_2, the degeneracy of certain levels increases, but, particularly if $L_1 = L_2$, the calculations become very complicated. We shall use a much more practical approximation method.

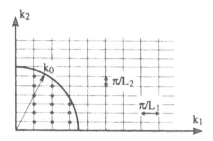

Let us count the number of orbital states (the spin will be introduced later) with energy between 0 and E_0. These states correspond in the plane k_1, k_2 to vectors $\overline{k}$ (with components k_1 and k_2) with modulus less than k_0 given by:

$$E_0 = \frac{\hbar^2 k_0^2}{2m}.$$

Now, the permissible values of k_1 and k_2 form a lattice of rectangles with surface area $\Delta S = \pi^2 / L_1 L_2$ in this plane. If we neglect the 'corners' (this is the basis of the approximation), the desired number of states is therefore:

$$\nu_0 = \frac{\pi k_0^2}{4} / \Delta S$$

or, replacing k_0^2 by the value $2m E_0 / \hbar^2$:

$$\nu_0 = \frac{E_0}{2\pi \hbar^2} L_1 L_2.$$

At low temperature, the Fermi level is obtained by putting two electrons (because of the two possible spin states) at each of these levels. Thus, it will be given by $\nu_F = N/2$, whence:

$$E_F = \frac{\pi \hbar^2}{m} \frac{N}{L_1 L_2}.$$

Since we have

$$N = N_1 N_2 \qquad L_1 = N_1 a \qquad L_2 = N_2 a$$

we find

$$\boxed{E_F = \frac{\pi \hbar^2}{m a^2}.}$$

The Fermi level is independent of the size of the platelets. For $a = 3$ Å, we find $E_F = 2.64$ eV.

The (orbital) state density is given by

$$n = \frac{\mathrm{d}\nu}{\mathrm{d}E} = -\frac{m}{2\pi \hbar^2} L_1 L_2.$$

It does not depend on the *surface area* $L_1 L_2$ of the platelet or on the cell size, which, in fact, is not known with accuracy. Unlike in the case of one- or three-dimensional boxes (see Course), it turns out that n is also *independent of the energy E*.

Because of the very small size of the platelets, the density of states n is very low. If we suppose that the order of magnitude of the distance between two consecutive energy levels is of the order of the mean distance $\overline{\Delta E} = 1/n$ (which is 'pessimistic', for this distance is larger if degeneracy occurs), we have

$$\overline{\Delta E} = \frac{2\pi\hbar^2}{mL_1L_2}.$$

Although this distance is exceedingly small for a macroscopic piece of metal, here we find finite values. For example, for $L_1 L_2 = 2000$ Å^2:

$$\overline{\Delta E} = 25 \times 10^{-3} \text{ eV}$$

i.e. an energy of the order of the thermal energy kT at room temperature.

3. Thus, we have a system whose 'characteristic temperature' $\theta = \Delta E/k$ is of the order of 300 K. For any excitation below this value, the system no longer 'responds'; whence 'low-temperature' quantum effects appear from room temperature.

Thus, the electronic specific heat of the platelets, which should vary like $\exp(-\theta/T)$, becomes very small below room temperature: the system is completely 'frozen'.

4. Here is another curious consequence. For $T < \theta$ (i.e. slightly below room temperature) the electrons pass to the lowest energy states. Electrons with opposite spins may be placed in each state. We then distinguish two cases:

(i)   The platelet has an even number of atoms (whence also of electrons): then the magnetic moments compensate each other pairwise and the platelet is not magnetic.

(ii) The platelet has an odd number of electrons: then *one* uncompensated spin remains and the platelet has a magnetic moment equal to *one* Bohr magneton.

It was by showing, by means of magnetic resonance, that this result is obtained from room temperature (and below) that C Taupin (see p. 277) was able to determine that the size of the platelets is about 1000 Å^2 (around 100 atoms).

INSTABILITY OF CAROTENE

(OR WHY CARROTS ARE RED)

Maurice Guéron and Maurice Bernard

Despite its extreme simplicity, the model of the linear chain of atoms is often useful: it provides an easy explanation of certain properties of solids.

We shall use this model to study certain properties of carotene, an organic substance with the chemical formula $R–(CH)_{2p}–R'$; each molecule is a long linear chain formed from a CH pattern repeated $2p$ times and terminated by two radicals R and R' which we shall neglect (p is of the order of 10).

We shall use the model of the linear chain of N atoms, with the assumption that each CH grouping is equivalent to an atom with a single external electron; we shall take cyclic boundary conditions (Born–von Karman) and assume that N is large.

I

We let $V(x)$ denote the potential seen by an electron from such an atom (assumed to be single), E_0 the energy eigenvalue of the ground state and $\varphi(x)$ the electron wavefunction corresponding to that state; we neglect the excited states which are assumed to have much higher energies and assume that $\varphi(x)$ is a real function. We shall first treat this problem using the tight binding approximation.

1. Let $x_n = na$ be the position of the nth atom, $V(x - x_n)$ the potential created by this atom, assumed to be single, and H the interaction which must be added to the potential of an atom to take into account the influence of the other atoms. We consider the electron wavefunction

$$\psi_k(x) = \sum_{n=1}^{N-1} \frac{1}{\sqrt{N}} e^{ikna} \varphi(x - x_n)$$

where k is a wavevector given by the cyclic boundary conditions. To simplify the notation, we shall write $\psi_k(x)$ in the form $|k\rangle$.

Verify that $|k\rangle$ is a Bloch function. Suppose that the function $\varphi(x)$ is chosen so that $\psi_k(x)$ is normalised; show that $|k\rangle$ and $|k'\rangle$ are orthonormal.

In the remainder of the problem, we shall consider the $|k\rangle$ basis.

2. *Calculate the matrix element*

$$\langle k|H|k'\rangle = \int_0^{Na} \psi^*(x) H \psi_{k'}(x) \, dx.$$

Show that this is zero if $k' \neq k$. If $k' = k$, show that:

$$\langle k|H|k \rangle = \sum_q A(qa)e^{+ikqa}$$

where q is a number which takes N consecutive integer values and $A(qa)$ is an integral, which should be given explicitly.

3. We suppose that the interaction H, considered as a pertubation, removes the degeneracy of the atomic levels E_0 and that each electron state is described approximately by a wavefunction $\psi_k(x)$ with energy eigenvalue:

$$E(k) = E_0 + \langle k|H|k' \rangle.$$

We retain only the terms $A(0) = -\alpha$ corresponding to $q = 0$ and $A(+a) = A(a) = -\beta$ corresponding to $q = \pm1$. We assume that α and β are positive and that the other terms $A(qa)$ are negligible.

Draw the curve $E(k)$ and show that there exists a band of allowed energies with width ΔE to be determined.

II

Because of the spin of the electron, the allowed energy band found above contains $2N$ places for N electrons. Thus, for N very large, the present model would lead to a crystal of metallic type. Now, carotene is an insulator and has a red colour which denotes a strong optical absorption for radiation of wavelength $\lambda \leq 0.6$ Å. To explain this result, we make the assumptiom that in the structure of carotene one in two atoms is slightly displaced, for example to the right, by a quantity εa (ε small); N is assumed to be even ($N = 2p$). This is called a Peierls distortion.

4. *What becomes of the new Brillouin zone?*

5. After the displacement εa, the interaction $H(x)$, which was initially periodic with period a, becomes $H'(x)$ with period $2a$; we shall assume that $H'(x)$ differs from $H(x)$ only by the term $2\varepsilon B \cos(\pi x/a)$:

$$H'(x) = H(x) + 2\varepsilon B \cos\left(\frac{\pi x}{a}\right).$$

Give a qualitative explanation for what happens in the neighbourhood of $k = \pm\pi/2a$.

Assuming that $E(0)$ is unchanged, sketch the new curve $E(k)$. What is the sign of the variation $\Delta_1 \mathcal{E}$ of the total energy of the N electrons of the system, assumed to be at room temperature? What is the order of magnitude of $\Delta_1 \mathcal{E}$?

6. In fact, it is natural to think that the displacement εa of the atoms of even rank a also has the effect of increasing the elastic energy of the N ions of the system. Suppose that this variation in energy $\Delta_2 \mathcal{E}$ may be written as $\Delta_2 \mathcal{E} = C\varepsilon^2 + A\varepsilon^4$, where C and A are positive constants. *What is the criterion for effective displacement of the even atoms (carotene-type instability)? Is this displacement limited?*

7. *If this condition is satisfied for carotene, explain why this model takes account of the fact that the body in question is an insulator and has a red colour. If we assume that B is of the order of magnitude of 10 eV, what is the size of the relative displacement ε?*

SOLUTION TO PROBLEM 2

I

By assumption, we have

$$\left(\frac{p^2}{2m} + v(x)\right)\varphi(x) = E_0\varphi(x) \qquad (p = \hbar k).$$

1. Consider the wavefunction:

$$\psi_k(x) = \sum_{n=0}^{N-1} \frac{1}{\sqrt{N}} e^{ikna}\varphi(x - x_n)$$

with $x_n = na$ and $k = (2\pi/Na)xr$ where r is an integer which may take N distinct values, for example, $r = 0, 1, \ldots, N - 1$ (cyclic boundary conditions).

$$\begin{aligned}
\psi_k(x + a) &= \sum_{0}^{N-1} \frac{1}{\sqrt{N}} e^{ikna}\varphi(x + a - x_n) \\
&= \sum_{0}^{N-1} \frac{1}{\sqrt{N}} e^{ik(n-1)a} e^{ika}\varphi(x - x_{n-1}) \\
&= e^{ika} \sum_{0}^{N-1} e^{ik(n-1)a}\varphi(x - x_{n-1}) \\
&= e^{ika}\psi_k(x).
\end{aligned}$$

Thus, the function $\psi_k(x)$ is a Bloch function. Let us calculate the scalar product

$$\langle k|k'\rangle = \frac{1}{N} \sum_n \sum_{n'} \int e^{-ikna}\varphi^*(x - x_n) e^{ik'n'a}\varphi(x - x_{n'})\,dx.$$

We set $n' = n + q$.

$$\begin{aligned}
\langle k|k'\rangle &= \frac{1}{N} \sum_n \sum_q \int_0^{Na} e^{i(k'-k)na} e^{ik'qa}\varphi(x - x_n)\varphi(x - x_n - x_q)\,dx \\
&= \frac{1}{N} \sum_n \sum_q e^{i(k'-k)na} e^{ik'qa} \int_{x_n}^{x_n+Na} \varphi(\xi)\varphi(\xi - x_q)\,d\xi
\end{aligned}$$

where we have made the change of variable $x = x_n + \xi$; the integral:

$$\int_{-x_n}^{-x_n+Na} \varphi(\xi)\varphi(\xi - x_q)\,d\xi = \Phi_q$$

does not depend on the integer q and represents the overlap of the functions $\varphi(x)$ centred at distance qa from each other. It follows that:

$$\langle k|k'\rangle = \frac{1}{N} \sum_q e^{ik'qa}\Phi_q\left(\sum_n e^{i(k'-k)na}\right).$$

Now, it is known that the expression in brackets is zero for $k' \neq k$ (modulo $2\pi/a$) and equal to N otherwise. We see that $|k\rangle$ and $|k'\rangle$ corresponding to different k are orthogonal. The norm of k is given by:

$$\langle k|k\rangle = \sum_q \Phi_q e^{ikqa}.$$

It is equal to 1 by construction.

Thus, the functions $|k\rangle$ can be used as an orthonormal basis for the wavefunctions for an electron of the crystal considered.

2. We calculate

$$\langle k|H|k'\rangle = \int_0^{Na} \psi_k^*(x) H \psi_{k'}(x)\, dx$$

$$= \frac{1}{N} \sum_N \sum_{n'} e^{ik'n'a - ikna} \int_0^{Na} \varphi(x - x_n) H \varphi(x - x_{n'})\, dx.$$

We set $n' = n + q$ and make the change of variable $x = x_n + \xi$. It follows that:

$$\langle k|H|k'\rangle = \frac{1}{N} \sum_n \sum_{n'} e^{i(k'n'a - kna)} \int_{-x_n}^{-x_n + Na} \varphi(\xi) H \varphi(\xi - x_q)\, dx$$

where the integral

$$\int_{-x_n}^{-x_n + Na} \varphi(\xi) H \varphi(\xi - x_q)\, d\xi = A(qa)$$

does not depend on x_n because H is periodic with period a. Thus, we have:

$$\langle k|H|k'\rangle = \sum_q A(qa) e^{ik'qa} \left(\frac{1}{N} \sum_n e^{i(k'-k)na} \right).$$

As earlier, the expression in brackets can be written as $\delta_{k,k'}$; it follows that:

$$\langle k|H|k\rangle = \sum_q e^{ikqa} A(qa).$$

3. From perturbation theory, if H is the difference between the crystal and the atom potentials, the Bloch function $|k\rangle$ corresponds to a characteristic value for the energy, given by:

$$E(k) = E_0 + \langle k|H|k\rangle.$$

If we suppose that the $A(qa)$ are negligible, except for:

$$A(0) = -\alpha \qquad A(\pm a) = -\beta$$

where α and β are positive, we see that for $E(k)$ we have:

$$E(k) = E_0 - \alpha + 2\beta \cos(ka).$$

The dispersion curve is the same as that of Figure 2, p. 270. It has a cissoidal appearance, lying between $E_0 - \alpha - 2\beta$ (minimum at the centre of the Brillouin zone) and $E_0 + \alpha + 2\beta$ (maximum on the edges of the Brillouin zone). The width of the energy band is $\Delta E = 4\beta$.

II

Taking into account the spin of the electrons, the total number of possible energy states in the above energy band $E(k)$ is $2N$; thus, at zero temperature, this band is half full and therefore we predict a 1D crystal of metallic type, i.e. an electrical conductor.

4.   The assumption that the atoms of even rank are displaced 'en bloc' by a very small quantity εa implies that the period becomes $2a$ instead of a; the Brillouin zone is half as large and its limits must be taken for $k = \pm\pi/2a$ rather than for $\pm\pi/a$ for the undeformed crystal. Thus, intutively, we would expect the Bragg reflections which occur for values of the wavevector equal to $\pm\pi/2a$ to modify the dispersion curve $E(k)$ in the neighbourhood of these points.

5.   The interaction $H(x)$ which is periodic with period a now becomes $H'(x)$ which is periodic with period $2a$. One can pass from $H(x)$ to $H'(x)$ by modifying the amplitude of the terms of the Fourier series expansion of $H(x)$: we retain the first of these terms and write:

$$H'(x) = H(x) + 2\varepsilon B \cos\left(\frac{\pi x}{a}\right).$$

The first question showed that we have:

$$\left(\frac{p^2}{2m} + U(x)\right)|k\rangle = E(k)|k\rangle$$

where $U(x)$ was the crystalline potential of the undisplaced linear crystal, now replaced by $U'(x) = U(x) + 2\varepsilon B \cos(\pi x/a)$. Assuming that εB is not too large, we may apply the method used in Chapter VI; we find that in the neighbourhood of $k = \pi/2a$, the energy $E(k)$ is modified:

$$E_+(k) = W_0 + \varepsilon B + \frac{\hbar^2(k - (\pi/2a))^2}{2m_+}$$

$$E_-(k) = W_0 - \varepsilon B \frac{\hbar^2(k - (\pi/2a))^2}{2m_-}$$

where $W_0 = E(\pi/2a) - E(0) = 2\beta$ and:

$$\frac{1}{m_+} = \frac{1}{m}\left(\frac{W_0}{\varepsilon B} + 1\right) \qquad \frac{1}{m_-} = \frac{1}{m}\left(\frac{W_0}{\varepsilon B} - 1\right).$$

Thus, a forbidden band $E_{for} = 2\varepsilon B$ appears.

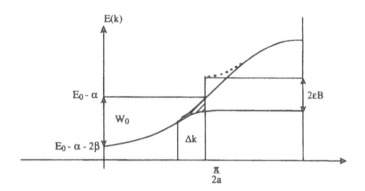

Moreover, the region Δk affected by a perturbation is approximately that for which we have

$$\varepsilon B = \frac{E(\pi/2a) - E(\Delta k)}{2} = \frac{E_0 - \alpha - E_0 + \alpha + 2\beta \sin(a\Delta k)}{2} = \beta a \Delta k.$$

Hence

$$\Delta k = \frac{\varepsilon B}{a\beta}.$$

In all this, we have assumed that the energy $E - \alpha - 2\beta$ was unmodified. It follows that the electrons at the top of the Brillouin zone have a decreased energy. The gain in energy is proportional to the shaded area; comparing this with a triangle, we find a decrease in the total energy of the system equal to:

$$\Delta_1 \mathcal{E} = -\frac{\varepsilon B}{2} \times \frac{2\Delta k}{\pi/a} = -\frac{\varepsilon^2 B^2}{\pi \beta}.$$

6. One might think that the 'carotene instability' should always occur since it leads to an increase in the binding. In reality, this conclusion is false; it is clear that the displacement εa of the atoms of even rank causes an increase in the energy of the system, written as:

$$\Delta_2 \mathcal{E} = C\varepsilon^2 + A\varepsilon^4 \qquad (C \text{ and } A > 0).$$

The criterion for the occurrence of instability is that:

$$W = \Delta_1 \mathcal{E} + \Delta_2 \mathcal{E} = \left(C - \frac{B^2}{\pi \beta}\right)\varepsilon^2 + A\varepsilon^4$$

has a minimum in the neighbourhood of $\varepsilon = 0$; this implies

$$\frac{B^2}{\pi \beta} > C.$$

The displacement is limited by the existence of the term in $A\varepsilon^4$: one finds a minimum in energy for a value of ε, such that:

$$\frac{dW}{d\varepsilon} = 4A\varepsilon^3 + 2\left(C - \frac{B^2}{\pi \beta}\right)\varepsilon = 0$$

that is:

$$\varepsilon_1 = \left(\frac{(B^2/\pi\beta) - C}{2A} \right)^{1/2}.$$

7. The present model gives a very simple explanation as to why the N electrons in such a crystal occupy an allowed band with N states; the upper band also contains N states and is found at an energy $E_{for} = 2\varepsilon B$ above. Such a crystal cannot be a conductor. If we assume that the red colour of carotene is due to the absorption of photons of energy $h\nu \geq E_f$, we see that the absorption threshold must lie just above the red, that is around 2 eV ($\lambda = 0.6 \; \mu m$); thus, we must have:

$$E_{for} = 2\varepsilon B \simeq 2 \text{ eV}.$$

If $B = 10$ eV, we see that we have:

$$\varepsilon = \frac{\Delta x}{a} \simeq 0.1.$$

ONE-DIMENSIONAL A–B COMPOUNDS

Jean-Noël Chazalviel

We consider a crystalline solid with chemical formula AB, where A and B represent two elements of Mendeleyev's classification. To simplify matters, we suppose that this solid is one dimensional and consists of the repetition of N patterns AB, where the distance between two consecutive atoms (A–B or B–A) is equal to d.

1. *For this crystal, give the length of the primitive cell and the size of the first Brillouin zone.*

2. We shall calculate the band structure of this crystal using the tight-binding approximation. We shall model the quantum states of the valence electrons of the isolated atom A (*resp.* B) by a single non-degenerate orbital state $|\phi_A\rangle$ (*resp.* $|\phi_B\rangle$) of energy ε_A (*resp.* ε_B). We suppose that the potential seen by an electron of the crystal is simply the sum of all the potentials V_{A_i} and V_{B_i} ($i = 1, 2, \ldots, N$) seen by the valence electrons of the supposedly isolated atoms A_i and B_i.

Give the Hamiltonian $\hat{H}$ of an electron of the crystal. Write down the general form $|\psi_k\rangle$ of an eigenstate of this Hamiltonian as a linear combination of atomic orbitals (tight-binding approximation, see Annex 4), taking into account the periodicity of the crystal, assuming periodic boundary conditions. Show, in particular, that the periodicity condition does not determine the eigenstates $|\psi_k\rangle$ completely but leaves one degree of hybridisation freedom which may be described by two parameters a_k and b_k (with $|a_k|^2 + |b_k|^2 = 1$).

3. In order to determine the $|\psi_k\rangle$ completely, together with their associated energies ε_k, we write down the energy eigenvalue equation $(\hat{H} - \varepsilon_k)|\psi_k\rangle = 0$ which we project alternately onto two atomic states $|\phi_{A_j}\rangle$ and $|\phi_{B_j}\rangle$ ($\forall j \leq N$). We neglect all the overlaps between atomic states ($\langle \phi_X | \phi_Y \rangle = \delta_{XY}$), and we retain only those matrix elements of type

$\langle \phi_X | V_Y | \phi_Y \rangle$ involving at most two nearest neighbours, thereby defining the four quantities (all negative) $\alpha_A = \langle \phi_A | V_B | \phi_A \rangle$, $\alpha_B = \langle \phi_B | V_A | \phi_B \rangle$, $\gamma_A = \langle \phi_A | V_A | \phi_B \rangle$, $\gamma_B = \langle \phi_A | V_B | \phi_B \rangle$, where A and B are nearest neighbours. (We suppose that the wavefunctions ϕ_A and ϕ_B are s functions, whence even, so that these matrix elements are identical for the pairs AB such that B is on the left or the right of A.)

Show that, with the above approximations, we now also have the identity $\gamma_A = \gamma_B$, *which quantity we henceforth denote by* γ. This can be seen by considering the matrix elements $\langle \phi_A | (\widehat{p}^2 / 2m) + \hat{V}_A + \hat{V}_B | \phi_B \rangle$, where A and B are nearest neighbours.

4. *Show that the projection procedure described above leads to a system of two equations in a_k and b_k. Obtain the following expression for one of these:*

$$(\varepsilon_A - \varepsilon_k + 2\alpha_A)a_k + \gamma(1 + e^{-2ikd})b_k = 0.$$

and give an expression for the other.

5. *Deduce the energy eigenvalues and sketch the band structure, giving the values at the centre and at the edge of the first Brillouin zone. For convenience, you may set $\delta = 2\sqrt{\gamma_A \gamma_B}$, and, for an appropriate choice of the zero energy, $\Delta = \varepsilon_A + 2\alpha_A = -(\varepsilon_B + 2\alpha_B)$.*

6. *How can we determine $|\psi_k\rangle$ completely? Do so at the centre and at the edge of the zone. Give an interpretation in terms of chemical bonding.*

7. *The atom A and the atom B each have one valence electron. Taking into account the spin degeneracy, is the material AB a conductor or an insulator? The same question if A has one valence electron and B has two.*

8. We now assume a situation in which the lowest energy band is filled exactly and we suppose that the forbidden band immediately above is sufficiently narrow that the material is a semiconductor.

Give the width of the forbidden band and an expression for the effective masses at the top of the valence band and at the bottom of the conduction band.

9. *What happens to the band structure in the limit A = B? Compare with the result of the Course (equation (11) and Figure 2, p. 270, brought to one dimension). Where does the difference come from?*

SOLUTION TO PROBLEM 3

1. The primitive cell, containing the basis AB has length $2d$ and the first Brillouin zone is a segment of length $2\pi/2d = \pi/d$.

2. According to the assumptions, the Hamiltonian may be written as

$$\hat{H} = \frac{\widehat{p}^2}{2m} + \sum_{i=1}^{N}(\hat{V}_{A_i} + \hat{V}_{B_i}).$$

The general form of an LCAO eigenstate is

$$|\psi\rangle = \sum_{l=1}^{N}(\xi_l |\phi_{A_l}\rangle + \eta_l |\phi_{B_l}\rangle).$$

The periodicity condition and the periodic boundary conditions imply that ξ_l and η_l are proportional to $\exp(2ikld)$ where k is an integer multiple of $2\pi/2Nd$.

Thus, the general form of a normalised eigenstate is given by

$$|\psi_k\rangle = \frac{1}{\sqrt{N}} \sum_{l=1}^{N} (a_k|\phi_{A_l}\rangle + b_k|\phi_{B_l}\rangle)e^{2ikld}$$

where $|a_k|^2 + |b_k|^2 = 1$.

3.

$$\left\langle \phi_A \left| \frac{\widehat{p}^2}{2m} + \hat{V}_A + \hat{V}_B \right| \phi_B \right\rangle = \left\langle \phi_A \left| \frac{\widehat{p}^2}{2m} + \hat{V}_B + \hat{V}_A \right| \phi_B \right\rangle = \langle \phi_A | \varepsilon_B + \hat{V}_A | \phi_B \rangle = \gamma_A$$

$$= \left\langle \phi_A \left| \frac{\widehat{p}^2}{2m} + \hat{V}_A + \hat{V}_B \right| \phi_B \right\rangle = \langle \phi_A | \varepsilon_A + \hat{V}_B | \phi_B \rangle = \gamma_B.$$

4. $\langle \phi_{A_j} | \hat{H} - \varepsilon_k | \psi_k \rangle = 0$ may be written as

$$\left\langle \phi_{A_j} \left| \frac{\widehat{p}^2}{2m} + \hat{V}_{A_j} + \sum_{i\neq j} \hat{V}_{A_i} + \sum_i \hat{V}_{B_i} - \varepsilon_k \right| \sum_{l=1}^{N}(a_k\phi_{A_l} + b_k\phi_{B_l})e^{2ikld} \right\rangle = 0.$$

Let us use the fact that $(\widehat{p}^2/2m + \hat{V}_{A_j})|\phi_{A_j}\rangle = \varepsilon_A|\phi_{A_j}\rangle$ and the orthogonality of the $|\phi\rangle$'s;

$$(\varepsilon_A - \varepsilon_k)a_k e^{2ikjd} + \sum_{l=1}^{N} e^{2ikld}\left(a_k\left\langle \phi_{A_j} \left| \sum_{i\neq j} \hat{V}_{A_i} + \sum_i \hat{V}_{B_i} \right| \phi_{A_l} \right\rangle \right.$$

$$\left. + b_k\left\langle \phi_{A_j} \left| \sum_{i\neq j} \hat{V}_{A_i} + \sum_i \hat{V}_{B_i} \right| \phi_{B_l} \right\rangle \right) = 0.$$

Retaining only the matrix elements between first neighbours leaves

$$(\varepsilon_A - \varepsilon_k)a_k e^{2ikjd} + e^{2ikjd}a_k 2\alpha_A + e^{2ikjd}b_k\gamma + e^{2ik(j-1)d}b_k\gamma = 0$$

or

$$(\varepsilon_A - \varepsilon_k + 2\alpha_A)a_k + \gamma(1 + e^{-2ikd})b_k = 0. \tag{1}$$

Analogously, projection of the energy eigenvalue equation on to $\langle \phi_{B_j}|$ gives

$$(\varepsilon_B - \varepsilon_k)e^{2ikjd} + \sum_{l=1}^{N} e^{2ikld}\left(a_k\left\langle \phi_{B_j} \left| \sum_i \hat{V}_{A_i} + \sum_{i\neq j} \hat{V}_{B_i} \right| \phi_{A_l} \right\rangle \right.$$

$$\left. + b_k\left\langle \phi_{B_j} \left| \sum_i \hat{V}_{A_i} + \sum_{i\neq j} \hat{V}_{B_i} \right| \phi_{B_l} \right\rangle \right) = 0$$

and finally

$$\gamma(1 + e^{2ikd})a_k + (\varepsilon_B - \varepsilon_k + 2\alpha_B)b_k = 0. \tag{2}$$

5. A necessary and sufficient condition for $|\psi_k\rangle$ to be an eigenvalue of $\hat{H}$ is that there exists a pair $(a_k, b_k) \neq (0, 0)$, satisfying (1) and (2); in other words, that the determinant of the system (1) + (2) is zero.

$$(\varepsilon_A - \varepsilon_k + 2\alpha_A)(\varepsilon_B - \varepsilon_k + 2\alpha_B) - \gamma^2(1 + e^{-2ikd})(1 + e^{2ikd}) = 0.$$

Thus, the energy eigenvalues are determined by solving for ε_k.

$$\varepsilon_k = \tfrac{1}{2}(\varepsilon_A + \varepsilon_B + 2\alpha_A + 2\alpha_B) \pm \left[\tfrac{1}{4}(\varepsilon_A + 2\alpha_A - \varepsilon_B - 2\alpha_B)^2 + 4\gamma^2\cos^2 kd\right]^{1/2}.$$

With the conventions of the statement, this reduces to

$$\varepsilon_k = \pm\sqrt{\Delta^2 + \delta^2\cos^2 kd}.$$

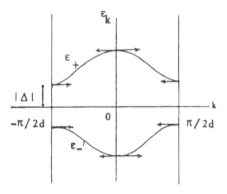

We obtain two symmetrical bands ε_+ and ε_-. The values in the centre of the zone are $\pm\sqrt{\Delta^2 + \delta^2}$ and those at the edge of the zone are $\pm|\Delta|$.

6. $|\psi_k\rangle$ may be determined completely by deriving a_k and b_k from the system (1) and (2). At $k = 0$, we have

$$\left(\Delta + \sqrt{\Delta^2 + \delta^2}\right)a_k - \delta b_k = 0$$

whence, for example

$$a_k = \delta\left[2\sqrt{\Delta^2 + \delta^2}\left(\sqrt{\Delta^2 + \delta^2} \mp \Delta\right)\right]^{-1/2}$$
$$b_k = \left(\Delta \mp \sqrt{\Delta^2 + \delta^2}\right)\left[2\sqrt{\Delta^2 + \delta^2}\left(\sqrt{\Delta^2 + \delta^2} \mp \Delta\right)\right]^{-1/2}.$$

a_k and b_k have opposite signs for ε_+ (antibinding orbital) and the same sign for ε_- (binding orbital).

At $k = \pi/2d$ the system (1) + (2) reduces to

$$(\Delta - \varepsilon_k)a_k = (-\Delta - \varepsilon_k)b_k = 0.$$

Thus, we have $(a_k, b_k) = (1, 0)$ for $\varepsilon_k = +\Delta$ and $(a_k, b_k) = (0, 1)$ for $\varepsilon_k = -\Delta$. The wavefunctions are purely type A or type B (non-binding orbitals).

7. Taking into account the spin degeneracy, each band can accommodate $2N$ electrons. If A and B each have one valence electron, the band ε_- will be filled exactly and the material will be an insulator.

If A has one valence electron and B has two, the $3N$ electrons will fill the band ε_- completely and half fill the band ε_+. The material will be a conductor.

8. The bottom of the band ε_+ and the top of the band ε_- lie at the edge of the zone; thus, the width of the forbidden band is $2|\Delta|$.

The effective masses are given by the expansion of ε_k in the neighbourhood of these extrema $(k = \pi/2d)$.

$$\varepsilon_k = \pm\sqrt{\Delta^2 + \delta^2 \cos^2 kd} \simeq \pm|\Delta|\left(1 + \frac{\delta^2}{2\Delta^2}\cos^2 kd\right) \simeq \pm|\Delta| \pm \frac{\delta^2 d^2}{2|\Delta|}\left(k - \frac{\pi}{2d}\right)^2.$$

Identifying this with $\varepsilon_k = \varepsilon_{\pi/2d} + (\hbar^2/2m^*)(k - (\pi/2d))^2$ we obtain:

$$m_c = \frac{\hbar^2|\Delta|}{\delta^2 d^2} \qquad \text{(conduction band)}$$

$$m_v = -\frac{\hbar^2|\Delta|}{\delta^2 d^2} \qquad \text{(valence band)}.$$

9. In the limit A = B, we have $\varepsilon_A = \varepsilon_B = \varepsilon_0$ and $\alpha_A = \alpha_B = \alpha$, whence it follows, in particular, that $\Delta = 0$. The band structure reduces to:

$$\varepsilon_k = \varepsilon_0 + 2\alpha \mp 2\gamma \cos kd = \pm\delta \cos kd.$$

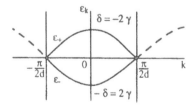

This is identical with the result of the Course ($\varepsilon_k = \varepsilon_0 + 2\alpha + 2\gamma \cos kd$) up to the detail that here the primitive cell is equal to $2d$ and the Brillouin zone is half as large. The upper band ε_+ is obtained simply by bringing the unique band of the Course $(-\pi/d < k < +\pi/d)$ back to the first Brillouin zone by translations with vectors $\pm\pi/d$.

EXTENDED STATES AND LOCALISED STATES ON A CHAIN

Jean-Noël Chazalviel

We know that the electron eigenstates of a perfect crystal are Bloch states delocalised over the whole crystal. We also know (see p. 149) that impurities may introduce localised states. Based on a band model for a one-dimensional crystal (**I**), the problem shows that localised states may also be created by violations of periodicity (even in the absence of foreign atoms), either on the surface of the crystal (**II**), or in the neighbourhood of 'stacking faults' (**III**). These localised states may play an important physical role and two concrete examples will be given in the problem.

I

1. We consider a periodic rectilinear chain of identical atoms. We suppose that this chain has been subject to the Peierls distortion (see **Problem 2**), which means that the distances between neighbouring atoms are successively shortened (a_1 between atom $(2j - 1)$ and atom $(2j)$) and lengthened (a_2 between atom $(2j)$ and atom $(2j + 1)$), as shown in the accompanying diagram ($a_2 > a_1$). We set $a = a_1 + a_2$.

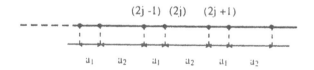

For this chain, give the length of the primitive cell and the size of the first Brillouin zone.

2. We shall calculate the band structure of this chain in the tight-binding approximation. We suppose that the electronic structure of each atom (assumed isolated) reduces to a single non-degenerate state $|\phi\rangle$ with s symmetry (even wavefunction) and energy ε.

Write down the general form for an eigenstate of an electron of the chain, as a linear combination of atomic orbitals (LCAO). For an infinite chain, what constraint does the periodicity impose on the coefficients of these linear combinations?

3. In the remainder of the problem, we suppose that the chain consists of $2N$ atoms (N large). We also suppose, in **I**, that the chain loops back on itself (periodic boundary conditions).

What additional condition do these assumptions impose on the electronic states of the chain? We shall denote these states by $|\psi_k\rangle$. Prove that there exist two quantities u_k and v_k ($|u_k|^2 + |v_k|^2 = 1$) characterising the contribution of atoms of odd and even rank, respectively, to $|\psi_k\rangle$. (Suppose for convenience that the $|\phi_l\rangle$ are orthonormal: $\langle\phi_l|\phi_m\rangle = \delta_{lm}$.)

4. In order to determine the $|\psi_k\rangle$ and their associated energies E_k completely, we write down the Hamiltonian $\hat{H}$ for an electron of the chain (assuming that the potentials V_i ($i = 1, 2, \ldots, 2N$) of the atoms of the chain are additive) followed by the energy eigenvalue equation $(\hat{H} - E_k)|\psi_k\rangle = 0$, which should be projected alternately on to two monatomic eigenstates $\langle\phi_{2j-1}|$ and $\langle\phi_{2j}|$ (where j is an arbitrary integer between 1 and N). Of the matrix elements of type $\langle\phi_l|V_m|\phi_m\rangle$, we retain only those of type $\langle\phi_p|V_p|\phi_{p+1}\rangle = \langle\phi_p|V_{p+1}|\phi_{p+1}\rangle$, which quantity is equal to $-\Gamma$ (respec. $-\gamma$) if the distance between (p) and $(p+1)$ is equal to a_1 (respec. a_2), where $0 < \gamma < \Gamma$. All the other matrix elements of potentials involving two different sites will be neglected, including, for example, $\langle\phi_p|V_{p\pm1}|\phi_p\rangle$.

Show that this procedure leads to a system of two equations in u_k and v_k, one of which is given by

$$(\varepsilon - E_k)u_k - (\Gamma + \gamma e^{-ika})v_k = 0.$$

Give the other equation.

5. Deduce the eigenvalues E_k from this system and draw the band structure and the form of the corresponding wavefunctions at the centre and the edge of the first Brillouin zone (this will involve calculating u_k and v_k at these points).

II

We are now interested in calculating the physical effects associated with the finite length of the chains; this requires the use of more realistic boundary conditions than the periodic ones. We clearly expect different results, according to whether the chains terminate with a 'long' (a_2) or a 'short' (a_1) interatomic distance.

6. We wish to determine the eigenstates $|\Phi_k\rangle$ ($0 < k < \pi/a$) of a finite chain as linear combinations $\alpha|\psi_k\rangle + \beta|\psi_{-k}\rangle$ of states previously calculated for the looped chain.

Show that the $|\Phi_k\rangle$ defined in this way are indeed eigenstates of the system provided that α, β and k are chosen so that the coefficient of $|\phi_i\rangle$ is zero for the two values of i corresponding to the positions of the first atoms excluded at each extremity of the chain (for example, for a chain of $2N$ atoms, $i = 1, 2, \ldots, 2N$, so that the coefficient of $|\phi_i\rangle$ is zero for $i = 0$ and $i = 2N + 1$).

7. For a chain terminated by short distances (such as the chain $i = 1, 2, \ldots, 2N$), one can write down the equations determining the triplets (α, β, k) satisfying the above condition and deduce that there exist $2N$ solutions of type $|\Phi_k\rangle$. Conversely, it can be shown that for a chain terminated by long distances (for example, the chain $i = 0, 1, 2, \ldots, 2N - 1$), this procedure gives only $2N - 2$ solutions of type $|\Phi_k\rangle$, which implies that two eigenstates of the system were not produced in this way. It is not required to write down these equations nor to carry out these calculations, the results of which will be assumed. To determine the missing solutions, which are conjectured to be associated with the extremities of chains terminated by a long distance, we begin with the general form for an eigenstate of the system, in the absence of boundary conditions (question **2**).

Show that the energy eigenvalue equation (analogous to the system determined in question 4) has solutions of the same type as $|\psi_k\rangle$, but where, now, k is a complex number, in the present case, $k = i\chi$ or $k = (\pi/a) + i\chi$, where χ is a real number, the domain of variation of which will be specified.

8. *We are interested in the states $|\psi_k\rangle$ with $k = (\pi/a) + i\chi$. What energy range do they describe as χ varies? What happens to the pair (u_k, v_k) when $E_k = \varepsilon$?*

9. *Deduce the two missing solutions of question 7. What can be said about the localisation of these states? Sketch their wavefunctions.*

10. *Assuming that $\gamma \ll \Gamma$, can you explain, in terms of chemical binding, why only the extremities of chains terminated by a long distance give rise to states of this type?*

11. The figure below shows the result of measurements of photoconductivity performed on a sample of silicon cleaved in a very good vacuum ($\sim 10^{-13}$ bar), then on the same sample after oxidation.

Can you intuitively explain the origin of the photoconductive effect for photon energies below the gap? Can you explain the important decrease in this effect after oxidation?

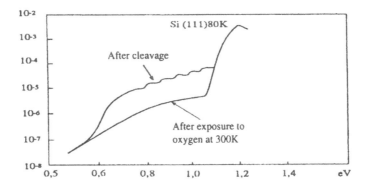

After **W Müller** and **W Mönch**, *Phys. Rev. Lett.* **27**, 250 (1971).

III

12. We now consider a 'stacking fault' in a chain involving the succession of two long distances.

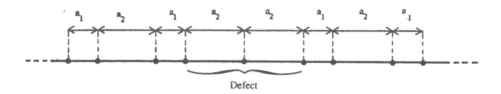

Write down the equation for the energy eigenvalues, projected onto the state $|\phi_i\rangle$ of the central site of the defect, for a state $|\psi\rangle$ with the general LCAO form. Then, using the states $|\psi_k\rangle$ determined in question 8, show that there exists an electron state localised in the neighbourhood of the defect; give its energy and draw its wavefunction.

13. The tight-binding model used in this problem gives a reasonably good description of the π electrons of trans-polyacetylene $(CH)_n$. The order of magnitude of Γ and γ is $\simeq 2-3$ eV.

Can you give a qualitative interpretation of the spectrum of optical absorption shown in the accompanying figure, in the case of material without defects (continuous curve) and in that of material in which defects of the above type were intentionally introduced (dashed curve). Deduce the value of $\Gamma - \gamma$.

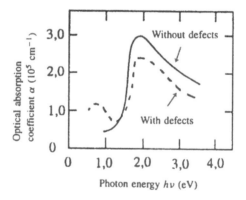

Trans-polyacetylene

After C R Fincher *et al., Phys. Rev.* **B-20**, 1589 (1979).

SOLUTION TO PROBLEM 4

I

1. a, $2\pi/a$ (two atoms per cell).

2. $|\psi\rangle = \sum_i \lambda_i |\phi_i\rangle$. Periodicity implies $\lambda_{i+2} = \xi \lambda_i$, where ξ is a constant, whence

$$|\psi\rangle = \sum_l \xi^{l-1} \left(\lambda_1 |\phi_{2l-1}\rangle + \lambda_2 |\phi_{2l}\rangle \right).$$

3. Looped chain of N cells, implies $\xi^N = 1$, whence $\xi = \exp(ika)$ with $k = 2n\pi/Na$ (n is an integer between, for example, $-N/2$ and $+N/2$). We may rewrite

$$|\psi_k\rangle = \sum_{l=1}^{N} \frac{1}{\sqrt{N}} e^{ikla} \left(u_k |\phi_{2l-1}\rangle + v_k |\phi_{2l}\rangle \right).$$

Normalising, we obtain $\langle \psi_k | \psi_k \rangle = |u_k|^2 + |v_k|^2 = 1$.

4.

$$\hat{H} = \frac{\hat{p}^2}{2m} + \sum_{i=1}^{2N} \hat{V}_i$$

$$\left\langle \phi_{2j-1} \Big| \frac{\hat{p}^2}{2m} + \hat{V}_{2j-1} + \sum_{i \neq 2j-1} \hat{V}_i - E_k \Big| \sum_{l=1}^{N} e^{ikla} \left(u_k |\phi_{2l-1}\rangle + v_k |\phi_{2l}\rangle \right) \right\rangle = 0.$$

We use $((\hat{p}^2/2m) + \hat{V}_{2j-1})|\phi_{2j-1}\rangle = \varepsilon |\phi_{2j-1}\rangle$, keeping only the non-negligible matrix elements

$$\langle \phi_{2j-1} | \varepsilon - E_k + \hat{V}_{2j-2} + \hat{V}_{2j} \left| e^{ikja} \left(e^{-ika} v_k |\phi_{2j-2}\rangle + u_k |\phi_{2j-1}\rangle + v_k |\phi_{2j}\rangle \right) \right\rangle = 0$$

$$(\varepsilon - E_k)u_k - \left(\Gamma + \gamma e^{-ika} \right) u_k = 0.$$

Similarly, for $\langle \phi_{2j} |$, we have

$$\langle \phi_{2j} | \varepsilon - E_k + \hat{V}_{2j-1} + \hat{V}_{2j+1} \left| e^{ikja} \left(u_k |\phi_{2j-1}\rangle + v_k |\phi_{2j}\rangle + e^{ika} u_k |\phi_{2j+1}\rangle \right) \right\rangle = 0$$

$$(\varepsilon - E_k)v_k - (\Gamma + \gamma e^{ika})u_k = 0.$$

5. This homogeneous linear system of two equations has a solution $(u_k, v_k) \neq (0,0)$ if and only if

$$(\varepsilon - E_k)^2 - (\Gamma + \gamma e^{-ika})(\Gamma + \gamma e^{ika}) = 0.$$

Whence

$$E_k = \varepsilon \pm (\Gamma^2 + \gamma^2 + 2\Gamma\gamma \cos ka)^{1/2} = \varepsilon \pm \left[(\Gamma - \gamma)^2 + 4\Gamma\gamma \cos^2 \frac{ka}{2} \right]^{1/2}.$$

$k = 0$

$\quad E_- = \varepsilon - (\Gamma + \gamma) \quad u_k = v_k = 1/\sqrt{2}$

$\quad E_+ = \varepsilon + (\Gamma + \gamma) \quad u_k = -v_k = 1/\sqrt{2}$

$k = \pi/a$

$\quad E_- = \varepsilon - (\Gamma - \gamma) \quad u_k = v_k = 1/\sqrt{2}$

$\quad E_+ = \varepsilon + (\Gamma - \gamma) \quad u_k = -v_k = 1/\sqrt{2}$

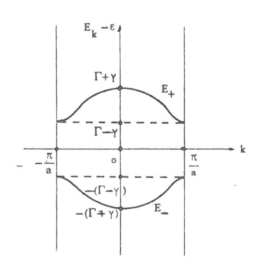

II

6. Starting once again with the general form $|\psi\rangle = \sum_i \lambda_i |\phi_i\rangle$, the equation

$$\langle \phi_i | \hat{H} - E_k | \psi \rangle = 0$$

may be written as

$$(\varepsilon - E_k)\lambda_i + \langle \phi_i | V_{i-1} | \phi_{i-1} \rangle \lambda_{i-1} + \langle \phi_i | V_{i+1} | \phi_{i+1} \rangle \lambda_{i+1} = 0.$$

This equation is satisfied by $|\psi_k\rangle$ and $\psi_{-k}\rangle$, whence also by $|\Phi_k\rangle$. Of the $2N$ equations of this type, the two equations corresponding to the atoms at the ends of the chain are again the exception. In fact, if i denotes, for example, the initial atom of the chain, then the atomic potential V_{i-1} does not exist and the corresponding term in the above equation is absent. The equation thus truncated will, nevertheless, be satisfied by $|\Phi_k\rangle$ if the latter is constructed in such a way that $\lambda_{i-1} = 0$. Analogous reasoning applies at the other end of the chain. (The λ_i are, of course, zero for values of i outside the chain.)

7. Outside the problem. In what follows, we justify the results which we asked readers to accept without proof.

For the two bands $E_\pm$, we have

$$\frac{u_k}{v_k} = \frac{\mp(\Gamma + \gamma e^{-ika})}{|\Gamma + \gamma e^{-ika}|}.$$

Setting $\tan 2\varphi_k = \gamma \sin ka / (\Gamma + \gamma \cos ka)$, we may write

$$u_k = \frac{1}{\sqrt{2}} e^{-i\varphi_k} \qquad v_k = \mp \frac{1}{\sqrt{2}} e^{+i\varphi_k}.$$

In the case of a chain ($i = 1, 2, \ldots, 2N$), the conditions of question **6** may be written as

$$0 = \lambda_0 = \alpha v_k + \beta v_{-k} = \mp \frac{1}{\sqrt{2}} (\alpha e^{i\varphi_k} + \beta e^{-i\varphi_k})$$

$$0 = \lambda_{2N+1} = \alpha u_k \exp(ik(N+1)a) + \beta u_{-k} \exp(-ik(N+1)a)$$

$$= \frac{1}{\sqrt{2}} (\alpha \exp(ik(N+1)a - i\varphi_k) + \beta \exp(-ik(N+1)a + i\varphi_k)).$$

Eliminating α and β, we obtain

$$2k(N+1)a - 4\varphi_k = 2n\pi.$$

The possible values of k are given by the equation

$$k = \frac{2\varphi_k}{(N+1)a} = \frac{n\pi}{(N+1)a} \quad \text{where } n = 1, 2, \ldots, N;$$

which implies N values $\times$ 2 bands $= 2N$ solutions.

Similarly, in the case of a chain ($i = 0, 1, \ldots, 2N - 1$), we have

$$0 = \lambda_{-1} = \alpha u_k + \beta u_{-k} = \frac{1}{\sqrt{2}} (\alpha e^{-i\varphi_k} + \beta e^{i\varphi_k})$$

$$0 = \lambda_{2N} = \alpha v_k e^{ikNa} + \beta v_{-k} e^{-ikNa} = \mp \frac{1}{\sqrt{2}} (\alpha \exp(ikNa + i\varphi_k) + \beta \exp(-ikNa - i\varphi_k)).$$

Whence $k + (2\varphi_k / Na) = n\pi / N$.

The only values of n leading to a solution which is not identically zero are $n = 1, 2 \ldots, N - 1$. Thus, we now have $(N - 1) \times 2 = 2N - 2$ solutions; and since we had $2N$ solutions at the beginning, there are two solutions missing.

We now return to the general form of $|\psi\rangle$ resulting from question **2**, which may be written as

$$|\psi\rangle = \sum_l \xi^l (u|\phi_{2l-1}\rangle + v|\phi_{2l}\rangle).$$

The energy eigenvalue equation may be rewritten, according to question **4**, replacing $\exp(ika)$ by ξ:

$$(\varepsilon - E)u - (\Gamma + (\gamma/\xi))v = 0$$
$$-(\Gamma + \gamma\xi)u + (\varepsilon - E)v = 0.$$

The equation obtained by setting the determinant to zero

$$(\varepsilon - E)^2 = \Gamma^2 + \gamma^2 + \Gamma\gamma\left(\xi + \xi^{-1}\right)$$

has real roots in E, subject only to the condition that $(\xi + \xi^{-1})$ is real and greater than $-(\Gamma^2 + \gamma^2)/\Gamma\gamma$. This condition is met if $\xi = 1$ (solutions $\xi = \exp(ika)$) and also for certain real values of ξ; the permitted values are $\xi > 0$ and $-\Gamma/\gamma < \xi < -\gamma/\Gamma$.

This reduces to writing $\xi = \exp(ika)$, but with $k = i\chi$ where χ is an arbitrary real number (for $\xi > 0$), or $k = (\pi/a) + i\chi$, where χ is real and $|\chi| < (1/a)\ln(\Gamma/\gamma)$ (for $-\Gamma/\gamma < \xi < -\gamma/\Gamma$).

8. As χ varies from $-(1/a)\ln(\Gamma/\gamma)$ to $+(1/a)\ln(\Gamma/\gamma)$, the energies

$$E_\pm = \varepsilon \pm (\Gamma^2 + \gamma^2 - 2\Gamma\gamma\cosh\chi a)^{1/2}$$

begin at ε, touch the bands at $\pm(\Gamma - \gamma)$ for $\chi = 0$, then return to ε. Thus, the energy range described corresponds to the gap of the band structure.

The pair (u, v) is given, for example, by

$$\frac{u}{v} = \frac{\Gamma + \gamma/\xi}{\varepsilon - E} = \mp\left(\frac{\Gamma - \gamma e^{\chi a}}{\Gamma - \gamma e^{-\chi a}}\right)^{1/2}$$

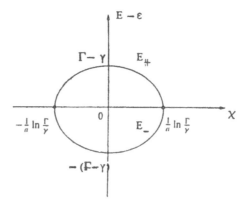

Thus, we have $E = \varepsilon$, either for

$$\chi = -\frac{1}{a}\ln\frac{\Gamma}{\gamma}$$

whence $u/v = \infty$, $(u, v) = (1, 0)$, or for

$$\chi = \frac{1}{a} \ln \frac{\Gamma}{\gamma}$$

whence $u/v = 0$, $(u, v) = (0, 1)$.

9. The two states above, which are not normalisable on an infinite chain may become so near an extremity and may, additionally, satisfy the condition of question **6**. The solution $\chi = -(1/a) \ln(\Gamma/\gamma)$ is appropriate for the right end of the chain ($\lambda_{2N} = v = 0$)

$$|\psi\rangle = \left[1 - \left(\frac{\gamma}{\Gamma}\right)^2\right]^{1/2} \sum_{n \leq N} \left(-\frac{\Gamma}{\gamma}\right)^{n-N} |\phi_{2n-1}\rangle.$$

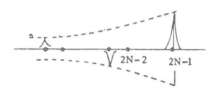

The solution $\chi = (1/a) \ln(\Gamma/\gamma)$ is suitable for the left end ($\lambda_{-1} = u = 0$)

$$|\psi\rangle = \left[1 - \left(\frac{\gamma}{\Gamma}\right)^2\right]^{1/2} \sum_{n \geq 0} \left(-\frac{\gamma}{\Gamma}\right)^{n} |\phi_{2n}\rangle.$$

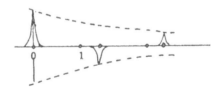

These states are localised on the first few atoms at the extremities of the chain.

10. In the limit $\gamma \ll \Gamma$, the chain is similar to a succession of weakly interacting diatomic molecules (Γ represents the binding energy of a molecule and γ the energy of interaction between two molecules). Only the *breaking of a bond* Γ may provoke the appearance of a localised state at an energy ε, while the states of the infinite chain are close to $\varepsilon + \Gamma$ and $\varepsilon - \Gamma$.

11. The bonds broken on the surface of the cleaved silicon provoke the appearance of electron states localised in the neighbourhood of the surface, with energies inside the forbidden band. One can observe the photoconductivity for photon energies $h\nu < E_g$ from electron excitations either from the surface states towards the conduction band or from the valence band (i.e. hole creation) towards the surface states. The superficial oxidation suppresses most of the broken bonds and thus the surface states, thereby diminishing the associated photoconductivity mechanism.

III

12. Calling $i = 0$ the site of the defect, the equation of question **6**, written for $i = 0$, becomes

$$(\varepsilon - E)\lambda_0 - \gamma(\lambda_{-1} + \lambda_1) = 0.$$

The connection of the state $(\chi = -(1/a)\ln(\Gamma/\gamma))$ to the left of 0 and the state $(\chi = (1/a)\ln(\Gamma/\gamma))$ to the right leads to a solution of the system. In fact, it satisfies the equation of question **6** for $i \neq 0$ (by construction) and also for $i = 0$, since $E = \varepsilon$ and $\lambda_{-1} = \lambda_1 = 0$. The normalised wavefunction is given by

$$|\psi\rangle = \left(\frac{\Gamma^2 - \gamma^2}{\Gamma^2 + \gamma^2}\right)^{1/2} \sum_{-\infty}^{\infty} \left(-\frac{\gamma}{\Gamma}\right)^n |\phi_{2n}\rangle.$$

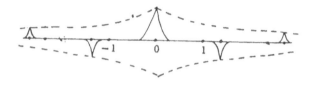

The energy of this state is ε.

13. The defect-free trans-polyacetylene has an optical absorption threshold for $h\nu \simeq$ 1.4 eV. Taking into account the band structure and the fact that there is one π electron per carbon atom, we expect the band E_- to be full and the band E_+ empty. Thus, the optical absorption threshold can be identified with the gap of the band structure.

We can verify that, in the presence of defects, an optical-absorption peak occurs at the half energy $\simeq 0.7$ eV, associated with optical transitions between the level of the defect and one or the other band (according to the state of occupancy of the level).

It follows that $\Gamma - \gamma \simeq 0.7$ eV.

N.B. Such defects, called **solitons**, may move: when the atom 1 moves by $(a_2 - a_1)$ to the left, the defect migrates from (0) to (2) and may contribute to the electrical conduction.

INSULATOR–METAL TRANSITION
(FROM HYDROGEN TO DOPED SEMICONDUCTORS)

Henri Alloul

The calculations of **I**, questions **4** and **5**, are useful for comprehension, without being necessary as far as following the problem is concerned. In **II**, it is possible to re-enter the problem in question **12** without working through the preceding questions in full.

III is independent of **IV**. The only requirement for **IV** is that one should have understood the physical sense of **I** and **II**.

Here, we shall exhibit parameters which determine whether a material is a conductor or an insulator. In **I**, we shall apply the tight-binding method to the case of a crystal of hydrogenoid atoms to determine the width of the band of electron states arising from the localised atomic state. In **II**, the important role played by the interactions between electrons will be introduced via the simple example of the hydrogen molecule. A condition for the solid described in **I** to be metallic will be obtained. This will be compared with experimental observations on hydrogen and alkalis in **III**. In **IV**, a complete analogy will be made with the case of doped semiconductors; this will provide an understanding of how the physical properties of these materials are modified when the concentration of donors is increased.

I

Tight-binding method for hydrogenoid orbitals

Let us consider a crystalline solid formed by a stacking of monovalent atoms according to a *body-centred cubic* structure whose conventional cubic cell has side b. We suppose that the electron-energy bands obtained from the internal electrons layers are excessively narrow and remote from that based on the external layer s of the atom. The latter may be described by the tight-binding approximation. If ε is the eigenenergy and $|\varphi(\overline{r})\rangle$ the wavefunction of the corresponding state s of the isolated atom, we shall seek an eigenfunction for the crystal in the form

$$|\psi_{\overline{k}}(\overline{r})\rangle = \sum_j A_{j\overline{k}}|\varphi(\overline{r} - \overline{R}_j)\rangle$$

where the sum over the j is taken over all the sites of the crystal.

1. *Describe a primitive cell of the lattice. What is its volume?*

2. We suppose that:

(i) the potential seen by an electron in the crystal is the sum of the atomic potentials $V(\bar{r} - \bar{R}_j)$.

(ii) $\langle \varphi(\bar{r} - \bar{R}_i) | \varphi(\bar{r} - \bar{R}_j) \rangle = \delta_{ij}$.

(iii) The only non-zero terms of the form $\langle \varphi(\bar{r} - \bar{R}_j) | V(\bar{r} - \bar{R}_l) | \varphi(\bar{r} - \bar{R}_m) \rangle$ are those which involve only two near-neighbour sites j and m, i.e.

$$-\alpha = \langle \varphi(\bar{r} - \bar{R}_j) | V(\bar{r} - \bar{R}_m) | \varphi(\bar{r} - \bar{R}_j) \rangle$$
$$-\gamma = \langle \varphi(\bar{r} - \bar{R}_j) | V(\bar{r} - \bar{R}_j) | \varphi(\bar{r} - \bar{R}_m) \rangle.$$

Give the expression for the energy $E_{\bar{k}}$ associated with the wavefunction $|\psi_{\bar{k}}(\bar{r})\rangle$.

In the case of hydrogen atoms, which is the only case considered up to the end of **II**, the corresponding values of $V(\bar{r})$, $\varphi(\bar{r})$ and ε are

$$V(\bar{r}) = -\frac{q^2}{4\pi \varepsilon_0 r} = -\frac{e^2}{r}$$

$$\varphi(\bar{r}) = \frac{1}{(\pi a_0^3)^{1/2}} e^{-r/a_0}$$

$$\varepsilon = -e^2/2a_0$$

where a_0 is Bohr's atomic radius and q is the elementary charge.

3. *Express α and γ in the form of integrals involving a_0 and the distances r_A and r_B separating an electron of two first-neighbour sites A and B.*

Analytic calculation of γ leads to

$$\gamma = \frac{e^2}{a_0} \left(1 + \frac{R}{a_0} \right) \exp(-R/a_0) \quad \text{(with } \bar{R} = \overline{AB}\text{)}.$$

4. α may be obtained analytically by observing that it corresponds to the potential energy of a point charge q placed at B, subjected to a spherical distribution of charges centred at A with electrostatic density $q\varphi(r_A)^2$, for which it is easy to calculate the electrostatic potential. This can be done by decomposing the distribution of charges into two judiciously chosen parts. *Give the expression for α as a function of R and a_0.*

5. *Write $-\alpha$ as the sum of two terms, one of which exactly annihilates the Coulomb interaction between the nuclear charges. Verify that the second term of α is negligible in comparison with γ, in the limit $R/a_0 \gg 1$, and that the assumption (ii), above, in fact only applies in this limit.*

6. *Express the width W of the band of electron states thus obtained as a function of R and a_0.*

7. *Where is the Fermi level located at $T = 0$ K? Is the solid a metal or an insulator? Does it depend on b?*

II

The interactions between electrons

The result obtained above is contrary to the physical reality. This is linked with the fact that we have completely neglected the Coulomb interaction between the electrons. We shall illustrate its effect in the simple case of a pseudo-molecule of hydrogen, consisting of two atoms whose protons are located at $\bar{r}_A$ and $\bar{r}_B$ and the electrons at $\bar{r}_1$ and $\bar{r}_2$. The overall electronic Hamiltonian of this system is given by

$$\hat{H} = \frac{p_1^2}{2m} + V_{1A} + V_{1B} + \frac{p_2^2}{2m} + V_{2B} + V_{2A} + V_{12}$$

where

$$V_{1A} = -\frac{e^2}{|\bar{r}_1 - \bar{r}_A|} \qquad V_{1B} = -\frac{e^2}{|\bar{r}_1 - \bar{r}_B|} \quad \text{etc} \dots$$

$$V_{12} = +\frac{e^2}{|\bar{r}_1 - \bar{r}_2|}.$$

Since $\hat{H}$ is independent of the electron spins, the eigenfunctions take the form of the product of spatial and spin wavefunctions. For the spatial wavefunction with two electrons, the following basis may be taken:

$$|\varphi_{1A}\rangle \otimes |\varphi_{2B}\rangle, \quad |\varphi_{1B}\rangle \otimes |\varphi_{2A}\rangle, \quad |\varphi_{1A}\rangle \otimes |\varphi_{2A}\rangle, \quad \text{and} \quad |\varphi_{1B}\rangle \otimes |\varphi_{2B}\rangle$$

with the obvious notation $|\varphi_{1B}\rangle = |\varphi(\bar{r}_1 - \bar{r}_B)\rangle$.

Nevertheless, it is more judicious to remark that $\hat{H}$ is invariant with respect to the exchange of protons A and B (operator $\hat{T}_{AB}$) which enables us to conclude that $\hat{H}$ and $\hat{T}_{AB}$ have a common basis of eigenfunctions. Thus, the eigenfunctions of $\hat{H}$ are either symmetric or antisymmetric with respect to the exchange of A and B, which leads us to choose the following basis for the spatial functions:

$$|\Phi_1\rangle = \frac{1}{\sqrt{2}} \{|\varphi_{1A}\rangle \otimes |\varphi_{2B}\rangle + |\varphi_{1B}\rangle \otimes |\varphi_{2A}\rangle\}$$

$$|\Phi_2\rangle = \frac{1}{\sqrt{2}} \{|\varphi_{1A}\rangle \otimes |\varphi_{2B}\rangle - |\varphi_{1B}\rangle \otimes |\varphi_{2A}\rangle\}$$

$$|\Phi_3\rangle = \frac{1}{\sqrt{2}} \{|\varphi_{1A}\rangle \otimes |\varphi_{2A}\rangle + |\varphi_{1B}\rangle \otimes |\varphi_{2B}\rangle\}$$

$$|\Phi_4\rangle = \frac{1}{\sqrt{2}} \{|\varphi_{1A}\rangle \otimes |\varphi_{2A}\rangle - |\varphi_{1B}\rangle \otimes |\varphi_{2B}\rangle\}.$$

Moreover, since the electrons are fermions, the Pauli principle implies that the overall wavefunction (spatial and spin) is antisymmetric with respect to the exchange of the two electrons 1 and 2.

8. With which functions $|\Phi_1\rangle$, $|\Phi_2\rangle$, $|\Phi_3\rangle$, $|\Phi_4\rangle$ is a spin singlet state associated? What about a triplet spin state? We shall use the notation $|\Phi_n^M\rangle$, where $M = S$ for a singlet and T for a triplet, to denote the basis vectors of all the wavefunctions obtained in this way.

9. Using the properties stated above in **II**, prove, without resorting to any calculations, that two of the states obtained in question 8 are eigenstates of $\hat{H}$.

10.   We shall investigate the eigenenergies and eigenstates of $\hat{H}$ in the limit $R \gg a_0$ (with $\overline{R} = \overline{r}_B - \overline{r}_A$), which will enable us, as we saw in **I**, to neglect the integrals of type α and retain only the integrals of type γ. In addition, we shall only retain the main matrix elements of V_{12} which involve two electrons on a single atom. The corresponding energy, called the intra-atomic Coulomb energy, is given by

$$U = \langle \varphi_{1A} | \otimes \langle \varphi_{2A} | V_{12} | \varphi_{1A} \rangle \otimes | \varphi_{2A} \rangle.$$

*Prove that in this limit, the eigenstates determined in question **9** have eigenenergies 2ε and $2\varepsilon + U$ and that the 2×2 matrix which represents $\hat{H}$ in the subspace defined by the two other states is given by*

$$\begin{pmatrix} 2\varepsilon + U & -2\gamma \\ -2\gamma & 2\varepsilon \end{pmatrix}.$$

11.   *Give the energies of the two-electron states. Note that in the limit as $R \to \infty$, only two levels persist. To which eigenstates do these correspond?*

12.   Thus, the introduction of the Coulomb interaction V_{12} between the electrons prevents a double occupation of a single atomic orbital in the ground state.

Thus, it is again possible to give a description in terms of *independent electrons*, as in **I**, provided one removes the spin degeneracy of the *one-electron* state of energy ε by introducing a state of energy ε and an excited state of energy $\varepsilon + U$. In the hydrogen molecule, the effect of the terms γ is, as in **I**, to remove the degeneracy within each of these states. By extending this remark to a crystal, one can show that the electron levels at ε and $\varepsilon + U$ give rise to energy bands having the width W, determined in question **6**.

Assuming this result, under what conditions is the solid an insulator? When does it become metallic?

13.   In order to obtain a quantitative estimate of when the insulator–metal transition will take place, we need an estimate of U.

*Express U in its integral form in the case of hydrogen. Use the result found for α in question **4** to show that $U = (5/8)(e^2/a_0)$.*

14.   *Verify that the metal–insulator transition takes place when the number of atoms per unit volume in the body-centred cubic solid, n_M, satisfies*

$$n_M^{1/3} a_0 = C$$

where C is a constant of the order of unity, to be determined.

15.   *Show, by considering the case of a simple cubic crystal, that C is only weakly dependent on the crystalline structure chosen.*

III

Alkalis and hydrogen

The method developed in **I** and **II** is applicable to the case of monovalent atoms such as alkalis, but the numerical result obtained in question **14** is not directly applicable. In fact, the valence electrons of alkalis are in states ns ($n = 2$ for Li, $n = 3$ for

Na ...). The corresponding radial wavefunctions φ_n decrease exponentially for large distances, but become small for r close to zero, since the s electron does not penetrate very far into the internal electron layers. As indicated in the Course, a pseudo-potential approximation may be used, in which the potential is infinite for $r < R_c$, where R_c, the ionic radius, increases with the atomic number. It can be shown that if the wavefunction is approximated by $\varphi_n(r) = \text{constant} \exp -[(r - R_c)/a_0]$ for $r > R_c$ then, to a first approximation, the condition of question **14** is modified by replacing R by $R - 2R_c$.

16. *Use the values of R_c given in the Course (Chapter IV) and the known sizes of the body-centred cubic crystalline cells of Li, Na, K, Rb and Cs (3.49, 4.23, 5.23, 5.59 and 6.05 Å, respectively) to show that these materials are normally metallic.*

17. Hydrogen solidifies at low temperature ($T \sim 15$ K) into an insulating *molecular* crystal. The hydrogen molecules are located at the nodes of a face-centred cubic lattice (edge: 5.4 Å).

Calculate the number of hydrogen atoms per unit volume in the solid state, n_0. How can one understand that this phase is an insulator?

18. *Can you imagine how one might obtain solid metallic hydrogen? Under what conditions do you think certain alkali materials become insulators?*

IV

Metal–insulator transition in Si–P

The model developed in **I** and **II** is scarcely easy to test in detail in the case of hydrogen or alkalis. On the other hand, we shall see that it is applicable, with minor modifications, to the case of doped semiconductors. For example, let us consider silicon doped with phosphorus. It has been seen (p. 151) that, at zero temperature, the phosphorus atom (valency 5) substitutes into a silicon site (valency 4) and traps its excess electron in a hydrogenoid orbit with an energy level below the conduction band. The ionisation energy of the donor is the energy of the electron in its enormous Bohr orbit. The introduction of an appreciable number N_D of P atoms per unit volume, results in a disordered system of hydrogenoid orbits. Assuming this disordered state can be modelled by a cubic crystal with cell size equal to the mean distance between the donors, we might expect the model of **I** and **II** to be applicable.

19. *Adapt the condition obtained in question **15** to the case of silicon doped with phosphorus.*

20. *What band structure should we expect, as a function of the concentration of phosphorus N_D? Represent the density of electron states as a function of the energy for the material Si–P in the three cases $N_D \ll n_M$, $N_D \simeq n_M$ and $N_D \gg n_M$. Indicate the position of the Fermi level at zero temperature.*

21. The electrical resistivity of Si–P has been measured for various concentrations of phosphorus. Figure 1 shows a brutal variation of behaviour as a function of N_D.

*Determine the concentration n_M from the experiment. Given that the dielectric constant of Si is $K = 11.75$ and that the effective mass is $m^*/m = 0.33$, compare the experimental value of C with that obtained in question **15**.*

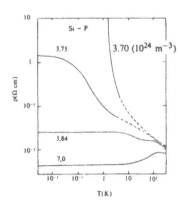

Figure 1 Electrical resistivity as a function of the temperature (log scale) for four samples of Si–P. N_D is given in units of 10^{24} atoms m^{-3}. After **T F Rosenbaum** *et al.*, *Phys. Rev.* **B27**, 7509 (1983).

We shall now attempt to understand the physical behaviours observed in the two extreme limits $N_D \ll n_M$ and $N_D \gg n_M$.

22. *Use Figure 1 to determine the order of magnitude of the collision relaxation time τ for electrons in the most metallic case at low temperature. What might the physical origin of the observed temperature dependence be? Why is it weaker than in pure metals?*

23. *What characteristic energy of the band structure of question 20 can be deduced from the analysis of ρ in the insulating case? Determine its order of magnitude and compare it with the ionisation energy of phosphorus.*

24. *What type of variation do you expect for this energy as a function of the concentration N_D?*

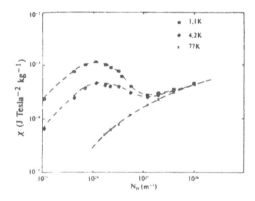

Figure 2 Magnetic susceptibility (log scale) as a function of the concentration N_D (in atoms m^{-3}, log scale), measured at three different temperatures. The susceptibility is given for 1 kg of silicon. After **D F Holcomb**, Summer School, St. Andrew's Edinburgh University, p. 316 (1986).

25. The insulator–metal transition is also accompanied by a profound modification of the magnetic properties of electrons. This is illustrated in Figure 2 by the variation of the electron magnetic susceptibility χ as a function of the temperature T. Very different behaviours are seen to occur in the limits $N_D \gg n_M$ and $N_D \ll n_M$.

What is the expected magnetic behaviour in the limit $N_D \gg n_M$? Give the expected expression for χ. Transfer the expression calculated for the three temperatures $T = 1.1$ K, $T = 4.2$ K and $T = 77$ K to Figure 2. In which limit do the results agree with the observations (density of Si: 2.33×10^3 kg m^{-3}; Bohr magneton $\mu_B = 0.927 \times 10^{-23}$ J/tesla)?

26. In the limit $N_D \gg n_M$, we may suppose that the impurity and conduction bands merge into a single parabolic band.

Give the expression expected for the magnetic susceptibility in this limit. How does it vary with N_D and T? From what concentration N_D is this assumption valid?

SOLUTION TO PROBLEM 5

I

1. The conventional cubic cell of the body-centred cubic lattice has a basis with two atoms. The Bravais lattice can be constructed in infinitely many ways. For example, by defining the three basis vectors

$$\bar{b}_1 = b\bar{x} \quad \bar{b}_2 = b\bar{y} \quad \bar{b}_3 = \frac{b}{2}(\bar{x} + \bar{y} + \bar{z}).$$

A more symmetric cell is given by

$$\bar{b}'_1 = \frac{b}{2}(\bar{y} + \bar{z} - \bar{x}) \quad \bar{b}'_2 = \frac{b}{2}(\bar{z} + \bar{x} - \bar{y}) \quad \bar{b}'_3 = \frac{b}{2}(\bar{x} + \bar{y} - \bar{z}).$$

The volume of the primitive cell is $b^3/2$.

2. E_k is calculated exactly as in **Annex 4** (p. 267), leading, for a crystal with N atoms, to

$$|\psi_{\bar{k}}(\bar{r})\rangle = N^{-1/2} \sum_j e^{i\bar{k}\cdot\bar{R}_j} |\varphi(\bar{r} - \bar{R}_j)\rangle$$

$$E_k = \langle \psi_k(\bar{r})|\hat{H}|\psi_k(\bar{r})\rangle = \varepsilon - \sum_{j',j} \alpha - \gamma \sum_{j',j} e^{i\bar{k}\cdot(\bar{R}_{j'} - \bar{R}_j)} \tag{1}$$

where the sums run over neighbouring pairs j' and j.

In the body-centred cubic lattice an atom has eight neighbours. The coefficient of γ is then given by

$$\sum_{\varepsilon_x = \pm, \varepsilon_y = \pm, \varepsilon_z = \pm} \exp\left[i\frac{b}{2}(\varepsilon_x k_x + \varepsilon_y k_y + \varepsilon_z k_z)\right] = 8\cos\left(k_x\frac{b}{2}\right)\cos\left(k_y\frac{b}{2}\right)\cos\left(k_z\frac{b}{2}\right)$$

whence

$$E_k = \varepsilon - 8\alpha - 8\gamma \cos\left(k_x\frac{b}{2}\right)\cos\left(k_y\frac{b}{2}\right)\cos\left(k_z\frac{b}{2}\right) \tag{2}$$

where $\bar{k}$ belongs to the first Brillouin zone. The reciprocal lattice of the body-centred cubic lattice is a face-centred cubic with conventional cell size $4\pi/b$.

3. $R = b\sqrt{3}/2$ is the distance between first neighbours. Let A and B be two atoms separated by this distance R.

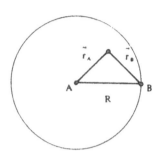

With the notation shown, we have

$$\alpha = \frac{1}{\pi a_0^3} \int \exp\left(-\frac{r_A}{a_0}\right) \frac{e^2}{r_B} \exp\left(-\frac{r_A}{a_0}\right) d^3 r_A$$

$$= \frac{e^2}{\pi a_0^3} \int \exp\left(-\frac{2r_A}{a_0}\right) \frac{1}{r_B} d^3 r_A \tag{3}$$

$$\gamma = \frac{e^2}{\pi a_0^3} \int \exp\left(-\frac{r_A}{a_0}\right) \frac{1}{r_A} \exp\left(-\frac{r_B}{a_0}\right) d^3 r_A. \tag{4}$$

4. α/q is the electrostatic potential created at B by a charge distribution $q\langle\varphi(\bar{r}_A)|\varphi(\bar{r}_A)\rangle$. The integral for α can be decomposed into an integral within the sphere of radius R and an integral outside this sphere

$$\alpha = \frac{e^2}{\pi a_0^3} \left(\int_{r_A < R} (\ ) d^3 r_A + \int_{r_A > R} (\ ) d^3 r_A \right).$$

To determine the former, according to Gauss's theorem, it is sufficient to concentrate the charge at A. The second integral corresponds to the potential created at B by the charge distribution outside the sphere. Since the corresponding electric field is zero inside the sphere by Gauss's theorem, the electrical potential is constant. It can be calculated at A. This gives:

$$\alpha = \frac{e^2}{\pi a_0^3} \left[\frac{1}{R} \int_0^R \exp\left(-\frac{2r_A}{a_0}\right) 4\pi r_A^2\, dr_A + \int_R^\infty \exp\left(-\frac{2r_A}{a_0}\right) \frac{1}{r_A} 4\pi r_A^2\, dr_A \right]$$

$$= \frac{e^2}{\pi a_0^3} \left[\frac{1}{R} \int_0^\infty \exp\left(-\frac{2r_A}{a_0}\right) 4\pi r_A^2\, dr_A + \int_R^\infty \exp\left(-\frac{2r_A}{a_0}\right) \left(\frac{1}{r_A} - \frac{1}{R}\right) 4\pi r_A^2\, dr_A \right]$$

$$= \frac{e^2}{R} + \frac{4e^2}{a_0^3} \int_R^\infty \exp\left(-\frac{2r_A}{a_0}\right) \left(\frac{1}{r_A} - \frac{1}{R}\right) r_A^2\, dr_A.$$

Setting $x = r_A/a_0$, we have

$$\alpha = \frac{e^2}{R} + \frac{4e^2}{R} \int_{R/a_0}^{\infty} x \left(\frac{R}{a_0} - x \right) \exp(-2x) \, dx.$$

Integrating by parts, with $u = ((R/a_0) - x) x$ and $dv = \exp(-2x) dx$, we obtain

$$\alpha = \frac{e^2}{R} + \frac{2e^2}{R} \int_{R/a_0}^{\infty} \left(\frac{R}{a_0} - 2x \right) \exp(-2x) \, dx.$$

A second integration by parts gives

$$\alpha = \frac{e^2}{R} - \frac{e^2}{R} \left[\left(1 + \frac{R}{a_0} \right) \exp\left(-\frac{2R}{a_0} \right) \right] = \frac{e^2}{R} - \frac{e^2}{a_0} \left[\left(1 + \frac{a_0}{R} \right) \exp\left(-\frac{2R}{a_0} \right) \right]. \tag{5}$$

5. The first term of $-\alpha$ is precisely the Coulomb interaction energy between two charges $+q$ and $-q$ located at A and B: in the total energy of the system of electrons plus nuclei it compensates the Coulomb interaction energy between the protons. Thus, for the modification of the electron levels, one has to retain only the second term of α in expression (2) for E_k.

As stated (see p.306)

$$\gamma = \frac{e^2}{a_0} \left(1 + \frac{R}{a_0} \right) \exp\left(-\frac{R}{a_0} \right). \tag{6}$$

For $R/a_0 \gg 1$, we have

$$\gamma \simeq \frac{e^2}{a_0} \frac{R}{a_0} \exp\left(-\frac{R}{a_0} \right) \qquad \alpha - \frac{e^2}{R} \simeq -\frac{e^2}{a_0} \exp\left(-\frac{2R}{a_0} \right)$$

whence $|\alpha - (e^2/R)| \ll \gamma$.

N.B. It is clear from formula (4) for γ that when $R/a_0 \gg 1$ the charge distributions are concentrated around A and B. The dominant contribution to γ comes from $r_A \sim r_B \sim R/2$, whence the dominant factor $\exp(-R/a_0)$ in the expression (6) for γ.

The same remarks leads to $\langle \varphi(\bar{r} - \bar{r}_A) | \varphi(\bar{r} - \bar{r}_B) \rangle \propto \exp(-R/a_0) \ll 1$, which enabled us to neglect the overlap integrals $\langle \varphi(\bar{r} - \bar{R}_j) | \varphi(\bar{r} - \bar{R}_i) \rangle$ in question **2**.

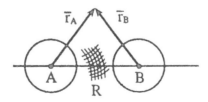

6. Since the term 8α in (2) is negligible, the atomic level ε gives rise in the solid to a band of energies essentially unshifted. The lowest level is $\varepsilon - 8\gamma$ (obtained for $\bar{k} = 0$) and the highest is $\varepsilon + 8\gamma$ (obtained for $\bar{k} = (2\pi/b)\bar{x}$, in other words, for the six vertices of the polyhedron representing the first Brillouin zone, namely those with four adjacent faces). The width of this band is $W = 16\gamma$.

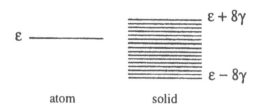

atom   solid

7. The band obtained in question **6** has N degenerate spin states and thus may accommodate $2N$ electrons. There are N electrons to be placed. Thus, the Fermi level at $T = 0$ is in the middle of this band, for all R. Although the band shrinks very rapidly when R increases, it is always half full and thus corresponds to a metallic state. This clearly contradicts the physical reality, since when the atoms are distant, the electrons are located on the atomic sites and conduction is impossible.

II

8. The Pauli principle implies that for fermions the total wavefunction is antisymmetric with respect to the exchange of electrons (1,2). Thus, a spatial wavefunction which is symmetric with respect to the exchange (1,2) corresponds to a spin wavefunction which is antisymmetric, that is, a singlet spin state. An antisymmetric spatial wavefunction corresponds to a triplet spin state. Thus, only $|\Phi_2\rangle$ is associated with a spin triplet; $|\Phi_1\rangle$, $|\Phi_3\rangle$ and $|\Phi_4\rangle$ are each associated with a singlet.

We shall denote the chosen basis by $|\Phi_1^S\rangle$, $|\Phi_2^T\rangle$, $|\Phi_3^S\rangle$ and $|\Phi_4^S\rangle$.

N.B. The Pauli principle has reduced the number of possible states to six (instead of 16).

9. The Hamiltonian $\hat{H}$ commutes with $\hat{T}_{AB}$ and with the total spin operator $S^2 = (\bar{s}_1 + \bar{s}_2)^2$. Thus, these three operators have a common eigenbasis. Since $|\Phi_2^T\rangle$ is an eigenstate of S^2 orthogonal to the three other states which are spin singlets, $|\Phi_2^T\rangle$ is also an eigenstate of $\hat{H}$.

Moreover, $|\Phi_1^S\rangle$ and $|\Phi_3^S\rangle$ are symmetric with respect to the exchange A, B. $|\Phi_2^T\rangle$ and $|\Phi_4^S\rangle$ are antisymmetric with respect to this exchange. Thus, the eigenstates of $\hat{H}$ are a linear combination of $|\Phi_1^S\rangle$ and $|\Phi_3^S\rangle$ or a linear combination of $|\Phi_2^T\rangle$ and $|\Phi_4^S\rangle$. Since $|\Phi_2^T\rangle$ is an eigenstate of $\hat{H}$, so too is $|\Phi_4^S\rangle$.

10. Let us investigate the eigenenergies of the states $|\Phi_2^T\rangle$ and $|\Phi_4^S\rangle$. For this, we set $\hat{H} = \hat{H}_0 + \hat{V}_{12}$, where $\hat{H}_0 = (p_1^2/2m) + V_{1A} + V_{1B} + (p_2^2/2m) + V_{2A} + V_{2B}$.

Let us consider, for example, $|\Phi_2^T\rangle$, whose eigenenergy is given by

$$E_2^T = \langle\Phi_2^T|\hat{H}_0 + \hat{V}_{12}|\Phi_2^T\rangle = \langle\Phi_2^T|\hat{H}_0|\Phi_2^T\rangle + \langle\Phi_2^T|\hat{V}_{12}|\Phi_2^T\rangle. \qquad (7)$$

With the assumption on the matrix elements of $\hat{V}_{12}$, the second term is zero because $|\Phi_2^T\rangle = (1/\sqrt{2})\{|\varphi_{1A}\rangle \otimes |\varphi_{2B}\rangle - |\varphi_{1B}\rangle \otimes |\varphi_{2A}\rangle\}$ does not contain any terms in which the two electrons are on the same site. To calculate the first term of (7), we note that

$$\hat{H}_0|\varphi_{1A}\rangle \otimes |\varphi_{2B}\rangle = \left\{\left(\frac{p_1^2}{2m} + V_{1A} + V_{1B}\right)|\varphi_{1A}\rangle\right\} \otimes |\varphi_{2B}\rangle$$

$$+ |\varphi_{1A}\rangle \otimes \left\{\left(\frac{p_2^2}{2m} + V_{2A} + V_{2B}\right)|\varphi_{2B}\rangle\right\}$$

$$= \{(\varepsilon + V_{1B})|\varphi_{1A}\rangle\} \otimes |\varphi_{2B}\rangle + |\varphi_{1A}\rangle \otimes \{(\varepsilon + V_{2A})|\varphi_{2B}\rangle\}. \qquad (8)$$

A non-zero matrix element in (7) is obtained by taking the 'bra' containing $\langle\varphi_{2B}|$ for the first term of (8) and $\langle\varphi_{1A}|$ for the second. This shows that in the calculation of E_2^T, the cross terms disappear and

$$E_2^T = 2 \times \tfrac{1}{2}\langle\varphi_{1A}| \otimes \langle\varphi_{2B}|\hat{H}_0|\varphi_{1A}\rangle \otimes |\varphi_{2B}\rangle$$
$$= 2\varepsilon + \langle\varphi_{1A}|V_{1B}|\varphi_{1A}\rangle + \langle\varphi_{2B}|V_{2A}|\varphi_{2B}\rangle$$
$$= 2\varepsilon - 2\alpha \simeq 2\varepsilon.$$

Similarly, for $|\Phi_4^S\rangle = (1/\sqrt{2})\{|\varphi_{1A}\rangle \otimes |\varphi_{2A}\rangle - |\varphi_{1B}\rangle \otimes |\varphi_{2B}\rangle\}$, we have

$$E_4^S = \langle\Phi_4^S|\hat{H}_0|\Phi_4^S\rangle + \langle\Phi_4^S|V_{12}|\Phi_4^S\rangle = 2\varepsilon - 2\alpha + U \simeq 2\varepsilon + U.$$

Applying the same arguments to $|\Phi_1^S\rangle$ and $|\Phi_3^S\rangle$, we immediately find

$$\langle\Phi_1^S|\hat{H}|\Phi_1^S\rangle = 2\varepsilon - 2\alpha \simeq 2\varepsilon \qquad \langle\Phi_3^S|\hat{H}|\Phi_3^S\rangle = 2\varepsilon - 2\alpha + U \simeq 2\varepsilon + U$$

and for the non-diagonal matrix element, we have $\langle\Phi_3^S|\hat{H}|\Phi_1^S\rangle = \langle\Phi_3^S|\hat{H}_0|\Phi_1^S\rangle$.

Taking the term $|\varphi_{1A}\rangle \otimes |\varphi_{2B}\rangle$ in $|\Phi_1^S\rangle$ and considering (8) and

$$\langle\Phi_3^S| = \tfrac{1}{\sqrt{2}}\{\langle\varphi_{1A}| \otimes \langle\varphi_{2A}| + \langle\varphi_{1B}| \otimes \langle\varphi_{2B}|\}$$

it is clear that the first term of (8) must be associated with the second term of $\langle\Phi_3^S|$, and *vice versa*. This leads to

$$\langle\Phi_3^S|\hat{H}|\Phi_1^S\rangle = \left\{\tfrac{1}{2}\langle\varphi_{1B}|V_{1B}|\varphi_{1A}\rangle + \tfrac{1}{2}\langle\varphi_{2A}|V_{2A}|\varphi_{2B}\rangle\right\} \times 2 = -2\gamma.$$

Thus, the matrix representing $\hat{H}$ in the subspace $|\Phi_1^S\rangle$, $|\Phi_3^S\rangle$ is given by

$$\hat{H} = \begin{pmatrix} 2\varepsilon & -2\gamma \\ -2\gamma & 2\varepsilon + U \end{pmatrix}.$$

11. i) The associated eigenenergies are given by

$$(2\varepsilon - E)(2\varepsilon - E + U) - 4\gamma^2 = 0$$

or

$$E_\pm = 2\varepsilon + \tfrac{1}{2}\left(U \pm \sqrt{U^2 + 16\gamma^2}\right).$$

We note that $E_+ > 2\varepsilon + U$ and $E_- < 2\varepsilon$, whence the structure of levels as shown in the accompanying figure.

a) $\gamma \neq 0$ b) $\gamma = 0$

ii) U is an intra-atomic quantity independent of the distance between the two atoms. On the other hand, γ decreases strongly when the atoms are moved away. In the limit $\bar{r}_B - \bar{r}_A \to \infty$, we have $\gamma \to 0$, $E_+ \to 2\varepsilon + U$ and $E_- \to 2\varepsilon$. The corresponding eigenstates are then $|\Phi_1^S\rangle$ for E_- and $|\Phi_3^S\rangle$ for E_+.

The energy 2ε is associated with the states $|\Phi_2^T\rangle$ and $|\Phi_1^S\rangle$ which only involve terms of the type $|\varphi_{1A}\rangle \otimes |\varphi_{2B}\rangle$ (in other words, two electrons in the same orbital).

Thus, in the ground state of the system, the two electrons are not simultaneously in the same orbital. In the excited state two electrons coexist on the same atom, thus constituting an ion H^-; U is the additional energy which must be supplied to put the two electrons on the same atom.

12. Under these conditions, as in **I**, the hybridisation due to the term γ produces a broadening W for the electron states at ε and at $\varepsilon + U$. The band due to ε can only accommodate N electrons. Thus, it is full, as long as $W < U$. Under this condition, the material is an insulator.

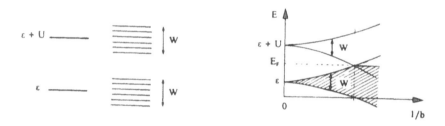

On the other hand, if $W > U$ the bands overlap and the system becomes metallic. This occurs if one varies the lattice cell size b. The insulator–metal transition occurs for $U = W$.

13. By definition:

$$U = \left(\frac{1}{\pi a_0^3}\right)^2 \int \int \exp\left(-\frac{2r_1}{a_0}\right) \frac{e^2}{|\bar{r}_2 - \bar{r}_1|} \exp\left(-\frac{2r_2}{a_0}\right) d^3r_1 \, d^3r_2$$

$$= \frac{1}{(\pi a_0^3)^2} \int d^3r_1 \left\{\exp\left(-\frac{2r_1}{a_0}\right) \int \exp\left(-\frac{2r_2}{a_0}\right) \frac{e^2}{r_{12}} d^3r_2\right\}.$$

The second integral in d^3r_2 corresponds exactly to that in question **3** for calculating α; whence, from (5) with $r_1 = R$, we have

$$U = \frac{e^2}{\pi a_0^3} \int \left[\frac{1}{r_1} - \frac{1}{a_0}\left(1 + \frac{a_0}{r_1}\right) \exp\left(-\frac{2r_1}{a_0}\right)\right] \exp\left(-\frac{2r_1}{a_0}\right) d^3r_1.$$

Thus,

$$U = \frac{4e^2}{a} \int_0^\infty \left[\frac{a_0}{r_1} - \left(1 + \frac{a_0}{r_1}\right) \exp\left(-\frac{2r_1}{a_0}\right) \right] \exp\left(-\frac{2r_1}{a_0}\right) \left(\frac{r_1}{a_0}\right)^2 \frac{dr_1}{a_0}$$

$$= \frac{4e^2}{a_0} \int_0^\infty [x \exp(-2x) - (x^2 + x) \exp(-4x)] \, dx$$

whence, after integrating by parts several times, we have

$$U = \frac{5e^2}{8a_0}. \tag{9}$$

14. The metal–insulator transition occurs when

$$U = W = 16\gamma$$

that is, when, according to the expressions (6) for γ and (9) for U

$$\frac{5}{8} = 16 \left(1 + \frac{R}{a_0}\right) \exp\left(-\frac{R}{a_0}\right) \qquad \exp\left(-\frac{R}{a_0}\right) = \frac{5}{128(1 + (R/a_0))}.$$

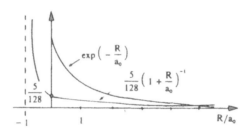

Graphically, we see that the solution corresponds to $R/a_0 \gg 1$. Numerically, we obtain $R/a_0 \simeq 5.5$. For the body-centred cubic lattice $R = b\sqrt{3}/2$ and $n_M = 2/b^3$ or $R = (\sqrt{3}/2^{2/3})n_M^{-1/3}$ and $R/a_0 = 5.5$ corresponds to $n_M^{1/3} a_0 = (\sqrt{3}/2^{2/3})/5.5 = 0.218 = C$.

The condition for insulator–metal transition is satisfied when

$$n_M^{1/3} a_0 = 0.218 = C_{\text{bcc}}.$$

15. In the case of a simple cubic crystal, there are only six first neighbours and the condition is $W = 12\gamma$, or $\exp(-R/a_0) = 5/[96(1 + (R/a_0))]$, which gives $R/a_0 = 4.7$. Since $n_M = 1/R^3$, we then obtain $n_M^{1/3} a_0 = (4.7)^{-1} = 0.213$.

$$n_M^{1/3} a_0 = 0.213 = C_{\text{sc}}.$$

For a face-centred cubic (12 neighbours), we find $n_M^{1/3} a_0 = 0.204 = C_{\text{fcc}}$.
The constant C is only weakly dependent on the crystalline structure

III

16. The condition $n_M^{1/3} a_0 = 0.218$ or $R/a_0 = 5.5$ for a body-centred cubic transforms into $(R_M - 2R_c)/a_0 = 5.5$ or $R_M/a_0 = (2R_c/a_0) + 5.5$ where $R/a_0 = b\sqrt{3}/2a_0$. We obtain the following table of values:

	Li	Na	K	Rb	Cs
b (Å)	3.49	4.23	5.23	5.59	6.05
R_c/a_0	1.13	1.79	2.51	2.79	3.19
R_M/a_0	7.76	9.08	10.52	11.08	11.88
R/a_0	5.70	6.91	8.55	9.13	9.89

We always have $R < R_M$ and the density is larger than the density needed to obtain a metallic state.

17. The atomic volume is $v = a^3/8 = 1/n_M = (5.4)^3/8 = 19.7$ Å^3. This gives $n_M^{1/3} a_0 = 0.196$, which is just less than the density needed for atomic hydrogen to give a metallic solid.

18. We evidently have to compress hydrogen to obtain a metallic state.

It is necessary to reach pressures of the order of 200 GPa (2 Mbar) in order to detect a metallic state. At present, such pressures can be obtained only by transient methods (shock wave produced by a violent explosion ...) for obtaining such a phase in the laboratory. Hydrogen with a very high density certainly exists on certain planets in the solar system. The magnetic field of Jupiter detected by Voyager I in 1975 can be explained only by the existence of metallic hydrogen on that planet at pressures of several hundred GPa.

As far as alkalis are concerned, to turn them into insulators, their density must be decreased (in the liquid state, near the critical point, Cs and Rb become insulators). Another way of obtaining a metal–insulator transition involves creating an alloy of alkalis and rare gas in order to increase the distance between the alkali atoms.

This effectively leads to a metal–insulator transition, but we are dealing with a disordered system of hydrogenoid atoms and not with a crystal and the transition in this case is actually linked to the existence of disorder (Anderson transition) rather than to an effect of interactions between electrons (Mott transition).

IV

19. The phosphorus nucleus and the additional electron which it gives to the silicon lattice constitute a large hydrogenoid atom. The interaction potential between the electron and the nucleus is decreased by the existence of a large dielectric constant K and the effective electron mass is reduced ($m^* < m$). The essential effect is to transform a_0 into $a_H = K a_0 m/m^*$ (see S.C.Ph.). If the phosphorus atoms are assumed to be distributed over a regular lattice, the arguments of **II** apply, leading to

$$n_M^{1/2} a_H \simeq 0.20.$$

20. The energies of states associated with the impurity are $E_D = E_C - \varepsilon$ where $\varepsilon = e^2/2a_H$ is the ionisation energy of the donor, and $E_{P^-} = E_D + U$, which corresponds to a phosphorus which traps two electrons (P^-). The latter is slightly larger than E_C, since $U = \frac{5}{8}(e^2/a_H) = \frac{5}{4}\varepsilon$. Thus, it lies in the conduction band of silicon.

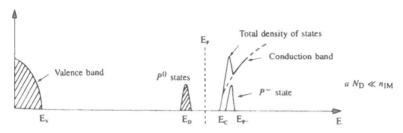

The case $N_D \ll n_M$ corresponds to the case described in the Course, with the P^- band superimposed on the conduction band. E_F lies between E_D and E_c at $T = 0$ (see S.C.Ph.) and the band of impurity states is full.

As N_D increases, the widths of the two impurity bands increase. Near the conduction band we obtain the scheme shown in the diagram below.

For $N_D > n_M$ the bands intersect and we obtain a metal.

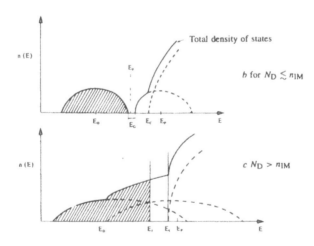

21. The experiment shows a metallic behaviour for $N_D \geq 3.75 \times 10^{24}$ m^{-3} (ρ is constant when $T \to 0$) and an insulating behaviour for $N_D \leq 3.70 \times 10^{24}$ m^{-3} (ρ diverges when $T \to 0$). Note the need to take measurements at very low temperature!

Taking $n_M \sim 3.74 \times 10^{24}$ m^{-3}, we obtain

$$C = n_M^{1/3} \frac{K}{m^*/m} a_0 = (3.74 \times 10^{24})^{1/3} \frac{11.75}{0.33} 0.53 \times 10^{-10} = 0.29.$$

This result is relatively close to the calculated value $C \sim 0.20\text{–}0.22$.

22. The number of electrons in the metallic band is N_D. The conductivity is given by $\sigma = N_D q^2 \tau / m^* = \rho^{-1}$. For $N_D = 7 \times 10^{24}$ m^{-3} we have $\rho \simeq 5 \times 10^{-5}$ Ω m, whence

$$\tau = \frac{m^*}{N_D q^2 \rho} = \frac{0.33 \times 9.1 \times 10^{-31}}{7 \times 10^{24} \times (1.6 \times 10^{-19})^2 \times 5 \times 10^{-5}} \simeq 3.3 \times 10^{-14} \text{ s}.$$

ρ increases slightly as the temperature increases, which is normal in a metal. This stems from the fact that $1/\tau$ is the sum of two contributions

$$\frac{1}{\tau} = \frac{1}{\tau}\bigg|_{\text{defects}} + \frac{1}{\tau}\bigg|_{\text{vibrations}}.$$

Although in a pure metal the second term dominates the first at high temperature, here, the spatially disordered donor atoms constitute centres where the electrons may be involved in collisions, and the first term dominates the second across practically the whole temperature range; this explains the weak variation of τ^{-1} with T. The value of the collision time at $T = 0$ is effectively comparable with $\tau \simeq 10^{-14}$ s obtained in a pure metal such as aluminium at $T \simeq 300$ K (see p. 20).

23. In the insulator case, the resistivity is due to thermally excited electrons in the conduction band (or the P$^-$ band). In the limit $N_D \ll n_M$ the number of these, n_e, satisfies (see S.C.Ph.)

$$n_e \propto \exp(-E_G/2kT) \quad \text{where} \quad E_G = E_C - E_D = \varepsilon.$$

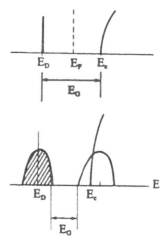

As N_D nears n_M the distance E_G between the full band and the empty band decreases (diagram (b), question **20**).

The resistivity is given by $\rho = (n_e e \mu)^{-1}$ and since $\mu = e\tau/m^*$ varies little with the temperature, the variation of ρ is dominated by that of n_e:

$$\rho \propto \exp(E_G/2kT).$$

In Figure 3, ρ diverges indeed exponentially as T decreases for $N_D = 3.7 \times 10^{24}$ m^{-3}.

We have $\ln \rho = (E_G/2k_B T) + \text{constant}$.

Thus, if we consider the case $N_D \simeq 3.7 \times 10^{24}$ m^{-3} we can plot $\ln \rho$ as a function of $1/T$ for $2K < T < 10$ K. This gives a straight line (Figure 3) with $\Delta(\ln \rho) = (E_G/2k_B)\Delta(1/T)$.

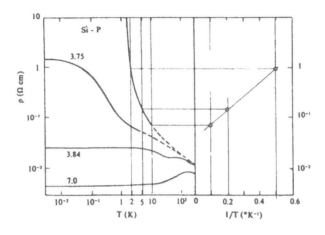

Figure 3   Determination of E_G (from $\ln \rho = E_G/2k_B T + \text{constant}$) in Si–P alloys (see Figure 1).

On the graph of Figure 3, for $\rho_1/\rho_2 = 10$, we obtain $(1/T_1) - (1/T_2) = 0.33$ K^{-1}, or

$$\frac{E_G}{k_B} = 2\frac{\ln 10}{0.33} = 14 \text{ K}.$$

This energy should be compared with $\varepsilon = E_1(m^*/m)/K^2$ where E_1 is the energy of the Bohr atom ($= 13.6$ eV). We obtain $\varepsilon = 13.6 \times 0.33/(11.7)^2 = 0.033$ eV or $\varepsilon/k_B = 400$ K.

We see that $E_G \ll \varepsilon$, since the concentration considered is very near the insulator–metal transition.

24. E_G should decrease from the value ε when N_D is very small, and become zero for $N_D \sim n_M$.

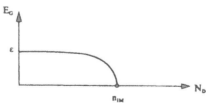

25. In the limit $N_D \ll n_M$, the electrons are localised on the phosphorus sites. This should give a Curie paramagnetism:

$$\chi_C = N_D \frac{\mu_B^2}{k_B T}.$$

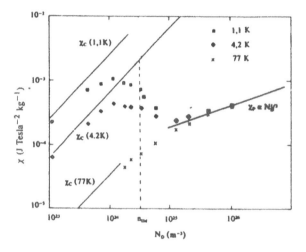

Figure 4   Magnetic susceptibility of Si–P versus N_D (see Figure 2).

The susceptibility χ_C increases linearly with N_D and is inversely proportional to T. There is such a rough tendency in Figure 4 for $N_D < 10^{24}$ m^{-3}. The susceptibility is given per kilogram of silicon. In this case, for example, for $N_D = 10^{23}$ m^{-3}, we have

$$\chi_C = \frac{10^{23}}{2.33 \times 10^3} \frac{(0.927 \times 10^{-23})^2}{1.38 \times 10^{-23}T} = \frac{2.67}{T} 10^{-4} \text{ SI}$$

which corresponds to the three straight lines for $\chi_C(1.1K)$, $\chi_C(4.2K)$ and $\chi_C(77K)$ shown on Figure 4.

26. For $N_D \gg n_M$, we are dealing with a metallic state. In this case, the susceptibility is of Pauli type (see Chapter III), independent of the temperature. This is in fact what we observe for $N_D > 10^{25}$ m^{-3}. The susceptibility is proportional to the electron density of states at the Fermi level:

$$\chi_P = n(E_F)\mu_B^2.$$

If we suppose that, for N_D sufficiently large, the conduction band may be considered to be simply parabolic, then taking the bottom of this band for the origin of the energies, we may write:

$$n(E) \propto E^{1/2} \text{ whence } n(E_F) \propto E_F^{1/2}$$

where $E_F = (\hbar^2/2m^*)(3\pi^2 N_D)^{2/3}$ (see p. 46).

We obtain $\chi_P \propto N_D^{1/3}$ and, more precisely,

$$\chi_P = \frac{3\mu_B^2 m^*}{\hbar^2 (3\pi^2)^{2/3}} N_D^{1/3}.$$

The experimental susceptibility acquires this asymptotic behaviour for $N_D > 10^{25}$ m^{-3} (see straight line in Figure 4).

N.B. The insulator–metal transition observed in Si–P does not have such a simple explanation. The effect of the interactions between electrons compounds with that of the disorder. Numerous experimental and theoretical attempts are currently being made to clarify the respective roles of these two effects in the various insulator–metal transitions.

BAND STRUCTURE OF A SUPERCONDUCTING OXIDE

Jean-Noël Chazalviel

It is not absolutely essential to answer question **1** before tackling the rest of the problem. Questions **3** to **7** are relatively short and should be dealt with quite quickly. If questions **8** or **9** should constitute a blockage, **II** may be studied in full, provided one has read the text of question **9**.

The discovery of two families of materials which are superconductors at abnormally high temperatures ($T_c \sim 40$ K for La_2CuO_4 and $T_c \sim 100$ K for $YBa_2Cu_3O_7$ has raised considerable interest in the scientific community (see p. 62). This is as much because of the potential applications as because of the fundamental problem of understanding the mechanisms capable of leading to such high values of T_c. Here, we shall be interested in the band structure of La_2CuO_4, which may provide a key to this problem, because of the existence of singularities in the electron density of states.

The La_2CuO_4 crystal has a *layered* structure. Each layer consists of CuO_6 octahedra located at the nodes of a square lattice and touching at their vertices (there is only one oxygen atom at the common vertices, which is responsible for the stoichiometry of CuO_4). The lanthanum atoms are interspersed between the layers. Figure 1 shows three views of the structure: (*a*) in perspective, (*b*) in elevation, (*c*) viewed from above a layer. Figure 2 shows a view of a single layer in perspective.

I

1. *Figure 1(a) shows a lattice cell with dimensions $a \times a \times c$. Is this cell primitive? Draw three vectors $\bar{a}_1$, $\bar{a}_2$ and $\bar{a}_3$ defining a primitive cell and circle the atoms which form a basis. List the number of atoms of each sort (La, Cu, O) in this basis and check that the stoichiometry is respected.*

2. Suppose that the layers are sufficiently far apart that the bonds between layers are negligible. Thus, in what follows, we are interested in the hypothetical *bidimensional* crystal of Figure 2, consisting of the periodic repetition of the basis (three dimensional) according to the lattice vectors lying in the plane of the layer. The lanthanum atoms, which do not appear in this model, will only be used to provide electrons for these layers (cf. question **10**).

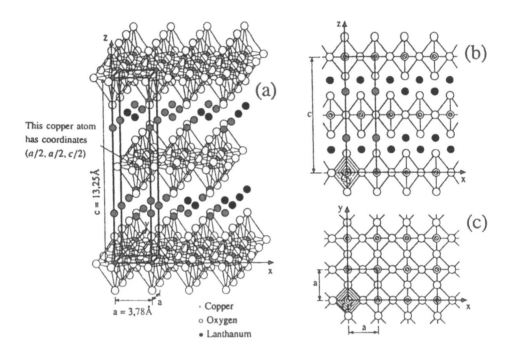

This copper atom has coordinates $(a/2, a/2, c/2)$

$c = 13.25$Å

(a)

(b)

(c)

$a = 3.78$Å

· Copper
o Oxygen
• Lanthanum

Figure 1   Crystalline structure of La_2CuO_4. The edges of the CuO_6 octahedra have been drawn to show the layers together with a cell (quadratic) with dimensions $a \times a \times c$. (a) View in perspective. (b) View parallel to the plane of the layers. The copper atoms lying at the centre of the octahedra are hatched and are not shown in black because they are in reality 'hidden' behind an oxygen atom. The octahedron at the origin is hatched, for clarity. (c) View from above a layer. Same remarks as in (b). The lanthanum atoms are not shown.

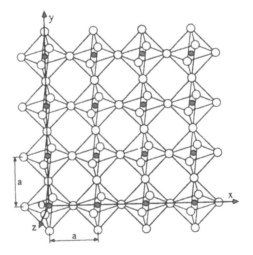

Figure 2   View of an isolated layer, in perspective. As in Figure 1, the lanthanum atoms are not illustrated.

    Check that the square primitive cell contains one copper atom and four oxygen atoms. Determine the reciprocal lattice (in two dimensions) and the first Brillouin zone of such a crystal. Sketch the latter.

3. We shall calculate the band structure of this crystal using the tight-binding approximation. To do this, we write the Hamiltonian for an electron in the crystal in the form $\hat{H} = (\hat{\bar{p}}^2/2m) + \sum_{n,a} V_{n,a}$ and we shall seek wavefunctions in the form of linear combinations of atomic orbitals $\psi = \sum_{n,a,\alpha} C_{na\alpha}\varphi_{na\alpha}$, where n denotes a primitive cell, a an atom of this cell and α an atomic wavefunction for this atom. *Prove that $C_{na\alpha}$ is of the form $\exp(i\bar{k} \cdot \bar{R}_n)u_{a\alpha}$, where $\bar{R}_n$ is a vector of the Bravais lattice.*

4. If one calculates the valency bands and the first conduction band only, it may suffice to express the Bloch functions in the form of linear combinations of Cu 3d and O 2p atomic orbitals. *Taking into account Bloch's theorem, the number of atoms per cell and the orbital degeneracy of the atomic states p and d, show that the set of coefficients $C_{na\alpha}$ reduces to a small number N of unknown coefficients $u_{a\alpha}$. Give the value of N (this value should lie between 15 and 20).*

5. To determine the $u_{a\alpha}$, we write down the equation for the energy eigenvalues $\hat{H}\psi = E\psi$ and project it onto each atomic function $\exp(i\bar{k} \cdot \bar{R}_n)\varphi_{na\alpha}$ for a certain value of n (fixed, arbitrary). *Show, without calculations, that this should give rise to a homogeneous linear system of N equations in N unknowns. What procedure could be used to deduce the band structure $\{E_m(\bar{k})\}$ of the crystal? Deduce that we expect to find N bands.*

6. In order to construct the matrix M for the above linear system, we make the following approximations: we assume that the atomic functions $\varphi_{na\alpha}$ are orthonormal ($\langle\varphi_{na\alpha}|\varphi_{n'a'\alpha'}\rangle = \delta_{nn'}\delta_{aa'}\delta_{\alpha\alpha'}$) and we retain only those matrix elements of type $\langle\varphi_{i\alpha}|V_j|\varphi_{l\alpha'}\rangle$ for which i and l are two first-neighbour atoms and $j = i$ or $j = l$ (here, i, j and l denote pairs of indices of the form (n, a); note that two first-neighbour atoms may or may not belong to the same primitive cell). We also neglect all terms of type $\langle\varphi_{i\alpha}|V_j|\varphi_{i\alpha'}\rangle$ with $i \neq j$.

    With these approximations, show that the diagonal elements of M reduce to $M_{a\alpha,a\alpha} = \varepsilon_{a\alpha} - E$, where $\varepsilon_{a\alpha}$ is the energy of the state $\varphi_{a\alpha}$ in the isolated atom a, and that the non-diagonal elements may be expressed in the form $M_{a\alpha,a'\alpha'} = \sum_m \exp(i\bar{k} \cdot \bar{R}_m)\gamma_m$ where $\bar{R}_m$ is either $\bar{0}$ or a simple vector of the Bravais lattice and γ_m is non-zero if and only if the atoms $(\bar{R}_n, a)$ and $(\bar{R}_n + \bar{R}_m, a')$ are first neighbours. Finally, show that under these conditions we have $\gamma_m = \langle\varphi_{na\alpha}|V_{n+m\,a'}|\varphi_{n+m\,a'\alpha'}\rangle$ (here, the index $n + m$ symbolically denotes the cell $\bar{R}_n + \bar{R}_m$).

7. *Writing the matrix element $\langle\varphi_{i\alpha}|(\hat{\bar{p}}^2/2m) + V_i + V_j|\varphi_{j\alpha'}\rangle$ in two different ways, show that $\langle\varphi_{i\alpha}|V_i|\varphi_{j\alpha'}\rangle = \langle\varphi_{i\alpha}|V_j|\varphi_{j\alpha'}\rangle$.*

8. To lighten the calculations, we shall make the following extreme simplification: we shall retain only those matrix elements $\langle\varphi_{na\alpha}|V_{na}|\varphi_{n'a'\alpha'}\rangle$ for which (n, a) and (n', a') are two first-neighbour atoms (one copper and one oxygen) *lying in the plane xOy* (Figure 1), with wavefunctions $(a\alpha)$ and $(a'\alpha')$ pointing towards each other ('σ bonds'). In particular, this leads us to assume that of the 3d functions of the copper, *only* the function[1]
$$\varphi_{Cu,x^2-y^2}(\bar{r}) = R_{ad}(r) \times \text{constant} \times (x^2 - y^2)/r^2$$
gives non-diagonal matrix elements.

---

[1] The algebraic form of these functions is given here only as a reminder of the symmetries and the signs.

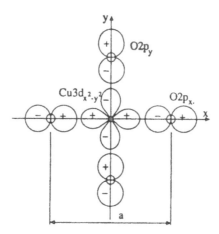

Figure 3 Schematic representation of the orbital Cu $3d_{x^2-y^2}$ and the orbitals $2p_x$ and $2p_y$ of neighbouring oxygen atoms, which may give rise to a non-zero matrix element. Note that these are different functions O 2p depending on whether the Cu–O bond is parallel to Ox or to Oy. Particular attention should be paid to the respective signs of the different matrix elements (we recall that the atomic potentials are *attractive*).

For each oxygen atom, we shall take for the functions 2p the basis $\{p_\beta\}$ ($\beta = x, y, z$) ($\varphi_{p\beta}(\overline{r}) = R_{2p}(r) \times \text{constant} \times \beta/r$). The shape and the signs of these 3d and 2p functions are shown schematically in Figure 3.

Show that, with this extreme approximation, the matrix M reduces to a 3×3 block complemented by a diagonal part. Give this matrix as a function of the only remaining parameter γ, which should not be calculated explicitly (choose $\gamma > 0$ and pay great attention to the respective signs of the different matrix elements, for example by considering Figure 3 in detail).

9. *Deduce the (approximate) band structure of La_2CuO_4. For convenience, you may choose the origin for the energies to be $\frac{1}{2}(\varepsilon_{Cu\,3d} + \varepsilon_{O\,2p})$ and set $\varepsilon_{Cu\,3d} = -\varepsilon_{O\,2p} = \Delta > 0$. Show that we obtain two non-degenerate bands $E_\pm = \pm\left[\Delta^2 + 2\gamma^2(2 - \cos k_x a - \cos k_y a)\right]^{1/2}$ and two flat bands, with energies $\pm\Delta$ and respective degeneracies 4 and 11.*

II

10. *Determine the number of electrons to be placed in these bands, knowing that, here, lanthanum is in the ion state La^{3+} and that the 4s and 4p Cu levels occur at energies far greater than Cu 3d and O 2p and thus are not close to the energies with which we are concerned (the atomic levels of lanthanum are not involved further here, however, the three electrons introduced by each lanthanum atom may populate the bands formed from Cu 3d and O 2p). We recall that the valency configuration of atomic oxygen is $2p^4$ and that of atomic copper is $3d^{10}\,4s^1$. Deduce that the Fermi level falls in the highest band calculated in question 9. What is the occupancy of this band?*

11. *Taking into account the formula $E_+(\overline{k})$ for the highest band found in question 9, determine the Fermi curve in this model and draw it (in two dimensions, the 'Fermi surface' becomes a curve). What is the value of E_F?*

12. *Show, by examining the curves of constant energy in the neighbourhood of $E = E_F$ that the density of states $n(E)$ has a singularity at the Fermi level $n(E) \simeq (2E_F/\pi^2\gamma^2 a^2)\ln(\gamma^2/E_F|E - E_F|)$.*

To do this, begin with the general formula for the density of states $n(E) = (2/(2\pi)^2)\int_S dk/|\overline{\nabla}_{\overline{k}}E|$, (see (24), p.135) where S is the curve of constant energy E (this formula is given here for a two-dimensional space) and note that the singularity is a consequence of the shape of the curves $E_+(\overline{k})$ in the neighbourhood of the points $(0, \pm\pi/a)$ and $(\pm\pi/a, 0)$. Use a limited expansion of $E_+(k)$ in the neighbourhood of one of these points $\overline{k}_0$ and calculate the singular contribution of this point *approximately*, considering a limited portion of S in its neighbourhood, for example such that $|\overline{k} - \overline{k}_0| \leq K = 1/a$, without attempting to optimise the choice of the value of K further.

III

13. In reality, the crystalline structure of Figure 1 is not always observed. At low temperature ($T \leq 200$ K), X-ray diffraction diagrams show additional spots which reveal a reduction in the symmetry of the crystal (distorted configuration). The precise structure of the distorted configuration has been much debated. Some authors have suggested that the configuration simply involves the transformation of the square lattice, on which a layer is constructed, into a rectangular lattice (sides of the rectangle $a_x = a(1 - \eta)$ and $a_y = a(1 + \eta)$, where η is small, of the order of 10^{-3}).

Take the matrix of question 8 and diagonalise it, taking this distortion into account. You will no longer have to consider a single parameter γ but a parameter $\gamma(1 + G\eta)$ corresponding to the shortened bonds and a second parameter $\gamma(1 - G\eta)$ corresponding to the extended bonds (here, G is a dimensionless constant of the order of unity).

14. *Noting that the singularities in the density of states are associated with the energies for which the Fermi curve touches a boundary of the first Brillouin zone, sketch the Fermi curve and the shape of the density of states in the distorted configuration, indicating the positions of the singularities and of E_F.*

15. *Can you explain qualitatively, based on energy-related arguments, why the La_2CuO_4 material 'prefers' the distorted configuration and why the undistorted configuration reappears above a certain temperature? Examine the decrease in energy introduced by the distortion and show that, in the limit of weak distortions, it behaves like $\eta^2 \ln\eta$.*

16. In fact, pure La_2CuO_4 material is neither a superconductor nor a metal, but a semiconductor, in contradiction to the results obtained in question **10**. *Taking your inspiration from the course (p. 120 and **Problem 2**) and the Fermi curve obtained in question 11, can you suggest another distortion which might explain this property?*

17. The superconducting materials are crystals with a modified composition (by abuse of terminology, these are said to be 'doped') of the type $La_{2-x}Sr_xCuO_{4-y}$ where $0 < x - 2y < 1$ (Sr = strontium, in the state Sr^{2+} here). *Why, for example, is the material $La_{1.85}Sr_{0.15}Cu\,O_4$ no longer a semiconductor but a metal?*

18. In the conventional theory of superconductivity (Bardeen, Cooper, Schrieffer, 1957), the critical temperature is given by the formula $T_c = \theta_D \exp[-1/Vn(E_F)]$, where θ_D is the 'Debye temperature' of the solid ($k_B\theta_D$ represents a mean energy of lattice vibration) and V is an effective *attractive* potential between electrons induced by the electron–phonon coupling. *What, in your opinion, is a possible reason for the abnormally high value of T_c for the materials $La_{2-x}Sr_xCuO_{4-y}$?*

SOLUTION TO PROBLEM 6

I

1. One answer is shown in the accompanying figure (there are many others). Number of atoms in a pattern:

$$\left.\begin{array}{l} 2\ La \\ 1\ Cu \\ 4\ O \end{array}\right\} La_2CuO_4 \qquad \text{(Figure below.)}$$

This works!

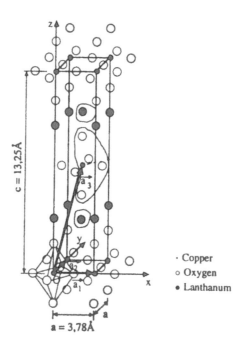

- · Copper
- ○ Oxygen
- • Lanthanum

2. The vertices of the CuO_6 octahedra lying in the plane xOy belong to two cells and thus count as half. Thus, the CuO_4 stoichiometry is respected. Square lattice with parameter $a \rightarrow$ reciprocal square lattice $(2\pi/a)$. Figure below.

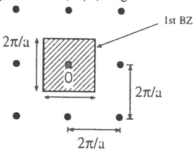

3. Bloch $\rightarrow \psi(\bar{r}) = u(\bar{r})\exp(i\bar{k} \cdot \bar{r})$ where $u(\bar{r})$ is periodic

$\rightarrow \psi(\bar{r} + \bar{R}_n) = \exp(i\bar{k} \cdot \bar{R}_n)\psi(\bar{r})$ for any $\bar{R}_n$, a vector of the Bravais lattice.

$\rightarrow C_{n a \alpha}$ proportional to $\exp(i\bar{k} \cdot \bar{R}_n)$.

4. By Bloch's theorem, we may write $C_{n a \alpha} = u_{a \alpha}\exp(i\bar{k} \cdot \bar{R}_n)$. The number N of coefficients $u_{a \alpha}$ to be determined is thus the number of pairs (a, α) under consideration, or, in this case

Cu 3d: 1 atom per cell, state d, 5 times degenerate $\rightarrow 1 \times 5 = 5$

O 2p: 4 atoms per cell, state p, 3 times degenerate $\rightarrow 4 \times 3 = 12$

Total number of coefficients $u_{a \alpha}$: $N = 5 + 12 = 17$.

5.

$$\left\langle e^{i\bar{k}\cdot\bar{R}_n}\varphi_{n a \alpha} \middle| \hat{H} - E \middle| \sum_{l,a',\alpha'} e^{i\bar{k}\cdot\bar{R}_l} u_{a'\alpha'}\varphi_{l a' \alpha'} \right\rangle = 0.$$

There are N possible pairs (a, α) and N unknowns $u_{a \alpha}$. Thus, we have a system of N equations in N unknowns. This system will have a solution (other than the trivial solution) if and only if its determinant is zero. This gives an equation of degree N in E; whence there are N solutions, that is, N bands $E_m(\bar{k})$.

6.

$$M_{a\alpha,a'\alpha'} = \langle\varphi_{n a \alpha}| \hat{H} - E \left| \sum_l e^{i\bar{k}\cdot(\bar{R}_l-\bar{R}_n)}\varphi_{l a' \alpha'} \right\rangle.$$

If $(a, \alpha) = (a'\alpha')$ (diagonal elements), the atoms (n, a) and (l, a) are either identical $(n = l)$ or more distant than first neighbours. Taking into account the approximations in the statement of the problem, the only contribution is for $n = l$, that is

$$M_{a\alpha,a\alpha} = \langle\varphi_{n a \alpha}| \left(\frac{\widehat{\bar{p}}^2}{2m} + V_{n a}\right) + \sum_{(j,b)\neq(n,a)} V_{j b} - E |\varphi_{n a \alpha}\rangle = \varepsilon_{a\alpha} - E$$

since $((\widehat{\bar{p}}^2/2m) + V_{n a})\varphi_{n a \alpha} = \varepsilon_{a\alpha}\varphi_{n a \alpha}$ and $\langle\varphi_{i\alpha}|V_j|\varphi_{i\alpha'}\rangle$ is negligible if $i \neq j$.

If $(a, \alpha) \neq (a', \alpha')$

$$\langle\varphi_{n a \alpha}| \left(\frac{\widehat{\bar{p}}^2}{2m} + V_{n a}\right) = \langle\varphi_{n a \alpha}|\varepsilon_{a\alpha},$$

whence

$$M_{a\alpha,a'\alpha'} = \langle\varphi_{n a \alpha}| \sum_{(n',b)\neq(n,a)} V_{n'b} \left| \sum_l e^{i\bar{k}\cdot(\bar{R}_l-\bar{R}_n)}\varphi_{l a' \alpha'} \right\rangle.$$

Taking into account the approximations in the statement of the problem, only the terms

$$M_{a\alpha,a'\alpha'} = \sum_m \langle\varphi_{n a \alpha}|V_{n+m\,a'}|\varphi_{n+m\,a'\alpha'}\rangle e^{i\bar{k}\cdot\bar{R}_m}$$

for which the atoms (n, a) and $(n + m, a')$ are first neighbours survive. In practice, this will be realised for $\overline{R}_m$ equal to zero (two first-neighbour atoms in the same cell) or small values of $\overline{R}_m$ (first-neighbour atoms belonging to two contiguous cells).

7.

$$\langle\varphi_{i\alpha}| \frac{\widehat{\overline{p}}^2}{2m} + V_i + V_j |\varphi_{j\alpha'}\rangle$$

$$= \langle\varphi_{i\alpha}| \left(\frac{\widehat{\overline{p}}^2}{2m} + V_i\right) + V_j |\varphi_{j\alpha'}\rangle = \langle\varphi_{i\alpha}|\varepsilon_i + V_j|\varphi_{j\alpha'}\rangle = \langle\varphi_{i\alpha}|V_j|\varphi_{j\alpha'}\rangle$$

$$= \langle\varphi_{i\alpha}| \left(\frac{\widehat{\overline{p}}^2}{2m} + V_j\right) + V_i |\varphi_{j\alpha'}\rangle = \langle\varphi_{i\alpha}|\varepsilon_j + V_i|\varphi_{j\alpha'}\rangle = \langle\varphi_{i\alpha}|V_i|\varphi_{j\alpha'}\rangle.$$

8.   Of the five 3d functions of copper, only the function (Cu, $x^2 - y^2$) is coupled to the 2p functions of the oxygen, the $2p_x$ functions of the atoms located at $(\pm a/2, 0, 0)$ ($\overline{R}_m = \overline{0}$ and $(-a, 0, 0)$) and the $2p_y$ functions of the atoms located at $(0, \pm a/2, 0)$ ($\overline{R}_m = \overline{0}$ and $(0, -a, 0)$). The matrix elements are $\pm\gamma$ and $\mp\gamma$, respectively (the potentials are attractive, whence negative). Thus, the matrix M consists of a diagonal part corresponding to the non-coupled atomic functions, that is four Cu 3d functions (diagonal elements $(\varepsilon_{Cu\,3d} - E)$) and 10 O 2p functions (diagonal elements $(\varepsilon_{O\,2p} - E)$) and a non-diagonal 3×3 part, given below

$$\begin{pmatrix} \varepsilon_{Cu\,3d} - E & \gamma(1 - e^{-ik_x a}) & -\gamma(1 - e^{-ik_y a}) \\ \gamma(1 - e^{+ik_x a}) & \varepsilon_{O\,2p} - E & 0 \\ -\gamma(1 - e^{+ik_y a}) & 0 & \varepsilon_{O\,2p} - E \end{pmatrix}.$$

9.   We find:
   4 bands $E_m(\overline{k}) \simeq \varepsilon_{Cu\,3d}$
   10 bands $E_m(\overline{k}) \simeq \varepsilon_{O\,2p}$
   3 bands obtained by setting the determinant of the 3×3 block to zero, namely

$$(\varepsilon_{Cu\,3d} - E)(\varepsilon_{O\,2p} - E)^2 = (\varepsilon_{O\,2p} - E)\gamma^2[(1 - e^{ik_x a})(1 - e^{-ik_x a})$$
$$+ (1 - e^{ik_x a})(1 - e^{-ik_x a})]$$
$$= (\varepsilon_{O\,2p} - E)2\gamma^2(2 - \cos k_x a - \cos k_y a).$$

This again gives an *eleventh* band $E_m(\overline{k}) \simeq \varepsilon_{O\,2p}$ and two bands

$$E_\pm = \tfrac{1}{2}(\varepsilon_{Cu\,3d} + \varepsilon_{O\,2d}) \pm \left[\tfrac{1}{4}(\varepsilon_{Cu\,3d} - \varepsilon_{O\,2p})^2 + 2\gamma^2(2 - \cos k_x a - \cos k_y a)\right]^{1/2}.$$

The energy E_+ is minimum for $\overline{k} = 0$ (when $E_+ = \varepsilon_{Cu\,3d}$) and maximum for $\overline{k} = (\pi/a, \pi/a)$, when $E_+ = \tfrac{1}{2}(\varepsilon_{Cu\,3d} + \varepsilon_{O\,2p}) + \sqrt{\tfrac{1}{4}(\varepsilon_{Cu\,3d} - \varepsilon_{O\,2p})^2 + 8\gamma^2}$. The band E_- is symmetric to the band E_+ about $\tfrac{1}{2}(\varepsilon_{Cu\,3d} + \varepsilon_{O\,2p})$.

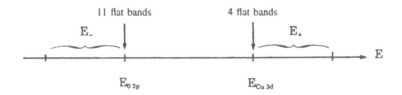

With the convention given in the statement of the problem, we may simply write

$$E_\pm = \pm\sqrt{\Delta^2 + 2\gamma^2(2 - \cos k_x a - \cos k_y a)}.$$

II

10. For each primitive cell there is an La_2CuO_4 basis. The lanthanum gives $2 \times 3 = 6$ electrons, since it is in the ion state La^{3+}. The copper gives 10 3d electrons plus one 4s electron, i.e. 11 electrons in total. The oxygen gives $4 \times 4 = 16$ electrons. Thus, we have a total of $6 + 11 + 16 = 33$ electrons to distribute among 17 bands, i.e. 34 places; whence, all the bands are full except the highest (E_+), which is exactly half full.

11. The energy curves $E_+(\vec{k}) = $ constant are given by $2 - \cos k_x a - \cos k_y a = C$, where the constant C may be between 0 and 4. Note the symmetry $C \rightarrow 4 - C$, $k_x \rightarrow (\pi/a) - k_x$, $k_y \rightarrow (\pi/a) - k_y$. Thus, the value $C = 2$ seems to correspond to half-filling. The corresponding curve $\cos k_x a + \cos k_y a = 0$ reduces to $|k_x| + |k_y| = \pi/a$: the Fermi curve is a square (see Figure below).

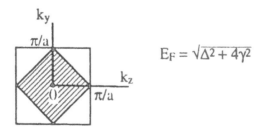

$$E_F = \sqrt{\Delta^2 + 4\gamma^2}$$

12. For $E_+ \simeq E_F$, we have $\cos k_x a + \cos k_y a \simeq 0$.

$$E_+ \simeq \sqrt{\Delta^2 + 4\gamma^2}\left[1 - \frac{\gamma^2}{\Delta^2 + 4\gamma^2}(\cos k_x a + \cos k_y a)\right]$$

$$= E_F - \frac{\gamma^2}{E_F}(\cos k_x a + \cos k_y a).$$

The symmetry $(E_+ - E_F \rightarrow E_F - E_+, k_x \rightarrow (\pi/a) - k_x, k_y \rightarrow (\pi/a) - k_y)$ indicates that, in this approximation, $n(E)$ is symmetric about E_F. For example, taking $E = E_F - \delta E$, the curve of energy E has equation $\cos k_x a - \cos k_y a = (E_F/\gamma^2)\delta E$. This curve is quasi-singular for k_x or $k_y = 0$. Let us consider, for example, the vertex A ($k_y = 0, k_x \simeq \pi/a$). In a neighbourhood of this vertex (see Figure below), we may write

$$2\frac{E_F}{\gamma^2 a^2}\delta E \simeq \left(k_x - \frac{\pi}{a}\right)^2 - k_y^2.$$

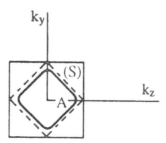

The Fermi curve is close to a hyperbola in the neighbourhood of this vertex. We have $n(E) = (2/(2\pi)^2) \int_S dk/|\overline{\nabla}_{\overline{k}} E|$.

The vertex A gives a singular contribution to $n(E)$

$$\frac{2}{(2\pi)^2}\int_{-K}^{+K}\frac{dk_y}{|\partial E/\partial k_x|} = \frac{2}{(2\pi)^2}\frac{E_F}{\gamma^2 a^2}\int_{-K}^{+K}\frac{dk_y}{\sqrt{k_y^2 + (2E_F\delta E/\gamma^2 a^2)}}$$

$$= \frac{E_F}{\pi^2\gamma^2 a^2}\,\text{arg sinh}\,\frac{\gamma K a}{\sqrt{2E_F\delta E}}$$

where K is a wavevector such that $Ka \simeq 1$ (Van Hove singularity).

Taking the four vertices into account, we have the approximation

$$n(E) \simeq \frac{2E_F}{\pi^2\gamma^2 a^2}\ln\frac{\gamma^2}{E_F\delta E}$$

(which is the density of states for a crystal with unit surface area).

N.B.   We expect a high density of states at the Fermi level to have consequences, for example for the magnetic susceptibility. When a magnetic field B is introduced, the energy of an electron with spin $\pm\frac{1}{2}$ is shifted by $\pm\mu_B B$. The induced magnetic moment μ is given by $2\mu_B$ times the number of electrons which change sub-band; in other words, at temperature T

$$\chi = \frac{\mu}{B} = \frac{2\mu_B}{B}\delta n = \frac{2\mu_B}{B}\int_{-\infty}^{+\infty}\frac{1}{2}n(E)(-\mu_B B)\frac{\partial f}{\partial E}\,dE$$

where $f(E)$ is the Fermi–Dirac distribution

$$\chi = \mu_B^2\int_{-\infty}^{+\infty}n(E)\left(-\frac{\partial f}{\partial E}\right)dE \qquad \text{(Course p. 65)}$$

$$= \mu_B^2 \frac{2E_F}{\pi^2 \gamma^2 a^2} \int_{-\infty}^{+\infty} \ln \frac{\gamma^2}{E_F |\delta E|} \frac{1/kT}{4\cosh^2(\delta E/kT)} \, d(\delta E)$$

$$= \mu_B^2 \frac{2E_F}{\pi^2 \gamma^2 a^2} \left[\underbrace{\ln \frac{\gamma^2}{E_F kT}}_{\text{term in } \ln(1/T)} + \underbrace{\int_{-\infty}^{+\infty} \ln \frac{1}{|x|} \frac{dx}{\cosh^2 x}}_{\text{constant term} \simeq 2} \right]$$

for a crystal with unit surface area.

Such curves have sometimes been observed for superconductors at high temperature. However, when the temperature falls below the critical temperature, the crystal becomes a superconductor and is thus highly diamagnetic, which makes it difficult to verify the above law.

III

13. Here, the 3×3 matrix of question **8** becomes

$$\begin{pmatrix} \Delta - E & \gamma(1+G\eta)(1-e^{-ik_x a(1-\eta)}) & -\gamma(1-G\eta)(1-e^{-ik_y a(1+\eta)}) \\ \gamma(1+G\eta)(1-e^{+ik_x a(1-\eta)}) & -\Delta - E & 0 \\ -\gamma(1-G\eta)(1-e^{+ik_y a(1+\eta)}) & 0 & -\Delta - E \end{pmatrix}$$

Only the two bands $E_\pm$ are modified (expression given to order 1 in η).

$$E_\pm = \pm[\Delta^2 + 2\gamma^2 (2 - (1 + 2G\eta)\cos k_x a(1 - \eta) - (1 - 2G\eta)\cos k_y a(1 + \eta))]^{1/2}.$$

14. The surface of constant energy touches the boundary $k_x a(1 - \eta) = \pi$ for

$$E_{+x} = [\Delta^2 + 2\gamma^2(2 + 4G\eta)]^{1/2}$$

and the boundary $k_y a(1 + \eta) = \pi$ for

$$E_{+y} = [\Delta^2 + 2\gamma^2(2 - 4G\eta)]^{1/2}.$$

Thus, we have two singularities in the density of states, not just one; the Fermi level lies between the two singularities.

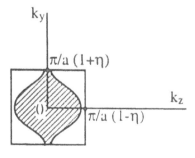

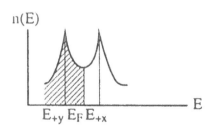

15. The electron energy is reduced by the distortion, as the accompanying figure shows. At zero temperature, the reduction in the energy is of the order of $\eta^2(E_F/a^2) \times \ln(1/\eta)$; this energy gain is important for small η and dominates the cost which may result for the electrons located far from E_F (elastic energy $\propto \eta^2$). When the temperature increases, the gain no longer varies as $\eta^2 \ln 1/\eta$, but, for small η, reaches a 'ceiling' of the order of $\eta^2 \ln(E_F/kT)$. The elastic terms dominate above a certain temperature and the distortion disappears.

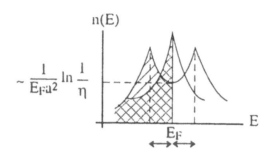

N.B. This interpretation of the distortion, first proposed by **J Labbé** and **J Bok** (Europhys. Lett., *3*, 1225 (1987)) seems to have been subsequently contradicted by experiments: the distorted structure seems to leave the copper–oxygen distances unchanged and the singularity seems not to be split (**J-P Pouget, C Noguéra** and **R Moret**). However, Labbé and Bok's mechanism may still provide the correct explanation for the analogous distortion observed in $YBa_2CU_3O_7$.

16. The one-dimensional systems with a half-full band are frequently subjected to a Peierls distortion: a doubling of the primitive cell results in the appearance in the reciprocal space of new Brillouin planes at exactly the Fermi level, which reduces the energy of the electron system and makes the material an insulator (see **Problem 2**).

Such a distortion is conceivable here, although the system considered is bidimensional, since the accidentally square shape of the Fermi curve may allow it to coincide with new 'Brillouin planes' for a hypothetical distortion such as that outlined in the next Figure (one octahedron in two is stretched or shrunk).

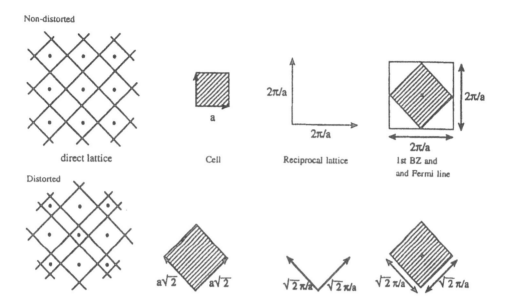

Non-distorted

direct lattice   Cell   Reciprocal lattice   1st BZ and and Fermi line

Distorted

17. For doped materials, the number of electrons available falls from 33 to $33 - x - 4y$. For example, for $La_{1.85}Sr_{0.15}CuO_4$ there are 32.85 electrons available for 34 places. The occupation rate for the last band is no longer 0.5 but 0.425. The Fermi curve is no longer square and can no longer coincide with the boundaries of the new Brillouin zone; the distortion disappears and the material becomes a conductor again. This hypothesis would enable us to understand the insulating or conducting nature of the material in association with its composition. However, despite this enticing aspect, it appears to be invalidated by current experiments (X-ray crystallography).

18. The reason for this abnormally high value of the critical temperature of materials of type La_2CuO_4 (and also $YBa_2Cu_3O_7$) is far from decided. Thus, there is no single answer to this question and all serious arguments will be marked positively by the examiner.

    Taking the BCS formula as starting point, the three parameters θ_D, V and $n(E_F)$ can be and have been invoked as responsible for the effect. The small mass of the oxygen atoms may lead to lattice vibrations of an abnormally high frequency in comparison with standard metals (θ_D large); the partially ionic nature of the bonds in an oxide may lead one to conjecture an abnormally efficacious electron–phonon coupling ($|V|$ large), while the problem shows that we may also hope to benefit from a high value of $n(E_F)$. Finally, many theoreticians think that the superconductivity of these materials should be explained in a completely different framework from BCS, for example an attraction between electrons generated other than through the intermediary of phonons (for example, spin waves). Even the delocalised character of the valence electrons (narrow bands, whence great importance of the correlation between electrons) may be called in question.

MAGNETIC PROPERTIES OF A CRYSTAL

Maurice Bernard

In order to study the magnetic properties of a crystal, it is viewed as consisting of quasi-pointlike distinguishable atoms, numbered from 1 to N and occupying fixed positions at the nodes of a three-dimensional lattice. These atoms, n per unit volume, are similar to particles with spin $S = \frac{1}{2}$, having a magnetic moment $\overrightarrow{M}$, the projection of which onto Oz for the atom i is

$$M_i = -2\mu S_{iz}$$

where μ denotes the Bohr magneton

$$\mu = \frac{e\hbar}{2m} = 0.93 \times 10^{-23} \text{ SI}.$$

We suppose that the canonical equilibrium is realised at the temperature T and we shall consider n as practically independent of T.

I

Assuming all the interactions between atoms are negligible, the crystal is placed in a uniform magnetic field parallel to Oz whose induction has a positive value B. Each atom is subjected solely to the action of this field and the variable part of its energy reduces to $-M_i B$.

1. *Calculate the mean value of M and deduce the relationship between the magnetisation intensity $J = n\langle M \rangle$ and B.*

2. *Given B, how does J vary as a function of T? Explain the behaviour of J for $T \to 0$ and for $T \to \infty$. Calculate the numerical value J_0 of J for $T = 0$. We shall take $n = 8 \times 10^{28} \text{ m}^{-3}$.*

3. *What does the relationship between J and B reduce to for weak fields ($B \to 0$)? What is the expression for the magnetic susceptibility*

$$\chi = \mu_0 \frac{\mathrm{d}J}{\mathrm{d}B}_{(B=0)} \qquad (\mu_0 = 4\pi \times 10^{-7})$$

in this case, and how does it vary with T? Calculate the numerical value of χ for $T = 300 \text{ K}$.

4. We denote the free energy per unit volume of the crystal by F.

Write down the equations which enable us to calculate, from F, the entropy S, the heat capacity per unit volume C^1, the magnetisation J and the susceptibility χ for weak fields.

5. *Calculate F as a function of the variables B and T and deduce the expressions for S and C^2. Check that we retrieve the values obtained in questions 1–3 for J and χ.*

6. *Examine what happens to J and S in the two limits $B \to \infty$ and $B \to 0$ and explain the results.*

7. *How does C vary as a function of T, for a given value of B? Calculate numerically the maximum of C in cal cm^{-3}.*

For what value of T is this maximum attained if $B = 1$ tesla?

Compare the contribution of the magnetic moments to the heat capacity per unit volume with that of the vibrations of the atoms C_V, calculated using the classical approximation.

II

From now on we take the spin interactions into account. We suppose that the interaction energy between two arbitrary atoms i and j is

$$u_{ij} = -a_{ij}\overline{S}_i \cdot \overline{S}_j.$$

We suppose that, as far as the spin $\overline{S}_i$ is concerned, we may, to a first approximation, replace the other spins in the interaction by their mean value at the thermal equilibrium. Thus, each of the atoms has a potential energy

$$-\overline{S}_i \cdot \left(\sum_j a_{ij} \langle \overline{S}_j \rangle \right)$$

which is added to that resulting from the magnetic field of external origin, in other words, to

$$-M_i B = S_{iz} 2\mu B.$$

We shall suppose that the mean values $\langle \overline{S}_i \rangle$ are all equal and directed along the axis Oz:

$$\langle S_{iz} \rangle = -\frac{\langle M \rangle}{2\mu}.$$

It is now as though every atom in thermal equilibrium were subjected to the action of an effective magnetic field

$$\mathcal{B} = B + a\mu_0 n \langle M \rangle$$

where a denotes the dimensionless constant

$$a = \frac{1}{4\mu^2 n \mu_0} \sum_j a_{ij}.$$

---

[1] C is the product of the specific heat and the density of the crystal.

[2] Since we have reduced the expression for the energy to its magnetic part, the thermodynamic quantities considered here are, in fact, only the contributions to these quantities made by the atomic magnetism.

8. Write down the new relation linking J and B.

9. Show, with the help of a graphical discussion, that if the temperature T is lower than a certain temperature T_c, to be determined, in the absence of any magnetic field of external origin ($B = 0$) the crystal takes on a spontaneous magnetisation J_S. Here, we shall suppose that if the equation determining J as a function of B has several solutions, then it is the largest value of J which corresponds to thermal equilibrium.

How does J_S vary as a function of T and what is its value for $T = 0$? Show that, provided T is not too close to T_c, J_S is not very different from this value.

Give the numerical value of T_c, assuming $a = 1590$ and, as before, $n = 8 \times 10^{28}\ m^{-3}$.

10. If the temperature T is greater than T_c what does the relation between J and B for weak fields ($B \to 0$) reduce to and what is the value of the susceptibility χ in this case? Compare the results with those obtained in the absence of interactions in question 1.

What is the numerical value of χ for $T = 1300\ K$?

SOLUTION TO PROBLEM 7

I

1. S_z may take the values $\pm\frac{1}{2}$, thus M may take the values $\pm\mu$ corresponding to the energies $\mp\mu B$. The partition function (see Stat. Ph.) is

$$Z = \mathrm{Tr}\, e^{-H/kT} = e^{\mu B/kT} + e^{-\mu B/kT} = 2\cosh\frac{\mu B}{kT}$$

and

$$\langle M \rangle = \frac{1}{Z}\mu e^{\mu B/kT} - \frac{1}{Z}\mu e^{-\mu B/kT} = \mu\tanh\frac{\mu B}{kT}$$

whence we have the magnetisation

$$J = n\mu\tanh\frac{\mu B}{kT}.$$

The crystal is paramagnetic, that is, the magnetisation has the same direction as the field.

2. For given B, J decreases from $J_0 = n\mu$ to zero as T increases from 0 to ∞. Numerical value of J_0: 7.44×10^5 A m^{-1} (Coulomb magnetisation; multiply by μ_0 to obtain the Ampère magnetisation, which gives 0.935 tesla or 9350 Gauss).

3. $B \to 0$

$$J = \frac{n\mu^2}{kT}B, \qquad \chi = \mu_0\frac{n\mu^2}{kT}.$$

The product $A = \chi T$ is constant (Curie's law).

$$A = \mu_0\frac{n\mu^2}{k} = 5 \times 10^5\mu_0 = 0.628.$$

For $T = 300$ K, $\chi = 0.00209$.

4.

$$S = -\frac{\partial F}{\partial T}, \qquad C = T\frac{\partial S}{\partial T} = -T\frac{\partial^2 F}{\partial T^2};$$

$$J = -\frac{\partial F}{\partial B}; \qquad \chi = -\mu_0 \frac{\partial^2 F}{\partial B^2}.$$

5. For n atoms,

$$F = -nkT \ln Z = -nkT \ln\left(2\cosh\frac{\mu\beta}{kT}\right)$$

whence, by the previous formulae,

$$S = nk\left[\ln\left(2\cosh\frac{\mu B}{kT}\right) - \frac{\mu B}{kT}\tanh\frac{\mu B}{kT}\right]$$

$$C = n\frac{\mu^2 B^2}{kT^2}\frac{1}{\cosh^2(\mu B/kT)}$$

and

$$J = n\mu\tanh\frac{\mu B}{kT},$$

as was found in question **1**.

6. $B \to \infty$: $J = n\mu$: all the spins are $|+\rangle$);

$S = 0$ (a single configuration).

$B \to 0$: $S = nk\ln 2 = k\ln 2^n$: each spin may be $|+\rangle$ or $|-\rangle$ and there are 2^n configurations.

$J \to 0$ (because $\langle M\rangle = 0$).

7. We set $u = -\mu B/kT$. Then

$$C = nk\frac{u^2}{\cosh^2 u} = nku^2(1 - \tanh^2 u).$$

C is zero for $u = 0$ ($T = \infty$) and $u = \infty$ ($T = 0$) and maximum for u a root of the equation $\tanh u = 1/u$, i.e. for $u \simeq 1.2$. Then

$$C = nk(u^2 - 1) = 0.44nk$$

or, numerically,

$$C = 8 \times 10^{23} \times 1.38 \times 10^{-23} \times 0.44 = 4.87 \times 10^5 \text{ Joule m}^{-3}$$

or $C = 0.116$ cal cm^{-3}.

As compared with the classical vibrational heat ($C_V = 3nk$ at high temperatures), this value is significant. Moreover, the temperatures for which C is maximum are very low. Even for $B = 1$ tesla, the maximum occurs for

$$T = \frac{\mu}{1.2k} = 0.56 \text{ K}.$$

At such temperatures, C_V is small and C is larger.

II

8.

$$J = n\mu \tanh \frac{\mu(B + \mu_0 a J)}{kT}.$$

9. For $B = 0$

$$J = n\mu \tanh \frac{\mu \mu_0 a J}{kT}.$$

Spontaneous magnetisation will occur if this equation in J has a non-zero solution J_S. Set

$$T_c = \frac{n\mu^2 \mu_0 a}{k} \qquad x = \frac{T_c}{T} \frac{J}{n\mu}.$$

The equation becomes:

$$\frac{T}{T_c} x = \tanh x.$$

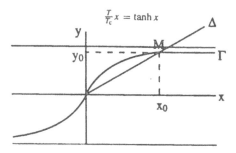

It has a non-zero root if the line Δ

$$y = \frac{T}{T_c} x$$

intersects the curve Γ

$$y = \tanh x$$

at a point $M(x_0, y_0)$ different from the origin; this occurs if $T < T_c$. T_c is a critical temperature. The spontaneous magnetisation is (the solution $J_S = 0$ is not stable)

$$J_S = n\mu \frac{T}{T_c} x_0 = J_0 y_0.$$

For $T = 0$, $x_0 = \infty$, $y_0 = 1$ and $J_S = J_0 = n\mu$.

Provided T does not become too close to T_c, y_0 remains near to 1 and J_S is little different from J_0, as the figure shows. Numerically, we have

$$T_c = Aa = 0.628 \times 1590 = 1000 \text{ K}.$$

The crystal is ferromagnetic: T_c is the Curie temperature.

10. For $T > T_c$ and $B \rightarrow 0$, we have

$$J = \frac{n\mu^2 B}{kT} = \frac{n\mu^2 B}{kT} + \frac{n\mu^2}{kT}\mu_0 a J = \frac{A}{T}\frac{B}{\mu_0} + \frac{T_c}{T}J$$

whence

$$J\left(1 - \frac{T_c}{T}\right) = \frac{A}{T}\frac{B}{\mu_0}$$

and

$$\chi = \frac{\mu_0 J}{B} = \frac{T}{T - T_c}.$$

Curie's law $\chi T = A$ is replaced by the *Curie–Weiss law* $\chi(T - T_c) = A$. The crystal is paramagnetic (paramagnetism with a Weiss molecular field). The ferromagnetism disappears for $T > T_c$.

For $T = 1300$ K, $T - T_c = 300$ and $\chi = 0.00209$ as in question **3**.

FERROELECTRICITY
OF BARIUM TITANATE

Roland Omnès

Barium titanate $BaTiO_3$ is a ferroelectric crystal whose electrical properties are analogous to the magnetic properties of ferromagnetic materials. We shall study these properties.

We shall first indicate the structure of this crystal. This structure will be little exploited in the calculations which follow, but it is the basis for the approximations made and provides us with a better understanding of the physical phenomena. The crystal has a cubic symmetry. It can be viewed as a stacking of cubes with a Ba^{++} ion at each vertex, an O^{--} ion in the centre of each face and a Ti^{++++} ion at the centre of the cube (*perovskite* structure). The titanium ions are small and have a certain freedom of movement, although the electrostatic forces tend to bring them back to the centre of the cube. The details of these forces will play an important role in what follows.

At the temperatures considered (between 0 and 200° C) the movement of the titanium ions can be viewed as that of *classical* particles.

We denote the *local* electric field experienced by the titanium ion by $\overline{E}$ and the charge of this ion by q. We shall consider a sample of unit volume V containing N titanium ions. We shall denote the polarisation (total dipolar moment for the unit volume V) by $\overline{P}$ and the *external* electric field to which the sample may be subjected by $\overline{E}_0$. The local field $\overline{E}$ is the sum of the field $\overline{E}_0$ and the electrostatic field created by the set of neighbouring ions. In what follows, we shall suppose that *all these vectors are directed along the axis Oz*.

A fairly simple, but legitimate electrostatic analysis, enables us to relate the polarisation (where only the contribution due to the displacement of the titanium ions from their equilibrium position is retained), the external field and the local field; this gives the equation

$$\overline{E} = \overline{E}_0 + \overline{P}/3\varepsilon_0.$$

We recall that when an ion with charge q is moved from its equilibrium position by a displacement $\overline{x}$ its dipolar moment is $q\overline{x}$.

1. *If the titanium ions form a perfect gas of the same volume V, what would the canonical partition function Z_0 for this gas be? $\overline{E}$ is assumed to be zero.*

2. Suppose that the forces pulling a titanium ion back to its equilibrium position are

345

given by the harmonic potential energy

$$\phi_0 = a(x^2 + y^2 + z^2)$$

when the ion is displaced from this position by the vector $\overline{x}$ with components (x, y, z). *Calculate the partition function Z_1 for the sample, still assuming $\overline{E}$ is zero and supposing that x, y and z are very small in comparison with the interionic distances.*

3. Since the local field $\overline{E}$ is zero (identical for all the ions and constant over the small distances which an ion is displaced), *calculate the new partition function Z_2 after determining the potential energy of each ion explicitly. Z_2 will be expressed as a function of the temperature and of the local field $\overline{E}$.*

4. *Give an expression for the polarisation $\overline{P}$ using a derivative of Z_2 and calculate this polarisation. Does it depend on the temperature?*

5. We further suppose that the crystalline potential is anharmonic and of the form

$$\phi = \phi_0 + \phi_1$$

where ϕ_1 is a homogeneous polynomial of degree four in x, y, and z. To facilitate the calculations, we shall restrict ourselves here to the case in which ϕ_1 is of the form

$$\phi_1 = b(x^4 + y^4 + z^4).$$

Calculate the partition function Z_3 in this new case, assuming the presence of a local electric field $\overline{E}$.

N.B. Integrals involving exponentials of polynomials will arise in this calculation. To calculate these, we assume that the coefficient b is sufficiently small that we can expand $\exp(-\beta b z^4)$ in series inside the integral and retain only the terms independent of b and those linear in b. We recall the value of the integral

$$\int_{-\infty}^{\infty} e^{-pz^2} z^{2n}\, dz = \frac{(2n-1)(2n-3)\ldots 3 \times 1}{(2p)^n}\sqrt{\frac{\pi}{p}} \qquad (n \geq 1)$$

$$= \sqrt{\frac{\pi}{p}} \qquad (n = 0).$$

6. *Calculate the polarisation of the sample in this case. Does it depend on the temperature?*

7. The dielectric susceptibility χ of the medium is defined by the equation

$$\chi = \lim_{E_0 \to 0} \frac{\partial P}{\varepsilon_0 \partial E_0}.$$

Calculate χ and show, taking into account the smallness of b, that it is approximately of the form

$$\chi = \frac{T_1}{T - T_0}.$$

Express T_1 and T_0 as a function of a and b.

8. *Knowing that $T_0 = 118°$ C, $T_1 = 150\,000$ K and the edge of a primitive cube of $BaTiO_3$ has length 4.00 Å, calculate a and b taking the eV and the Å as units of energy and length, respectively.*

9. Next, we shall use the polarisation $\overline{P}$ as a thermodynamic variable in order to study the ferroelectric properties of the material. To this end, we introduce the new thermodynamic function $A(T, \overline{P})$ defined by

$$A = F + \overline{P} \cdot \overline{E}_0 + \overline{P}^2/6\varepsilon_0$$

where F is the free energy. *Find the equation which determines $\overline{E}_0$ using the derivative $(\partial A/\partial \overline{P})_T$.*

10. *Calculate $A(T, \overline{P})$ to the first order in b. Show that A is of the form*

$$A(T, \overline{P}) = A_0(T) + \lambda \overline{P}^2(T - T_0) + \mu(\overline{P}^2)^2.$$

It is not worthwhile giving an expression for $A_0(T)$. Express λ and μ as a function of a and b.

11. The polarisation is said to be *spontaneous* (and the crystal *ferroelectric*) if $\overline{P}$ may be non-zero in the presence of an external field $\overline{E}_0$ equal to zero. The expression for the susceptibility obtained in question **7** is not valid when the polarisation is spontaneous. *Calculate this spontaneous polarisation (which may be expressed in terms of λ and μ) and calculate $\chi(T)$ in the various possible temperature ranges.*

12. *Taking into account the numerical values obtained in question 8, can the approximations made be considered legitimate?*

SOLUTION TO PROBLEM 8

1. The partition function for a classical gas of indistinguishable particles is given by

$$Z_0 = \frac{1}{N!}\left[\int \frac{V\,d^3p}{(2\pi h)^3}e^{-(p^2/2mkT)}\right]^N$$

or

$$Z_0 = \frac{1}{N!}\left[-\frac{V(2\pi mkT)^{3/2}}{(2\pi h)^3}\right]^N.$$

2. Since each classical ion occupies a potential cage defined by a cell of the crystalline lattice, there is no $N!$. In fact, the factor $(N!)^{-1}$ due to the indistinguishability is compensated by a factor $N!$ due to the permutation of the different atoms among the different potential cages. The partition function is then given by

$$Z_1 = \left[\int \frac{d^3p}{(2\pi h)^3}e^{-\beta p^2/2m} \times \int d^3x e^{-\beta\phi(x)}\right]^N \quad \left(\text{where } \beta = \frac{1}{kT}\right).$$

Now we have

$$\int d^3 p e^{-\beta p^2/2m} = (2\pi m k T)^{3/2}$$

and

$$\int d^3 x e^{-\beta\phi} = \int d^3 x e^{-\beta a(x^2+y^2+z^2)} = \left(\frac{\pi}{\beta a}\right)^{3/2}.$$

Thus,

$$Z_1 = \left[\left(\frac{2\pi^2 m (kT)^2}{a}\right)^{3/2} \frac{1}{(2\pi h)^3}\right]^N.$$

N.B. The presence or absence of a factor $N!$ is of no further consequence in what follows.

3. The potential energy of an ion is

$$\phi_0 - q E z$$

whence

$$Z_2 = \left[\frac{(2\pi m k T)^{3/2}}{(2\pi h)^3}\int d^3 x e^{-\beta(\phi-q E z)}\right]^N$$

$$\int d^3 x e^{-\beta(\phi-Ez)} = \left[\int dx e^{-\beta a x^2}\right]^2 \int dz e^{-\beta(a z^2 - q E z)}.$$

But, setting $z' = z - q E/2a$, we have

$$\int dz\, e^{-\beta(a z^2 - Ez)} = \int dz'\, e^{-\beta a z'^2} e^{\beta E^2 q^2/4a}.$$

Whence

$$Z_2 = Z_1\left(e^{q^2 \beta E^2/4a}\right)^N.$$

4. The dipole moment of an ion is

$$\bar{p} = q\bar{x}$$

whence

$$\bar{P} = \int \prod_{i=1}^{N}\frac{d^3 p_i\, d^3 x_i}{(2\pi h)^s} e^{-\sum_i \beta(p_i^2/2m + \phi(x_i) - q\bar{E}\cdot\bar{x}_i)} \times \left(\sum_{i=1}^{N} q\bar{x}_i\right) \Bigg/ Z_2$$

or

$$\overline{P} = \frac{\partial Z_2}{\partial E} \frac{1}{\beta Z_2} = \frac{1}{\beta} \frac{\partial}{\partial E} \ln Z_2.$$

Since $\partial Z_1/\partial E$ is zero, this gives

$$P = \frac{N}{\beta} \left(\frac{q^2 \beta E}{2a} \right) = \frac{Nq^2 E}{2a}.$$

P does not depend on T.

5. We have

$$Z_3 = \left[\frac{2\pi mkT}{(2\pi h)^2} \right]^{3N/2} \left[\int d^3 x \, e^{-\beta(\phi - qEz)} \right]^N.$$

The integral may be written as

$$\int d^3 x \exp\left\{ -\beta \left[a(x^2 + y^2 + z^2) + b(x^4 + y^4 + x^4) - qEz \right] \right\}$$

$$I_x = \int dx \, e^{-\beta(ax^2 + bx^4)} \simeq \int dx \, e^{-\beta ax^2}(1 - \beta b x^4)$$

$$= \sqrt{\frac{\pi}{\beta a}} \left(1 - \frac{3\beta b}{(2\beta a)^2} \right)$$

$$I_z = \int dz \, e^{-\beta(az^2 + bz^4 - qEz)} = \int dz' \, e^{\beta q^2 E^2/4a} e^{-\beta az'^2} \left[1 - \beta b \left(z' + \frac{qE}{2a} \right)^4 \right]$$

$$I_z = e^{\beta q^2 E^2/4a} \sqrt{\frac{\pi}{\beta a}} \left[1 - \frac{3\beta b}{(2\beta a)^2} - \frac{6\beta b}{2\beta a} \left(\frac{qE}{2a} \right)^2 - \beta b \left(\frac{qE}{2a} \right)^4 \right]$$

whence

$$Z_3 = Z_1 \left(e^{\beta q^2 E^2/4a} \right)^N \left[1 - \frac{9b}{4\beta a^2} - \frac{3b}{a} \left(\frac{qE}{2a} \right)^2 - \beta b \left(\frac{qE}{2a} \right)^4 \right]^N$$

$$= Z_1 \left(e^{\beta q^2 E^2/4a} \right)^N H^N.$$

6. Recalling that

$$P = \frac{1}{\beta} \frac{\partial \ln Z_3}{\partial E}$$

it follows that

$$P = \frac{N}{\beta} \frac{\partial}{\partial E} [\ln H + \beta q^2 E^2/4a] \qquad \frac{\partial}{\partial E} \ln H \simeq -\frac{3bq^2 E}{2a^3} - \frac{\beta b q^4 E^3}{4a^4}$$

whence

$$P = N \left(\frac{q^2 E}{2a} - \frac{3bq^2 E}{2a^3 \beta} - \frac{bq^4 E^3}{4a^4} \right) = \rho E - \sigma E^2.$$

We surmise that P depends on T via the second term of ρ.

7. We have

$$P = \chi \varepsilon_0 E_0 = \rho E - \sigma E^3$$

and

$$E = E_0 + P/3\varepsilon_0$$

whence it follows immediately that

$$E = \frac{P + \sigma E^3}{\rho} \simeq \frac{P}{\rho} + \sigma \frac{P^3}{\rho^4}$$

and so

$$P \left(\frac{1}{\rho} + \frac{\sigma P^2}{\rho^4} - \frac{1}{3\varepsilon_0} \right) = E_0 = \frac{P}{\chi \varepsilon_0}.$$

Neglecting the terms nonlinear in E or P which are still known to be very small and are, moreover, proportional to b here, we have

$$\chi = \frac{1}{\varepsilon_0} \left(\frac{1}{\rho} - \frac{1}{3\varepsilon_0} \right)^{-1} = \frac{3\rho/\varepsilon_0}{3 - \rho/\varepsilon_0}.$$

Now

$$3 - \rho/\varepsilon_0 = 3 - \frac{Nq^2}{2a\varepsilon_0} \left(1 - \frac{3bkT}{a^2} \right)$$

is of the form

$$\frac{3Nq^2 bk}{2a^3 \varepsilon_0} (T - T_0)$$

where

$$kT_0 = \frac{2a^3}{3Nq^2 b} \left(3\varepsilon_0 - \frac{Nq^2}{2a} \right).$$

In the numerator, we find

$$\frac{3\rho}{\varepsilon_0} = \frac{3Nq^2 bkT_1}{2a^3 \varepsilon_0}$$

where we neglect the terms involving b in ρ, to obtain

$$kT_1 = \frac{2a^3}{Nq^2 b} N \frac{q^2}{2a} = \frac{a^2}{b}$$

and

$$\chi = \frac{T}{T - T_0}.$$

8. Since ϕ is an energy (dimension M L^2 T^{-2}), the dimensions of a and b are, respectively MT^{-2} and ML^{-2}: a^2/b is an energy.

We have seen that

$$kT_1 = \frac{a^2}{b} \qquad kT_0 = \frac{2\varepsilon_0}{Nq^2} \frac{a^3}{b} - \frac{a^2}{3b}$$

whence

$$\frac{Nq^2}{2\varepsilon_0} \left(kT_0 + \frac{kT_1}{3} \right) = \frac{a^3}{b}.$$

It follows that

$$a = \frac{Nq^2}{2\varepsilon_0} \left(\frac{T_0}{T_1} + \frac{1}{3} \right).$$

Now

$$\frac{Nq^2}{2\varepsilon_0} = \frac{4\pi (4 \times 1.602 \times 10^{-19})^2}{2(4 \times 10^{-10})^3} 10^{-7} (3 \times 10^8)^2$$

$$= 22.6 \text{ eV Å}^{-2}.$$

Let

$$a = 22.6 \left(\frac{391}{150\,000} + \frac{1}{3} \right) = 7.59 \text{ eV Å}^{-2}$$

and

$$b = \frac{a^2}{kT_1} = \frac{(7.59)^2 \times 11\,605}{150\,000} = 4.46 \text{ eV Å}^{-4}.$$

9. Considering E as a function of P given by

$$E = E_0 + P/3\varepsilon_0$$

and E_0 as a parameter, we have

$$dA = dF + E_0 dP + P dP/3\varepsilon_0.$$

Now, we have

$$dF = -P dE - S dT$$

whence

$$dA = -\frac{P\,dP}{3\varepsilon_0} - S\,dT + E_0\,dP + \frac{P\,dP}{3\varepsilon_0}$$
$$= -S\,dT + E_0\,dP$$

whence

$$E_0 = \left(\frac{\partial A}{\partial P}\right)_T.$$

10. We have seen that

$$-\beta F = \ln Z_1 + \frac{N\beta q^2 E^2}{4a} + N\ln H.$$

Since $\ln Z_1$ does not depend on E, we do not retain it in F, and, up to this constant independent of E, we have

$$F = -\frac{Nq^2 E^2}{4a} - NkT\ln H$$

$$\ln H \simeq -\frac{9b}{4\beta a^2} - \frac{3b}{a}\left(\frac{qE}{2a}\right)^2 - \beta b\left(\frac{qE}{2a}\right)^4.$$

Therefore, still up to a constant, we have

$$F = -\frac{Nq^2 E^2}{4a} + \frac{3Nb}{\beta a}\left(\frac{qE}{2a}\right)^2 + Nb\left(\frac{qE}{2a}\right)^4 = -\frac{1}{2}\rho E^2 + \frac{1}{4}\sigma E^4$$

and

$$P = \rho E - \sigma E^3$$

whence

$$A = F + P\left(E - \frac{P}{3\varepsilon_0}\right) + \frac{P^2}{6\varepsilon_0} = F + PE - \frac{P^2}{6\varepsilon_0}.$$

Using (see question **7**)

$$E = \frac{P}{\rho}\left(1 + \sigma\frac{P^2}{\rho^3}\right)$$

we find that

$$A = \frac{1}{2}P^2\left(\frac{1}{\rho} - \frac{1}{3\varepsilon_0}\right) + \frac{1}{4}\sigma\frac{P^4}{\rho^4}.$$

But we know that

$$\frac{1}{\rho} - \frac{1}{3\varepsilon_0} = \frac{1}{\chi\varepsilon_0} = \frac{T - T_0}{T_1\varepsilon_0}$$

and, neglecting the terms of degree two in b

$$\frac{\sigma}{\rho^4} = \frac{Nbq^4}{4a^4}\left(\frac{2a}{Nq^2}\right)^4 = \frac{4b}{N^3q^4}.$$

Thus, $A = \lambda P^2(T - T_0) + \mu P^4$, where

$$\lambda = \frac{1}{2}\frac{1}{T_1\varepsilon_0} = \frac{kb}{4a^2\varepsilon_0} \qquad \mu = \frac{b}{N^3q^4}.$$

11. For $E_0 = 0$, P is given by $\partial A/\partial P = 0$, whence

$$2\lambda P(T - T_0) + 4\mu P^3 = 0$$

which always has the solution $P = 0$.

For $T < T_0$, two other solutions exist

$$P = \pm\left(\frac{\lambda(T_0 - T)}{2\mu}\right)^{1/2} = \pm P_0.$$

To see which solution arises, we must determine which corresponds to a minimum of F or A. Now, we have

$$\frac{\partial^2 A}{\partial P^2} = 2\lambda(T - T_0) + 4\mu P^2.$$

We see that, for $T < T_0$, the solution $P = 0$ gives $(\partial^2 A/\partial P^2) < 0$, which is thermodynamically impossible. For $P^2 = P_0^2$, we surmise that $(\partial^2 A/\partial P^2) > 0$. Thus, *spontaneous polarisation occurs*.

For $T > T_0$ there is no spontaneous polarisation. The susceptibility is then given by

$$\chi = \frac{T_1}{T - T_0}.$$

For $T < T_0$, the polarisation is given by

$$E_0 = \frac{\partial A}{\partial P} = 2\lambda P(T - T_0) + 4\mu P^3$$

while the spontaneous polarisation is given by

$$0 = 2\lambda P_0(T - T_0) + 4\mu P_0^3.$$

We set $P = P_0 + \Delta P$ and retain only the terms of first order in ΔP. It follows that

$$E_0 = 2\lambda(T - T_0)(P_0 + \Delta P) + 4\mu(P_0^3 + 3P_0^2\Delta P)$$

whence

$$E_0 = (2\lambda(T - T_0) + 12\mu P_0^2)\Delta P.$$

Replacing $4\mu P_0^2$ by $-2\lambda(T - T_0)$, we have

$$E_0 = -4\lambda(T - T_0)\Delta P$$

whence

$$\Delta P = \frac{1}{4\lambda(T_0 - T)}E_0 = \frac{1}{2}\frac{T}{T_0 - T}\varepsilon_0 E_0,$$

i.e.

$$\chi = \frac{1}{2}\frac{T_1}{T_0 - T} \quad \text{for} \quad T < T_0.$$

12.   We note that for $kT \sim kT_0 \sim k \times (400 \text{ K}) \sim 3.4 \times 10^{-2}$ eV we have

$$\beta a = 223 \text{ Å}^{-2}.$$

Thus, the titanium ion is displaced by a distance of the order of

$$l = \frac{1}{\sqrt{223}} \simeq 0.07 \text{ Å}$$

and we have

$$bl^4 \simeq 9 \times 10^{-5} \ll al^2 \sim 8 \times 10^{-2}$$

which justifies the approximations made.

ORDER AND DISORDER IN ALLOYS

Maurice Bernard

We shall study the transition between order and disorder for cubic alloys of type AB (for example CuZn).

For this, we consider a set N of points in space forming the nodes of a cubic lattice; this lattice is denoted by (a). We consider another set of N points in space forming the nodes of another cubic lattice (b). The two lattices are arranged in such a way that each node of the lattice (b) is the centre of a cube of the lattice (a), and conversely (Figure 1). We observe that each node of (a) is surrounded by eight nearest-neighbour nodes belonging to (b), and conversely. We also assume that the number N is sufficiently large that the side effects due to the limits of the lattices are negligible. We suppose that the A and B atoms are arranged, in equal numbers, at the nodes of the lattices (a) and (b), so as to form a solid alloy AB, whose constituents are in equal atomic proportions.

We could suppose that the alloy is perfectly *ordered*, that is, the A atoms occupy the nodes of one lattice, (a) for example, while the B atoms evidently occupy the nodes of the lattice (b). We could also suppose that the A and B atoms occupy the nodes of the two lattices at random, whence the alloy is perfectly *disordered*. Between these two extremes, there exist intermediate, partially ordered situations.

We shall study the order of this alloy as a function of the temperature; we shall assume that the alloy is incompressible, whence the pressure plays no role.

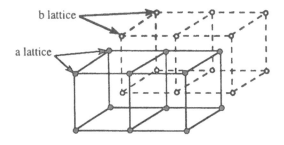

Figure 1 Lattices (a) and (b) considered here. Note that this is the CsCl structure (Figure 4, p. 7).

We define a given configuration of the alloy, that is a microstate of the system in question, by placing n A atoms on n well-defined nodes of the lattice (a) and $(N - n)$ A atoms on $(N - n)$ well-defined nodes of the lattice (b); the sites to be occupied by the B atoms are then completely determined. We assume that the range of the interatomic forces is sufficiently small that the energy state of each configuration depends only on the binding energies of the nearest neighbours, which will therefore be the only ones taken into consideration. The A atoms are indistinguishable among themselves as are the B atoms.

We consider the set $\{\lambda\}$ of configurations corresponding to a given value of n $(0 \leq n \leq N)$; we denote the statistical distribution such that all the configurations of the set $\{\lambda\}$ are equiprobable by $\mathcal{D}_\lambda$. We define the 'order parameter' λ by the equation

$$n = \frac{N}{2}(1 + \lambda)$$

where, *a priori*, λ lies between -1 and $+1$. In fact, changing λ to $-\lambda$ reduces to transforming the lattice (a) into the lattice (b) and *vice versa*; thus, we shall only consider values $\lambda \geq 0$. Let n_{Aa} denote the number of A atoms situated on the lattice (a), n_{Ab} the number of A atoms situated on the lattice (b), n_{Ba} the number of B atoms situated on the lattice (a) and n_{Bb} the number of B atoms situated on the lattice (b). We see immediately that

$$n_{Aa} = n_{Bb} = n = \frac{N}{2}(1 + \lambda)$$

$$n_{Ab} = n_{Ba} = N - n = \frac{N}{2}(1 - \lambda).$$

For a given configuration, we denote the number of bonds between nearest neighbours of type AA, BB and AB by N_{AA}, N_{BB} and N_{AB}, respectively.

1. Suppose we are given a value of λ between 0 and 1. *Show that the numbers N_{AA}, N_{BB} and N_{AB} are not defined by the specification of λ alone. On the other hand, the mean values $\overline{N}_{AA}$, $\overline{N}_{BB}$ and $\overline{N}_{AB}$ of the numbers N_{AA}, N_{BB} and N_{AB}, respectively, calculated on the macrostate $\mathcal{D}_\lambda$ depend only on λ. Calculate these mean values and show that*

$$\overline{N}_{AA} = \overline{N}_{BB} = 2N(1 - \lambda^2) \qquad \overline{N}_{AB} = 4N(1 + \lambda^2).$$

In preparation for the last question, *give simple examples of microstates which are very different from the mean configuration, in the particular case $\lambda = 0$.*

2. We denote the binding energy between two nearest-neighbour A atoms by W_{AA}; like on p. 186, this is the energy which would have to be supplied to break the bond and separate these two atoms, which are assumed to be linked by this single bond. Similarly, we denote the B–B and A–B binding energies by W_{BB} and W_{AB}, respectively. We suppose that these energies are known and that they are independent of the temperature.

A given value of λ corresponds to a large number of microstates of the systems, with energies $E_i(\lambda)$. *Write down the canonical partition function for the system in the form*

$$Z = \sum_\lambda z(\lambda).$$

What is the form of $z(\lambda)$? Is this function calculable?

Since it is impossible to use the Boltzmann–Gibbs distribution directly, we seek to describe the system by the sampling distribution $\mathcal{D}_\lambda$. We shall ultimately determine the parameter λ to simulate the thermal equilibrium distribution as well as possible. *Calculate the mean energy $U(\lambda)$ associated with the sampling distribution $\mathcal{D}_\lambda$ as a function of W_{AA}, W_{BB}, W_{AB}, N and λ. Show that $U(\lambda)$ is of the form:*

$$U(\lambda) = U_0 - \lambda^2 U_1$$

and give expressions for U_0 and U_1.

3. *Calculate the entropy $S(\lambda)$ associated with the sampling distribution $\mathcal{D}_\lambda$. Sketch the shape of the curve $S(\lambda)$.*

4. *Give a variational criterion for selecting that distribution of $\mathcal{D}_\lambda$ corresponding to the best possible approximation to the canonical partition function Z.*

5. *For each temperature T, deduce a value $\lambda_0(T)$ which gives an approximate description of the state of the alloy as a function of the temperature. Distinguish between the cases $U_1 > 0$ and $U_1 < 0$. What can be said about the order of the alloy as a function of the temperature? What happens at the temperature $T_c = U_1/Nk$?*

6. *Assuming the same approximation, for $U_1 > 0$, sketch the shape of the graphs of the entropy and the heat capacity against the temperature.*

What simple physical measurement would enable us to obtain the experimental value of T_c?

7. We now suppose that U_1 is negative and we examine what happens at temperature zero. *Show that there exist stable configurations which escape the previous analysis. Comment on this fact.*

SOLUTION TO PROBLEM 9

1. To take a value of λ is to say that, of the N A atoms, n occupy nodes of (a) and $(N - n)$ occupy nodes of (b). For a given value of λ, there is a very large number of distinct configurations, that is, ways of arranging n A atoms on n of the N sites of A and $(N - n)$ A atoms on $(N - n)$ of the N sites of (b); each of these microstates is associated with specific numbers of A–A, B–B and A–B bonds, N_{AA}, N_{BB} and N_{AB}, respectively.

The probability that a site of the lattice (a) is occupied by an A atom is equal to $n/N = n_{Aa}/N$; this site is linked to eight surrounding sites of the lattice (b) and in the approximation of the model, these are the only sites to which it is linked. The probability that one of the eight sites is occupied by an A atom belonging to the lattice (b) is n_{Ab}/N and these two events are statistically independent in the macrostate in question, since the sites of the two lattices (a) and (b) are occupied independently. The probability that the bond considered is of type AA is therefore $n_{Aa}n_{Ab}/N^2$. The total number of bonds is $8N$. Thus,

$$\overline{N}_{AA} = 8N n_{Aa}n_{Ab}/N^2 = 2N(1 - \lambda^2).$$

Similarly, we have

$$\overline{N}_{BB} = 8N n_{Bb}n_{Ba}/N^2 = 2N(1 - \lambda^2).$$
$$\overline{N}_{AB} = 8N(n_{Aa}n_{Bb} + n_{Ab}n_{Ba})/N^2 = 4N(1 + \lambda^2).$$

It is easy to check that $\overline{N}_{AA} + \overline{N}_{BB} + \overline{N}_{AB} = 8N$.

We may examine a particular case, for example $\lambda = 0$. We have

$$n_{Ab} = n_{Bb} = n_{Ab} = n_{Ba} = \frac{N}{2}.$$

A microstate may correspond to the configuration in which all the A atoms occupy all the sites of (a) and (b) in the left half of the crystal and all the B atoms occupy all the sites of (a) and (b) in the right half (see Figure 2). Then (neglecting the central interface), we have the approximation:

$$N_{AA} = 4N = N_{BB} \qquad N_{AB} = 0.$$

For the same value $\lambda = 0$ another microstate corresponds to a configuration in which the A atoms occupy sites of (a) and (b) uniformly distributed in the crystal volume (see Figure 3); then we have:

$$N_{AA} = 2N = N_{BB} \qquad N_{AB} = 4N.$$

It is apparent that we are dealing with completely different physical situations in these two cases.

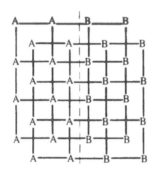

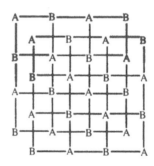

Figure 2   (See text).   **Figure 3**   (See text).

2.   The energy $E_i(\lambda)$ of an arbitrary microstate of the set $\{\lambda\}$ is given by

$$E_i(\lambda) = N_{AA}W_{AA} + N_{BB}W_{BB} + N_{AB}W_{AB}$$

where N_{AA}, N_{BB} and N_{AB} are the effective numbers of bonds of the form A–A, B–B and A–B, respectively for this microstate. These numbers are not *a priori* calculable since they depend on the actual microscopic arrangement of n A atoms on the lattice (a) and $(N - n)$ A atoms on the lattice (b), where n and $(N - n)$ have order of magnitude 10^{23}! We may view N_{AA}, N_{BB} and N_{AB} as random numbers whose mean values, calculated in the previous question, are known.

The partition function for the system is given by

$$Z = \sum_{i,\lambda} e^{-\beta E_i(\lambda)} \qquad (\beta = 1/kT)$$

where the sum runs over all the microscopic states; isolating the terms corresponding to a single value of λ, we may write

$$Z = \sum_{\lambda} \sum_{i} e^{-\beta E_i(\lambda)} = \sum_{\lambda} z(\lambda)$$

with

$$z(\lambda) = \sum_{i} e^{-\beta E_i(\lambda)}$$

where the sum runs over all microscopic states of the set $\{\lambda\}$. The function $z(\lambda)$ is no more calculable than $E_i(\lambda)$.

The average energy $U(\lambda)$ is given by

$$U(\lambda) = \overline{N}_{AA} W_{AA} + \overline{N}_{BB} W_{BB} + \overline{N}_{AB} W_{AB}$$

where $\overline{N}_{AA}$, $\overline{N}_{BB}$ and $\overline{N}_{AB}$ are the means calculated in the first question. Introducing these values, we have:

$$U(\lambda) = 2N(W_{AA} + W_{BB} + 2W_{AB}) - \lambda^2 2N(W_{AA} + W_{BB} - 2W_{AB})$$
$$= U_0 - \lambda^2 U_1$$

where

$$U_0 = 2N(W_{AA} + W_{BB} + 2W_{AB}) \qquad U_1 = 2N(W_{AA} + W_{BB} - 2W_{AB}).$$

3. The entropy $S(\lambda)$ is equal to $k \ln W(\lambda)$ where $W(\lambda)$ is the number of microstates in $\{\lambda\}$, i.e. the number of microstates corresponding to a given value of λ.

Thus, we have:

$$W(\lambda) = C_N^{n_{Aa}} C_N^{n_{Ab}} = \left(\frac{N!}{n_{Aa}! n_{Ab}!} \right)^2.$$

Whence

$$S(\lambda) = k \ln \frac{(N!)^2}{\left[((N/2)(1 + \lambda))! \, ((N/2)(1 - \lambda))! \right]^2}$$
$$\simeq - kN \left[(1 + \lambda) \ln \frac{1 + \lambda}{2} + (1 - \lambda) \ln \frac{1 - \lambda}{2} \right].$$

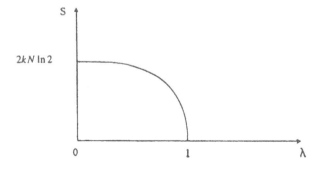

Figure 4   Variation of entropy versus order parameter.

4. We recall (see Stat. Ph.) that, in the special case in which the mean energy is given (canonical set), the function λ defined by

$$S(\lambda) - k\beta U(\lambda)$$

is less than $k \ln Z$ for all λ. Thus, the best possible choice of λ is that which maximises this function.

5. The best approximation corresponds to λ_0 which maximises $S(\lambda) - k\beta U(\lambda)$, i.e.

$$\frac{\partial}{\partial \lambda}[U(\lambda) - T S(\lambda)] = 0.$$

After simple calculations, we obtain:

$$-2\lambda_0 U_1 + NkT \ln \frac{1 + \lambda_0}{1 - \lambda_0} = 0 \tag{1}$$

which defines λ_0 as an implicit function of N, T and U_1.

(i)   If U_1 is negative ($W_{AA} + W_{BB} < 2W_{AB}$), the two terms of the first member of (1) are positive for all λ_0. Thus, there is no maximum other than $\lambda_0 = 0$ and the alloy remains disordered all all temperatures; the A–B binding energy is sufficiently large in comparison with the A–A and B–B binding energies that disorder is preferred at all temperatures.

(ii) If U_1 is positive ($W_{AA} + W_{BB} > 2W_{AB}$), we set $T_c = U_1/Nk$ and the equation becomes

$$\lambda_0 = \tanh \frac{\lambda_0 T_c}{T}. \tag{2}$$

For $T > T_c$, the sampling function is maximal for $\lambda_0 = 0$. For $T < T_c$, $\lambda = 0$ is a minimum and the maximum we seek is achieved for a non-zero value $\lambda_0(T)$ satisfying equation (2). Thus, the alloy is totally disordered when its temperature is greater than the 'transition temperature' T_c; it becomes partially ordered as soon as it is cooled below T_c and totally ordered at $T = 0$. Graphical solution of equation (2) shows that $\lambda_0(T)$ is a decreasing function of T (see Figure 5).

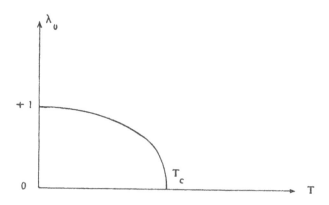

Figure 5   Variation of the order parameter versus temperature.

When T approaches T_c from below, λ_0 is small, and expanding the expression (1) in a series we obtain:

$$\lambda_0 \simeq \sqrt{3\left(\frac{T_c}{T} - 1\right)}.$$

For T small, λ_0 is close to 1, and setting $\lambda_0 = 1 - \varepsilon$, the series expansion of (1) gives

$$\lambda_0 \simeq 1 - 2e^{-2T_c/T}.$$

6. Retaining the assumptions of the previous question and supposing that, at all temperatures T, λ takes the value $\lambda_0(T)$ given above, we have:

$$S(\lambda_0) = -Nk\left[(1 + \lambda_0)\ln\frac{(1 + \lambda_0)}{2} + (1 - \lambda_0)\ln\frac{(1 - \lambda_0)}{2}\right].$$

It follows that

$$\frac{\mathrm{d}S}{\mathrm{d}T} = \frac{\partial S}{\partial \lambda_0}\frac{\mathrm{d}\lambda_0}{\mathrm{d}T} = -Nk\ln\frac{(1 + \lambda_0)}{(1 - \lambda_0)}\frac{\mathrm{d}\lambda_0}{\mathrm{d}T}.$$

The function $\partial S/\partial \lambda_0$ remains constantly negative as λ_0 decreases from 1 to 0; thus, $S(\lambda_0)$ is a decreasing function of λ_0, whence, as one might expect, it is an increasing function of temperature (Figure 6).

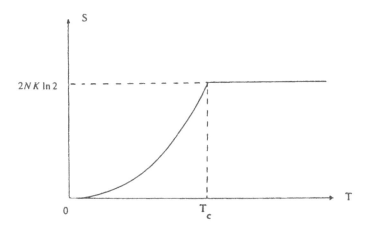

Figure 6 Entropy versus temperature.

The specific heat is given by:

$$C = \frac{\partial U}{\partial T} = \frac{\partial U}{\partial \lambda_0}\frac{\mathrm{d}\lambda_0}{\mathrm{d}t} - 2U_1\lambda_0\frac{\mathrm{d}\lambda_0}{\mathrm{d}T}.$$

This quantity is positive, since $\lambda_0(T)$ being a decreasing function of T, $d\lambda_0/dt$ is negative. For T near zero, we have:

$$C \simeq 8Nk\left(\frac{T_c}{T}\right)^2 e^{-2T_c/T}.$$

For T slightly less than T_c, we have

$$C \simeq 3Nk\left(1 - \frac{8}{5}\frac{T_c - T}{T_c}\right).$$

For $T = T_c$, we have

$$C(T_c) \simeq 3Nk.$$

For T greater than T_c, we clearly have $C = 0$.

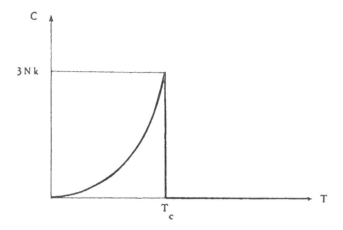

Figure 7   Order-disorder specific heat versus temperature.

Measurement of the specific heat of an alloy as a function of the temperature (which also involves contributions from other degrees of freedom, such as vibration of the atoms, which are continuous around T_c), including, in particular, investigation of the peaks of the curve $C(T)$, constitutes a classical method for determining the transition temperatures (see Figure 7).

Let us also mention X-ray (or neutron) diffraction which exhibits new spots when order occurs, due to the fact that the Bravais lattice becomes larger. Finally, all physical properties are more or less sensitive to the appearance of order. For example, the electrical conductivity is, most generally, significantly increased when one passes from disorder to order.

7.   At zero temperature, the most stable configuration of those described by the sampling distributions $\mathcal{D}_\lambda$ is that which corresponds to a minimal value of $U(\lambda)$. When U_1 is negative, λ_0 is zero for all temperatures, whence we have

$$U_{\min} = U_0.$$

In fact, the configuration described in the first question, in which the A atoms occupy half of the crystal and the B atoms occupy the other half, corresponds to an energy $U_{A/B}$ less than U_0; indeed, we find that

$$U_{A/B} = 4N(W_{AA} + W_{BB}) = U_0 + U_1 < U_0.$$

Thus, at low temperature, this configuration has a far greater probability than most other configurations $\{\lambda\}$ for $\lambda = 0$ and the sampling distributions $\mathcal{D}_\lambda$ were thus not suitable for describing the equilibrium at low temperature for $U_1 < 0$. We conclude that when U_1 is negative, the alloy at low temperature will tend to exhibit a phase segregation. A new form of order may appear in this way. In fact, the evolution to the stable state in which the A and B atoms are separated in space may actually be too slow to be observed effectively.

N.B. From this point of view, an irradiation at a temperature $< T_c$ may help (by creation of point defects which enhance the diffusion) to reach a state of order which could not otherwise be obtained. This is the case of the *irradiation-induced ordering* of Fe Ni alloys (**J Paulevé, D Dautreppe, J Laugier** and **L Néel**, *Comptes Rendus* **254**, 965 (1962)).

CRYSTALLINE SURFACES AND STEPS

Roger Balian

We study here a number of geometric characteristics of the faces of a crystal in equilibrium with its vapour. For this, we model the crystal as a stacking of N cubes with edge a, arranged in a simple lattice. Each cube represents an atom of mass m and spin 0 whose internal structure we neglect. The centres of the cubes may occupy the *sites* with coordinates $x = (q + \frac{1}{2})a$, $y = (q' + \frac{1}{2})a$, $z = (q'' + \frac{1}{2})a$, where q, q' and $q'' \in \mathbb{Z}$. The energy of each atom is taken to be 0 when it is isolated and at rest. In particular, in the vapour, the energy reduces to the sum of the kinetic energies of the atoms $p^2/2m$. In the crystalline phase, we neglect the movement of the atoms, but we take their mutual interaction into account. At very short range, the atoms repel each other and in the model, this repulsion is very simply represented by the impenetrability of the atomic cubes. At a greater range, they attract each other, which gives rise to the cohesion of the crystal. In the present model, this attraction is represented by a potential energy $-V$ associated with *each pair of neighbouring* cubes attached by a face. To evaluate the energy of a configuration, characterised by the presence or absence of a cube at each site, it will be convenient to attribute the energy $-V$ to *each face* separating two occupied sites (not forgetting that such a face is common to two cubes). The interaction between two cubes is neglected when they only touch on an edge or at a vertex and when they do not touch at all.

For *numerical applications*, we shall take an atomic mass 100 g mol^{-1}, a cell size $a = 2.5$ Å and an attractive energy $V = 0.2$ eV. Only the orders of magnitude of the results are important.

Although the questions follow consecutively, there is no need to solve each one before progressing. In particular, in **I** and **II**, the only indispensable questions for what follows are questions **2** and **7**; **III** is used in **IV**, but **V** and **VI** may be tackled without having found the solutions to **III** and **IV**. The questions within each part are also to some extent independent.

I

Volume-related properties of the solid

1. *Show that the cohesion energy ρ per unit volume defined as the ratio of the energy of the ground state of a crystal to its volume Ω in the limit $\Omega \to \infty$ is equal to $-3Va^{-3}$.*

2. *Deduce the chemical potential μ_0 of the crystal at temperature zero (and pressure zero).*

3. At non-zero temperature, the crystal (still considered to be infinite) at equilibrium contains vacancies. We shall assume that there are so few vacancies that they practically never touch each other. *What energy must be supplied to create a vacancy by suppressing the corresponding cube?*

We consider a fixed volume $\Omega = \mathcal{N}a^3$, which we imagine to be cut out in an infinite crystal (to avoid side effects) and which is assumed to contain a large number of sites $\mathcal{N}$, some of which may be vacant. Let l be the number of vacancies and $N = \mathcal{N} - l$ the number of atoms occupying the $\mathcal{N}$ sites. *Write down the energy as a function of l and N.*

4. *Assuming a low density of vacancies, use the above result to evaluate the overall partition function for the region Ω as a function of β $(= 1/kT)$ and of $\alpha = \beta\mu$, where μ is the chemical potential of the atoms. Deduce the pressure P_s of the solid as a function of T and μ.*

5. *Write down the probability that at equilibrium there exists a vacancy at a given site, at temperature T and pressure P_s. How does this vary with T and P_s?*

6. *Justify the assumption of question 4, giving a numerical upper limit for this probability at $T = 300$ K.*

In what follows, we shall systematically neglect the possibility that vacancies are created within the crystal.

II

Saturated vapour

The crystal is assumed to be surrounded by its vapour, with which it may exchange atoms and energy, thus leading to an equilibrium. The system is systematically controlled by acting on the pressure P and the temperature T of the vapour. The latter is considered to be a perfect monatomic gas.

7. *Give the expression for the chemical potential μ of the vapour as a function of P and T. Assuming that when the temperature and the pressure increase the chemical potential of the crystal remains fixed with value μ_0, and that the vapour is at equilibrium with the crystal, write down the expression for its pressure $P_0(T)$. The latter is the saturated vapour pressure.*

8. *Deduce the form of the diagram of the solid–vapour phases in the plane T, P. Calculate the saturated vapour pressure $P_0(300$ K$)$ at room temperature.*

9. *Try to justify the assumption made in question 7 that the chemical potential can be replaced by μ_0.*

III

Smooth face of a crystal

We consider a very large crystal, one of whose faces is in the plane $z = 0$. The solid is on the side $z < 0$ and the vapour on the side $z > 0$. At temperature zero the faces are perfectly smooth so that the energy is minimal. In what follows, as a reference for the crystal energy, we shall take this configuration in which all the sites $z < 0$ are occupied and all the sites $z > 0$ are empty. At temperature T, the exchanges of atoms and energy with the vapour may give rise to irregularities on the surface of the crystal. In this part, we shall consider only the two simplest types of surface defect, namely the *surface vacancy*, represented by a missing cube whose centre is a site in the plane $z = -a/2$ and the *adsorbed atom*, represented by an extra cube whose centre is a site in the plane $z = a/2$.

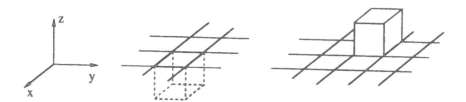

10. *Write down the respective energies of the two types of defect, related to the energy of the perfectly smooth face.*

11. The crystal is in grand canonical equilibrium with its saturated vapour at pressure $P_0(T)$. We let p_0, p_{-1} and p_1 denote, respectively, the probabilities that at a given point on the surface there is no defect, there is a vacancy or there is an adsorbed atom. *Write down p_{-1}/p_0 and p_1/p_0 and deduce that the probabilities p_{-1} and p_1 are equal in this case. We shall assume that the defects considered are sufficiently rare that they are almost never adjacent; check this assumption by calculating p_{-1} and p_1 at $T = 300$ K.*

12. At a fixed temperature T, the vapour is brought to a pressure P different from $P_0(T)$ (either higher or lower). The surface of the crystal then moves into equilibrium with its vapour, which process involves a variation in its chemical potential. *Write down the new probabilities p_{-1} and p_1 as a function of the saturation ratio $\lambda \equiv P/P_0(T)$ and give the numerical value of p_1 for $\lambda = 10$.*

13. *What can you conclude about the sublimation of the crystal (i.e. its evaporation) and the condensation of its vapour?*

IV

Growth of a step on a face

The adsorption of isolated atoms on a smooth surface is not sufficient to explain why a crystal in the presence of its supersaturated vapour must grow. The theory of crystal growth due to Burton, Cabrera and Frank (1951) is based on the prior existence of a *monatomic layer, bounded by a step* of height a on the face of the crystal. In our figures, we shall shade cubes reaching the height $z = a$, while the smooth face with reference height $z = 0$ will be shown in white.

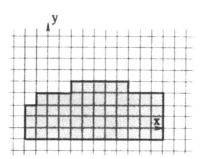

14. We shall consider a step consisting of two rectilinear parts ($x < 0$, $y = 0$ and $x > 0$, $y = a$), separated by a *kink* ($\Delta y = a$) at $x = 0$ (see figure). We denote sites on the side $z = a/2$ situated far from the step, on the edge of the step, and on the kink by A, B and C, respectively. The surface is in contact with the supersaturated vapour at temperature $T = 300$ K and pressure $P = \lambda P_0(T)$ where $\lambda = 10$. We neglect the formation of surface vacancies. We denote the probabilities of the three configurations containing an extra atom at A, B or C by p_A, p_B and p_C, respectively (for $\lambda \gg 1$, p_A is identical to the probability denoted by p_1 in question **12**). *Write down the ratios $p_A/(1 - p_A)$, $p_B/(1 - p_B)$ and $p_C/(1 - p_C)$ and deduce the numerical values of p_A, p_B and p_C.*

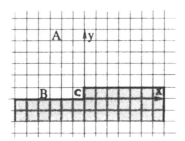

15. In order to determine whether or not the kinks are numerous along a step, we begin by estimating the energy of a kink. For this, we compare the two configurations shown in the next figure, which were constructed with the same number of atoms but differ as far as the kinks are concerned. The configuration (1) consists of two alternating smooth steps. In configuration (2) each step has acquired a kink at $x = 0$. *Calculate the difference in energy between these two configurations. Deduce the energy which must be attributed to a kink.*

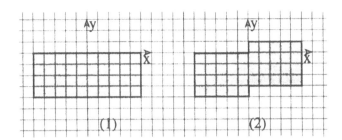

16. We consider a step parallel to Ox on a face in equilibrium with its saturated vapour $P_0(T)$. For simplicity, we assume that this step involves only 'simple' kinks ($\Delta y = +1$ or $\Delta y = -1$). We also assume that these kinks are few in number and that they may be handled using the methods of statistical physics as independent objects whose energies were obtained in question **15**. *Give a numerical estimate of the probability p_D that there is a kink (of one or other of the types $\Delta y = \pm 1$) at a given point of the step.*

17. When it collides with the face of the crystal, an atom emanating from the vapour has little chance of meeting the edge of a step. On the other hand, it is natural to suppose that the atoms are very mobile on the surface of the crystal when they move at a constant energy. *Using the values for p_A, p_B, p_C and p_D obtained above, and taking particular account of the high value of p_C, try to envisage an efficient mechanism for crystal growth at the expense of the supersaturated vapour (for a saturation ratio $\lambda = 10$).*

Try to interpret the photograph of a face of a crystal (of NaCl) in Figure 11(a), p. 214.

V

Shape of a step

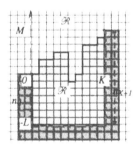

We shall study the average shape of a step on the face $z = 0$ in the general case where this step *is not parallel* to the axis Ox and the saturation ratio $\lambda = P/P_0(T)$ is *different from 1*. In the x, y plane, the step must run from the point $(0,0)$ to the point (Ka, Ma) (see figure above, left). To fix these extremities, we suppose that all the sites in the column $x = -a/2$, $y < 0$, $z = a/2$ and all the sites in the column $x = (K + \frac{1}{2})a$, $y < Ma$, $z = a/2$ are occupied, as are all the sites far behind the step $0 < x < Ka$, $y < -La$, $z = a/2$ (densely shaded areas). The fixed number L is chosen sufficiently large that the step practically never reaches an ordinate $y < -La$. We shall characterise the shape of the step by a function $y(x)$ defined on discrete values. Thus, we neglect all the configurations (see previous figure, right) which would include lateral projections of the step (1), backwards subsidences of this step (2) or rises in a forward direction (3). This is justified at low temperature and for a step with a gentle inclination ($M < K$), in which case such accidents are not very likely. Thus, each possible configuration is uniquely identified by a set of K integers n_k, $1 \leq k \leq K$, each of which represents the *number of cubes stacked in the column* $x = (k - \frac{1}{2})a$, from $y = -La$ to $y = (-L + n_k)a$. Equivalently, this configuration is also identified uniquely by the set $\{m_k\}$ of the $K + 1$ integers $m_k = n_{k+1} - n_k$, $(0 \leq k \leq K)$, $m_k = 0, \pm 1, \pm 2, \ldots$ (see figure below), each of which represents the *algebraic value* of the y-shift at the point of the step with abscissa $x = ka$ (at the extremities, we have $n_0 = L$ and $n_{K+1} = L + M$).

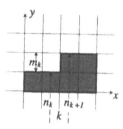

The region $\mathcal{R}$ whose properties we intend to study is that defined by $0 \leq x \leq Ka$, $y \geq -La$ and $z \leq a$ and bounded on the figure at the beginning of this section (left) by dash-dotted lines. The system $\mathcal{R}$ includes not only the atoms situated in this region, but also the faces in the planes $x = 0$, $x = Ka$, $y = -La$ and $z = 0$, namely the faces to which we attribute the energy $-V$ when they separate two atoms. Thus, the reference energy is that corresponding to the region $\mathcal{R}$ devoid of atoms.

18. *Show that, for a given configuration $\{m_k\}$, the number of atoms N situated in the region $\mathcal{R}$ is*

$$N = KL + \sum_{k=0}^{K}(K - k)m_k.$$

19. *Show that, for the configuration $\{m_k\}$, the energy associated with the set of faces situated in the region $\mathcal{R}$ is*

$$E = -3NV + \frac{V}{2}\sum_{k=0}^{K}|m_k| - V\left(L + \frac{M}{2}\right),$$

where $\sum_{k=0}^{K}|m_k|a$ is the total length of the y- shifts.

This formula for the energy makes it natural to use the $K + 1$ variables $\{m_k\}$ rather than the K independent variables $\{n_k\}$. However, the calculation of the overall partition function for the region $\mathcal{R}$ is complicated by the existence of the constraint

$$\sum_{k=0}^{K}m_k = M$$

expressing the fact that the ordinate of the arrival point of the step is Ma for $x = Ka$ (in principle, there are additional constraints arising from $n_k \leq 0$, but these do not come into play if L is sufficiently large, so that the m_k may be taken from $-\infty$ to $+\infty$). So that we can sum freely over each of the $K + 1$ numbers m_k ($0 \leq k \leq K$) we introduce a constraint on their sum, in the *mean statistical* sense only, in the form

$$\sum_{k=0}^{K} \langle m_k \rangle = \langle M \rangle$$

together with a Lagrange multiplier γ to take account of this constraint. Then the corresponding Boltzmann–Gibbs distribution gives a probability proportional to $e^{\alpha N - \beta E - \gamma M}$ for each configuration $\{m_k\}$.

20. *Show that the partition function $Z(\alpha, \beta, \gamma)$ associated with the set which we have just defined, is equal (up to a multiplicative factor) to the product of $K + 1$ functions $Z_k(\alpha, \beta, \gamma)$, each of which is associated with an abscissa $x = ka$ where a kink may take place. Give an expression for Z_k. You may use the identity ($\xi > |\eta|$)*

$$\sum_{m=-\infty}^{+\infty} e^{-\xi |m| - \eta m} = \frac{\sinh \xi}{\cosh \xi - \cosh \eta}.$$

21. *Deduce $\langle m_k \rangle$ (as a function of α, β and γ), the average slope in the (x,y) plane of the step at the point with abscissa $x = ka$.*

22. *Discuss the concavity of the step depending on the values of the saturation ratio λ.*

VI

Islands of condensation

On a face in equilibrium with a lightly supersaturated vapour ($\lambda \gtrsim 1$), we consider an island consisting of a condensed monatomic layer bounded by a step in the form of a closed curve (see figure, left). We shall study the geometry of these islands.

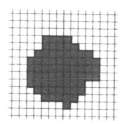
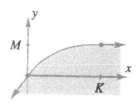

23. *Using the result of question 21, determine γ and K as a function of λ and T so that the step has an average slope of $45°$ at the point $x = y = 0$ and slope 0 at $x = K$, as shown in the figure (right).*

24. *Calculate $\langle M \rangle$ assuming that at the scale in question, the edge of the step behaves like a continuum so that the sums over k reduce to integrals.*

25. *Deduce the shape and average dimensions of an island for* $T = 300\ K$ *and* $\lambda = 1.02$. *How do these geometric characteristics vary with the overpressure* λ *and with the temperature?*

Comment briefly on the micrography of a face of an NaCl crystal, shown in Figure 5 of p. 160.

SOLUTION TO PROBLEM 10

1. The ground state is obtained for a compact stacking. In a large volume Ω, $\mathcal{N} = \Omega/a^3$ cubes will then be occupied with $3\mathcal{N}$ faces (six faces per cube, each face being common to two cubes), whence the (negative) cohesion energy per unit volume is $\rho = -3Va^{-3}$ (numerically, $\rho = -6\ kJ\ cm^{-3}$).

2. Because $T = P = 0$, the equation $dU = T\,dS - P\,d\Omega + \mu\,dN$ reduces to $dU = \mu\,dN$. We have $U = -3VN$, whence $\mu_0 = -3V$.

3. The energy $6V$ is necessary in order to suppress a cube (since six faces common to two adjacent atoms disappear from the overall expression for the energy). Thus, a configuration with l vacancies has energy $-3\mathcal{N}V + 6lV = -3NV + 3lV$.

4. The previous question leads us to attribute the energy $+3V$ to a vacant site and $-3V$ to an occupied site. Since each of the $\mathcal{N}$ sites can be treated independently of the others we obtain

$$Z_G(\alpha, \beta) = (e^{-3\beta V} + e^{3\beta V + \alpha})^{\mathcal{N}}.$$

The pressure is derived from the equation

$$P_s = \frac{kT}{\Omega} \ln Z_G = \frac{kT}{a^3} \ln\left[e^{-3V/kT} + e^{(3V+\mu)/kT}\right]. \tag{1}$$

5. According to the grand canonical Boltzmann–Gibbs distribution for a site, the probability that this site is vacant is $\exp(-3\beta V)/[\exp(-3\beta V) + \exp(3\beta V + \alpha)]$. Eliminating the chemical potential using (1), this probability of having a vacancy may be written as

$$\exp\left(-\frac{3V + P_s a^3}{kT}\right).$$

It increases with the temperature (increase in energy) and decreases with the pressure (decrease in volume).

6. At zero pressure, when the desired probability is maximal, it is equal to $\exp(-3V/kT) = 10^{-10}$ at 300 K. This value is negligible. It would still only be equal to 10^{-5} at $T = 600$ K.

7. This is given (see Stat. Ph.) by:

$$P = -\frac{A}{\Omega} = kTe^{\mu/kT}\left(\frac{2\pi mkT}{\hbar^2}\right)^{3/2} \tag{2}$$

whence

$$\mu = kT \ln\left[P(kT)^{-5/2}(2\pi\hbar^2/m)^{3/2}\right].$$

At equilibrium, the chemical potential of the vapour is equal to $\mu_0 = -3V$, consequently

$$P_0(T) = (kT)^{5/2} e^{-3V/kT} \left(\frac{m}{2\pi\hbar^2} \right)^{3/2}.$$

8. The expression for $P_0(T)$ leads to the diagram

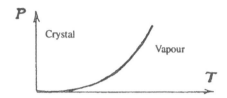

and to the value $P_0(300\ \text{K}) = 300$ Pa. The curve rises very quickly, since we already find 2×10^5 Pa $= 2$ atmospheres at 400 K.

9. The low concentration of vacancies implies first that the entropy per unit volume remains almost zero and second that the energy and the volume vary almost proportionally with the number of atoms. Thus, for the crystal we have

$$dU \simeq -3V\, dN = T\, dS + \mu\, dN - P\, d\Omega \simeq (\mu - Pa^3)\, dN$$

so that $\mu \simeq \mu_0 + Pa^3$. Microscopically, by inverting (1), we obtain the more precise formula

$$\mu = -3V + Pa^3 + kT \ln \left(1 - e^{-(3V+Pa^3)/kT} \right).$$

This expression should be inserted in (2), identifying both the chemical potentials and the pressures of the two phases, to obtain the saturation curve $P_0(T)$. But the last term (which even at 600 K is only equal to 5×10^{-7} eV) and the term Pa^3 (equal to 10^{-5} eV at atmospheric pressure) are negligible in comparison with $\mu_0 = -0.6$ eV along this curve, which is therefore obtained by replacing μ by μ_0 in (2).

10. As far as the smooth surface is concerned, the removal of a cube suppresses five faces which carried an attraction $-V$, so that the surface vacancy has the energy $5V$. Adding a cube adds a facial attraction, so that the atom adsorbed has the energy $-V$.

N.B. The total energy of the crystal does not only include the volume term calculated in question **1** and these contributions of surface defects. Let us use a parallelepipedal crystal with sides $L_x a$, $L_y a$ and $L_z a$ and perfectly smooth faces as a reference. In the direction $x = 0$, it contains $L_x - 1$ lattice planes each giving a contribution $-V L_y L_z$ to the cohesion energy. Noting that the volume is $\Omega = L_x L_y L_z a^3$ and that the area of the faces is $S = 2(L_y L_x + L_x L_z + L_x L_y)$, the reference energy is not equal to $\rho\Omega$, but to $\rho\Omega + \sigma S$ where $\sigma = V/2a^2 \simeq 0.25$ J m^{-2}. Since the surface contribution σS increases with the area of the faces, the latter should be minimal at equilibrium at zero temperature. This is why the surfaces are perfectly smooth then.

11. According to the Boltzmann–Gibbs distribution for grand canonical equilibrium, the ratio of the probabilities p and p' for two configurations with respective energies ε and $\epsilon + \Delta\epsilon$ and respective numbers of atoms N and $N + \Delta N$ is given by

$$\frac{p'}{p} = e^{-\beta\Delta\epsilon + \alpha\Delta N}. \tag{3}$$

Here, we find $\Delta\epsilon + \mu\Delta N = 2V$ for both types of defect, so that $p_1/p_0 = p_{-1}/p_0 = \exp(-2V/kT)$. This number is small ($2\times 10^{-7}$ at $T = 300$ K) so that $p_0 = 1 - p_1 - p_{-1} \simeq 1$ and the desired probabilities $p_1 = p_{-1}$ are equal to 2×10^{-7}. Moreover, this justifies the implicit assumption we have just made, namely that the sites can be treated independently of each other (there are practically never two neighbouring defects).

12. According to (2), the saturation ratio is

$$\lambda \equiv \frac{P}{P_0(T)} = e^{(\mu - \mu_0)kT} = e^{\alpha - \alpha_0} = e^\alpha e^{3V/kT}.$$

Carrying this new value of α over to (3) and using $p_0 \simeq 1$ we obtain the respective probabilities $p_{-1} = \exp(-2V/kT)/\lambda$ for surface vacancies and $p_1 = \lambda\exp(-2V/kT) = 2 \times 10^{-6}$ for adsorbed atoms.

13. If the vapour is supersaturated ($\lambda > 1$, $P > P_0(T)$) the number of atoms adsorbed is λ^2 times the number of surface vacancies. This excess of atoms indicates a tendency for the vapour to condense on the smooth face of the crystal. On the other hand, undersaturation ($P < P_0(T)$) implies a tendency to sublimation. However, the number of atoms essentially *remains small* in the present theory, which does not explain why the equilibrium with super- or under-saturated vapour cannot be stable. In fact, there exist more complex configurations than surface point defects which, as we shall see later, have large probabilities at equilibrium but are more difficult to create. Thus, the equilibrium considered here (with smooth faces and point defects) is *metastable*; the only truly stable equilibria are the crystal alone for $\lambda > 1$ or the vapour alone for $\lambda < 1$.

14. The binding energies at A, B and C are $-V$, $-2V$ and $-3V$, respectively, while the chemical potential is $-3V + kT\ln\lambda$. Thus, we have $p_A/(1 - p_A) = \lambda\exp(-2\beta V)$, $p_B/(1 - p_B) = \lambda\exp(-\beta V)$, $p_C/(1 - p_C) = \lambda$. Whence $p_A \simeq \lambda\exp(-2V/kT) \simeq 2 \times 10^{-6}$, $p_B \simeq \lambda\exp(-V/kT) \simeq 4 \times 10^{-3}$ and $p_C = \lambda/(1 + \lambda) \simeq 0.9$.

15. As far as the number of faces in contact is concerned, the configuration (2) (see lower Figure on p. 369) only differs from (1) in the faces in the plane $x = 0$: there is one less. Thus, the energy of (2) exceeds that of (1) by V and a kink has the energy $V/2$.

16. For each type of kink the probability is proportional to $\exp(-\beta V/2)$, with a normalisation factor $Z = 1 + 2\exp(-\beta V/2) \simeq 1$. Thus, we have $p_D = 2e^{-V/2kT} = 0.04$.

17. For a vapour which is just saturated ($\lambda = 1$), we have $p_C = 1/2$. The site C adjacent to the kink has as great a chance of being empty as of being occupied, which means that the position of the kink is immaterial, as one might expect. But if the vapour is supersaturated, the site adjacent to the kink has a chance of being occupied λ times that of its being vacant if the vapour reaches equilibrium. The large value $p_C = 0.9$ therefore indicates that a step with a kink is unstable in the presence of supersaturated vapour, for the kink tends to move and cause the step to advance. But the vapour atoms must still reach as far as C, which requires several stages and some time, so that the step is metastable. In fact, the vapour atoms can only be deposited on the crystal when they collide with its face; there is practically no chance that such collisions will occur at a point C or even at a point B on the edge of the step. However, they rapidly lead to the establishment of traps at A sites. The thermal agitation may redistribute them, but in the resulting metastable equilibrium, a proportion $p_A = 2 \times 10^{-6}$ per surface site will remain.

The two-dimensional diffusion of these atoms on the smooth face enables them to find a step, most commonly at a point B, but more rarely at a point C (in 4% of cases, since $p_D = 0.04$). Thus, on average a proportion $p_B = 4 \times 10^{-3}$ is fixed. The one-dimensional diffusion of these atoms along the step finally leads them to the points C where they have 9 chances in 10 of becoming fixed. The efficiency of this process rests on the large number of kinks (4% of the length of the step). The populations at A and B would tend to become depleted, but new exchanges with the vapour and amongst themselves maintain them at 2×10^{-6} and 4×10^{-3} as the step grows. An oblique step involves more kinks and thus grows more rapidly; therefore, it tends to straighten out parallel to an axis Ox or Oy. This mechanism explains dynamically why macroscopic crystals have simple geometric forms.

The steps may be created spontaneously on a face as the edges of islands of condensation (which grow and may merge). But there is another mechanism leading to the *automatic* creation of steps. In fact, the crystals have *screw dislocations*. There is one in the middle of Figure 11(a), p. 214, perpendicular to the face. The existence of this dislocation implies a step on the surface, starting from it. In a supersaturated regime, this step moves forward at an almost constant rate perpendicular to the local direction. This forces it to curl up and gives it the spiral form which we observe. We also see two other screw dislocations with the same sign Δz. In addition, we see flattening of the spiral in the directions Ox and Oy which is a consequence of the higher growth rate of the step when it is not parallel to these directions. There are a score of steps between the centre and the edge of the micrography, amounting to a drop of 5 nm.

18. The number of atoms in $\mathcal{R}$ is

$$
\begin{aligned}
N &= n_1 + n_2 + \ldots + n_K \\
&= (n_0 + m_0) + (n_0 + m_0 + m_1) + (n_0 + m_0 + m_1 + m_2) + \ldots \\
&\quad + (n_0 + m_0 + m_1 + \ldots + m_{K-1}) \\
&= KL + \sum_{k=0}^{K} (K - k) m_k.
\end{aligned}
$$

19. For the directions parallel to the planes $z = 0$ and $y = 0$, the region $\mathcal{R}$ contains as many faces carrying the interaction $-V$ as there are atoms in this region; this gives the contribution $-2NV$. In each vertical plane $x = k$ ($0 \le k \le K$), the numbers of cubes on either side of which are n_k and n_{k+1}, respectively, the number of faces carrying $-V$ is the smaller of the two numbers n_k and n_{k+1}, i.e.

$$
\tfrac{1}{2}(n_k + n_{k+1}) - \tfrac{1}{2}|n_{k+1} - n_k| = \tfrac{1}{2}(n_k + n_{k+1}) - \tfrac{1}{2}|m_k|.
$$

Summing over k, we obtain

$$\frac{1}{2}\sum_{k=0}^{K}(n_k + n_{k+1}) = \frac{1}{2}n_0 + N + \frac{1}{2}n_{K+1} = N + L + \frac{1}{2}M$$

whence the given formula for the total energy of the configuration.

20. Using the expressions for N, E and M and summing over all the configurations $\{m_k\}$, we have

$$Z(\alpha, \beta, \gamma) \equiv \mathrm{Tr}\, e^{\alpha N - \beta E - \gamma M}$$

$$= \sum_{\{m_k\}} \exp\left\{(\alpha + 3\beta V)\left[KL + \sum_{k=0}^{K}(K - k)m_k\right]\right.$$

$$\left. - \frac{1}{2}\beta V \sum_{k=0}^{K}|m_k| + \beta V L + \left(\frac{1}{2}\beta V - \gamma\right)\sum_{k=0}^{K}m_k\right\}.$$

Apart from the factor

$$\exp\left[\alpha K L + \beta V L(3K + 1)\right]$$

this expression is the product of partition functions

$$Z_k(\alpha, \beta, \gamma) \equiv \sum_{m=-\infty}^{+\infty} \exp\left\{\left[(\alpha + 3\beta V)(K - k) + \left(\tfrac{1}{2}\beta V - \gamma\right)\right]m - \tfrac{1}{2}\beta V|m|\right\}$$

associated with each point $(0 \le k \le K)$ at which the step may have a kink.

Each function Z_k is of the form

$$\sum_{m=-\infty}^{+\infty} e^{-\xi|m|-\eta m} = \sum_{m=0}^{+\infty}\left[e^{-(\xi+\eta)m} + e^{-(\xi-\eta)m}\right] - 1 = \frac{1}{1 - e^{-\xi-\eta}} + \frac{1}{1 - e^{-\xi+\eta}} - 1$$

$$= \frac{1 - e^{-2\xi}}{1 + e^{-2\xi} - e^{-\xi+\eta} - e^{-\xi-\eta}} = \frac{\sinh \xi}{\cosh \xi - \cosh \eta}$$

so that

$$\ln Z(\alpha, \beta, \gamma) = \alpha K L + \beta V L(3K + 1) + (K + 1)\ln \sinh \beta V/2$$

$$- \sum_{k=0}^{K} \ln[\cosh(\beta V/2) - \cosh \eta_k]$$

$$\eta_k \equiv \gamma - \beta V/2 - (\alpha + 3\beta V)(K - k)$$

$$= \gamma - \beta V/2 - (K - k)\ln \lambda.$$

21. From the definition of Z_k, $\langle m_k \rangle$ is obtained by differentiation:

$$\langle m_k \rangle = -\frac{\partial}{\partial \gamma}\ln Z_k = \frac{\partial}{\partial \gamma}\ln[\cosh(\beta V/2) - \cosh \eta_k]$$

$$= -\frac{\sinh \eta_k}{\cosh(\beta V/2) - \cosh \eta_k}.$$

22. The slope $\langle m_k \rangle$ is constant when $\lambda = 1$, since η_k does not then depend on k. In the presence of a vapour which is just saturated, the step then (on average) links its two extremities (assumed fixed) in a straight line.

As a function of η_k, $\langle m_k \rangle$ is decreasing, since its derivative is equal to

$$-\frac{\cosh(\beta V/2)\cosh \eta_k - 1}{[\cosh(\beta V/2) - \cosh \eta_k]^2}.$$

Since η_k increases with k for $\lambda > 1$ and decreases for $\lambda < 1$, the step is convex $(d^2y/dx^2 < 0)$ when the vapour is supersaturated and concave $(d^2y/dx^2 > 0)$ when it is under-saturated. The rectilinear steps are unstable if $\lambda \neq 1$. The direction of the concavity reflects the tendency of the step (whose extremities are assumed fixed here) to advance for $\lambda > 1$ and retreat for $\lambda < 1$. The curvature increases with $|\ln \lambda|$.

N.B. The present formalism assumes that $|\eta_k| < \beta V/2$, which constrains the possible values of γ by

$$(K - k)\ln \lambda < \gamma < \beta V + (k - k)\ln \lambda.$$

These inequalities must be satisfied for all $0 \leq k \leq K$, giving

$$|2\gamma - \beta V - K \ln \lambda| < \beta V - K|\ln \lambda|.$$

The multiplier γ should be adjusted so that M takes the desired value, but this is impossible if

$$K > \frac{V}{kT|\ln \lambda|}.$$

This limits the length of the possible steps when the vapour is over- or under-saturated. In fact, if $|\ln \lambda|$ increases, the curvature of the metastable steps which we have just studied increases, so that they cannot reach a great length.

We also note that the existence of step-edge irregularities, which is not taken into account here (lateral projections, subsidence, relief), is essential to explain that the step is not stable, but only metastable for $\lambda \neq 1$.

Finally, for a step parallel to Ox in equilibrium with the saturated vapour ($\lambda = 1$), we have $\langle m_k \rangle = 0$, whence $\eta_k = 0$ and $\gamma = \beta V/2$. The corresponding Boltzmann–Gibbs distribution gives the following probability of a shift $\Delta y = ma$:

$$p_m = \frac{\cosh(\beta V/2) - 1}{\sinh(\beta V/2)}\exp(-\beta V|m|/2).$$

Using the fact that $\exp(-\beta V/2) \simeq 2 \times 10^{-2} \ll 1$, we rediscover the properties assumed in question **16** and the value $p_D = 0.04$.

23. We must have

$$1 = \langle m_0 \rangle = \frac{\sinh(\beta V/2 + K \ln \lambda - \gamma)}{\cosh(\beta V/2) - \cosh(\beta V/2 + K \ln \lambda - \gamma)}$$

and

$$0 = \langle m_k \rangle = \frac{\sinh(\beta V/2 - \gamma)}{\cosh(\beta V/2) - \cosh(\beta V/2 - \gamma)}$$

which implies that $\gamma = \beta V/2$ and

$$1 = \frac{\sinh(K \ln \lambda)}{\cosh(\beta V/2) - \cosh(K \ln \lambda)}$$

that is,

$$K = \frac{\ln \cosh(\beta V/2)}{\ln \lambda}.$$

24. Treating k as a continuous variable, we obtain

$$M = \sum_k \langle m_k \rangle \simeq -\int_0^K dk \frac{\sinh \eta_k}{\cosh(\beta V/2) - \cosh \eta_k}$$

in which η_k has to be replaced by its value

$$\eta_k = -\ln \cosh(\beta V/2) + k \ln \lambda.$$

The integration is immediate, using the variable $t = \cosh \eta_k$, which gives

$$M = \frac{1}{\ln \lambda} \int_1^X \frac{dt}{\cosh(\beta V/2) - t} = \frac{1}{\ln \lambda} \ln \frac{\cosh(\beta V/2) - 1}{\cosh(\beta V/2) - X}$$

where $X = \cosh[\ln \cosh(\beta V/2)] = \frac{1}{2}[\cosh(\beta V/2) + \cosh^{-1}(\beta V/2)]$, whence

$$M = \frac{1}{\ln \lambda} \ln \frac{2 \cosh(\beta V/2)}{\cosh(\beta V/2) + 1}.$$

N.B. This result could have been obtained directly. We note that Z_k only depends on the variables α, β, γ and k through β and the combination η_k. A first consequence of this is that

$$\frac{\partial}{\partial \alpha} \ln Z_k = -(K - k)\frac{\partial}{\partial \gamma} \ln Z_k = (K - k)\langle m_k \rangle$$

which, when we sum over k, leads us back to the expression of question **18** for N. A more remarkable consequence is obtained if we consider k as the continuous variable x/a. Then we have

$$\frac{\partial}{\partial x} \ln Z_k = \frac{\ln \lambda}{a} \frac{\partial}{\partial \gamma} \ln Z_k = -\frac{\ln \lambda}{a}\langle m_k \rangle = -\frac{\ln \lambda}{a} \frac{\partial y}{\partial x}$$

which implies (if we determine the constant of integration at $x = 0$) that

$$y = -\frac{a}{\ln \lambda} \ln(Z_k/Z_0) \qquad x = ka.$$

For $\lambda \neq 1$, the logarithm of the *partition function* $Z_k(\alpha, \beta, \gamma)$ is thus interpreted, up to constants, as the *equation for the edge of the step* in the plane $y(x)$. Explicitly, we have

$$\frac{y}{a} = \frac{1}{\ln \lambda} \ln \frac{\cosh(\beta V/2) - \cosh \eta_k}{\cosh(\beta V/2) - \cosh \eta_0} \tag{4}$$

which again gives M for $k = K$. Lastly, integration of the equation $y(x)$ from $k = 0$ to $k = K$ shows that $\ln Z(\alpha, \beta, \gamma)$ is *itself* directly linked to the area behind the step, that is, to the *number of atoms* N. The thermodynamic interpretation of the multiplier γ follows from the identification of the Legendre transform

$$-kT(\ln Z + \gamma M) = U - TS - \mu N \equiv A$$

with the grand potential (function of T, μ and M), so that

$$-kT\gamma = \left.\frac{\partial A}{\partial M}\right|_{T,\mu}$$

is the deformation energy for the step edge (in equilibrium with a vapour at given T and P), when, for K fixed, M varies by one unit.

25. Neglecting $\exp(-\beta V/2) \simeq 2 \times 10^{-2}$ in comparison with 1, we may replace K and M by

$$K = \frac{\beta V}{2\ln\lambda} - \frac{\ln 2}{\ln\lambda} \qquad M = \frac{\ln 2}{\ln\lambda} \simeq 35$$

giving

$$K + M = \frac{\beta V}{2\ln\lambda} \simeq 200.$$

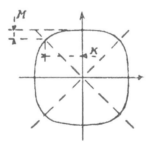

The curve $y(x)$ determined above begins with gradient 1 near $x = 0$ and soon bends downwards ending up with a fairly long almost horizontal section (for $x = Ka/2$, the gradient

$$\langle m_k \rangle = \frac{\sinh(K\ln\lambda/2)}{\cosh(\beta V/2) - \cosh(K\ln\lambda/2)} \simeq 0.1$$

is already small and

$$M - \frac{y}{a} \simeq \frac{1}{\ln\lambda} \ln \frac{\cosh(\beta V/2) - 1}{\cosh(\beta V/2) - \cosh(K\ln\lambda/2)} \simeq 3.5$$

is small in comparison with $M \simeq 35$).

To describe an island, we link eight such step segments deduced from each other by the symmetries of the square. Thus, the island is shaped like a square with rounded corners and side $2(K + M)a \simeq 1000\,\text{Å} = 0.1\ \mu\text{m}$ (this value is the same as that we found at the end of **V** as an upper bound for K for a stable step, which bound is attained by the step delimiting half the island).

As the supersaturation increases, the size of the island, which may also be written as $V/2(\mu - \mu_0)$, decreases as $\ln \lambda$ or as $\mu - \mu_0$. Although the instability increases, the metastable islands become smaller and smaller. However, the shape of the islands, characterised by the ratio $K/M = \beta V/2\ln 2 - 1$ does not vary. Moreover, equation (4) for their edge $y(x)$ depends only on the coordinates and on λ via the combinations $x \ln \lambda/a$ and $y \ln \lambda/a$. An increase in the temperature (for constant λ) causes K/M to decrease and also results in a decrease of $K + M$: so the islands become rounder and shrink as the temperature increases. At very low temperature ($\beta V \gg 1$), they are almost square (see Figure 5(b), p. 160).

We observe the various characteristics of islands studied above on the photo, including the square shape with rounded corners and the almost identical dimensions. The sides of the squares mark the directions Ox and Oy of the cubic lattice; the rounded contours represent numerous kinks on the scale $a \simeq 2.5$ Å. In fact, the photo corresponds to $\lambda < 1$, since the closed curves delimit basins cut into the face and not projecting islands; the preceding theory gives the same shape for λ as for $1/\lambda$ with convex steps replaced by concave steps.

N.B. If the free surface of a crystal has been cut nearly parallel to a major crystallographic plane, it then consists of terraces separated by straight parallel steps. The following micrography obtained by scanning tunnel microscopy (STM) shows a square portion (0.15 nm × 0.15 nm) of such a surface in copper. The surface is close (within 7.33°) to a (001) plane. Atomic kinks may be observed along the steps (temperature = 293 K).

Terraces and steps on a (1, 1, 11) surface of copper. After **L Barbier**, **L Masson** and **J Cousty**, CEA Saclay (1994).

COLOUR CENTRES IN IONIC CRYSTALS

Let us consider the diatomic solid NaCl. We know (see Chapters I and V) that it is an ionic solid of faced-centred cubic structure. The stacking, which is compact, is recalled in the accompanying figure on the left. The arrangement of the ions, viewed parallel to a cube face is shown on the right of the figure.

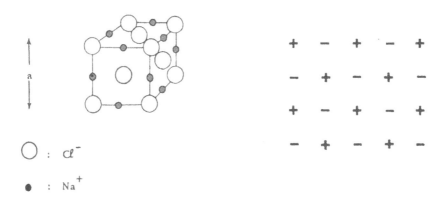

$\bigcirc$: Cl^-

$\bullet$: Na^+

This structure, called the *rock-salt structure*, is very frequent in nature. In particular, it is that of all alkali halides.

Sufficiently pure, the halides are *transparent*. If they are 'irradiated' with photons with sufficient energy (for example X-rays or γ-rays), they *become coloured*. The reason for this colouring is the following. A photon is capable of ejecting an anion from its site, leaving an unoccupied site or *vacancy* in its place. This anionic vacancy, which is surrounded by positive ions, is electronegative and tends to *trap an electron* (which restores the neutrality of the crystal locally). Such a trapped electron may be excited from its ground state and thus *absorb light*; whence the colouring of the crystal. The whole (vacancy plus electron) is called an F centre (see Chapter VII). Here is a sketch of the F centre.

Figure 1

I

Mollwo–Ivey law

Let us denote the lattice parameter by a.

Measurements of the wavelength λ or of the energy ε of light-absorption lines on various alkali halides by Mollwo and Ivey have shown that this energy varies in a simple manner as a function of the lattice parameter. The experimental results are shown in Figure 2.

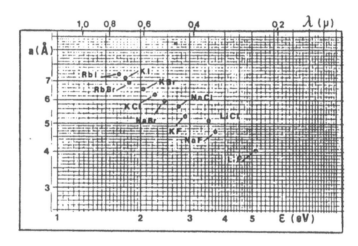

Figure 2 Energy of the light-absorption peak for various alkali halides, as a function of their lattice parameter a.

1. *Express the experimental law which emerges from these results in the form*

$$\varepsilon = Ka^n. \tag{1}$$

K will be given in units such that ε is in eV if a is in Å.

This law is called the *Mollwo–Ivey law*. It immediately suggests a model. The observation that the transition energy ε depends on the nature of the crystal only via the intermediary of a, the size of its crystalline cell, indicates that the structure of the F centre should be the same for all these crystals.

Thus, the simplest model of the F centre is the following. The Z positive ions which are immediate neighbours of the F centre (at distance $a/2$) form a shell in which the electron of the F centre is trapped. This shell may be taken as a *cubic box* in which there is a constant potential, taken, for example, to be zero. To a first approximation, the potential will be considered to be infinite outside the box (infinitely deep well). Initially, it is sensible to take the edge of the crystalline cell as the edge of the cubic box (see Figure 1). Taking the origin of the coordinates on a vertex of the cube, the potential is thus:

$$V = \begin{cases} 0 & \text{for } 0 < x < a, \, 0 < y < a \text{ and } 0 < z < a \\ \infty & \text{for } x < 0, \, y < 0 \, z < 0, \, x > a, \, y > a \text{ or } z > a. \end{cases}$$

2. *How many positive ions (Z) are near neighbours of the F centre?*

3. *Write down the energy eigenfunctions: (i) for the ground state; (ii) for the first excited state.*

Give the energies (E_1 and E_2, respectively) of these states.

What is the degeneracy of the states (i) and (ii)?

Give an example to show that they are not eigenstates of the momentum. Why are they not?

4. *Assuming that the light absorption is due to the transition $E_1 \rightarrow E_2$ of the electrons of the F centres, express the energy ε of the absorption peak as a function of a.*

Note that the model leads to an expression of type (1).

Compare[1] the theoretical and experimental values of the coefficient n and the constant K.

5. It is clear that the above comparison is satisfactory as far as *n* is concerned but is scarcely favourable as far as *K* is concerned.

We then recall that there was a certain arbitrariness in our choice of the size of the cube. Thus, we shall introduce an *effective size $a_0 = \alpha a$* and choose α so that the theoretical formula obtained in question **4** best describes the experimental results.

Write down this value of α. Give a brief physical commentary on the new value a_0 for the size of the well. Sketch the corresponding physical law in Figure 2.

II

Jahn–Teller Effect

If a quantum state of a nonlinear molecule is found to be degenerate, it can be shown that there exists a distortion of the molecule which leads to a lower symmetry for the latter and thus removes the degeneracy and stabilises the molecule. This distortion corresponds to the *Jahn–Teller effect*[2].

The set comprising the F centre and the neighbouring ions may be viewed as a pseudo molecule liable to undergo a Jahn–Teller distortion. We shall verify this for a simple model with the F centre.

As in **I**, we continue to take the F centre as an infinitely deep cubic well with edge a_0 and potential zero. Here, we shall again take the x, y and z axes along the edges of the cube and the origin on a vertex (see diagram).

---

[1] Here we shall use the system of atomic units (see p. 36).
[2] **H A Jahn** and **E Teller**, Proc. Royal Soc. **A 161**, 220, 1937.

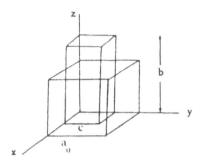

6. We then deform the vacancy by giving it a parallelepipedal form with a square base (edges of length c along Ox and Oy, height b along Oz, see diagram). Taking into account the global rigidity of the crystal, it is reasonable to assume that this change in shape of the vacancy takes place for constant volume ($a_0^3 = c^2 b$). The deformation is characterised by the parameter $\eta = b/c$.

Show how the deformation removes the degeneracy of the level E_2 (see question 3).

For the above, calculate the effect of the deformation η on the level E_2.

Note that the energy has a minimal value E_2^0 for one (which?) of the excited states. What is the value η_0 of the deformation corresponding to this energy minimum? Is the F centre stretched or flattened?

7. *Calculate the energy of the ground level E_1 as a function of the deformation η. Calculate the value $E_1^0 = E_1(\eta_0)$.*

8. *Sketch the variations of E_1 and E_2 as a function of η.*

III

Stokes Shift

We are now able to give a simple description of the overall absorption and emission of an F centre.

For this, we return to experiments.

In **I**, we gave a summary description of the *absorption* of light by the F centres. Experiments show that after a lifetime of the order of 10^{-6} s, the excited state becomes de-excited, emitting a 'luminescence' photon. This is the *emission*.

Experimental study of the emission lines shows that they are systematically shifted towards larger wavelengths (or smaller energies) in comparison with the absorption lines. This shift, an example of which is shown in Figure 3, is called the *Stokes shift*.

9. We shall suppose that initially (and we shall prove this in questions **11** to **13**), in most cases, this shift moves the emission line *into the infrared* (invisible).

If this is the case, by what simple mechanism do the F centres colour a crystal viewed in ordinary light?

10. Each of the sachets provided[3], labelled A, B and C, contains a few crystals of one of the following halides: Kl, KCl and NaCl.

---

[3] Three sachets, each containing coloured crystals, were distributed at this point.

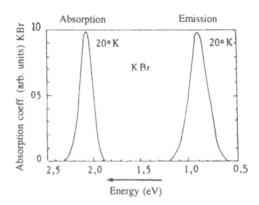

Figure 3 Absorption and emission spectra of the F centre in KBr at low temperature. The maxima of the two lines are located at 2.06 and 0.92 eV, respectively. After **W Gebbart** and **A Kuhnert**, Physica Status Solidi, **14**, 157 (1966).

These crystals have been irradiated by X-rays. In their initial state they were colourless but have become coloured with the creation of F centres (yellow in sachet A, green in B and mauve in C).

Assuming that they satisfy the assumption of question 9 (emission line in the invisible), indicate the contents of each sachet.

N.B. Here, we recall that the colours of the spectrum of white light, are, in order of increasing energy: red ($\sim$ 1.65–2.0 eV), orange (2.0–2.1 eV), yellow (2.1–2.3 eV), green ($\sim$2.3–2.55 eV), blue (2.55–2.65 eV), violet (2.65–3.1 eV).

We also recall that 'complementary colours' are colours which when associated reconstitute white light, and that the main complementary colours are yellow–violet, red–green and blue–orange. Thus, an absorption in the blue gives rise to an orange colour.

11. In what follows, we shall try to give a simple description of the Stokes shift.

We shall assume that the electron excitation and de-excitation times are negligible in comparison with the local deformation times for the crystal, which are themselves negligible in comparison with the lifetimes of the excited states[4] (namely $\sim 10^{-6}$ s).

For these conditions, give a very simple description of the overall absorption and emission of an F centre. Use the results of II for this.

12. *More quantitatively, show that the results of questions 6 and 7 take reasonable account of the experimental result of Figure 3.*

13. *Finally, justify the assumption made in question 9 by showing that for most of the crystals mentioned in Figure 2 (say which) the emission line is located in the infrared[5].*

SOLUTION TO PROBLEM 11

1. The experimental points are practically aligned in a log log representation. Thus, the experimental law has the form

$$\varepsilon = K a^n \tag{2}$$

---

[4] This assumption, which can be proved, is called the *'Franck–Condon' principle*.
[5] Here, we recall that the visible domain lies between 0.4 and 0.75 μm or 3.1 and 1.65 eV.

where $K \simeq 68$ and $n \simeq -1.85$.

2. There are six positive ions at distance $a/2$ from the F centre.

3. Choosing the origin to be the vertex of the potential cube, the energy eigenfunctions are:

$$\psi_1 = \left(\frac{2}{a}\right)^{3/2} \sin\frac{\pi}{a}x \sin\frac{\pi}{a}y \sin\frac{\pi}{a}z \qquad (i)$$

for the ground state, with eigenenergy

$$E_1 = \frac{\hbar^2}{2m}\frac{3\pi^2}{a^2}$$

(this state is non-degenerate), and the functions ψ_{2x}, ψ_{2y}, ψ_{2z} (triple degeneracy), where, for example

$$\psi_{2z} = \left(\frac{2}{a}\right)^{3/2} \sin\frac{\pi}{a}x \sin\frac{\pi}{a}y \sin\frac{2\pi}{a}z \qquad (ii)$$

with

$$E_2 = \frac{\hbar^2}{2m}\frac{6\pi^2}{a^2}.$$

Let us consider, for example, the operator $\hat{L}_z$.
Applied to ψ_1 it gives

$$\hat{L}_z\psi_1 = -i\hbar\left(x\frac{\partial}{\partial y} - y\frac{\partial}{\partial x}\right)\psi_1$$

$$= -i\hbar\frac{\pi}{a}\left(\frac{2}{a}\right)^{3/2} \sin\frac{\pi}{a}z\left(x\sin\frac{\pi}{a}x\cos\frac{\pi}{a}y - y\cos\frac{\pi}{a}x\sin\frac{\pi}{a}y\right).$$

Thus, ψ_1 is not an eigenstate of $\hat{L}_z$.

This result is not surprising because the potential used is not a central potential.

4. The transition $E_1 \rightarrow E_2$ corresponds to the absorption of the energy

$$\varepsilon = E_2 - E_1 = \frac{\hbar^2}{2m}\frac{3\pi^2}{a^2} \qquad (3a)$$

where a is just the lattice parameter. This expression is of the type (2) with $K = 112$ and $n = -2$.

This theoretical value of n is reasonably close to the experimental value (-1.85). The value of K is clearly overestimated.

5. Thus, the experimental results may be accounted for reasonably correctly by a law of the type (3a), that is, such that $n = -2$ provided the edge of the potential-well cube, a, is replaced by a greater value $a_0 = \alpha a$. The line with slope -2 shown in Figure 12, p. 168 has equation

$$\varepsilon = \frac{\hbar^2}{2m}\frac{3\pi^2}{\alpha^2}\frac{1}{a^2} \qquad (3b)$$

where $\alpha = 1.13$.

It corresponds to $K = 87$ (and $n = -2$). It accounts reasonably well for the experimental results.

The size of the 'effective cube' is therefore 13% greater than the crystal cell. Here, therefore, the effective potential well takes in the first-neighbour positive ions. This result is not surprising; it is associated with the fact that the six neighbouring positive ions attract the electron of the F centre. In other words, if we use a more realistic potential than this infinite potential, the probability that an electron is present should be non-zero outside the anionic vacancy itself.

6. Let us consider the state (see question 3, (ii))

$$\psi_{2z} = \left(\frac{2}{a}\right)^{3/2} \sin\frac{\pi}{a_0}x \sin\frac{\pi}{a_0}y \sin\frac{2\pi}{a_0}z.$$

During the deformation it becomes:

$$\psi'_{2z} = \frac{2}{c}\left(\frac{2}{b}\right)^{1/2} \sin\frac{\pi}{c}x \sin\frac{\pi}{c}y \sin\frac{2\pi}{b}z.$$

The energy

$$E_2 = \frac{\hbar^2}{2m}\frac{6\pi^2}{a_0^2}$$

becomes

$$E'_{2z} = \frac{\hbar^2\pi^2}{2m}\left(\frac{2}{c^2} + \frac{4}{b^2}\right).$$

If we set $\eta = b/c$ and deform with a constant volume ($a_0^3 = c^2 b$), we have:

$$c = a_0\eta^{-1/3} \quad\text{and}\quad b = a_0\eta^{2/3}$$

whence

$$E'_{2z} = \frac{\hbar^2}{2m}\frac{\pi^2}{a_0^2}\left(2\eta^{2/3} + 4\eta^{-4/3}\right).$$

Similarly, we find immediately that

$$E'_{2x} = E'_{2y} = \frac{\hbar^2}{2m}\frac{\pi^2}{a_0^2}(5\eta^{2/3} + \eta^{-4/3}).$$

It is clear that both E'_{2z} and E'_{2x} ($= E'_{2y}$) are different from E_2. Thus, the deformation removes the degeneracy, at least partially.

Study of the variations of E'_{2z} and E'_{2x} as a function of η shows that these two energies pass through a minimum. The energy E'_{2z} is minimised for the value $\eta = 2$ and the energy E'_{2x} is minimised for the value $\eta = \sqrt{2/5} \simeq 0.63$. The minimum of E'_{2z} is $E'_{2z}(\eta = 2) = m(\hbar^2/2m)(\pi/a_0^2)4.76$, while the minimum of E'_{2x} is $E'_{2x}(\sqrt{2/5}) = (\hbar^2/2m)(\pi^2/a_0^2)5.52$ (see Figure 13, p. 169).

These two minima are lower than the value of E_2. The first is the minimum minimorum. Thus, the energy passes through the minimum

$$E_2^0 = \frac{\hbar^2}{2m}\frac{\pi^2}{a_0^2}4.76$$

for the deformation value $\eta_0 = 2$. Since this value of η is greater than 1, the F centre is stretched along Oz.

7. When the F centre is deformed, the energy of the ground state is given by

$$E_1' = \frac{\hbar^2}{2m}\pi^2\left(\frac{2}{c^2} + \frac{1}{b^2}\right)$$

that is

$$E_1' = \frac{\hbar^2}{2m}\frac{\pi^2}{a_0^2}\left(2\eta^{2/3} + \eta^{-4/3}\right).$$

This function is minimal for $\eta = 1$ (centre not deformed). Thus, any deformation adds to the energy of the ground state. In particular, we have

$$E_1'(\eta_0) = E_1'(2) = \frac{\hbar^2}{2m}\frac{\pi^2}{a_0^2}3.57.$$

8. The variations in these energy levels of the F centre as a function of the deformation η are shown in Figure 13 of p. 169 with the correspondence $E_{2z}'(\eta) \leftrightarrow E_{112}$, $E_{2x}' \leftrightarrow E_{211}$ and $E_1' \leftrightarrow E_{111}$.

9. If the luminescence emission is located in the infrared, it does not contribute to the colouring, which is solely due to the absorption. The colour observed is then the complementary colour of the colour absorbed.

10. The crystals in sachet A are yellow. Thus, the colour absorbed is violet. Therefore, the absorption takes place in the band 2.65–3.1 eV. Of the crystals provided, only NaCl absorbs in this band. Similarly, one can show that sachet B (green) contains KI and sachet C (mauve) KCl.

11. If an F centre is deformed after absorption (whence excitation to the state ψ_2) its energy decreases to E_2^0. If it becomes de-excited, the energy of the emitted photon will necessarily be lower than the energy of the absorbed photon; whence the Stokes shift. Assuming the Franck–Condon principle, the absorption and emission may then be represented as shown in Figure 4.

12. Taking the above into account, the energy of the emission peak in this model is:

$$\varepsilon' = E_2^0 - E_1'(\eta_0)$$
$$= \frac{\hbar^2}{2m}\frac{\pi^2}{a_0^2}(4.76 - 3.57) = \frac{\hbar^2}{2m}\frac{\pi^2}{a_0^2}1.19.$$

This emission energy is less than the absorption energy (see question 4) with ratio $(1.19/3) \simeq 0.4$. The experimental ratio (see Figure 3) for KBr is $0.92/2.06 = 0.44$. Thus, the agreement between this experiment and the theoretical model is satisfactory, taking into account the extreme simplicity of the latter.

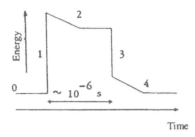

Figure 4   0: F centre in its ground state.

1:   Absorption of a photon with energy $\varepsilon = E_2 - E_1$. Instantaneous transition to the degenerate state ψ_2.

2:   Deformation of the F centre ($\eta \rightarrow \eta_0$, stretching). The electron energy decreases to E_2^0. The corresponding energy ($E_2 - E_2^0$) is yielded to the crystal in the form of thermal vibrations ('phonons').

3:   De-excitation. Since this is instantaneous, it takes place in a deformed centre. Emission of a photon with energy $E_2^0 - E_1'(\eta_0)$.

4:   The centre returns to its initial undeformed state. The corresponding energy $E_1'(\eta_0) - E_1$ is yielded to the crystal in the form of thermal vibrations.

13. The ratio ε'/ε calculated at present does not depend on a_0 whence it does not depend on the crystal. For an absorption energy corresponding to the extreme of the visible spectrum, i.e. 3.1 eV, the calculated emission energy is therefore $3.1 \times 0.4 = 1.24$ eV. This energy lies in the infrared. We deduce that *if the light absorbed is in the visible* (crystals from RbI to KF in Figure 2), then *the emitted light is outside the visible domain.* This was the assumption made in question **9**.

Complementary Remarks

1   A plausible mechanism for the formation of the F centres by irradiation with X- or γ-rays is presented in the course (p. 170).

2   The colouring may also be obtained by adding impurities (for example, a few Ca^{++} ions in NaCl). It is for this reason that many minerals with a very marked ionic nature, which are transparent when pure (for example, quartz), take on a coloured appearance if they are contaminated by 'foreign' atoms at the time of their crystallisation. This is the origin of the beautiful colours of 'pierres précieuses'.

3   The model developed in **I** provides a reasonable explanation of the Mollwo–Ivey law. It is, of course, too simple. Normally, the real potential is not infinitely deep, whence the wavefunctions may have range far greater than $a_0/2$. Experiments (using electron resonance measurements) have detected wavefunctions extending as far as the eighth ionic layer around the F centre, which is far beyond $a_0/2$.

4   The F centres may move by diffusion (see Chapter VIII, diffusion of vacancies).

   This mobility tends to lead to their ultimate disappearance (for example, when it drives them to a free surface). This disappearance (whence the mobility) could be observed by the students by gently heating the crystals provided in the sachets and noting the disappearance of the colouring.

   There is also a possible disappearance of the colouring when the crystals are exposed to light. This is associated with the ionisation of the F centres by ultraviolet photons (ejection of the electron outside its vacancy).

DIFFUSION OF HYDROGEN ON INTERSTELLAR DUSTS

Alfred Vidal Madjar, Pierre Petiau
and Jean-Louis Basdevant

Parts **II** and **III** *are independent.*

Atomic hydrogen is the main component of the very sparse matter found between the stars (around 1 atom cm^{-3}). However, paradoxically, whenever this material is a little more concentrated (> 10 atoms cm^{-3}) atomic hydrogen no longer appears except in the trace state. In this case, it is essentially transformed into molecular hydrogen. The mechanism for this transformation has long been an enigma. One might think that it took place directly during collisions between the hydrogen atoms in the gas phase of the interstellar medium. However, selection rules prohibit evacuation in the form of electromagnetic emission of the energy corresponding to the molecular binding energy (4.5 eV).

In 1963, Gould and Salpeter proposed another mechanism which permits this association on the surface of dust particles present in the interstellar medium. In fact, in the clouds of the interstellar medium, around 1% of the mass occurs in the form of dust particles of diameter approximately 1 μm, which are probably enveloped in a coat of ice.

Thus, to evaluate the orders of magnitude, we shall suppose that the atoms of the interstellar medium may be adsorbed on the surface of a crystal formed from regularly spaced atoms, with the kinetic energy of the atom being transmitted to the crystal in the form of vibrational energy during the absorption. The dust particles then serve as catalysers for the formation of H_2 molecules.

N.B. In what follows, we shall use the term 'energy of the hydrogen atom' to refer to the energy of this atom as an object in the crystal potential. We shall not concern ourselves with the internal energy levels of the atom which is assumed to be in its ground state throughout the problem.

I

Assuming that the force between the hydrogen atom and each of the atoms of the crystal (at distance r) is a Van der Waals-type force, the corresponding interaction potential is described by a so-called Lennard–Jones expression

$$\varphi(r) = 4\varepsilon \left[\left(\frac{\sigma}{r}\right)^{12} - \left(\frac{\sigma}{r}\right)^{6} \right] \quad \varepsilon > 0.$$

1. *Show that this potential has a minimum for a distance r_0, to be calculated. Sketch the shape of $\varphi(r)$. We shall take $\varepsilon = 7 \times 10^{-2}$ eV and $\sigma = 3$ Å.*

2. Suppose that the surface of the crystal is planar and that the atoms of the surface are distributed at the nodes of a square lattice with cell size p, where the crystal lattice is simple cubic. We shall take $p = 3$ Å.

Using simple qualitative arguments, indicate the possible positions for the minima of the interaction potential between the hydrogen atom and the crystal, in a plane parallel to its surface.

3. This interaction potential with the crystal is described in the neighbourhood of the surface of the crystal (defining the plane $z = 0$) by the following expression

$$V(x, y, z) = f(x) + f(y) + \varphi(z)$$

with

$$f(x) = \frac{\varepsilon}{4}\left(1 - \cos\frac{2\pi x}{p}\right) \qquad f(y) = \frac{\varepsilon}{4}\left(1 - \cos\frac{2\pi y}{p}\right)$$

and where $\varphi(z)$ is a potential of Lennard–Jones type:

$$\varphi(z) = 4\varepsilon\left[\left(\frac{\sigma}{z}\right)^{12} - \left(\frac{\sigma}{z}\right)^{6}\right].$$

We introduce a further approximation by replacing the potential in the neighbourhood of a minimum by the following expression involving three harmonic potentials:

$$W(x, y, z) = C + \tfrac{1}{2}mw_x^2 x^2 + \tfrac{1}{2}mw_y^2 y^2 + \tfrac{1}{2}mw_z^2(z - z_0)^2.$$

Calculate C, w_x^2, w_y^2, w_z^2 and z_0 as a function of the data for the problem.
Find the energy E_0 of the ground state of the hydrogen atom adsorbed on the crystal. Take $\hbar c = 2 \times 10^3$ eV Å and $mc^2 = 10^9$ eV (where c is the velocity of light).

4. *What energy must be supplied to the atom for it to reach the first excited state in the crystal potential? Given that the temperature of the dust particles in the interstellar medium is of the order of 20 K, can an atom of the crystal supply the hydrogen atom with this energy by virtue of its thermal vibration?*
We recall that, for $T = 300$ K, we have $kT \simeq 1/40$ eV.

II

To form hydrogen molecules on the surface of the particles, the atoms must meet, whence they must move horizontally in x and y.

5. *What mechanism accounts for the mobility of the atoms on the surface of the particles?*

6. To assess the efficiency of this mechanism, we shall evaluate an order of magnitude for the time taken for a hydrogen atom to pass from one site to a neighbouring site.

For this, we assume that the hydrogen atoms, situated at time $t = 0$ at site s_0 (see figure) can only jump to the adjacent sites s_1, s_2, s_3 and s_4. The existence of other sites in the lattice is not taken into account.

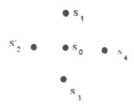

If this atom is in the ground state calculated in question **3** and situated on the sites s_0, $s_1, \ldots, s_4$, we denote its five states by $|\varphi_0\rangle, |\varphi_1\rangle, \ldots |\varphi_4\rangle$. If one neglects the probability that the atom may jump from one site to another, the five states $|\varphi_i\rangle$ are eigenstates of its Hamiltonian $\hat{H}_0$ with the same eigenvalue E_0 calculated above.

The coupling between the state $|\varphi_0\rangle$ and the states $|\varphi_1\rangle, \ldots, |\varphi_4\rangle$ modifies the Hamiltonian: so we must add a further term $\hat{H}_1$, defined as follows:

$$\hat{H}_1|\varphi_0\rangle = -a(|\varphi_1\rangle + |\varphi_2\rangle + |\varphi_3\rangle + |\varphi_4\rangle)$$
$$\hat{H}_1|\varphi_i\rangle = -a|\varphi_0\rangle \quad i = 1, \ldots, 4.$$

The term a is a positive real quantity.

We neglect other possible couplings. *Calculate the energy levels of the Hamiltonian* $\hat{H} = \hat{H}_0 + \hat{H}_1$.

7. *Write down the corresponding normalised eigenstates of $\hat{H}$ as a function of the states* $|\varphi_i\rangle$.

8. *At time $t = 0$, the hydrogen atom is at site s_0. Write down its state $|\psi(t)\rangle$ at time t as a function of the eigenstates of $\hat{H}$. Discuss the location of the atom at time t. After what time T can we say that the atom has changed site?*

We take $4a = 5 \times 10^{-5}$ eV.

III

9. The order of magnitude of the time of transfer from one site to another can be calculated in another way by simplifying the problem to the extreme. For this, we consider a one-dimensional problem and replace the real potential by the simplified potential shown in the accompanying figure.

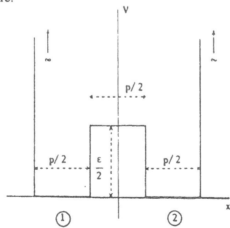

At time $t = 0$ the hydrogen atom is located in the well on the left. We shall suppose that its energy with respect to the bottom of the well is equal to that of the ground level of a harmonic oscillator with the frequency ω_x determined in question **3**.

Calculate the time T after which the atom has left the initial site.

IV

10. *Knowing that the dust particle has dimensions of the order of a micron, give the order of magnitude of the time after which two atoms adsorbed at arbitrary points of the crystal surface meet.*

11. *What mechanism do you think they might use to evacuate the energy (4.5 eV) corresponding to the binding energy of the hydrogen molecule in order to form this molecule? What will happen to this molecule then?*

SOLUTION TO PROBLEM 12

1. We find $\varphi' = 4\varepsilon(-12\sigma^{12}r^{-13}+6\sigma^6 r^{-7})$ whence $\varphi'(r) = 0$ for $r_0 = \sigma \times 2^{1/6} = 1.12\sigma$. The curve $\varphi(r)$ has a very sharp minimum for $r = r_0$ and we have $\varphi(r_0) = -\varepsilon$.

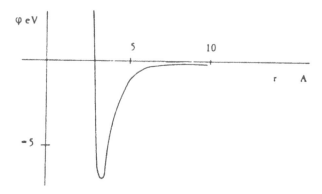

2. For symmetry reasons, the minima of the potential in a plane parallel to the surface of the crystal can only occur in three types of position: above atoms (positions A); above the centre of the square cell (position S); above the centre of a side of the cell (position C).

N.B. When we sum the potentials due to all the sites of the lattice, we find, in general, that the case S gives the minimal energy, the position A gives the maximal energy and the position C corresponds to a saddle-point. This is illustrated in the accompanying figure which shows an example of equipotential lines in a cell.

3. $\varphi(z)$ is minimal for $z_0 = \sigma 2^{1/6}$ and we have $\varphi(z_0) = -\varepsilon$. Moreover, $f(x)$ is minimal for $x = np$ (integer n): $f(np) = 0$. It follows that $C = -\varepsilon$.

Expansion of $f(x)$ about $x = 0$ gives

$$\frac{\varepsilon}{4} \times \frac{1}{2}\left(\frac{2\pi}{p}\right)^2 x^2 = \frac{1}{2}mw_x^2 x^2$$

whence

$$w_x^2 = w_y^2 = \frac{\varepsilon}{4m}\left(\frac{2\pi}{p}\right)^2.$$

Calculation of the second derivative of $\varphi(z)$ for $z = z_0$ gives

$$\varphi''(z_0) = \frac{72\varepsilon}{\sigma^2 2^{1/3}} = mw_z^2$$

whence

$$w_z^2 = \left(\frac{1}{2}\right)^{1/3}\frac{72}{m}\frac{\varepsilon}{\sigma^2}.$$

Thus, the energy of the atom in its ground state is

$$E_0 = -\varepsilon + \frac{1}{2}\hbar(w_x + w_y + w_z).$$

Numerical calculation:

$$\hbar^2 w_x^2 = \varepsilon\frac{\hbar^2 c^2}{4mc^2}\left(\frac{2\pi}{p}\right)^2 = \frac{7 \times 10^{-2} \times 4 \times 10^6}{4 \times 10^9}\left(\frac{2\pi}{3}\right)^2 = 3 \times 10^{-4} \text{ eV}^2$$

whence

$$\tfrac{1}{2}\hbar w_x = \tfrac{1}{2}\hbar w_y = 0.9 \times 10^{-2} \text{ eV}$$
$$\tfrac{1}{2}\hbar w_z = 2.1 \times 10^{-2} \text{ eV}$$

and consequently

$$E_0 = -3.1 \times 10^{-2} \text{ eV}.$$

4. The energy of the first excited state is equal to $E_0 + \hbar w_x$ (or $E_0 + \hbar w_y$), that is $E_0 + 1.8 \times 10^{-2} = -1.3$ eV.

The average thermal energy of an atom of the crystal at 20 K is $kT \sim 0.2 \times 10^{-2}$ eV. Thus, it cannot bring the atom to the first excited state. In reality, the average energy of a vibration mode of the crystal is even lower.

5. The hydrogen atoms must move from one site to another by tunnelling.

6. In the subspace of the $|\varphi_i\rangle$ the time-independent Schrödinger equation is given by

$$\hat{H}|\psi\rangle = (\hat{H}_0 + \hat{H}_1)|\psi\rangle = E|\psi\rangle.$$

We set $|\psi\rangle = \sum_i c_i |\varphi_i\rangle$ $i = 0, 1, \ldots, 4$. In the basis of the $|\varphi_i\rangle$, the Hamiltonian $\hat{H}$ is described by the following matrix:

$$H = \begin{vmatrix} E_0 & -a & -a & -a & -a \\ -a & E_0 & 0 & 0 & 0 \\ -a & 0 & E_0 & 0 & 0 \\ -a & 0 & 0 & E_0 & 0 \\ -a & 0 & 0 & 0 & E_0 \end{vmatrix}.$$

Thus, the energy eigenvalues are solutions of the equation

$$\begin{vmatrix} E_0 - E & -a & -a & -a & -a \\ -a & E_0 - E & 0 & 0 & 0 \\ -a & 0 & E_0 - E & 0 & 0 \\ -a & 0 & 0 & E_0 - E & 0 \\ -a & 0 & 0 & 0 & E_0 - E \end{vmatrix} = 0.$$

Setting $E - E_0 = \lambda$, and after a number of simple transformations, we find:

$$\lambda^3(\lambda^2 - 4a^2) = 0.$$

Thus, there are three eigenvalues of E:

$$E = E_0 \text{ triply degenerate}$$
$$E = E_0 \pm 2a$$

7. We decompose the eigenstates $|\psi\rangle$ in the basis of the $|\varphi_i\rangle$

$$|\psi\rangle = c_n |\varphi_0\rangle + c_1 |\varphi_1\rangle + \ldots + c_4 |\varphi_4\rangle.$$

For the triply degenerate eigenvalue $E = E_0$, we may seek an orthonormal basis $|\psi_1^0\rangle$, $|\psi_2^0\rangle$, $|\psi_3^0\rangle$ of the corresponding eigensubspace. We see that

$$c_0 = 0 \qquad c_1 + c_2 + c_3 + c_4 = 0.$$

We may choose the following basis:

$$|\psi_1^0\rangle = \tfrac{1}{2}(|\varphi_1\rangle + |\varphi_2\rangle - |\varphi_3\rangle - |\varphi_4\rangle)$$
$$|\psi_2^0\rangle = \tfrac{1}{2}(|\varphi_1\rangle - |\varphi_2\rangle + |\varphi_3\rangle - |\varphi_4\rangle)$$
$$|\psi_3^0\rangle = \tfrac{1}{2}(|\varphi_1\rangle - |\varphi_2\rangle - |\varphi_3\rangle + |\varphi_4\rangle).$$

For the eigenvalue $E = E_0 + 2a$, we have:

$$c_0 + 2c_i = 0 \quad \text{whence} \quad c_i = -\tfrac{1}{2}c_0$$

and

$$|\psi^+\rangle = \frac{1}{\sqrt{2}}\left[|\varphi_0\rangle - \frac{1}{2}(|\varphi_1\rangle + |\varphi_2\rangle + |\varphi_3\rangle + |\varphi_4\rangle)\right].$$

For the eigenvalue $E = E_0 - 2a$, we find:

$$c_0 - 2c_i = 0 \quad \text{whence} \quad c_i = \tfrac{1}{2}c_0$$

and the corresponding state is given by:

$$|\psi^-\rangle = \frac{1}{\sqrt{2}} \left[|\varphi_0\rangle + \frac{1}{2} \left(|\varphi_1\rangle + |\varphi_2\rangle + |\varphi_3\rangle + |\varphi_4\rangle \right) \right].$$

8. At time $t = 0$, we may write:

$$|\psi(0)\rangle = |\varphi_0\rangle = \frac{1}{\sqrt{2}} \left(|\psi^+\rangle + |\psi^-\rangle \right).$$

Since $|\psi^+\rangle$ and $|\psi^-\rangle$ are eigenstates of $\hat{H}$, their evolution with time is known. At time t, we will have

$$|\psi(t)\rangle = \frac{1}{\sqrt{2}} \left\{ |\psi^+\rangle \exp\left[-\frac{i(E_0 + 2a)t}{\hbar} \right] + |\psi^-\rangle \exp\left[-\frac{i(E_0 - 2a)t}{\hbar} \right] \right\}$$

whence:

$$|\psi(t)\rangle = \frac{1}{\sqrt{2}} \exp\left(-\frac{iE_0 t}{\hbar} \right) \left[|\psi^+\rangle \exp\left(-\frac{2iat}{\hbar} \right) + |\psi^-\rangle \exp\left(\frac{2iat}{\hbar} \right) \right].$$

Replacing $|\psi^+\rangle$ and $|\psi^-\rangle$ by their expressions as a function of the localised states $|\varphi_i\rangle$ determined in the previous question, we obtain:

$$|\psi(t)\rangle = \exp\left(-\frac{iE_0 t}{\hbar} \right) \left[\cos\frac{2at}{\hbar} |\varphi_0\rangle + i \sin\frac{2at}{\hbar} \left(|\varphi_1\rangle + |\varphi_2\rangle + |\varphi_3 + |\varphi_4\rangle \right) \right].$$

We see that the probability of observing a hydrogen atom at the site s_0 at time t is equal to $\cos^2(2at/\hbar)$. The probability that it is present is identical for each of the four neighbouring sites and is equal to $\frac{1}{4} \sin^2(2at/\hbar)$.

Thus, we can say that it has left its site s_0 after a time T such that

$$\frac{2aT}{\hbar} = \frac{\pi}{2} \quad \text{whence} \quad T = \frac{\pi\hbar}{4a}.$$

This model is clearly imperfect, since it does not allow the atom to diffuse on the surface of the crystal: subsequently, it can only return to s_0, which explains the periodic function found.

Numerical calculation.

$$T = \frac{\pi\hbar}{4a} = \frac{\pi\hbar c}{4ac} = \frac{\pi \times 2 \times 10^3}{5 \times 10^{-5} \times 3 \times 10^{18}} = 0.4 \times 10^{-10} \text{ s}.$$

9. From the results concerning the behaviour of a particle in a double potential well (see Q.M.), we know that

$$T = \frac{\pi\hbar}{2A} \quad \text{where} \quad A = \frac{1}{2}\hbar w_x \frac{4e^{-K\Delta}}{Ka}.$$

Here, we have

$$\Delta = a = \frac{p}{2} \qquad K = \frac{1}{\hbar}\sqrt{2m\left(\frac{\varepsilon}{2} - \frac{1}{2}\hbar w_x\right)}.$$

Numerical calculation.

$$\frac{1}{2}\hbar w_x\left(\simeq \frac{\hbar^2\pi^2}{2m(p^2/4)}\right) = 0.9 \times 10^{-2} \text{ eV}$$

$$K = 3.6 \text{ Å}^{-1} \quad \text{whence} \quad \frac{Kp}{2} = 5.4.$$

It follows that

$$A = 3 \times 10^{-5} \text{ eV} \quad \text{whence} \quad T = 0.35 \times 10^{-10} \text{ s}.$$

10. The transition time from one state to another is very short. On a cubic dust particle of dimension 1 μm, there are approximately 10^8 possible sites if the lattice spacing is 3 Å. Two hydrogen atoms should meet after a time of the order of 10^{-3} seconds.

11. They may then form a hydrogen molecule by yielding the binding energy to the crystal lattice in the form of vibration energy. A very small fraction of this energy will also be used to draw the molecule away from the crystal's attractive potential.

DIFFUSION IN THE PRESENCE OF A FORCE

(FROM PERRIN GRANULES TO COTTRELL CLOUDS)

Although this problem forms a whole, it is constructed in such a way that each of the three parts can be studied independently.

It involves an analysis of two cases of *diffusion in the presence of a force*, namely: the diffusion of *granules subjected to gravity* (**II**) and that of *atoms subjected to the field of stresses of a dislocation* (**III**).

But first (**I**), we shall *generalise the theory of diffusion* described in the Course (Chapter VIII). In the latter, we only consider a very special case, that of objects (vacancies) for which the elementary jump takes place in identical *steps* (r_0) in discrete *directions* (those of the edges of the crystalline cells). We wish to generalise this theory to the case of objects which we shall call particles (these may be electrons or holes in a solid, neutrons in a nuclear reactor, granules in a liquid, etc) whose *directions* of movement are *arbitrary*. We shall assume that their velocity v is constant and that they are involved in *collisions* characterised by an exponential probability distribution (mean time τ). No spatial direction is preferred after a given collision.

I

First, we calculate the number of particles per unit time crossing the *unit surface area* dS (situated at 0 in the plane xOy) which emanate from the upper half space ($z > 0$). Let us consider an element dV of this half space situated at (r, θ, φ) and defined by $dr, d\theta$ and $d\varphi$.

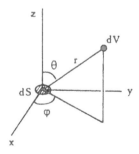

1. *Write down the expression for* dV.

2. *Denoting the concentration (number of particles per unit volume) by* c $(= c(r, \theta, \varphi))$, *indicate the number of collisions occurring in* dV *per unit time.*

3. *Give an expression for the number of particles per unit time, which emanate from* dV, *where they underwent their last collision, and cross the element* dS.

You should, of course, take into account the 'losses' due to collision on the path dV–dS.

4. *Use the above expression to determine, in the form of a triple integral, the total number of particles* j^+ *per unit time emanating from the upper half-space* ($z > 0$) *and crossing* dS.

5. Because of the losses referred to in question **3** the contribution of regions remote from dS to j^+ is small. Thus, we shall assume that it is legitimate to expand c in the form:

$$c = c_0 + x\left(\frac{\partial c}{\partial x}\right)_0 + y\left(\frac{\partial c}{\partial y}\right)_0 + z\left(\frac{\partial c}{\partial z}\right)_0. \tag{1}$$

Under these conditions, the expression obtained in question **4** can be integrated.

Deduce j^+ *as a function of* c_0, $(\partial c/\partial z)_0$, v *and* τ.

6. *As a function of the above, write down an expression for the number of particles* j^- *crossing* dS *and emanating from the lower half space* ($z < 0$).

Deduce the excess number of particles $j_z = j^- - j^+$ *per unit time crossing the element* dS *in the direction* $z > 0$.

7. j_z *is the* z *component of the flux* $\overline{j}$ *at point 0; express* j_z *as a function of* $\overline{\nabla}c$.

Show that the diffusion law thus established generalises and incorporates Fick's first equation described in the Course for the diffusion of vacancies. Establish the identicalness of the two expressions.

8. From now on we suppose that the random movement as characterised above is supplemented by a movement described by a uniform drift velocity V_z due to the application of an external force F_z assumed to be uniform.

Say why the application of a uniform force manifests itself through a constant drift velocity. Is there a contradiction to Newton's law?

Give an expression for the projection J_z *of the total flux of particles crossing* dS.

9. *Deduce the equation for the evolution of the concentration* $\partial c/\partial t$ *at the origin 0 (second Fick's equation).*

II

Here, we shall apply the evolution law determined in question **9** to the case of the *diffusion* (in the sense of the introduction) of identical *solid granules* immersed *in a liquid* at rest at temperature T.

10. In principle, in question **9** we found a law of the form

$$\frac{\partial c}{\partial t} = \alpha\frac{\partial^2 c}{\partial z^2} - \beta\frac{\partial c}{\partial z}. \tag{2}$$

If you were unable to solve I, determine the order of magnitude of α and β, using dimensionality arguments, as a function of the problem data (velocity v, collision time τ, drift velocity V_z). This will enable you to proceed with the problem.

11. Here, the drift velocity is that due to the force of gravity $-mg$ parallel to the z-axis.

Find the form of $c(z)$ (c = volume concentration of granules) in the stationary regime (or 'equilibrium').

12. The accompanying photograph taken from an article by Jean Perrin[1] shows a vertical equilibrium distribution of identical granules (of gamboge) arising from the desiccation of a latex, contained in water. The magnification corresponds to a total height of approximately 100 μm. The temperature is 20° C.

The granules in different sections of constant thickness dz at different heights z can clearly be counted. Let us listen to Jean Perrin.

'A very careful series of measurements was taken using particles of gamboge with radius 0.21 μm. Cross readings were taken in a 100-μm deep tank in four equidistant horizontal planes at the levels 5 μm, 35 μm, 65 μm and 95 μm. From these readings, counting 13 000 particles, we obtained concentrations for these levels proportional to the numbers 100, 47, 22.6 and 12, respectively.'

Are these experimental concentrations compatible with the distribution found in question 11? N.B. Possibility of sketching a graph of $c(z)$.

13. This distribution gives rise to a characteristic equilibrium distance z_0. *What is the value of z_0 in the experiment described by J Perrin?*

14. *Give an approximate (question 10) or, better still, an exact expression for z_0 as a function of v, τ and V_z.*

15. *Assuming that the granules are in thermal equilibrium with the water molecules (temperature T), give an expression for z_0 as a function of m, g and kT (k = Boltzmann's constant).*

16. *Assuming the density of latex is $\rho = 1.2$ g cm^{-3}, compare the experimental and theoretical values of z_0.*

17. *Give a qualitative description of what the 'equilibrium' of these granules involves.*

Indicate how II is similar to (or differs from) the problem of the atmospheric rarefaction of air (i.e. the variation of pressure with altitude).

---

[1] **J Perrin**. *Les preuves de la réalité moléculaire*, in Conférence Solvay 1911, p. 153.

III

The general ideas of the above discussion will now be applied to the case of *Cottrell clouds*. This is the name given to groups of foreign atoms which generally surround dislocations in solids.

Suppose A is a solid containing the foreign atom B as an impurity (low concentrations, for example $c_B \leq 10^{-2} c_A$), for example A = iron, B = carbon, or A = NaCl, B = calcium. Let us consider a rectilinear edge dislocation (Δ) in A, along the axis Oz with Burgers vector b parallel to Ox. We shall use cylindrical coordinates (r, θ), see the accompanying figure.

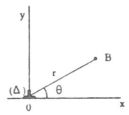

An atom B in solution in A creates an elastic perturbation whose intensity can be characterised by a positive or negative volume δv called the *size effect*[2]. This perturbation interacts with the stress field (Δ). It may be shown that the elastic interaction energy has the form:

$$W = K \frac{\sin \theta}{r} \tag{3}$$

with $K \simeq (\mu b / 3\pi) \delta v$ where μ is the shear modulus.

18. *Generalising without proof what was said (see p. 176) about vacancies subjected to a force, write down the form of the drift velocity V of the B atoms under the effect of the presence of (Δ). We shall denote the diffusion coefficient of B in A by D_B.*

In the plane (r, θ) perpendicular to (Δ) draw the equipotential lines and the lines of flux of B atoms.

19. *Taking into account both the random diffusion of the B atoms and the drift movement due to (Δ), express the flux $\overline{J}$ of the B atoms in its vectorial form. Deduce the components J_r and J_θ of this flux.*

20. *Find the form of the 'equilibrium' concentration $c(r, \theta)$ of B atoms. For this, we assume that in the stationary regime the flux is zero everywhere. The change of variable $u = \sin \theta / r$ may be used. We let c_0 denote the concentration of B far from (Δ).*

In the formula for $c(r, \theta)$, indicate why, whatever the sign of K (that is, that of δv), a 'cloud' of B atoms forms around (Δ) provided only that D_B has a sufficiently high value.

Give the size of this cloud, that is, a distance r_0 to the dislocation where c is significantly greater than c_0 (for example, $c = 2c_0$) and the numerical value of r_0 for current values of μ and b. We shall take $T \sim 300\ K$ and $\delta v \simeq b^3$ (the atomic volume).

---

[2] To the first order, δv is the difference between the 'volume' of the B atom and the volume available to it in A. Thus, the value of δv depends, in particular, on the fact that the B atom is substitutional or interstitial in the A crystal.

21. The expression found for $c(r, \theta)$ diverges when $K \sin \theta < 0$ as $r \to 0$.

Give the reason (based on elasticity) why, in this case, this expression is a large overestimate of the concentration c near (Δ).

22. *Say what should happen if B is not soluble in all proportions in A (limit of solubility c_s at temperature T).*

23. *Explain, in a few words, why the Cottrell clouds tend to harden materials.*

24. The accompanying optical micrograph (by transmission) shows a KCl crystal previously doped by Ag atoms and annealed at 700° C.

In your opinion, what can be seen on this micrograph and why can one see that?

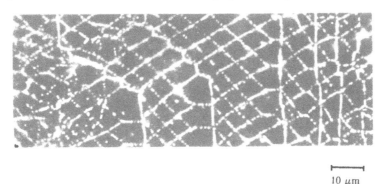

10 µm

After **S Amelinckx**, *Acta Metal.*, **6**, 34 (1958).

SOLUTION TO PROBLEM 13

1. $dV = r^2 \sin \theta \, d\theta \, dr \, d\varphi$.

2. This number is $c \, dV / \tau$.

3. A point of dV sees the unit element dS with an apparent surface area $\cos \theta$. The probability that a particle emanating in dV crosses dS is thus $\cos \theta / 4 \pi r^2$. Taking into account collisions on the path dV–dS, for which there is a probability of 'survival' $\exp(-r/v\tau)$ ($v\tau$ = mean free path of the particle of velocity v), we have

$$dc = \frac{c}{4 \pi \tau} \cos \theta \sin \theta e^{-r/v\tau} \, dr \, d\theta \, d\varphi. \qquad (4)$$

4.

$$j^+ = \int_{r=0}^{\infty} \int_{\theta=0}^{\pi/2} \int_{\varphi=0}^{2\pi} dc. \qquad (5)$$

5. If in (4) and (5) we replace c by its expansion (1), the integrals in φ concerning x ($= r \sin\theta \cos\varphi$) and y ($= r \sin\theta \sin\varphi$) are zero. Thus, we are left with:

$$j^+ = \frac{c_0}{2\tau} \int\int \sin\theta \cos\theta d\theta e^{-r/v\tau} dr + \frac{1}{2\tau}\left(\frac{\partial c}{\partial z}\right)_0 \int\int \sin\theta \cos^2\theta d\theta r e^{-r/v\tau} dr$$

whence

$$j^+ = \frac{c_0 v}{4} + \frac{1}{6}\left(\frac{\partial c}{\partial z}\right)_0 v^2\tau.$$

6. Similarly, we have

$$j^- = \frac{c_0 v}{4} - \frac{1}{6}\left(\frac{\partial c}{\partial z}\right)_0 v^2\tau$$

whence, the excess number of particles in the direction $z > 0$ is

$$j_z = -(j^+ - j^-) = -\frac{v^2\tau}{3}\left(\frac{\partial c}{\partial z}\right)_0. \tag{6}$$

7. j_z is simply the z-component of a flux of particles $\bar{j}$, which, according to (6), is proportional to $\bar{\nabla}c$:

$$\bar{j} = -\frac{v^2\tau}{3}\bar{\nabla}c$$

in conformity with the law (8) (p. 175). The coefficient $v^2\tau/3$, applied to the case of vacancies, may be written as $\frac{1}{6}v\Lambda$ (where Λ is the free path and 2τ is the time between the last and the next collision, see p. 18) or

$$\frac{1}{6} \underbrace{Z\bar{v}\exp(-E_m/kT)r_0}_{\text{velocity}} \times \underbrace{r_0}_{\text{free path}} = \frac{Z}{6}\bar{v}r_0^2\exp(-E_m/kT)$$

which is equal to the diffusion coefficient of vacancies.

8. The diffusive flux (6) is now supplemented by the drift flux

$$J_z = -\frac{v^2\tau}{3}\left(\frac{\partial c}{\partial z}\right)_0 + Vc.$$

The relaxation, as in the case of the movement of electrons in Drude's theory, gives rise to a constant drift velocity, without any contradiction with Newton's law.

9.

$$\left(\frac{\partial c}{\partial t}\right)_0 = \frac{v^2\tau}{3}\left(\frac{\partial^2 c}{\partial z^2}\right)_0 - V_z\left(\frac{\partial c}{\partial z}\right)_0 \tag{7}$$

in which expression we can, of course, omit the indices $(\ )_0$.

10. The dimensions of α and β in (2) are $l^2 t^{-1}$ and $l t^{-1}$, respectively. Taking into account the data (v, τ and V_z) and their incidence in the problem, α involves v and τ, while β involves V_z. The dimensions then imply $\alpha \sim v^2 \tau$ and $\beta \sim V_z$.

11. Equilibrium is reached when we $\partial c / \partial t = 0$ in (7) or in (2). It then follows immediately that

$$c = c(0) e^{-z/z_0}. \tag{8}$$

12. The accompanying figure shows $\ln c_{\text{exp}}$ as a function of z. The linearity observed matches the exponential function (8) well.

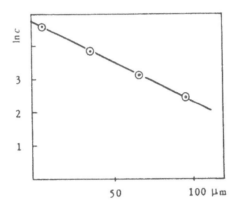

13. From the previous graph, we read off $z_0 = 42$ μm.

14. From (7) (or from (2)), we deduce that

$$z_0 = v^2 \tau / 3 V_z \quad \text{or} \quad z_0 = \alpha / \beta \sim v^2 \tau / V_z \quad \text{(see question 10)}.$$

15. The thermal equilibrium implies that we have $mv^2 = 3kT$. On the other hand, the force F_z of question **8** is equal to $m'g$ (m' is the apparent mass of a granule in water, g denotes the acceleration due to gravity). Thus, the drift velocity V_z is given by:

$$\text{acceleration} \times \tau = m'g\tau/m.$$

Whence, finally, we have

$$z_0 = \frac{v^2 \tau}{3 V_z} = \frac{kT}{m'g}.$$

16. At room temperature, $kT = 1/40$ eV $= 1/40 \times 1.6 \times 10^{-12}$ CGS and $m' = \frac{4}{3}\pi(0.21 \times 10^{-4})^3 (1.2-1)g$. Whence $z_0 = 52$ μm agrees very well with the experimental value (see question **13**), taking into account the uncertainty relating to the shape (and the radius) of the granules.

17.   'Equilibrium' here is clearly not the same thing as 'rest'. The granules migrate upwards under the effect of the concentration gradient and downwards under the effect of gravity, where these two fluxes are equal.

Formally, in a first analysis, this problem is the same as that of the molecules forming the air in the atmosphere, except that, in the latter case, taking into account the values of m, the values of z_0 are measured in kilometres rather than in microns.

18.   An atom B is subjected to the force

$$\overline{F} = -\overline{\nabla}W.$$

due to (Δ).

This results in a drift velocity (see p. 177)

$$\overline{V} = -\frac{D_B \overline{\nabla}W}{kT}.$$

The lines $W = $ constant are circles (shown by dashes), as are the lines of flux perpendicular to these (shown by continuous curves, with arrows for $\delta v > 0$).

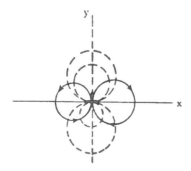

19. $\overline{J}$ is given by

$$\overline{J} = -D_B \overline{\nabla} c - \frac{D_B \overline{\nabla} W}{kT} c$$

whence it follows that

$$J_r = -D_B \frac{\partial c}{\partial r} - \frac{D_B c}{kT} \frac{\partial W}{\partial r} \qquad J_\theta = -\frac{D_B}{r} \frac{\partial c}{\partial \theta} - \frac{D_B c}{kT} \frac{1}{r} \frac{\partial W}{\partial \theta}. \qquad (9)$$

20.   In the steady state we have $J_r = J_\theta = 0$. If we introduce the variable $\sin\theta/r$, the corresponding system is easily solved giving

$$c = c_0 \exp\left(-\frac{K}{kT} \frac{\sin\theta}{r}\right). \qquad (10)$$

If we have $K > 0$ (*resp.* < 0) and $\sin\theta < 0$ (*resp.* > 0), then c is greater than c_0 and increases as r decreases: a supersaturation (or cloud) of B atoms forms below (*resp.* above) the dislocation, that is in the region under tension (*resp.* under compression). This is normal, taking into account the corresponding sign of the size effect δv. The establishment of this cloud only requires the temperature (i.e. D_B here) to be sufficiently high that the diffusion of B towards the dislocation takes place within a reasonable time.

It goes without saying that, since the direction Oz is free, the supersaturation extends uniformly along (Δ), creating a 'cloud' with cylindrical symmetry around (Δ).

N.B. Taking the sign of $\sin\theta$ into account in (10), an 'anticloud' with concentration $c < c_0$ forms on the other side of (Δ).

We obtain a characteristic size $r = r_0$ by writing, for example for $\sin\theta = 1$, $2 = \exp(K/r_0 kT)$ or $r_0 = K/kT \ln 2$. We write $K \simeq (\mu b/3\pi)b^3$. Taking $\mu b^3 \simeq 5$ eV (see p. 210) and $kT = 1/40$ eV, we find $r_0 \simeq 60b \simeq 125$ Å.

N.B. The expression (3) is approximate. A more complete calculation gives $K = [(\mu b/3\pi)(1 + v)/(1 - v)]\delta v$, where v is the Poisson ratio for the solid A. Taking the current value $v = \frac{1}{3}$ or $(1 + v)/(1 - v) = 2$, we obtain $r \simeq 120b \simeq 250$ Å.

21. When the concentration of B increases near (Δ) as an application of (10) (for example, accumulation of small B atoms $\delta v < 0$ in the region under compression, above the dislocation) the system of stresses developed by (Δ) decreases since the compression (in the given case) decreases. Thus, the energy W decreases and tends to decrease the driving force. Therefore, in (9), we should consider W as a decreasing function of c, which would tend to suppress the divergence of c for $r \to 0$.

22. If B is not completely soluble in A, the atoms B precipitate in the cloud from the concentration $c(r, \theta) = c_s$. The cloud then consists of small precipitates with a phase rich in B near the centre and a supersaturation c (such that $c_s > c > c_0$) of the type (10) towards the periphery.

23. The presence of the cloud around (Δ) tends to anchor the latter in its position. In fact, a glide or a climb of (Δ) from this position, requires the supply of the cloud–dislocation binding energy: this corresponds to 'alloy hardening'. If precipitates have formed (question **22**) they generally contribute to a further increase of this hardening.

24. As shown above, the silver atoms cluster along the edge-type dislocations present initially. It is plausible for them to precipitate (see question **22**) forming a chaplet of precipitates along the dislocations. This is what one sees using an optical microscope (low magnification) in the (transparent) KCl crystal; it is these bracelets which decorate the dislocations.

In the 1950s, this *decoration* technique was one of the first methods which enabled us to 'see' dislocations (or, as here, networks of dislocations) in the sense that we observe not the dislocations themselves (the magnification is too weak) but the clouds of precipitates which they induce *via* their stress field. Another method comprised a chemical etching of the surface of the sample. The points of emergence of the dislocations were preferentially etched, due to the impurities collected along the dislocations creating very characteristic '*etch-pits*' on the surfaces[3].

[3] **P Lacombe**, *Conference on Strength of Solids*, The Physical Society, London (1948)

OXIDATION OF METALS

If one introduces a metallic surface (metal M) into an oxygen atmosphere, an oxide film (OX, for example: M_2O), whose thickness $L(t)$ (Figure 1) generally increases with time t, forms. This is the phenomenon of the *oxidation of metals*.

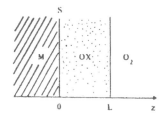

Figure 1 Formation of an oxide layer OX of thickness L on a metal M.

Frequently, but not systematically, the 'oxidation law' observed experimentally is of the form

$$L(t) = Kt^{1/2} \tag{1}$$

where K is a constant which depends on the temperature, among other things. Figure 2 gives examples of this empirical law.

The aim of the problem is to show that the different conceivable oxidation mechanisms lead to different laws $L(t)$. Thus, conversely, experimental determination of L as a function of t gives valuable information about the mechanism in play in the case studied.

One of the simplest mechanisms imaginable involves the following stages:

(i)   Dissolution of atoms of metal M in the oxide at the interface M/OX.

(ii) Diffusion of the M atoms in the oxide OX, the aim of the problem being to determine the kinetics of this diffusion.

(iii) Reaction $M + O_2 \rightarrow OX$ at the OX/O_2 interface, i.e. formation of a stable oxide.

Throughout the problem, we shall assume that stages (i) and (iii) are sufficiently fast that the corresponding chemical equilibria are reached, so that the concentrations of metal M in the oxide OX at the two interfaces of the oxide remain constant with time.

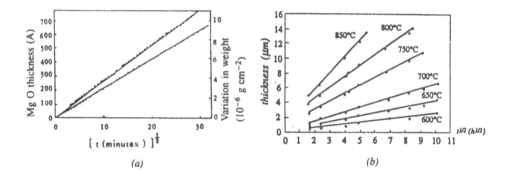

(a)   (b)

Figure 2 (a) **Oxidation of magnesium.** For two magnesium single crystals with different crystallographic orientations, a microbalance was used to measure the increase in weight as a function of time t when the metal was introduced into oxygen at 400° C. After **R Schwoebel**, *J. Appl. Phys.*, **34**, 2776 (1963).
(b) **Oxidation of zirconium by pure oxygen (normal pressure) at various temperatures.** The thickness of the oxide layer is plotted as a function of $\sqrt{t}$. After **G Béranger**.

Let z be the distance measured from S where the M/OX interface is assumed to be a plane perpendicular to S (Figure 1). $C(z)$ will denote the volume concentration of M atoms in OX, J_D will denote the flux of M atoms parallel to Oz in OX and D will be the diffusion coefficient (assumed to be independent of C) of M in OX.

In what follows, we shall assume that the absolute temperature T remains constant and that the bulk of the oxide OX does not contain either sources of M atoms or sinks for these atoms, which is equivalent to saying that the flux J_D does not depend on z: $J_D = J_0$. For the initial time of oxidation we shall take the time $t = 0$ (thus, let $L(0) = 0$, assuming that the initial thickness of the oxide is negligible).

I

1. Give the expression for the gradient dC/dz as a function of J_D $(= J_0)$ and D.

2. Denote the (constant) concentrations of the metal in the oxide at $z = 0$ and $z = L$ by C_0 and C_1, respectively. Express the flux J_0 as a function of these quantities.

3. Recalling that the thickness of the oxide layer increases with the arrival of M atoms at the OX/O$_2$ interface and denoting the volume of oxide formed per atom of metal M by Ω, give the expression for dL/dt and deduce the law $L(t)$.

4. Does this law correspond to the experimental observations of Figure 2? How might one try to explain qualitatively (in two or three lines) the difference between the oxidation rates for the two different crystallographic faces of magnesium? What variation in the kinetics with the temperature T does this law imply?

II

The oxidation mechanism is rarely as simple as described above. In particular, the possible accumulation of charges (cations, anions, ...) on the surfaces may create an *electric field* in the oxide.

Initially (questions **5** to **9**), we shall suppose that this electrical field is constant, i.e. uniform in the space between $z = 0$ and $z = L$, parallel to the Oz axis and constant in time: $E = E_0$. We shall again assume we are in the case in which the oxidation is due to the diffusion of the metal M in the oxide OX, but now in the form of M^+ cations (with charge $+e$).

5. *Give the expression for the drift flux J_{Drift} of the M^+ cations, created by field E_0.*

N.B. For this, use Einstein's formula (see (13), p. 177) which was established for the diffusion of vacancies, but which may be generalised directly to the diffusion of any other object subjected to a force and characterised by a coefficient of diffusion D (see also **Problem 13**). We shall denote the diffusion coefficient of the M^+ cations in OX at the temperature T in question by D.

6. For the total flux of the cations J, we then take the sum of the diffusive flux J_D and the drift flux J_{Drift}; as in the introduction, we shall suppose that this flux J is uniform in OX: $J = J_0$.
   Show that the concentration $C(z)$ of M^+ in OX is a solution of a differential equation of the form

$$\alpha = \beta \frac{dC}{dz} + \gamma C$$

where α, β and γ are constants whose values should be determined.
   Deduce the expression for $C(z)$.

7. *Again with the assumption that C_0 and C_1 are constants, write down the expression for the flux J_0 together with the differential equation which describes the variation of L as a function of t.*

8. We shall solve this equation for the particular case in which the concentration gradient is high: $C_1 \ll C_0$. More precisely, and for convenience, we set $C_1 = 0$. *Give the expression describing the dependence of L on t.*

9. In the general case, E_0 may equally well be positive or negative.
   Study the asymptotic behaviour of $L(t)$ for small t and for large t. In the latter case, study the effect of the sign of E_0 using the expression for dL/dt determined in question 7. How does $L(\infty)$ behave when E_0 is negative and C_1 is small in comparison with C_0, but non-zero?
   Sketch the variations of L as a function of t in three graphs corresponding to $E_0 > 0$, $E_0 = 0$ and $E_0 < 0$, respectively. Comment in simple terms on the differences between these three graphs.

10. In the last question, we shall continue to study the oxidation resulting from the diffusion of metallic cations in OX in the presence of an electric field. However, we shall now assume that the electric field E, which is still uniform in space and parallel to Oz, varies with time (with the thickness of the oxide layer) so that the potential difference between the two surfaces of the oxide $V_0 = EL(t)$ remains constant with time.
   Under these conditions write down the differential equation describing the evolution $dL(t)/dt$ (found in question 7) again and deduce the form of $L(t)$.
   What do you notice in comparison with the formula found in question 3? What experiment can be used (at least in principle) to separate the corresponding mechanisms?

SOLUTION TO PROBLEM 14

I

1. According to Fick's first equation, we have:

$$\frac{dC}{dz} = -\frac{J_0}{D}.$$

2. Since the flux J_0 is constant from $z = 0$ to $z = L$, we have:

$$J_0 = -\frac{D}{L}(C_1 - C_0).$$

3. In unit time, the thickness of OX increases by $J_0\Omega$. Thus, we have:

$$dL = J_0\Omega dt = \frac{C_0 - C_1}{L}D\Omega dt$$

whence

$$L(t) = [2(C_0 - C_1)D\Omega t]^{1/2}. \tag{2}$$

4. Here, the oxidation law is of the form (1). It corresponds to the experimental data of Figure 2. The kinetic constant K varies with temperature as $D^{1/2}$, that is as $\exp(-\Delta E/2kT)$ where ΔE is the diffusion energy of the M atoms in the oxide.

The following may be proposed as possible explanations of the difference of kinetics between two crystallographically distinct faces of magnesium:

(i) different epitaxy conditions of OX on the metal (in particular, different concentrations of dislocations) which modify the value of C_0;
(ii) an anisotropy of the coefficient of diffusion D of M in OX associated with a different crystalline orientation of the oxide OX.

II

5. The M$^+$ cations are subjected to the uniform force

$$F = eE_0.$$

This gives rise (Einstein's formula) to a drift velocity parallel to this force, of magnitude

$$v = \frac{DeE_0}{kT}$$

whence to a drift flux:

$$J_{\text{Drift}} = C(z)v = \frac{DeE_0}{kT}C(z)$$

which depends on z.

6. The total flux of cations is

$$J = J_D + J_{Drift} = -D\frac{dC(z)}{dz} + \frac{DeE_0}{kT}C(z).$$

Without loss or creation of cations in the bulk of the oxide OX, J is constant: $J = J_0$. Thus, $C(z)$ satisfies the equation

$$\alpha = \beta\frac{dC(z)}{dz} + \gamma C(z)$$

where

$$\alpha = J_0 \qquad \beta = -D \qquad \gamma = \frac{DeE_0}{kT}.$$

It follows immediately that

$$C(z) = C_0 e^{(\gamma/D)z} + \frac{J_0}{\gamma}\left(1 - e^{(\gamma/D)z}\right).$$

7. From the above equation for $z = L$, it follows that

$$J_0 = \gamma\frac{C_1 - C_0 e^{(\gamma/D)L}}{1 - e^{(\gamma/D)L}}.$$

The differential equation describing the variations of L is again given by $dL = J_0\Omega dt$, whence here

$$\frac{dL}{dt} = \gamma\Omega\frac{C_1 - C_0 e^{(\gamma/D)L}}{1 - e^{(\gamma/D)L}}. \tag{3}$$

8. Let us consider the case $C_1 = 0$. Equation (3) becomes

$$\frac{dL}{dT} = -\gamma\Omega C_0 e^{(\gamma/D)L}/(1 - e^{(\gamma/D)L}). \tag{4}$$

Taking into account the condition $L(0) = 0$, L is therefore defined by the equation:

$$e^{-(\gamma/D)L} + \frac{\gamma}{D}L = 1 + \frac{\gamma^2}{D}\Omega C_0 t. \tag{5}$$

9. *Short times.* For t (and L) small, we have (see (5)):

$$\frac{\gamma^2}{2D^2}L^2 \simeq \frac{\gamma^2}{D}\Omega C_0 t$$

whence

$$L^2 \simeq 2D\Omega C_0 t.$$

Thus, whatever the sign of γ (i.e. that of E_0) the oxidation initially follows a parabolic law of the form (1), or, more precisely (2), which is the same as in a zero field.

Intermediate times. The thickness L varies according to the law:

$$\sum_{p=z}^{\infty}(-1)^p\frac{1}{p!}\frac{\gamma^{p-2}}{D^{p-1}}L^p = \Omega C_0 t.$$

Long times. If E_0 (i.e. γ) is *positive*, the oxidation rate dL/dt (see (4) or (3)) tends to the constant value $\gamma\Omega C_0$.

If E_0 is negative, the oxidation rate dL/dt decreases continuously (see (4)) and even tends to 0 if C_1 is non-zero (see (3)). Thus, in this case, the oxide layer has a limiting thickness L_∞ equal to

$$L_\infty = -\frac{D}{\gamma}\ln\frac{C_0}{C_1} = -\frac{kT}{eE_0}\ln\frac{C_0}{C_1}.$$

The corresponding variations of the thickness are illustrated in the accompanying diagram.

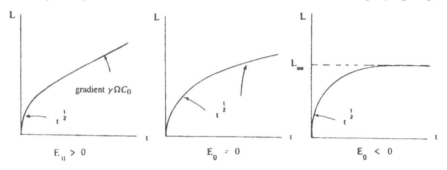

In comparison with the purely diffusive case ($E_0 = 0$; $L_\infty \propto t^{1/2}$) a *positive* electrical field *stimulates* the diffusion of the cations in the direction favouring oxidation: as the thickness of the oxide increases the concentration gradient decreases, whence so too does the diffusive flux. The latter progressively becomes negligible in comparison with the drift flux, which then ensures that the oxidation has a constant rate. In return, a *negative* field *slows down* the diffusion of cations. It creates a flux opposite to the diffusive flux. When the absolute values of these two fluxes become equal the oxidation stops.

10.   If we replace γ by its value DeE/kT (or $-(DeV_0/LkT)$ in this case) equation (3) is transformed to

$$\frac{dL}{dt} = -\frac{DeV_0\Omega}{LkT}\times\frac{C_1 - C_0e^{-eV_0/kT}}{1 - e^{-eV_0/kT}}$$

and this equation integrates to

$$L(t) = Kt^{1/2}.$$

Here we again find the characteristic parabolic law for a diffusion mechanism (question **3**). However, in this case, the kinetic oxidation constant K depends on v_0. Thus, to remove the ambiguity between these two possible mechanisms, one could measure the oxidation under the application of a constant potential and study the variations of the measured constant K as a function of the potential applied.

SNOEK EFFECT AND MARTENSITIC STEELS

N.B.   Although **I** and **II** are related, they may be studied independently. It is advisable to study them in order, but, in case of difficulties in **I**, readers may move to **II**.

III is more qualitative. While it requires a good understanding of **I** and **II**, it also calls for intuition and and physical instinct.

Short answers will suffice for most questions.

Carbon is a classical 'impurity' of iron. Up to contents c_0 of the order of 8×10^{-2} (c_0 = number of carbon atoms $\div$ number of iron atoms) Fe–C alloys constitute *steels*. Beyond this order, one enters the region of *cast irons*.

Here, we shall use an experiment conceived by Snoek to study the evolution of carbon in steels with low carbon contents ($c_0 \leq 10^{-2}$).

At room temperature, the iron atoms occupy all the sites of a body-centred cubic lattice (edge of the primitive cube: 2λ). In such a lattice, the vertices of the primitive cubes and their centres are equivalent, in that a translation by the vector $(\lambda, \lambda, \lambda)$ interchanges the roles of the centres and the vertices.

The carbon atoms may be located at sites either at the centres of edges (α) or in the centres of faces (β) of primitive cubes (Figure 1). Here again, a translation by the vector $(\lambda, \lambda, \lambda)$ interchanges the roles of the centres of the edges and the centres of the faces; each site where a carbon atom may be located is situated half-way between two iron atoms at distance 2λ apart. Thus, we shall distinguish between three families of carbon sites (x), (y) and (z), according to the direction in which the neighbouring iron atoms are found. For example, in Figure 1, α is an (x) site located between two iron atoms lying on the same parallel to Ox and β is a (y) site.

In the absence of stresses, these three families clearly play the same role. At equilibrium the corresponding concentrations are equal:

$$c_x = c_y = c_z = \frac{c_0}{3}. \tag{1}$$

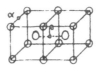

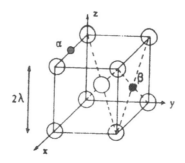

Figure 1   Two possible carbon sites α and β in iron.

Note, for what follows, that there are four carbon sites which are nearest neighbours of a given (z) site. These lie at distance λ from that site. Two of them are (x) sites, the other two are (y) sites.

Note also that a macroscopic volume Ω of metal contains $\Omega/8\lambda^3$ primitive cubes, $N = 2(\Omega/8\lambda^3)$ iron atoms and $6(\Omega/8\lambda^3)$ carbon sites.

Snoek's experiment

We shall now try to break the equidistribution manifest by (1). For this, we take a single crystal of steel cut in the form of a prism of volume Ω, with edges a, a and b parallel to Ox, Oy and Oz (Figure 2), which is initially unstressed and in equilibrium ($c_x = c_y = c_z = c_0/3$). At time t, we subject it to a *uniaxial tensile* stress σ along Oz ($\sigma a^2 =$ applied force). This stress then remains constant. We measure the length of the sample along Oz as a function of time. We shall denote the extension $\Delta b/b$ by ε.

At time $t = 0$, we observe the normal *elastic* extension

$$\varepsilon_1 = \frac{\Delta b}{b} = \frac{\sigma}{Y} \tag{2}$$

where Y is the (Young's) modulus of elasticity of iron.

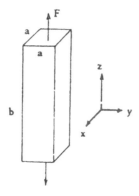

Figure 2   Snoek's experiment.

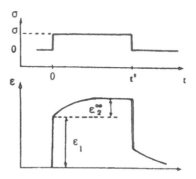

Figure 3   Snoek's results.

Then, over time, an additional, so-called *inelastic* extension ε_2 appears, which saturates to a value ε_2^∞ (Figure 3).

If we remove the stress applied after a long time t', two successive contractions $\varepsilon_1' = -\varepsilon_1$ and $\varepsilon_2' = -\varepsilon_2$ restore the initial size of the sample. Thus, the phenomenon known as the *Snoek effect* is reversible. The experiment shows that

i) ε_2 varies exponentially with t. The higher the temperature T is, the faster the equilibrium ε_2^∞ is attained.

ii) ε_2^∞ is proportional to the carbon concentration c_0.

iii) ε_2^∞ is proportional to the stress σ and inversely proportional to T.

I

Statistical thermodynamics approach

The idea is as follows. By moving the iron atoms apart along Oz the tension σ tends to stabilise the carbon atoms at (z) sites to the detriment of the (x) and (y) sites. Indeed, the latter are more 'compressed' so that the energy of a carbon atom located at a (z) site becomes lower (by an amount $2w$) than if it were located at an (x) site or a (y) site. This energy variation creates an asymmetry so that the populations c_x, c_y and c_z no longer have any reason to be equal. We shall characterise the distance from the isotropic situation (1) by a parameter

$$\eta = c_z - \frac{c_0}{3} \tag{3}$$

called the *polarisation* of the system. At equilibrium, this parameter reaches the value η^∞.

We shall show later (question 5) that the polarisation η of the carbon atoms in the direction Oz gives rise to an extension ε_2 which adds to the elastic extension ε_1 defined by (2) and is proportional to η:

$$\varepsilon_2 = B\eta \tag{4}$$

where, in particular,

$$\varepsilon_2^\infty = B\eta^\infty.$$

This will enable us to explain the Snoek effect.

First, we shall calculate the equilibrium value η^∞ of η at temperature T for a given value of the overall extension ε. For this, we shall study the free energy $F(\varepsilon, \eta, T)$ of quasi-equilibrium states of the sample considered at various stages of the polarisation phenomenon. Apart from temperature, these states are characterised by a given overall extension $\varepsilon = \varepsilon_1 + \varepsilon_2$ and a given polarisation η, considered to be independent variables. The equilibrium polarisation η^∞ is a function of ε which will be determined as a reflection of the fact that at equilibrium the two systems comprising the (z) sites and the (x) and (y) sites may exchange carbon atoms (for fixed ε).

We now consider the case where ε_2 is *negligible in comparison with ε_1*, which condition is satisfied if c_0 is sufficiently small.

We assume that the only notable variations of the free energy come from the following contributions.

a)   The contribution from the *elastic free energy* associated with the interactions between iron atoms, which has the same expression as for pure iron:

$$\Omega \tfrac{1}{2} Y \varepsilon^2.$$

b) The contribution from the *decrease in internal energy* $2w$ produced by each jump (x) or $(y) \rightarrow (z)$, since the carbon at (z) is in a more favourable situation than at (x) or (y). The energy $2w$ will be taken to be proportional to the deformation ε (practically constant and equal to ε_1 during the polarisation):

$$2w = A\varepsilon. \tag{5}$$

c) The contribution due to the *variation of configurational entropy* S_m of the carbon atoms.

We let N denote the number of iron atoms contained in the sample of Figure 2, N_c the number of carbon atoms and ρ the number of iron atoms per unit volume.

1. *Determine the entropy S_m for arbitrary initial concentrations c_x, c_y and c_z (use Stirling's formula $\ln n! \simeq n \ln n - n$).*

Write down the expression S_m^0 for S_m in the case of equilibrium in the absence of stress $(c_x = c_y = c_z = c_0/3)$.

2. *What happens to this entropy S_m when the carbon atoms are polarised? Here, we suppose that σ is sufficiently small that we have $\eta \ll c_0$ and we recall that $c_0 \ll 1$. Calculate $S_m - S_m^0$ as a function of η to order two in η.*

3. *Using the assumptions b) and c) above, determine the value η^∞ at equilibrium as a function of ε. Express this value as a function of the parameters σ, c_0 and T which can be controlled experimentally.*

4. *Use the expression for $F(\varepsilon, \eta, T)$ resulting from assumptions a), b) and c) above to give the relationship between the stress σ and the overall extension ε for a given polarisation η.*

5. *Deduce that equation (4) is a consequence of these assumptions and that the parameter B is given as a function of the quantities A and Y (defined by (5) and (2)) by*

$$B = \frac{A\rho}{Y}.$$

6. *Give the expression for the inelastic extension at equilibrium. Which experimental results (of questions 1 to 3 above) are taken into account by the model?*

7. *Evaluate w numerically in eV, taking $\sigma = 1000$ bars $(= 100$ MPa$)$. For B, we shall take the experimentally determined value 0.8. Iron has atomic mass 56 and density 7.8 g cm^{-3}.*

Evaluate the inelastic extension ε_z^∞ at room temperature for a steel with content $c_0 = 10^{-3}$.

II

Kinetics approach

We shall study the way in which the directional order establishes itself with time; in other words, the evolution of the polarisation η.

This clearly requires jumps of carbon atoms from (x) (or (y)) sites to (z) sites. We shall assume that these jumps of carbon only occur between near-neighbour sites. Each jump is taken along a diffusion path of length λ and joins two near-neighbour sites of different types (the accompanying diagram represents a jump $(x) \rightarrow (z)$).

We shall also assume that during the jump the carbon atom passes through an energy saddle-point half-way between the two sites (departure and arrival, see Figure 4). Let ΔE be the height of this saddle-point. We denote the characteristic frequency with which the carbon may vibrate on each site by ν.

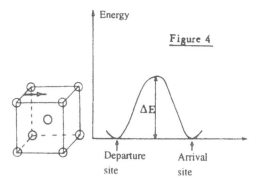

Figure 4

Figure 4 Jump of a carbon atom.   **Figure 5** Influence of a stress.

The jump of a carbon atom from a *given* departure site to a *given* (neighbouring) arrival site is characterised by a probability per unit time depending on the temperature T and the height of the saddle-point. This probability per unit time is equal to $\nu \exp[-(E_m - E_0)/kT]$, where E_m is the energy of the top of the col and E_0 is the energy of the departure site.

We recall that the concentration c_0 is very small.

8. *What is the average number of jumps ν_j per unit time performed by a given carbon atom?*

At room temperature, what is the average time τ_s a carbon atom stays at a given site? Use the experimental values $\nu = 1.5 \times 10^{14}$ s^{-1} and $\Delta E = 0.87$ eV.

9. The stress applied in Snoek's experiment modifies the energy profile of Figure 4 into that of Figure 5: under stress, the (z) sites have an energy $2w$ lower than that of the (x) or (y) sites (see the introduction to **I** and formula (5)). However, the energy E_M of the top of the saddle-point is unchanged.

The stress σ is taken to be sufficiently small that w is small in comparison both with ΔE and with kT. The variation of w during the experiment is negligible. The carbon atoms at (z) sites (concentration c_z) may jump to (x) or (y) sites. Conversely, (x) or (y) atoms may jump to (z) sites.

Write down the differential equation describing the overall variations dc_z of c_z at time t during time dt.

10. *Deduce the general form of the functions $c_z(t)$ and $\varepsilon_2(t)$ from $t = 0$ for Figure 3, to order one in w/kT, for the initial conditions $c_{z0} = c_{x0} = c_{y0} = c_0/3$.*

In particular, evaluate the quantities c_z^∞ and ε_2^∞ (check that this value is identical to that found in question 6) together with the characteristic time τ of the evolution of the inelastic extension. How does this time vary with T?

Which experimental results (points i) to iii) of the introduction) does this theoretical result explain?

Using the numerical values calculated previously, give the order of magnitude of τ. Verify that the assumptions $w \ll \Delta E$ and $w \ll kT$ were justified.

11. The numerical values of the frequency v and the energy ΔE were given above without justification (question **8**).

What experimental procedure would you adopt to measure them?

III

Martensitic steels

Figure 6 shows the Fe–C equilibrium diagram. The α phase (or ferrite) is the body-centred cubic phase studied above. The γ phase (or *austenite*) is a face-centred cubic phase in which the carbon is more soluble than in the α phase. The Fe_3C phase is called *cementite*.

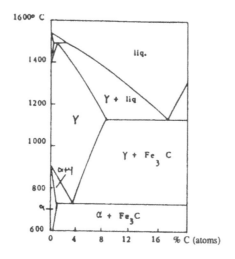

Figure 6.

Suppose we heat a steel ($c_0 = 5 \times 10^{-2}$ atom) to 1100° C and cool it rapidly (quench) to room temperature.

12. *What is the equilibrium structure of this steel at 1100° C? What would its equilibrium structure normally be at room temperature (as at 600° C)?*

13. The experiment shows that, in fact, neither of these structures is that observed after quenching. The quenched steel consists of a pseudo-ferrite analogous to the phase studied earlier, but in which, locally, *all* the carbon atoms occupy *a single* type of site, for example (z).

This structure, which is said to be *martensitic*, is not in thermodynamic equilibrium: it is metastable. The formation of the martensite, that is the *complete* polarisation of the carbon atoms in (z) positions, is a dynamic process due to the quenching which we shall not describe here.

What is the value of the polarisation? In your opinion, what is the shape of the primitive cell of the martensite? What are its dimensions compared with the edge 2λ of the ferrite cube of Figure 1? Does experimental measurement (using X-rays) give access to the value of the coefficient B of (4)?

14. Although it is metastable, martensite may survive a very long time: martensite has been found in prehistoric tools.

Can you give a qualitative explanation of the reasons for this longevity of martensite?

15. The axis of polarisation of martensite may equally well be x, y or z. In any region of a piece of quenched steel, one may observe small zones of martensite (from several hundred Å to several microns) of an (x), (y) or (z) nature.

Taking into account the answers to question 13, what do you deduce about the internal state of stress of the piece?

Do you sense a link between this answer and the hardness of quenched steels? Do you understand why the hardness of these steels increases with the carbon content?

16. *Using what has gone before (including, II in particular), how would you propose to decrease the hardness of a piece of martensitic steel?*

What sort of treatment do you think the more ductile of the two accompanying nails (i.e. the one which is easier to deform) has undergone [1]*?*

SOLUTION TO PROBLEM 15

1. There are N (x) sites $(x$, for example) in the crystal. Of these, $c_x N$ are occupied and $(1 - c_x)N$ are empty. Thus, we have

$$S_m = k \ln \frac{N!}{(c_x N)![(1 - c_x)N]!} \frac{N!}{(c_y N)![(1 - c_y)N]!} \frac{N!}{(c_z N)![(1 - c_z)N]!}$$

whence, applying Stirling's formula:

$$S_m = -kN[c_x \ln c_x + c_y \ln c_y + c_z \ln c_z + (1 - c_x) \ln(1 - c_x) + \ldots]. \tag{6}$$

This entropy is positive since the c_i and $(1 - c_i)$ are < 1.

If we have isotropy $(c_i = c_0/3)$, S_m becomes:

$$S_m^0 = -3kN \left[\frac{c_0}{3} \ln \frac{c_0}{3} + \left(1 - \frac{c_0}{3}\right) \ln \left(1 - \frac{c_0}{3}\right) \right]$$

$$\simeq kNc_0 \left(\ln \frac{3}{c_0} + 1 \right) \simeq kNc_0 \ln \frac{3}{c_0}.$$

2. In (6) we now replace c_z by $(c_0/3) + \eta$ and $c_x = c_y$ by $(c_0/3) - (\eta/2)$. Whence

$$S_m \simeq -kN \left[2 \left(\frac{c_0}{3} - \frac{\eta}{2} \right) \ln \left(\frac{c_0}{3} - \frac{\eta}{2} \right) + \left(\frac{c_0}{3} + \eta \right) \ln \left(\frac{c_0}{3} + \eta \right) \right],$$

which, after expansion to order two in η becomes

$$S_m = -kN \left(c_0 \ln \frac{c_0}{3} + \frac{9}{4} \frac{\eta^2}{c_0} \right) = S_m^0 - kN \frac{9}{4} \frac{\eta^2}{c_0}.$$

---

[1] **N.B.** Two nails, identical, except for the fact that one had been heat treated for one hour at 500° C, were given to the students.

The directional order (η) decreases the entropy by $kN(9/4)\eta^2/c_0$.

3. With the assumptions b) and c), the polarisation of the carbon atoms implies the following variation of the free energy

$$-A\varepsilon N\eta + NkT\frac{9}{4}\frac{\eta^2}{c_0}$$

because $N\eta$ carbon atoms have moved from (z) sites to (x) or (y) sites. Equilibrium is attained when this expression is a minimum, which condition corresponds to the maximum of the entropy taking the stresses into account. This gives:

$$\eta^\infty = \frac{2A\varepsilon c_0}{9kT} = \frac{2A\sigma c_0}{9kTY}.$$

4. The total free energy resulting from assumptions b) and c) is

$$F(\varepsilon, \eta, T) = \Omega\frac{1}{2}Y\varepsilon^2 - A\varepsilon N\eta + NkT\frac{9}{4}\frac{\eta^2}{c_0} - TS_m^0.$$

The work applied during the tensile strain, $\delta\varepsilon$ is $\sigma a^2 b\delta\varepsilon = \Omega\sigma\delta\varepsilon$, so that the stress may be deduced from F by equation

$$\sigma = \frac{1}{\Omega}\frac{\partial F}{\partial \varepsilon},$$

which gives

$$\sigma = Y\varepsilon - A\rho\eta \qquad (\rho = N/\Omega).$$

5. It follows from (2) that

$$\sigma = Y\varepsilon_1 = Y(\varepsilon_1 + \varepsilon_2) - A\rho\eta$$

so that

$$\varepsilon_2 = \frac{A\rho}{Y}\eta.$$

This proves (4), with $B = A\rho/Y$.

6. At equilibrium

$$\varepsilon_2^\infty = B\eta^\infty = \frac{2AB\sigma c_0}{9kTY} = \frac{2B^2}{9\rho k}\frac{\sigma c_0}{T}.$$

This formula takes account of the results i) and iii) (proportionality of ε to σ, of c_0 and of $1/T$).

7. By virtue of question **5**, (2) and (5), we have

$$w = \frac{A\varepsilon}{2} = \frac{A\sigma}{2Y} = \frac{B\sigma}{2\rho}.$$

The numerical application gives $\rho = 8 \times 10^{28}$ m^{-3}, whence

$$w = 5 \times 10^{-22} \text{ J} = \frac{5}{1.6} \times 10^{-22} \times 10^{19} \text{ eV} = 3 \times 10^{-3} \text{ eV}.$$

The extension ε_2^∞ is given by $\varepsilon_2^\infty = B\eta^\infty = (4Bc_0w/9kT)$ where $w = 3 \times 10^{-3}$ eV and $kT = 1/40$ eV. We obtain $\varepsilon_2^\infty \simeq 4 \times 10^{-5}$ (or $\Delta b \simeq 0.4$ μm for a sample of length $b = 1$ cm).

8. For each oscillation, the probability that a carbon atom has sufficient energy to pass over the saddle-point is $\exp(-\Delta E/kT)$, so that the probability per unit time of a jump is $\nu \exp(-\Delta E/kT)$. Given that there are four possible jumps from a given site, we obtain

$$\nu_j = 4\nu \exp(-\Delta/kT) \quad \text{and} \quad \tau_s = \frac{1}{4\nu} \exp(\Delta E/kT).$$

At room temperature $kT \simeq 1/40$ eV, whence

$$\tau(300 \text{ K}) = \frac{e^{-0.87 \times 40}}{4 \times 1.5 \times 10^{14}} = 2.2 \text{ s}.$$

At room temperature, a carbon remains several seconds at each site, but this time decreases rapidly if the temperature increases.

9. We have the balance:

$$N \, dc_z = \left[-\text{number of atoms per unit time making a jump } (z) \to (x) \text{ or } (y) \right.$$

$$\left. + \text{number of atoms per unit time making a jump } (x) \text{ or } (y) \to (z) \right] \times dt.$$

Whence

$$\frac{dc_z}{dt} = -4\nu c_z \exp\left(-\frac{\Delta E + w}{kT}\right) + 2\nu c_x \exp\left(-\frac{\Delta E - w}{kT}\right) + 2\nu c_y \exp\left(-\frac{\Delta E - w}{kT}\right).$$

Clearly, we have $c_x + c_y = c_0 - c_z$, which gives

$$\frac{dc_z}{dt} = -4\nu e^{-(\Delta E + w)/kT} c_z + 2\nu e^{-(\Delta E - w)/kT}(c_0 - c_z),$$

whence

$$\frac{dc_z}{dt} = n\left[c_0 e^+ - (e^+ + 2e^-)c_z\right]$$

where we have set $n = 2\nu \exp(-\Delta E/kT)$ and $e^\pm = e^{\pm(w/kT)}$.

10. Making the approximations indicated in the statement of the problem $\left(e^+ \simeq 1 + (w/kT)\right)$, the above differential equation integrates to:

$$c_z = \frac{c_0}{3} + \frac{4c_0 w}{9kT}[1 - \exp(-3nt)]$$

and $c_z^\infty = (c_0/3)(1 + (4w/3kT))$; whence

$$\varepsilon_2 = B\eta = B\left(c_z - \frac{c_0}{3}\right) = B\frac{4c_0w}{9kT}[1 - \exp(-3nt)]$$

and $\varepsilon_2^\infty = 2AB\sigma c_0/9kTY$.

The parameter η and the extension ε_2 increase with time and tend exponentially to the values determined in question **6**.

The characteristic time of this variation is $\tau = 1/3n = 2/3\tau_s$ (of the order of a second), or $(1/6\nu)\exp(\Delta E/kT)$ which decreases if the temperature increases. This accounts well for the experimental results i).

Thus we have shown that $w = 3 \times 10^{-3}$ eV is much smaller than $\Delta E = 0.87$ eV and than $kT = 1/40$ eV.

11. Thus, in order to access ν and ΔE experimentally it is sufficient to carry out a Snoek experiment such as that described in the introduction (measurement of length using capacitative or optical measurements ...) at *constant temperature* T; to measure the value of τ (for example from $\simeq -50°$ C to $+50°$ C); to sketch the log of the value of τ as a function of $1/T$; to check that the points obtained lie on a line; and thus deduce ν and ΔE.

12. According to the diagram, the equilibrium structure at 1100° C is austenite (γ phase, of fcc structure).

At room temperature, the equilibrium structure should be a mixture of the two phases of *ferrite* (or α iron containing a concentration $c_0 < 1\%$ of carbon in solution) and *cementite*.

13. If all the carbon atoms are at (z) sites, the polarisation is equal to

$$\eta^{(M)} = c_0 - \frac{c_0}{3} = \frac{2c_0}{3} \qquad (M = \text{martensite}).$$

The primitive cube is deformed by extension along Oz and contraction along Ox and Oy; its previous cubic structure becomes (centred) quadratic, and the cell has sides p, q and q (see figure below). The volume is preserved ($pq^2 = 8\lambda^3$) to the first order.

The extension of the cell along Oz is given by

$$\varepsilon_2^{(M)} = B\eta^{(M)} = \frac{2Bc_0}{3}.$$

Whence we have

$$p = 2\lambda\left(1 + \frac{2Bc_0}{3}\right) \qquad q = 2\lambda\left(1 - \frac{Bc_0}{3}\right).$$

Measurement of p (and/or q) for various values of the carbon content enables us to determine B: the value of B is the gradient of the experimental line $p/2\lambda = f(\frac{2}{3}c_0)$.

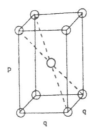

14. The extension of the axis of polarisation (z) of the martensite $\varepsilon^{(M)} = 2Bc_0/3 \simeq 3 \times 10^{-2}$ is very much greater (by a factor $\simeq 10^3$) than the inelastic extension calculated in question **7**. This implies a much larger stabilisation of the carbon atoms, whence a much larger value of w. The activation energy $\Delta E + w$ needed for a jump $(z) \rightarrow (x)$ or (y) becomes such that the carbon atoms may find themselves fixed for a very long time in a metastable position (z). Multiplying the value w calculated in question **7** by 100 would multiply the lifetime by $\exp(w/kT) \simeq e^{12} \simeq 10^5$; multiplying by 1000 would give 10^{50}!

15. The quadratic domains (x), (y) and (z) exert strong stresses on each other, the extension of one, (i), being opposed by the presence of its neighbours (j) and (k).

The interactions, of an elastic nature, between these internal stresses and the dislocations responsible for the deformation inhibit the mobility of these defects in the crystal, thereby increasing the hardness.

The intensity of these stresses, whence the effect of hardening, increases with the ratio p/q, thus with c_0 (see question **13**). Thus, we expect the hardness of quenched steels to increase with the carbon content.

16. To implement the martensite $(z) \rightarrow$ ferrite transformation, or at least to trigger it (which decreases the ratio p/q), the jumps $(z) \rightarrow (x)$ or (y) must be preferred. For this, it suffices to heat the sample in order to increase the term $\exp[-(\Delta E + w)/kT]$.

One of the two nails is hard because it is (at least partially) martensitic. The other, which was initially identical, was heated for an hour at 500° C, which process, as it were, eliminated the martensite, permitted equilibrium to be reached (ferrite + cementite) and thus removed many of the internal stresses referred to in question **15**.

SWELLING OF NUCLEAR FUELS

Solution of questions **3, 4, 5** and **7** is not an absolute prerequisite for a general understanding of the problem.

In the 'fuel' which forms the central part (or core) of a nuclear reactor[1], the fission of uranium 235 gives birth to new nuclei ('fission fragments'). A proportion of approximately 10% of these fragments consists of nuclei of rare gases (Xe and Kr). These rare gases, which are generally not very soluble in solids, collect into precipitates, which, in this case, are gas bubbles. Here, a *bubble* consists of a cavity (formed by agglomeration of vacancies of the solid, referred to in what follows as the *crystal*) containing gas atoms. The accumulation and growth of these bubbles over time clearly results in an increase in the external volume of the fuel, called *swelling*, which may have a major effect on the smooth running of the reactor. The aim of this problem is to study some of the swelling mechanisms.

In what follows we shall make the following simplifying assumptions.

(i) The surface energy Γ of the crystal is isotropic; thus, the bubbles are spherical cavities. A given bubble (radius R) contains n gas atoms at pressure p.

(ii) The solubility of the gas is low and its diffusion in the crystal is sufficiently fast that at any given time all the gas atoms are contained in the bubbles. A gas atom in a bubble is assumed never to leave that bubble.

(iii) The temperature T of the crystal is constant and is sufficiently high that the bubbles are in equilibrium (this point will be discussed in more detail in questions **1** and **7**).

(iv) The gas in the bubbles is a *perfect gas*.

(v) The external pressure is zero.

(vi) The crystal has a simple cubic structure (lattice parameter b).

I

State of the bubbles

1. Suppose that a particular bubble contains n gas atoms. We shall identify its free energy with the sum of the free energy of the (perfect) gas and the (internal) surface

---

[1] These fuels are generally oxides, essentially UO_2.

energy of the (spherical) cavity. Assuming that T and n are fixed and varying R by a quantity dR, *show that at equilibrium the size of the bubble and the gas pressure are related by the (Laplace) equation*

$$p = 2\Gamma/R.$$

Numerical application. We shall take $\Gamma = 1\ J\ m^{-2}$ *(see p. 156). Calculate the pressure at equilibrium in a bubble of radius 100 Å. Do the same for a bubble of radius 1 μm.*

2. *Use the state equation for perfect gases to show that it is the surface area (S) of the bubble, not its volume (V), which is proportional to the number of gas atoms which it contains.*

3. If the crystal contains dislocations, the bubbles tend to settle on these defects (Figure 1). *Explain the reason for this tendency and give the binding energy* E_{bd} *between the bubble (R) and a screw dislocation (length of the Burgers vector b, shear modulus μ).*

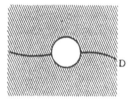

Figure 1   Bubble situated on a dislocation D.

Comment briefly on the micrograph on p. 157 of the Course, where the dislocations form a lattice with triple or quadruple points.

4. *In the spirit of the previous question, check that the bubbles have a similar tendency to settle on grain boundaries. Give the binding energy* E_{bb} *of a bubble (R) and of a grain boundary of surface energy* γ.

5. *By comparing* E_{bd} *and* E_{bb}, *show that a bubble is more stable on a boundary than on a dislocation when its radius becomes greater than a critical value* R_c. *Give an expression for* R_c *as a function of* μ, b *and* γ.
   Numerical application. Calculate R_c. *We take* μ *such that* $\mu b^3 = 5\ eV$, $b = 2.7$ Å *and* Γ *such that* $\Gamma b^2 = 0.2\ eV$ *(which corresponds to* $\Gamma \simeq 0.5\ J\ m^{-2}$*).*

II

Growth of the bubbles

We shall now consider a bubble situated at the centre of a region (ξ), which we assume to be cubic with edge h (see Figure 2). We also assume that the bubble collects all the fission gas produced in this region. The latter constitutes the 'basin' which feeds the bubble in question.

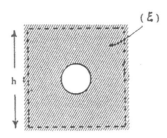

Figure 2 Bubble at the centre of its basin (ξ).

The neutrons in the reactor produce K gas atoms per unit volume per unit time. Via collisions with the atoms of the crystal they also produce a very large number ($\gg K$) of point defects, including vacancies in particular.

6. When dn additional gas atoms join the bubble (R, n) which is assumed to be initially at equilibrium, the bubble $(R, n + \mathrm{d}n)$ is no longer at equilibrium. The return to equilibrium requires a flux of vacancies to the bubble, which increases its size. Suppose the new bubble at equilibrium is $(R + \mathrm{d}R, n + \mathrm{d}n)$. *Show that we have*

$$\mathrm{d}R = A \frac{\mathrm{d}n}{R}$$

where A is a constant to be determined. Taking the volume of a vacancy to be the atomic volume (b^3), give the expression for the number of vacancies α needed to re-establish equilibrium when one gas atom joins the bubble (R, n).
 Numerical application. Calculate α for $R = 0.1$ μm. Do the same for $R = 10$ μm. We take $b^3 = 20$ $\overset{\circ}{A}{}^3$, $T = 1160$ K (or $kT = \frac{1}{10}$ eV) and $\Gamma = 1$ J m^{-2}. What can you say about these values of α?

7. *Calculate the order of magnitude of the time τ necessary for an irradiation-induced vacancy to join the bubble, that is to cross a distance of the order of $h/4$. We shall let E_m be the migration energy of the vacancies in the crystal and v the mean atomic vibration frequency.*
 Since the vacancies are provided in unlimited number by the irradiation, the equilibrium of the bubbles implies that this time τ is shorter than (or of the same order as) the mean time t_g between the arrival of two successive gas atoms in the bubble. *Deduce the condition that the temperature must satisfy for the assumption that 'bubbles are always at equilibrium' to be respected.*
 Numerical application. Calculate τ under the following conditions: $E_\mathrm{m} = 0.4$ eV, $v = 10^{13}$ s^{-1}, $T = 1160$ K, $h = 10$ μm. What can you say about the equilibrium for $K \simeq 2 \times 10^{18}$ m^{-3} s^{-1} (operational conditions for a typical PWR (Pressurised Water Reactor)?
 In what follows, we shall assume that this condition is always satisfied.

8. Assuming that the fuel represents the juxtaposition of small basins (ξ) with *one* bubble each, where each basin is identical with that of Figure 2, *write down the 'law of swelling' for this fuel, in other words, the law for the variation of the volume $\Delta V/V = f(t)$ (where*

$\mathcal{V}$ is the external volume) as a function of the time t and as a function of the number of bubbles per unit volume B $(B = h^{-3})$. For this, all the change in volume will be attributed to the bubbles, the density of the rest of the crystal being assumed to remain constant.

Is it more suitable to design a fuel with a large or a small number of sites of bubble nucleation?

III

'Breakaway' swelling

The swelling law found in question **8** does in general account satisfactorily for the experimental results *at the beginning* of the irradiation. This can be seen qualitatively from the first four or five points of Figure 3. However, the same figure also shows that the swelling begins to increase dangerously ($\Delta \mathcal{V}/\mathcal{V} = 1$ means a doubling of the volume!) after a certain period of irradiation. We then speak of *breakaway swelling*. It is clear that this phenomenon should be avoided at all costs in nuclear reactors and therefore should be understood first of all.

In this respect, we note that as the bubbles increase in size the probability that two of them enter into contact increases. We first examine the consequence of such a meeting.

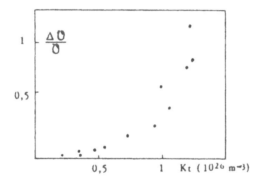

Figure 3   Example of measurement of the swelling of a fuel (here, a U – 3% Mo alloy). *Ordinates*. Change in volume of the samples. *Abscissae*. Irradiation time, or, more precisely, the number of atoms of fission gas produced per m³, Kt. The figure shows a brutal acceleration of the swelling from $Kt \simeq 1 \times 10^{26}$ m⁻³. After **W A Holland**, Trans. Amer. Nucl. Soc. 4, 337, 1961, *plus* data from **H Mikailoff**.

9. *At its equilibrium, calculate the volume V_3 and the surface area S_3 of the bubble (R_3, n_3) resulting from the meeting of two bubbles (R_1, n_1) and (R_2, n_2). For this, one should return to the results of question 2 and give V_3 as a function of V_1 and V_2 and S_3 as a function of S_1 and S_2. What is there to remark?*

For $R_1 = R_2 = R$, calculate the number of vacancies β needed for the equilibrium of the new bubble.

10. *Use the above result to describe, in one sentence, the effect the meeting of the bubbles should have on the swelling law for the fuel.*

11.   We shall model this effect as follows.

We suppose that the nucleation of the bubbles is highly inhomogeneous. More precisely, we assume that, at time $t = 0$, on the grain boundaries, there exist bubble nuclei with a small non-zero volume v_0, numbering G per unit volume. These ultimately correspond to bubbles which have grown faster than the rest. Thus, we shall distinguish between two classes of bubbles: these *large bubbles* with concentration G which essentially triggered the process described in questions **9** and **10** and a larger number of *small bubbles*, which are also assumed to lie on the boundaries, according to question **5**.

Suppose then that such a large bubble with radius R is situated at the centre of a cylindrical basin (η) of volume $\eta = \sigma a$ (see Figure 4, a is the mean distance between boundaries and σ is the surface area of the boundary concerned) with $G = \eta^{-1} = (\sigma a)^{-1}$. In what follows, we shall assume that the gas atoms produced in the basin diffuse homogeneously towards the boundary J reaching the surface of area s in the proportion s/σ. In particular, a fraction $\pi R^2/\sigma$ of the gas formed joins the bubble of radius R, while the remainder reaches the small bubbles in proportion to the surface area of the boundary in question; this amounts to a uniform redistribution of the gas on the boundary in quantity $K s a t$ over the element s.

If the central bubble is (R, n) at time t, it becomes $(R + dR, n + dn)$ at time $t + dt$.

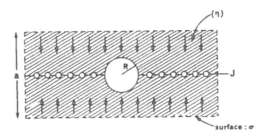

Figure 4 The basin (η) supplies the gas atoms for all the bubbles situated on the boundary J in a homogeneous manner (as indicated by the arrows).

Determine the differential equation describing the evolution of this central bubble, which is still assumed to be in equilibrium. For this, write dn as the sum of the dn_1 'fresh' atoms produced in the time dt which join the bubble immediately and the dn_2 atoms stored on the boundary J throughout the irradiation which are absorbed by the bubble during its growth from R to $R + dR$.

Deduce the law of swelling for the basin (η), whence that for the fuel $\Delta V/V = \Delta \eta/\eta = g(t)$. For this, we assume that practically all the swelling is due to the central bubbles which all have the same evolution with time.

12. The above model describes an avalanche phenomenon where the swelling accelerates with the absorption of small bubbles by a central bubble (as when a liquid boils) and ends by diverging. *Show that this divergence takes place at time*

$$t_\infty = (2\pi A K a)^{-1}.$$

Does this model seem to describe the experimental results of Figure 3 correctly?

13. *Try to summarise the swelling of nuclear fuels in a few sentences (without formulae or calculations). Assuming that breakaway swelling is unacceptable in a reactor and that* the temperature is fixed by efficiency considerations, say which physical parameters of the fissile material one can play with, and in what sense, to avoid this phenomenon or at least delay it as long as possible.

SOLUTION TO PROBLEM 16

1. Let S be the surface area of the bubble and V its volume. The equilibrium corresponds to

$$\Gamma \, dS - p \, dV = 0$$

or, for a sphere:

$$8\pi \Gamma R \, dR - p 4\pi R^2 \, dR = 0 \tag{1}$$

which gives $p = 2\Gamma/R$.
    For $R = 100$ Å, $p = 2 \times 10^8$ Pa = 2000 atm.
    For $R = 1$ μm, $p = 2 \times 10^6$ Pa = 20 atm.

2. Here, the law of perfect gases is written

$$\frac{2\Gamma}{R} \frac{4}{3}\pi R^3 = nkT$$

whence

$$n = \frac{8\pi \Gamma}{3kT} R^2 = \frac{2\Gamma}{3kT} S \tag{2}$$

while

$$n = \frac{2\Gamma}{kT} \left(\frac{4\pi}{3}\right)^{1/3} V^{2/3}. \tag{3}$$

3. A bubble of radius R situated on a dislocation eliminates (at the most) a length $2R$ of the latter and thus *decreases* its energy by

$$E_{bd} \simeq -2\mu b^2 R \quad \text{(see (13), p. 210)}.$$

This energy is $-3\mu b^2 R$ or $-4\mu b^2 R$ for a triple or quadruple point, respectively. This explains the tendency for the bubbles to anchor themselves to such points (micrograph on p. 157).

4. In the same way, a bubble eliminates a part of the boundary on which it is located. The maximum energy gained (boundary on the equatorial plane) is

$$E_{bb} = -\pi R^2 \gamma.$$

5. E_{bb} is less than E_{bd} for $R > R_c$ such that

$$R_c = 2\mu b^2/\pi\gamma \quad \text{(Figure 5)}.$$

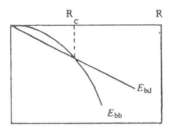

Figure 5   Binding energy of a bubble with a dislocation (E_{bd}) or a boundary (E_{bb}).

Numerically.

$$R_c = \frac{2}{\pi} \frac{\mu b^3}{\Gamma b^2} b \simeq 43 \text{ Å}.$$

6.   From equation (2) it follows that

$$dn = \frac{16\pi \Gamma R \, dR}{3kT}$$

whence

$$dR = A \frac{dn}{R} \tag{4}$$

where

$$A = \frac{3kT}{16\pi \Gamma}.$$

For *one* gas atom ($dn = 1$) the change in volume is $\delta V = 4\pi R^2 \, dR = 4\pi A R$. This change in volume requires the arrival of

$$\alpha = \frac{\delta V}{b^3} = \frac{4\pi A R}{b^3} = \frac{3}{4} \frac{kT}{\Gamma} \frac{R}{b^3}$$

vacancies at the bubble.

Numerically. For $R = 0.1 \ \mu m$, we have $\alpha = 60$ and for $R = 10 \ \mu m$, $\alpha = 6000$.

Note that the arrival of a gas atom triggers the inflow of a considerable number of vacancies, which increases with the size of the bubble.

7.   The time τ is such that we have

$$(6D\tau)^{1/2} \simeq \frac{h}{4} \quad \text{(see p. 174)}$$

where D is the diffusion coefficient for the vacancies, or

$$D = \nu b^2 \exp(-E_m/kT)$$

whence

$$\tau \simeq \frac{h^2}{96\nu b^2} \exp(E_m/kT).$$

The equilibrium of the bubble (in the sense of (ii) and (iii) of the introduction) requires τ to be of the order of (better still, less than) the mean time t_g between the arrival of two successive gas atoms, $t_g \simeq (Kh^3)^{-1}$. The condition $\tau \leq t_g$ then becomes

$$T \geq \frac{E_m}{k}\left(\ln \frac{96vb^2}{Kh^5}\right)^{-1}.$$

Numerical application. For the values proposed, we find $\tau = 0.08$ ms and $t_g = 0.5$ ms; thus, the vacancies created by the irradiation are able to flow towards the bubble faster than the gas atoms are created.

8. Returning to (3) with $n = Kh^3t$, we find

$$V = \left(\frac{3}{4\pi}\right)^{1/2}\left(\frac{kT}{2\Gamma}\right)^{3/2}(Kh^3t)^{3/2}$$

whence we have the law of swelling

$$\frac{\Delta V}{V} = \frac{V}{h^3} \simeq \frac{1}{2}\left(\frac{KkT}{2\Gamma}\right)^{3/2}\frac{t^{3/2}}{B^{1/2}}.$$

To ensure a low swelling it is important (variation in $B^{-1/2}$) to design a fuel with a maximum number of nucleation sites and thus as many bubbles as possible.

9. Following (2) and (3), we write $n_3 = n_1 + n_2$, whence

$$V_3 = (V_1^{2/3} + V_2^{2/3})^{3/2} \qquad \text{and} \qquad S_3 = S_1 + S_2.$$

The volume V_3 is larger than the sum $V_1 + V_2$ while the surface areas of the bubbles are additive when the two bubbles merge.

For $R_1 = R_2 = R$, we have $V_3 = (2V^{2/3})^{3/2} = 2^{3/2}V$. The equilibrium of the bubble (R_3) requires the arrival of β vacancies, where

$$\beta = \frac{V(2^{3/2} - 2)}{b^3} = 0.83\frac{V}{b^3}.$$

10. The above result shows that the meeting of the two bubbles is a 'catastrophic' event for the fuel because it is accompanied by a sharp increase in volume: when two equal bubbles with volume V meet this increase is almost equal to V (namely $0.83V$).

11. $dn_1 = K\sigma a\,dt(\pi R^2/\sigma)$ and $dn_2 = K\sigma at(2\pi R\,dR/\sigma)$.
    Using (4), we find

$$dR = \pi AKa(R\,dt + 2t\,dR).$$

It follows that

$$R = \frac{r_0}{(1 - 2Ct)^{1/2}}$$

where $C = \pi A K a$; whence

$$\frac{\Delta \mathcal{V}}{\mathcal{V}} = \frac{\Delta \eta}{\eta} = \frac{4\pi R^3}{3\sigma a} \simeq \frac{v_0}{\eta}(1 - 2Ct)^{-3/2}. \tag{5}$$

r_0, v_0: size of the central bubble at $t = 0$.

12. The law of swelling (5) predicts a divergence of the volume at time

$$t_\infty = \frac{1}{2C} = (2\pi A K a)^{-1}. \tag{6}$$

It accounts quite well for the experimental results of Figure 3 (see Figure 6) if we take the values $a = 1.8$ μm and

$$\frac{v_0}{\eta} = 2 \times 10^{-2}.$$

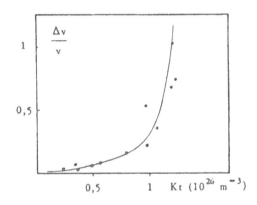

Figure 6   Sketch of the law (5) versus the experimental points of Figure 3.

13.   To summarise: in this problem, we have seen that by collecting into bubbles the fission gas causes the fuel to swell. As long as the bubbles do not meet, a relatively unalarming law in $t^{3/2}$ applies. When the bubbles begin to meet in a significant manner, giving rise to a massive inflow of vacancies (whence to an additional important increase in volume) the swelling accelerates and a regime of 'breakaway swelling' is entered which leads to a divergence of the volume after a critical irradiation time t_∞.

To delay the breakaway swelling, that is to increase t_∞, we need to decrease a (formula (6)) or, in other words, increase the number of bubble nucleation sites. This can be done by seeding the fuel with small precipitates.

Of course, we also slow the growth of the bubbles by increasing the external pressure P, where formula (1) becomes $2\Gamma/R = p - P$. For this, the fuel is enclosed in a rigid metallic container ('cladding'). At the start of the irradiation the swelling puts the fuel in contact with the container, and the external pressure P begins to increase. The effect is further reinforced if the cooling fluid in which the container is immersed is under pressure. This is the case for pressurised water reactors (PWR). Moreover, a free 'expansion volume' is left, in the top of the cladding tube, collecting the gas which escapes from the bubbles via the thermal stress-induced cracks in the fuel.

Complementary remarks

(i)   The irradiation creates an equal number of vacancies and interstitials. As we have
      seen, the former contribute to the growth of bubbles. The latter diffuse, in equal
      numbers, in the crystal towards the dislocations which they cause to *climb* (see
      p. 215), thereby creating additional atomic planes (Figure 7). It is these which, in
      the final analysis, cause the external volume to increase by an amount equal to the
      total volume of the bubbles.

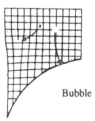

Bubble

Figure 7   The interstitial (·) formed by the irradiation causes an edge dislocation to climb while the corresponding vacancy $\left(-\dfrac{|}{|}- \right)$ migrates towards the bubble.

(ii) Of course, the conditions for bubble nucleation vary greatly from one crystal to
      another. This can be seen in Figure 8(*a*) which shows an electron micrograph
      ('replica' technique) of a two-phase fuel element, after irradiation. The density and
      size of the bubbles differ greatly between the two phases.

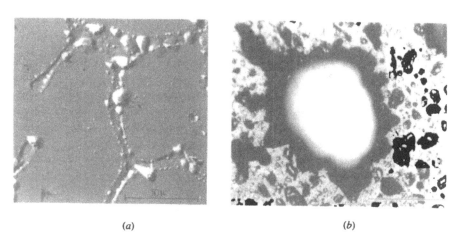

(*a*)   (*b*)

Figure 8   (*a*) Example of bubbles seen by electron microscopy (shadowed replica technique) in a solid
with two phases (mixture of α uranium and UC carbide) irradiated at 680° C. Note the very strong
inhomogeneity of the sizes of these bubbles: a very large number of very small bubbles in the carbide
and a small number of large bubbles in the metal. After H Mikailoff. (*b*) Example of a giant bubble
absorbing smaller neighbouring bubbles in a U–6% (weight) Nb alloy irradiated at 620° C. Optical
microscopy. Bubble diameter $\simeq 0.1$ mm. After H Mikailoff.

(iii) In the regime of breakaway swelling, certain bubbles, referred to as giant bubbles in the above, may effectively attain considerable sizes. An example of this can be seen in the micrograph of Figure 8(*b*), which shows such a bubble in its phase of absorption of smaller neighbouring bubbles.

PROBLEM 17

EVOLUTION OF IRRADIATED STEELS

In a nuclear reactor, the solid materials (metallic in particular) which form the core structure are subjected to bombardment by a flux of ambient neutrons. In a fast neutron reactor this flux may reach $\simeq 10^{18}$ neutrons m^{-2} s^{-1}, where the neutrons have energies between $\simeq 1$ and 10 MeV (1 MeV = 10^6 eV). By displacing the atoms of solids, this bombardment by neutrons modifies most of their properties. Here, we shall study an important dimensional modification, namely an increase in volume, referred to as *cavity swelling*[1]. The discovery of this phenomenon, which is now well understood, was a major challenge for the development of the fast-neutron reactors[2].

N.B. *I, II and III are independent of each other. IV depends on III.*

I

Creation of point defects

We consider a solid $\mathcal{A}$, for example a steel. The atoms with atomic mass A are assumed to be identical. When an atom which is normally situated at a crystalline site L receives an energy E from a neutron–nucleus collision, it may leave its site, leaving a vacancy (v) at L and creating an interstitial (i) at a different point $\dot{\mathrm{I}}$ (see Figure 1). This event only occurs if E is greater than a critical value, or displacement threshold (or just 'threshold') E_T.

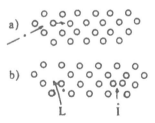

Figure 1 (*a*) Initial event: a neutron (·) collides with an atom. (*b*) Final situation: a vacancy created at L and an interstitial at $\dot{\mathrm{I}}$.

---

[1] This *swelling*, unlike in **Problem 16**, does not relate to fission gases, and concerns the cladding, not the fuel.
[2] In fact, this swelling of the solid parts dangerously decreases the cross-sectional area of passage of the heat-carrying fluid (liquid sodium for these reactors).

1. The passage from the initial state (perfect crystal, energy assumed to be zero) to the final state ((i) + (v): see Figure 1(b)) is accompanied by an evolution of the energy for all the atoms concerned. *Indicate which of the four evolution schemes illustrated is correct and add, on the sketch, the energy E_T and the sum $E_f = E_{fv} + E_{fi}$ together with a brief comment.*

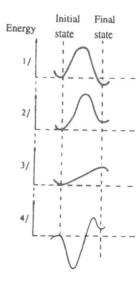

2. *Indicate a plausible value for E_f (in eV), according to the Course. Use question 1 to deduce a condition for E_T (in the form of an inequality).*

3. A collision between a neutron (n) with energy W and a nucleus (N) of $\mathcal{A}$ is represented as a hard-spheres collision (HS: interaction potential *zero* if the spheres are not in contact; *infinite* if they are in contact). Here, we recall a formula for the HS[3] which gives the energy E yielded from the mobile sphere (n) to the initially fixed sphere (N):

$$E = \frac{4 M_n M_N \cos^2 \theta}{(M_n + M_N)^2} W$$

(see diagram), where M_n and M_N are the masses of the spheres (n) and (N), respectively.

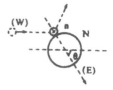

What is the maximum energy E_M yielded in the case of a neutron with energy W (assuming $A \gg 1$)? Show that the average energy yielded over all possible impacts is $\overline{E} = \frac{1}{2} E_M$.

4. In what follows, we shall use the value $E_T = 25$ eV for E_T, which is a common value in most metals. *Show that for an 'average' neutron with energy $\overline{W} = 3$ MeV the average energy yielded $\overline{E}$ is far greater than E_T.*

5. Under these conditions, the atom hit by a neutron (call it a_1: primary knocked-on atom, or *primary*, with energy $\overline{E}_1$) then hits another atom (a_2), which hits a third, etc. This produces a cascade of collisions, whose genealogical structure is illustrated in the accompanying diagram.

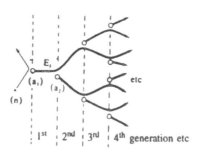

---

[3] Obtained using the conservation of energy and impulse.

The simplest cascade model is one in which: (i) all atom–atom collisions are HS collisions; (ii) all collisions are 'average' collisions (in the sense of question 3).

With these assumptions, determine how the energy is redistributed at the time of an atom–atom collision.

Why does the cascade stop? Give the number of vacancies p and interstitials created at the end of the cascade as a function of E_T and $\overline{E}_1$.

Numerical application. Calculate p for a cascade created by a neutron with energy $\overline{W} = 3$ MeV in a solid with atomic mass $A = 60$ and threshold $E_T = 25$ eV.

6. Knowing p, we can then evaluate the concentration K of pairs (v)+(i) produced *per unit time* in the solid subjected to the flux ϕ of (average) neutrons. For this we denote the classical radius of a neutron (or a proton) by r_0 and represent the nucleus (N) as a spherical aggregate of A nucleons. *Give an approximate value for the radius R of (N). Deduce the surface area $\sigma \simeq \pi R^2$ (or effective cross section) of the nucleus (N) involved in the impact with a neutron.*

Numerical application. Calculate σ for $A = 60$, given $r_0 = 1.2 \times 10^{-15}$ m.

7. In a reactor, the number of neutrons is characterised by a *flux* (expressed in terms of m^{-2} s^{-1}). *Knowing the value of the flux ϕ and the effective collision cross section σ, write down the concentration c_k of primaries per unit time (c_k = number of first atoms hit/total number of atoms).*

8. *Using question 5, evaluate the concentration K of pairs (v) + (i) created per unit time in the solid A, assuming that the neutrons all have the average energy $\overline{W}$.*

Numerical application. Calculate K for the numerical values of questions 5 and 6 for a neutron flux at 10^{18} m^{-2} s^{-1}.

II

Nucleation of cavities

In **I**, we showed that in a solid A subjected to a flux of fast neutrons, a concentration K (proportional to ϕ) of point defects ((v)+(i)) is created per unit time. Before we study the evolution of these defects globally, we first examine the question of the *agglomeration of the vacancies created*.

In fact, at the temperatures at which reactors operate (for example, $\simeq 600°$ C), the vacancies (and interstitials) migrate rapidly. Of the numerous possible reactions on meeting, we shall conceptually isolate here the reactions of type

$$v + v \rightarrow 2v \qquad 2v + v \rightarrow 3v \qquad \text{etc} \ldots \qquad (n-1)v + v \rightarrow nv \ldots$$

which denote the agglomeration of vacancies into clusters (for example, of n vacancies). We shall assume that the inverse reactions (for example, $3v \rightarrow 2v + v$) are impossible. Throughout what follows, we shall assume that A has a *simple cubic* crystalline structure (cube with edge b, atomic volume b^3).

We shall retain only two of the simplest and most classical types of vacancy clusters, namely the *cavity* and the perfect vacancy dislocation loop (referred to as a *vacancy loop* in what follows). The latter forms when vacancies agglomerate in a dense atomic plane (here, the planes {100}) and the free surfaces thus formed knit together again.

9. *Draw, at atomic scale, the section of a vacancy loop intersected by a plane containing the Burgers vector. Draw a section of a cavity (assumed spherical).*

10. *Assuming that the same number n of vacancies has agglomerated in both cases, evaluate the energy of the loop $E_l(n)$ (assumed circular) and that of the cavity $E_c(n)$ (assumed spherical). We denote the shear modulus of the metal $\mathcal{A}$ by μ and its surface energy by Γ. We assume that the formula of the Course for the energy of a rectilinear screw dislocation holds for an edge dislocation.*

11. *Show that above a certain number of agglomerated vacancies n_c, the vacancy loop is always more stable than the cavity. Give the expression for n_c.*

12. Experiments show that after a cavity is born it grows as such by absorption of vacancies, without turning into a vacancy loop at the critical size n_c, despite the result of question **11**.

Explain in simple terms why this is; but first, say how you think the transformation from a cavity of size $\gg n_c$ to a loop might take place.

III

Evolution of point defects

In this part, we shall study the evolution of point defects created by irradiation, with a view to describing the increase in size of the cavities which are assumed to have been nucleated.

We shall simplify this problem by considering only the following reactions (in the chemical sense) which are illustrated in the accompanying diagram:

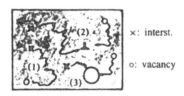

$$v + i \rightarrow 0 \tag{1}$$
$$v + d \rightarrow d \qquad i + d \rightarrow d \tag{2}$$
$$v + C_n \rightarrow C_{n+1} \qquad i + C_n \rightarrow C_{n-1}. \tag{3}$$

(1) denotes the *mutual annihilation* which occurs when (v) and (i) meet.

(2) denotes the annihilation of (v) and (i) when these defects meet a dislocation. In fact, $\mathcal{A}$ initially contains a *dislocation density* ρ which we shall assume to be *constant* in what follows.

(3) describes the purely random *annihilation* of (v) and (i) on *cavities* of size n (in the sense of question **10**). In what follows, we shall assume, in accordance with question **12**, that at the start of operation of the reactor (in practice, at time $t = 0$) N cavities per unit volume form, and that N *subsequently remains constant*.

We then denote the concentrations of vacancies and interstitials (at time t) by c_v and c_i, respectively, where the evolution of c_v is described by the following equation

$$\frac{dc_v}{dt} = K - \alpha(D_i + D_v)c_i c_v - \beta_v \rho D_v c_v - \gamma N D_v c_v \tag{4}$$

in which D_v and D_i denote the diffusion coefficients for vacancies and interstitials.

13. *Comment on this equation and, in particular, say what each of the terms on the right describes. Give the dimensions of the constants K, α, β and γ and, if possible, say what they represent physically.*

14. *Write down the equation describing the evolution of the concentration of interstitials*

$$\frac{dc_i}{dt} = \tag{5}$$

and comment on it as above. Say why certain constants which occur in (4) and (5) are equal.

15. Equations (4) and (5) form a system of nonlinear equations which are not mutually symmetric in v and i.

We shall not attempt to solve this system. For our purposes, it will suffice first to give the form of c_i and c_v for short times and second to calculate the ratio c_i^∞ / c_v^∞ for the steady state which becomes established after a sufficiently long time.

IV

Growth of cavities: swelling

16. In what follows, we shall assume that *this steady state is reached* and that the concentrations of point defects are c_v^∞ (we neglect the concentration of vacancies in thermodynamic equilibrium) and c_i^∞. *In (4) and (5), we also neglect the term in α* which has an insignificant role in what follows.

Assuming that, at any given time, the cavities all have the same average radius $r(t)$, adapt the theory of the growth of precipitates (see Course) to the problem of the 'precipitation' of the point defects (v) and (i) (reaction (3)) on the cavities. In particular, show that the constant γ of equations (4) and (5) is equal to $4\pi r$.

17. If, from that time, the constants β_v and β_i of (4) and (5) are taken to be equal (this is a legitimate assumption to the first order), *show that the N cavities (per unit volume) do not increase in size during the irradiation: they remain in the nucleus state and there is practically no swelling (except that due to the nuclei themselves).*

18. If, on the other hand, we introduce[4] a small disequilibrium into (4) and (5) arising from an attraction of the dislocations which is slightly stronger for interstitials than for vacancies, we may then write:

$$\beta_v = \beta_0 \quad \text{and} \quad \beta_i = \beta_0(1 + \varepsilon) \qquad (\varepsilon > 0).$$

Give a qualitative justification for the existence of such a disequilibrium and the positive sign of ε. Indicate what this disequilibrium enables the cavities to grow into (to the great displeasure of those responsible for reactors!).

19. In what follows, the ratio $g = \Delta V / V$ of the volume increase due to the cavities to the initial volume will be referred to as the 'swelling' of the solid $\mathcal{A}$. ΔV is just the total volume of the cavities.

Express g as a function of N and r.

20. *Show that we may write:*

$$\frac{dg}{dt} = N\gamma (D_v c_v^\infty - D_i c_i^\infty).$$

[4] After **R Bullough**.

By eliminating $D_i c_v c_v^\infty$ and $D_i c_i^\infty$ using the equations previously determined, show that

$$\frac{dg}{dt} \simeq \frac{K\gamma N\beta_0\rho\varepsilon}{(\beta_0\rho + \gamma N)^2}. \qquad (6)$$

21. *Taking into account the fact that g and γ are functions of r (see questions 16 and 19) use (6) to deduce the evolution g(t) of the swelling in the following two extreme cases*

(i) $4\pi Nr \le \beta_0\rho$ *(small cavities, short times)*
(ii) $4\pi Nr \ge \beta_0\rho$ *(large cavities, long times).*

Use a graph to illustrate the general evolution of the swelling with time.

Deduce a (happy) tendency towards saturation and comment on the physical origin of this, in simple terms.

Numerical application. Determine the value of g obtained after two years (t ≃ 6×10^7 s, case (ii) above) for the following values of the constants: K = value similar to that obtained in question 8, for example, K = 2×10^{-7} s^{-1}; N = 10^{21} m^{-3}; ρ = 10^{13} m^{-2}; ε = 5×10^{-2}.

22. *If you were responsible for materials for fast neutron reactors, in which directions would you initiate a research programme designed to decrease the swelling?*

SOLUTION TO PROBLEM 17

1. The correct scheme is shown in panel 2 of the figure on p. 440. E_f is the difference in energy between the final state and the initial state. E_T is an 'activation energy' (in the chemical sense).

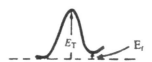

2. We may take $E_{fv} \simeq 1$ eV and $E_{fi} \simeq 4$ eV, or $E_f \simeq 5$ eV. Thus, we have

$$E_T \gtrsim 5 \text{ eV}.$$

3. The maximal energy ($\cos\theta = 1$, head-on collision) is

$$E_M = \frac{4M_n M_N}{(M_n + M_N)^2} W = \frac{4A}{(1 + A)^2} W \simeq \frac{4}{A} W.$$

The probability that the impact occurs between θ and $\theta + d\theta$ is

$$\frac{2\pi (R_0 + r_0)^2 \sin\theta \cos\theta \, d\theta}{\pi (R_0 + r_0)^2} = 2\cos\theta \, d\cos\theta.$$

Thus, the average energy yielded is

$$E_M \int_0^{\pi/2} 2\cos^3\theta \, d\cos\theta = \frac{E_M}{2} = \frac{2}{A} W.$$

4. A is always less than 250. Thus, for $\overline{W} = 3$ MeV, $\overline{E}$ is always greater than $\frac{2}{250} \times 3 \times 10^6 = 2.4 \times 10^4$ eV, which is itself far greater than $E_T = 25$ eV.

5. In assumptions (i) and (ii) of question 5 each atom of the cascade yields half of its energy (since the masses are equal). The two atoms of the second generation thus have energy $E_2 = E_1/2 \dots$ and the 2^{n-1} atoms of the nth generation have energy $E_n = E_1/2^{n-1}$. By the nth generation, 2^{n-1} vacancies (and as many interstitials) are formed. When the atoms have an energy slightly less than $2E_T$ new defects can no longer be created and the cascade stops. In its wake it leaves

$$p = \frac{E_1}{2E_T} \quad \text{vacancy–interstitial pairs}$$

(Kinchin and Pease formula).

Numerical application.

$$\overline{E_1} = \frac{2}{A} W = \frac{2}{60} \times 3 \times 10^6 = 10^5 \text{ eV}$$

whence

$$p = \frac{10^5}{50} = 2000 \quad \text{pairs.}$$

N.B. The energy stored by these p pairs is pE_f, or, here, $2000 \times 5 = 10^4$ eV. The energy introduced into this crystal is $\overline{E_1} = 10^5$ eV. Thus, most of this energy (90%) is dissipated as heat. It is this energy which passes into the heat-carrying fluid then into the turbines.

6.

$$\tfrac{4}{3}\pi R^3 \simeq A \tfrac{4}{3}\pi r_0^3 \rightarrow \sigma = \pi R^2 = \pi r_0^2 A^{2/3}.$$

Numerical application. $\sigma = \pi (1.2 \times 10^{-15})^2 \rightarrow \sigma \simeq 7 \times 10^{-29}$ m^2 ($= 0.7$ barn).

7. A disc with unit cross section and thickness dx contains dx/Ω atoms (Ω = atomic volume). The probability that a neutron penetrating this disc collides with a nucleus is thus $\sigma dx/\Omega$. The number of atoms hit per unit time is $\phi\sigma dx/\Omega$. Their concentration is $\phi\sigma$ per unit time.

8. For neutrons with energy $\overline{W}$, the first atoms hit have energy $\overline{E_1} = (2/A)\overline{W}$. Their concentration is $\phi\sigma$ per unit time and they produce on average $p = \overline{E_1}/2E_T$ vacancy–interstitial pairs. Thus, we have

$$K = \phi\sigma \frac{\overline{W}}{AE_T} \quad \text{per unit time}$$

Numerical application. $K = 10^{18} \times 7 \times 10^{-29} \times 3 \times 10^6/(60 \times 25) = 1.5 \times 10^{-7}$ s.

9.

Vacancy loop

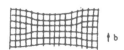

Cavity

10. A vacancy loop which has accumulated n vacancies has a surface area $\pi r^2 = nb^2 \rightarrow r = b\sqrt{n/\pi}$. Thus, it has an energy of the order of $E_1 \simeq 2\mu b^3 (\pi n)^{1/2}$ (see p. 210).

A cavity of n vacancies has a volume

$$\frac{4}{3}\pi r'^3 = nb^3 \rightarrow r' = b\left(\frac{3n}{4\pi}\right)^{1/3}.$$

Thus, it has an energy of the order of $E_c \simeq \Gamma b^2 (4\pi)^{1/3}(3n)^{2/3}$.

11. The functions $E_1 \alpha n^{1/2}$ and $E_c \alpha n^{2/3}$ have the behaviour shown in the accompanying diagram.

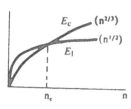

Thus, we have

$$E_c < E_1 \text{ for } n < n_c \qquad n_c = \frac{4\pi}{81}\left(\frac{\mu b}{\Gamma}\right)^6.$$
$$E_c > E_1 \text{ for } n > n_c$$

Numerical application. Taking $\mu b^3 = 5$ eV (see p. 210) or $\simeq 0.2$ a.u. and $\Gamma b^2 = \frac{1}{1000} \times (5)^2$ a.u. (see p. 156), we find

$$n_c = \frac{4\pi}{81}\left(\frac{0.2 \times 1000}{25}\right)^6 \simeq 4 \times 10^4.$$

N.B. This value is overestimated, since the energy of a loop is less than μb^2 per unit of length, which is the value determined for a *rectilinear* dislocation.

12. During the transformation from a spherical cavity to a circular loop, the cavity takes on the following flattened intermediate shapes:

Since the transformation takes place for constant volume (except for the final collapse), the surface area S increases greatly during the stages 2, 3, etc, as does the energy ΓS. For $n > n_c$ the energy diagram has the following form:

The transformation, which is energetically favourable, therefore requires a generally prohibitive activation energy ΔE. Thus, in the above numerical example, $n_c = 4 \times 10^4$ or $R \simeq 10^2$ a.u. and a simple doubling of the surface area leads to $\Delta E = 4\pi \Gamma R^2 \simeq 120$ a.u. or $\simeq 3000$ eV!

13. K describes the production of vacancies (see question **8**). It is a concentration (i.e. a number) divided by a time.

The term in α describes the 'bimolecular' annihilation reaction between vacancies and interstitials both migrating in the solid. α has dimension (surface area)$^{-1}$.

More precisely, a vacancy (*resp.* an interstitial) moves the distance $v \exp(-E_m/kT)b\, dt \simeq (D_v/b)\, dt$ (*resp.* $(D_i/b)\, dt$) in time dt. If the average distance to annihilation with an interstitial is λ, it sweeps out in dt the effective volume $\simeq (\lambda^2/b)(D_v + D_i)\, dt$ containing $(\lambda^2/b^4)(D_v + D_i)\, dt$ sites. During dt the concentration of vacancies which annihilate with interstitials is thus $(\lambda^2/b^4)(D_v + D_i)c_v c_i\, dt$. Therefore, the constant $\alpha = \lambda^2/b^4$ essentially describes the square of the distance to annihilation.

The term in β_v describes the annihilation of vacancies on the dislocations of the solid in concentration ρ (ρ in m^{-2}, see p. 210). The constant β_v is a number.

During dt, the number of sites visited by a vacancy is $(1/b^2)D_v\, dt$. The number of sites which permit annihilation on the dislocations is z/b per unit length of dislocation (z = numerical constant; $1/b$ = number of sites on the dislocation). Thus, the concentration of vacancies which are annihilated on the dislocations is $(D_v/b^2)\, dt \times (z/b)\rho b^3 = z\rho D_v\, dt$. Therefore, β_v is just z, the number of sites favourable to annihilation around a dislocation, per elementary length b of the dislocation.

The term in γ describes the elimination of vacancies on cavities in volume concentration N. γ has the dimension of a length (evidently associated with the radius of the cavity, see question **16**).

14. We write directly:

$$\frac{dc_i}{dt} = K - \alpha(D_i + D_v)c_i c_v - \beta_i \rho D_i c_i - \gamma N D_i c_i. \qquad (7)$$

The production K is the same as that for vacancies (see Figure 1). The mutual recombination term (in α) is clearly the same as in (4). The (purely random) elimination on the cavities must involve the same constant γ. The term in β_i describes the elimination of the interstitials on the dislocations, in density ρ.

N.B. If the temperature is very low ('academic' situation which is not representative of the operation of a nuclear reactor), we may write $D_i = D_v = 0$ and there is no longer any migration of the point defects. However, there is still a recombination term, corresponding to the case in which an interstitial (*resp.* a vacancy) is created in a certain volume v, called the *recombination volume*, around an existing vacancy (*resp.* an interstitial): the pair (interstitial in the volume v) + (vacancy) is unstable and is annihilated instantly in an athermal manner. Therefore, we have

$$dc_i/dt = K(1 - c_v v)\,dt \quad \text{and} \quad c_i = c_v$$

whence it follows that

$$c_i \quad (\text{or } c_v) = v^{-1}[1 - \exp(-Kvt)].$$

Thus, as a function of time, a physical parameter such as the electrical resistivity ρ_{el} (see p. 58) which is sensitive to the presence of defects, should first grow linearly then tend towards a *saturation*. This is observed experimentally (see Figure). The value of $\Delta\rho_{el}$ at saturation can be used to deduce the value of the volume v which is generally found to equal several hundred atomic volumes. It goes without saying that the above expression for the constant α includes this volume v directly, corrected for its possible variation with the temperature.

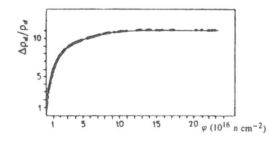

Figure 2 Increase and saturation of the electrical resistivity of uranium irradiated by neutrons at $T \simeq 5$ K. Ordinates: relative variation in resistivity. Abscissae: 'fluence' of thermal neutrons (proportional to the time t). J C Jousset, Y Quéré and H Wenzl.

15. At the beginning, we have a linear creation of defects: $c_v = c_i = Kt$. In the steady state $(dc_v/dt = dc_i/dt = 0)$, we have $(\beta_v\rho + \gamma N)D_v c_v^\infty = (\beta_i\rho + \gamma N)D_i c_i^\infty$. Whence

$$\frac{c_i^\infty}{c_v^\infty} = \frac{(\beta_v\rho + \gamma N)D_v}{(\beta_i\rho + \gamma N)D_i}. \tag{8}$$

16. The vacancies (and the interstitials) migrate randomly and meet (at random) cavities of radius r. This problem is essentially described in the Course (p. 196). For each defect ($j = $ i or v), we have (formula (33))

$$\frac{dc_j}{dt} = -\frac{3D_j}{R^3}(c_R - c_\alpha)r$$

where, here, $c_R = c_j^\infty$, $c_\alpha = 0$ and $\frac{4}{3}\pi R^3 = 1/N$, whence

$$\frac{dc_j}{dt} = -4\pi N r D_j c_j^\infty. \tag{9}$$

Thus, $\gamma = 4\pi r$.

17. If we have $\beta_i = \beta_v$, then, by virtue of expression (7), we may write

$$D_i c_i^\infty = D_v c_v^\infty. \tag{10}$$

Under these conditions, as many vacancies as interstitials are eliminated, in a given time, on the cavities[5] (see (8)). The latter do not grow or shrink.

18. On the other hand, if β_i and β_v are different then the previous conclusion does not apply. If β_i is (even slightly) greater than β_v, slightly more interstitials than vacancies are eliminated on the dislocations. Thus, the balance is perturbed in favour of the cavities. These grow.

It is understandable that β_i, which is associated with the attraction exerted by a dislocation on an interstitial (see questions **13** and **14**) is greater than β_v, since the elastic distortions around an interstitial are greater than those due to a vacancy. We then write: $\beta_v = \beta_0$ and $\beta_i = \beta_0(1 + \varepsilon)$, $\varepsilon > 0$.

19. $\Delta V/V$ is just the volume of a cavity divided by the volume $1/N$ surrounding a cavity:

$$\Delta V/V = g = \frac{4}{3}\pi N r^3. \tag{11}$$

20. According to (4) and (5), the excess vacancies (over the interstitials) reaching the cavities give rise to an increase in swelling:

$$\frac{dg}{dt} = N\gamma(D_v c_v^\infty - D_i c_i^\infty),$$

or, by (8)

$$\frac{dg}{dt} = N\gamma D_i c_i^\infty \left(\frac{\beta_i\rho + \gamma N}{\beta_v\rho + \gamma N} - 1\right) = N\gamma D_i c_i^\infty \frac{\beta_0\varepsilon\rho}{(\beta_0\rho + \gamma N)}.$$

[5] And on the dislocations (see (4) and (5)).

Moreover, following (7), where we neglect the term in α (see the statement of question 16), we have $K = D_i c_i^\infty (\beta_0(1 + \varepsilon)\rho + \gamma N)$. Whence:

$$\frac{dg}{dt} \simeq \frac{K\gamma N\beta_0 \rho \varepsilon}{(\beta_0 \rho + \gamma N)^2}. \tag{12}$$

21. The differential equation (12) describes the evolution of the swelling g. Here, g and γ are simple functions of the average radius of the cavities r: $g = \frac{4}{3}\pi N r^3$ and $\gamma = 4\pi r$. We solve this equation in the two extreme cases:

$$4\pi Nr \ll \beta_0 \rho \simeq \rho : g = 12N\left(\varepsilon \frac{K}{\rho}\right)^{3/2} t^{3/2} \tag{13}$$

and

$$4\pi Nr \gg \beta_0 \rho : g \simeq 0.25(\varepsilon K\rho)^{3/4} N^{-1/2} t^{3/4}. \tag{14}$$

The following diagram illustrates the schematic evolution of the swelling, which tends to a saturation over long periods. This saturation is achieved when the radius of the cavities is such that for point defects, the probability of reaching the dislocations becomes negligible in comparison with that of reaching cavities. The role of 'bias' played by the dislocations towards the interstitials (term ε) then becomes inoperative and the cavities stop growing when all the interstitials and all the vacancies reach them (in equal numbers).

Numerical application. Using (14), we obtain $\Delta V/V \simeq 3 \times 10^{-2}$.

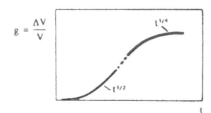

22. The very existence of swelling depends on the following two conditions: (i) initial nucleation of cavities and (ii) existence of a bias represented by the term ε.

A research programme should have the following aims:

- It should attempt to gain a better understanding of nucleation (e.g. by studying the shape and stability of the cascade-induced vacancy clusters[6]) and to favour the initial *cavity → loop* transformation. One should aim to understand (and combat) the factors stabilising the cavities, such as the presence of gas atoms in the solid, the effect of the surface energy Γ, etc.

[6] These clusters of individually resolved vacancies have been observed experimentally: **D N Seidman, R S Averback, R Benedek** Phys. Stat. Sol (b) **144** (1987) 85.

- It should attempt to gain a better understanding of the mechanism behind the bias, and, in particular, study the relationship between the factor ε and the crystalline structure of the alloys used. This should involve the following parallel research.

  — *Theoretical* research into the interaction between an interstitial and a dislocation. Particular attempts should be made to determine the effect of the Burgers vector (BV) on the value of ε. Thus, for alloys of body-centred cubic structure (for example, for those steels known as *ferritic*) one should try to calculate the effect for the two possible types of perfect dislocations (those with BV $\langle 100 \rangle$ and those with BV $\frac{1}{2}\langle 111 \rangle$, where the latter may *a priori* give rise to a smaller (whence more favourable) factor ε than the former).

    We also expect a decrease in ε if *dissociation* occurs (see p. 214). Thus, all the parameters favouring dissociation (in particular, the composition of the alloy) should be studied.

  — *Experimental* research (in particular by electron microscopy) to provide a quantitative appreciation of the elimination of the interstitials and the vacancies on the dislocations for different structures and different alloy contents.

  Finally, some work should be devoted to the simple idea that anything that favours direct (i)–(v) recombination is welcome and can only decrease swelling. In particular, the presence of certain *impurities* (foreign atoms), leading to the trapping of a defect (i, for example), may allow its ultimate annihilation by a defect of the other type, thereby removing both from the processes described above (questions 18–20). Thus, one might carry out a systematic study of the effect of such impureties (in particular, those of strong size effect) on swelling.

N.B. The problems arising in alloys irradiated in nuclear reactors are much more complex than indicated by this problem which tackles only the question of swelling. In particular, the equilibrium of phases is modified due to the continuous creation of point defects (see **G Martin** and **P Bellon**, *Static and Dynamics of Alloy Phase Transformation*, Plenum Press, New York (1994); *Solid State Physics*, **50**, 189 (1997)).

ATOMIC VIBRATIONS
IN SOLIDS

In a solid, the atoms oscillate about their equilibrium position. We first establish a simple oscillation model (**I**). Then, (**II**), we describe an experiment bringing these oscillations to light.

N.B. Questions **7** and **8** are independent of **I**.

I

Suppose a crystal $\mathcal{A}$ contains N identical atoms M_α of mass μ. Each atom is repelled by its neighbours and oscillates about its equilibrium position α, identified by $\bar{r}_\alpha$.

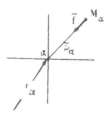

We suppose that it is subjected to a force $\bar{f}$ directed towards α and proportional to $\overline{\alpha M_\alpha} = \bar{\rho}_\alpha$:

$$\bar{f} = -B\bar{\rho}_\alpha$$

(where B is a constant independent of α). This amounts to treating the movement of M_α as that of a classical harmonic oscillator in three dimensions. Under these conditions, all the atoms oscillate with the same frequency ν_E. This is Einstein's model. ν_E is Einstein's frequency.

1. *Express ν_E as a function of B and μ.*

2. In questions **2** and **3**, we shall evaluate B in the case where $\mathcal{A}$ is a 'good' monovalent 'metal'.

Each atom is assigned a spherical atomic volume (radius R) whose centre represents the nucleus. The definition of this volume (namely:

$$[\text{total volume of } \mathcal{A}] = N \times \tfrac{4}{3}\pi R^3;$$

see p. 68) clearly excludes an arrangement in which these spheres are exactly tangential. However, we shall suppose that at any time

$$b = 2R$$

(b is the distance between near-neighbour atoms), which is a very reasonable approximation in the case of compact stackings of atoms.

To estimate B, we mentally isolate a small cluster of atoms, for example, in the case of a face-centred cubic crystal, the regular tetrahedron, shown in the accompanying figure, with edge b and comprising four atoms (we neglect all interaction between this cluster and the other atoms).

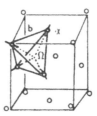

We then consider a particular oscillation mode in which the four atoms oscillate in phase, with frequency ν_E along axes of type $\Omega\alpha$ (Ω is the centre of the tetrahedron). This so-called 'respiration' mode preserves the tetrahedral shape of the cluster, whose volume varies with the frequency ν_E. The radii of the four atomic spheres oscillate in phase around the equilibrium value R_0.

Write down the relationship between the distance ρ of the four atoms from their equilibrium position and the variation $\Delta R = R - R_0$ of the atomic radii.

N.B. We recall the relationship in a regular tetrahedron: $\Omega\alpha = (b/2)\sqrt{3/2}$.

3. *Recalling that the expression for the energy of an atomic sphere for a monovalent metal was given in the Course, write down the energy $E(R)$ of the tetrahedral cluster (in atomic units a.u.; denote the ionic radius by R_c). Deduce the harmonic approximation $V(\rho)$ for this energy E (with V such that $V(0) = 0$), whence the constant B.*

4. *Using this estimate of B, calculate ν_E numerically (in s^{-1}) for any one of the alkali metals Li, Na or K.*

N.B. We suggest you work in the atomic system (see p. 36). The numerical values of R_0 are given on p. 71. The atomic masses of the alkali metals are: Li: $\simeq 7$; Na: 23; K: 39.

We recall that: mass of proton $\simeq$ mass of neutron $\simeq 1840 \times$ mass of electron.

5. The amplitudes of oscillation are, on average, greater the greater the temperature T is. The average displacement is not characterised by $\langle \rho \rangle$, which is zero, but by $\langle \rho^2 \rangle$ (mean square amplitude).

Taking into account the fact that the oscillators here obey classical statistics, express $\langle \rho^2 \rangle$ as a function of T and ν_E.

6. The quantity $\rho_T = \sqrt{\langle \rho^2 \rangle}$ then characterises an average displacement of the atoms. *Show numerically that, in the case chosen in question 4, ρ_T remains small in comparison with the interatomic distances, even at the melting point of the metal (melting points: Li: 179° C; Na: 98° C; K: 63° C).*

II

The diffraction of X-rays by a crystal is described (in **Annex 1**) under the assumption that the atoms are situated at ideal crystalline positions. We have just shown that, in reality, they are almost always *elsewhere* rather than at these ideal crystalline positions. Could this fact not lead to a blurring of the constructive interferences producing diffraction? Does this become impossible, or is it only modified, and if so, how? In this part, we shall attempt to answer these questions and at the same time give access to the measurement of ν_E.

However fast it may be, the oscillation frequency ν_E is slow in comparison with the frequency of X-rays. Thus, we may suppose that at each instant the scattering of the X-ray is produced by a set of fixed atoms located at the points $M_\alpha(\overline{r}_\alpha + \overline{\rho}_\alpha)$. In what follows, the different $\overline{\rho}_\alpha$ will be assumed to be entirely uncorrelated. The intensity scattered in a given direction is therefore the average of the intensities scattered by all the possible configurations.

We shall adopt the notation of the Course: $\overline{a}_1 \ \overline{a}_2 \ \overline{a}_3$ form the primitive cell (with $|\overline{a}_j| = a$) and $\overline{a}_1^* \ \overline{a}_2^* \ \overline{a}_3^*$ the reciprocal lattice. N_j is the number of atoms along $\overline{a}_j$ (where $N_1 N_2 N_3 = N$), $\overline{k}$ is the incident wavevector ($\overline{k}$ is fixed), and $\overline{k}'$ is the scattered wavevector with $\overline{\Delta k} = \overline{k}' - \overline{k}$ and $|\overline{k}| = |\overline{k}'|$. The direction of observation is $\overline{k}'$. The coordinates of the vectors $\overline{\rho}$ are denoted by x_j: $\overline{\rho} = \sum_j x_j \overline{a}_j$.

7. *For an instantaneous configuration of the atoms M_α, give the general formula for the amplitude scattered along $\overline{k}'$ in the form*

$$A(\overline{\Delta k}) = f \times \sum_\alpha \ldots \tag{1}$$

(here f is the atomic-scattering factor for which we shall not give an explicit expression and which we shall assume to be constant).

8. Experimentally, we choose $\overline{k}'$ so that $\overline{\Delta k}$ is a vector $\overline{L}$ of the reciprocal lattice, for example, $\overline{L} = p\overline{a}_1^*$ (for some integer p).

Using the fact (see question 6) that $x_1 \ll 1$, give an approximate formula for $A(p)$ involving the mean values $\langle x_1^n \rangle$ up to $n = 2$ (n is an integer > 0).

Deduce the intensity of X-rays, $I(p)$ in the direction $\overline{k}'$ in the form

$$I(p) = F \times I_0 \tag{2}$$

where I_0 is the intensity diffracted by the ideal crystal (where all the ρ are zero, case handled in the Course) and F is a factor which depends on $\langle x_1^2 \rangle$.

9. We are now in a position to respond to the concerns and queries raised at the head of **II**.

   For this, express the factor F as a function of the temperature T and the Einstein frequency (see question 5).

   Then describe, in simple terms, the effect of the thermal vibrations on the diffraction of X-rays by crystals and more precisely, on the existence of diffracted beams, their position and their intensity as a function of the temperature and the distance p from the centre of the reciprocal lattice. Summarise these conclusions using a scheme analogous to that of p. 255, but corrected for the effect of thermal vibrations.

10. The accompanying diagram shows experimental values of the intensity I of X-rays diffracted by a copper crystal as a function of temperature T for the diffraction corresponding to $p = 8$.

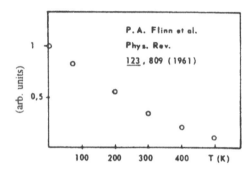

   Show that these data 'broadly' confirm the model described above. Deduce the value of ν_E (in s^{-1}) for copper (atomic mass: 66.5, distance between near neighbours $a = 2.55$ Å). Indicate the reason why, in the above approximations, the experimental values $I(T)$ differ from the theoretical predictions of questions 8 and 9.

SOLUTION TO PROBLEM 18

1. The movement of M_α decomposes into three sinusoidal movements with pulsatance $\omega_E = \sqrt{B/\mu}$. Whence $\nu_E = (1/2\pi)\sqrt{B/\mu}$.

2. In the mode considered here, the distances ρ are measured along the axes $\Omega\alpha$. Thus, taking into account the fact that $b = 2R$, we have

$$\rho = \sqrt{3/2}\Delta R.$$

3. Since we are dealing with a 'good metal', the energy of a cluster of four atoms may be written (see (8), p. 70) in the form:

$$E = 4\left(\frac{1.1}{R^2} - \frac{1.36}{R} + \frac{1.5R_c^2}{R^3}\right) \text{ a.u.}$$

   This formula remains valid when the cluster oscillates while remaining homothetic with itself, since the atomic volumes are still spheres as R varies about its equilibrium R_0. We then have (harmonic approximation):

$$V(\rho) = V(0) + \frac{1}{2}\frac{d^2E}{dR^2}\bigg|_{R_0}(\Delta R)^2$$

or, per atom:

$$V(\rho) = \frac{2}{3}\left[\frac{3.3}{R_0^4} - \frac{1.36}{R_0^3} + \frac{9R_c^2}{R_0^5}\right]\rho^2.$$

This potential is of the form $\frac{1}{2}B\rho^2$ (force in $-B\rho$) with $B = \frac{4}{3}[\ldots]$.

4. We can then calculate $\nu_E = (1/2\pi)\sqrt{B/\mu}$ numerically, since R_0 is given (in a.u.) in Chapter IV (p. 71) and μ is equal (in a.u.) to (atomic mass) $\times 1840$. Using the value of the time unit in a.u. (see p. 36), for sodium we have

$$\nu_E = \frac{1}{2\pi}\sqrt{\frac{0.0203}{42320}} \times 4.13 \times 10^{16}\ \text{s}^{-1};$$

whence, finally

	Li	Na	K
ν_E (s^{-1})	1.3×10^{13}	0.45×10^{13}	0.24×10^{13}

5. The mean potential energy of a classical harmonic oscillator is $\frac{1}{2}kT$ per degree of freedom (k is the Boltzmann constant), giving a total of $\frac{3}{2}kT$ here. In the potential $V = \frac{1}{2}B\rho^2$, an atom has an average potential energy $\langle V \rangle = \frac{1}{2}B\langle\rho^2\rangle$. Thus, we have

$$B\langle\rho^2\rangle = 3kT$$

whence

$$\langle\rho^2\rangle = \frac{3kT}{\mu(2\pi\nu_E)^2}.$$

6. Numerical calculation gives, at the melting points:

	Li	Na	K
ρ_T (Å)	0.16	0.22	0.30

ρ_T is always small in comparison with the interatomic distances (3.72 Å for Na).

7. Here, adding the phase factors (see p. 256) gives

$$A(\overline{\Delta k}) = f\sum_\alpha \exp[i\overline{\Delta k} \cdot (\overline{r}_\alpha + \overline{\rho}_\alpha)]. \tag{3}$$

8. If we have $\overline{\Delta k} = p\overline{a}_1^*$ and $\overline{r}_\alpha = n_1\overline{a}_1 + \ldots$, the terms $\exp(i\overline{\Delta k} \cdot r_\alpha)$ are all equal to 1 and we are left with

$$A = f\sum_\alpha \exp(i\overline{\Delta k} \cdot \overline{\rho}_\alpha) = f\sum_\alpha \exp[ip\overline{a}_1^* \cdot (x_{1\alpha}\overline{a}_1 + \ldots)].$$

Since $\overline{a}_1^* \cdot \overline{a}_1 = 2\pi$ and $\overline{a}_1^* \cdot \overline{a}_2 = \overline{a}_1^* \cdot \overline{a}_3 = 0$, we are left with

$$A = f\sum_\alpha \exp(i2\pi p x_{1\alpha}) \simeq f\sum_\alpha (1 + i2\pi p x_{1,\alpha} - 2\pi^2 p^2 x_{1,\alpha}^2)$$

$$= N_1 f(1 + i2\pi p\langle x_1\rangle - 2\pi^2 p^2\langle x_1^2\rangle)$$

or, since $\langle x_1 \rangle = 0$:

$$A = N_1 f(1 - 2\pi^2 p^2 \langle x_1^2 \rangle)$$

whence

$$I(h) \simeq N_1^2 f^* f(1 - 4\pi^2 p^2 \langle x_1^2 \rangle). \tag{4}$$

The case handled in the Course corresponds to $\langle x_1^2 \rangle = 0$ or $I_0 = N_1^2 f^* f$.

The factor F, called the *Debye–Waller factor* is therefore

$$F = 1 - 4\pi^2 p^2 \langle x_1^2 \rangle. \tag{5}$$

9. The isotropy of the problem enables us to write

$$\langle x_1^2 \rangle = \langle x_2^2 \rangle = \langle x_3^2 \rangle = \frac{1}{3a^2} \langle \rho^2 \rangle.$$

Thus, by virtue of question **5**, we have

$$F = 1 - \frac{p^2 kT}{\mu v_E^2 a^2}. \tag{6}$$

The above (expressions (4)–(6)) shows that the thermal vibrations do not eliminate the diffraction. The latter is produced in the same spatial directions $(\overline{\Delta k} = \overline{L}$, vector of the reciprocal lattice) as for an ideal crystal. The intensity of the diffracted beams is decreased by the thermal vibrations. This decrease is linear in T and parabolic in p, the distance from the centre of the reciprocal lattice (see diagram). It varies as v_E^{-2} (where v_E is the Einstein frequency).

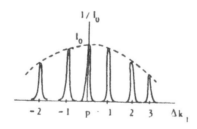

10. As predicted by formulae (4) and (6), the diffracted intensity decreases as the temperature increases. Approximately this decrease is linear, in accordance with (6). From the measured gradient $(\simeq -2.2 \times 10^{-3} \ \mathrm{K}^{-1})$, which according to (6) is equal to $-p^2 k / \mu v_E^2 a^2$ (with $p = 8$), we deduce $v_E = 7.6 \times 10^{12} \ \mathrm{s}^{-1}$.

The experimental decrease in I is actually slower than linear. This is a consequence of stopping the expansion of A in question **8** at the second order. An extension to order four would introduce a positive term in $\langle x^4 \rangle$.

N.B. 1   Calculation of the full development of A shows that, for an harmonic oscillator, the factor F expands as an exponential series of which the expression in (6) is only the first term. Thus, under these conditions, we have

$$I = I_0 F = I_0 \exp\left(-\frac{p^2 kT}{\mu v_E^2 a^2}\right) \tag{7}$$

which expression corresponds better to the experimental results than the linear expression (6).

N.B. 2   At very low temperature, the above expansion is not valid. The Debye–Waller factor is not equal to 1 for $T = 0$ since the atoms cannot be at rest (zero-point vibration).

At $T = 0$ K, all the oscillators are in their ground state and have the energy $\frac{1}{2} h v_E$. The mean value of the potential energy is then $\frac{3}{4} h v_E$.

Thus, we have

$$\tfrac{1}{2} B \langle \rho^2 \rangle = \tfrac{3}{4} h v_E$$

where $B = 4\pi^2 \mu v_E^2$. Formula (7) then becomes

$$I(0) = I_0 \exp\left(-\frac{p^2 h}{2\mu v_E a^2}\right)$$

where $I(0)$ is the intensity diffracted at $T = 0$ K and I_0 is the intensity which would be diffracted if the atoms were at rest at their ideal crystalline site.

Returning to the specific case of copper, we find numerically that for $p = 1$, $I(0)/I_0 = 0.97$ while for $p = 8$, $I(0)/I_0 = 0.16$.

STRUCTURE AND ELASTICITY OF POLYMERS

This problem gives a more precise description of the structure (**I**) and the elasticity (**II** and **III**) of polymeric chains than in the Course, together with a very simplified overview of the influence of the 'excluded volume' on the size of polymers in solution (**IV**).

Each part may be tackled independently of the others.

The following are provided: metallic kit models of the CH_4 structure, a rubber band, a small board, a nut, some liquid nitrogen and an elastic balloon.

I

Structure of polyethylene

We consider a polymeric chain as a sequence of N monomers $\bar{b}_1, \bar{b}_2, \ldots, \bar{b}_N$, each of length b. Initially, we suppose that the $\bar{b}_i$ are constrained to remain *in the plane*, each randomly making an angle $\pm\alpha$ with its two neighbours (Figure 1). We ignore all other constraints (of the 'excluded volume' type). We propose to show that, for N large, the formula of the Course, $\langle r^2 \rangle = Nb^2$, which is valid for $\alpha = \pi/2$, is replaced by

$$R^2 = \langle r^2 \rangle = Nb^2 \frac{1 + \cos\alpha}{1 - \cos\alpha}. \tag{1}$$

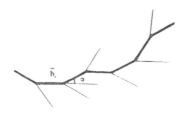

Figure 1.

1. *Expand the square $r^2 = (\bar{b}_1 + \ldots + \bar{b}_N) \cdot (\bar{b}_1 + \ldots + \bar{b}_N)$.*

2. *Show that we have*

$$\langle \bar{b}_i \cdot \bar{b}_j \rangle = b^2 \cos^{(j-i)} \alpha \quad (j > i; \langle \, \rangle = \text{mean value}). \tag{2}$$

N.B. For that, one might calculate $\langle \bar{b}_i \cdot \bar{b}_{i+1} \rangle$, $\langle \bar{b}_i \cdot \bar{b}_{i+2} \rangle$ and then generalise.

3. *Deduce the formula (1).*

4. *The expression (2) describes the angular correlation between the monomers $\bar{b}_i$ and $\bar{b}_j$. How does this correlation behave when $(j - i)$ increases? Determine the number v of successive monomers such that this correlation (expressed by $\cos^{(j-i)} \alpha$) becomes less than a given number η.*
    Numerical application: $\eta = 0.1$ and $\alpha = 45°$.

5. This progressive decorrelation leads us to represent the initial chain by an idealised chain of N completely decorrelated (that is, freely articulated: see Figure 2) segments of length l. *Write down the value of $\langle r^2 \rangle$ for this chain (N, l). What length l must the segments of the idealised chain have for it to be of the same 'size' R as the previous real chain (N, b, α).*
    Numerical application for $\alpha = 45°$ and $\alpha = 30°$.

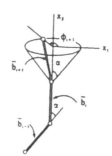

6. Generalise the above to the case in which the monomers retain the angle α between themselves (so-called valence angle) without being constrained to remain in a plane (Figure 3): the position of $\bar{b}_{i+1}$ is then defined on the cone with axis $\bar{b}_i$ and angle α by the angle of rotation ϕ_{i+1}.
    Show that if all ϕ are equally likely, formulae (2) and (1) remain valid.

7. To be specific, consider the case of polyethylene $(CH_2)_N$. In this, each carbon atom (i) is linked to two hydrogen atoms (together forming the monomer i) and to two carbon atoms $(i - 1)$ and $(i + 1)$. We shall assume that this configuration is similar (angularly) to the configuration of the methane molecule (CH_4) (hybridisation sp$_3$), metallic models of which were distributed. By manipulating two successive elements (i) and $(i + 1)$ and

Figure 2.   **Figure 3.**

linking them to form the bond $C_{(i)}$–$C_{(i+1)}$, and with the knowledge that the angles (all identical here) between bonds are equal to $\simeq 104°$, *identify the valence angle (above), denoted by α, and give its value together with that of the corrective term of (1).*

8. *Manipulating the rotation ϕ, reflect intuitively on the possible variation of the energy E of the group (i)–$(i + 1)$ as a function of the angle ϕ. Sketch the graph of $E(\phi)$ $(0 < \phi < 2\pi)$. How many stable (or metastable) positions in ϕ do you detect in this way?*

9. *Does the existence of these positions call in question the result of question 6.*

II

The elasticity of rubber

In this part, we analyse the origin of the *elasticity of rubber*. We assume that the latter consists of 'Gaussian' polymeric chains obeying equation (2) of the Course (Chapter V, p. 90), and that the behaviour of a chain accurately describes that of a set of chains, that is of a macroscopic sample.

The probability $p(\bar{r})$ is proportional to the number of configurations (equally likely) which give rise to the 'path' $\bar{r}$.

10. *Writing the entropy of the chain configuration in the form $S = S_0 + S(\bar{r})$, give the expression for $S(\bar{r})$.*

11. Suppose that the internal energy (essentially that of the bonds between monomers) does not depend upon the path followed. Then the free energy takes the form $F = F_0 + F(\bar{r})$. *Give an expression for $F(\bar{r})$ as a function of the temperature T.*

12. At the two ends of the chain, which is initially assumed to be 'closed' $(r = 0)$ a force f is applied and thus these two ends are separated by a distance r.

Give the expression for r as a function of f. What happens if the force f is suppressed? Can we speak of elasticity? If yes, is it of Hooke type?

13. Here, we suppose that the *vulcanisation* process which involves fastening the chains together by sulphur atoms, reduces, with the fixing of certain points, to decreasing their size R (defined in (1)) globally.

What is then the consequence of vulcanisation on the elastic rigidity of rubber?

14. *With the chain being subjected to a constant force f, how does r vary when the temperature decreases? What do you deduce about the sign of the coefficient of thermal expansion of a polymer, such as rubber, subjected to elongation?*

By fixing one end of the rubber band provided on the small board and hanging the nut on the other end, cool the band using a small amount of liquid nitrogen (without hardening it). *Does the change in length you observe have the sign indicated above?*

15. Carry out the following test using the rubber band: stretch it quickly and place it in contact with your cheek. You should then observe a slight variation ΔT in the temperature of the band.

What is the sign of ΔT?

16. We wish to evaluate ΔT in the form

$$\Delta T = \int \left(\frac{\partial T}{\partial r} \right)_s dr.$$

Express $(\partial T/\partial r)_s$ as a function of $(\partial S/\partial r)_T$ and of $(\partial S/\partial T)_r$. Then show that one can write

$$\left(\frac{\partial T}{\partial r} \right)_s = \frac{T}{C_r} \left(\frac{\partial f}{\partial T} \right)_r \tag{3}$$

where $C_r = T(\partial S/\partial T)_r$ is the specific heat (at constant length) of the rubber band. Deduce the expression for ΔT and, in particular, check that its sign is as you experienced it in question 15.

III

Inflation of a balloon

We now proceed to inflate the elastic balloon. This experiment should lead you to infer, as you continue to blow, the existence of three successive phases φ_1, φ_2 and φ_3. The first provides the balloon with its spherical equilibrium form without constraints. At the end of this first phase φ_1, the pressure p in the balloon is equal to the atmospheric pressure p_0, the radius of the balloon is ρ_0 and the thickness of the rubber is h_0. During φ_2, the pressure p increases, as does the radius ρ, while h decreases.

17. *Measure ρ_0 (approximately). Identify (through blowing) the transition from φ_2 to φ_3. Measure ρ ($= \rho_c$) at the time of this transition and state what appears to happen during φ_3.*

18. In what follows, we shall assume that the volume V ($= 4\pi\rho^2 h$) of the rubber (material) remains constant, that its behaviour is elastic (if you did not answer question **12**, you may simply write the force–elongation proportionality in the form $f = A\Delta r$, where Δr is the elongation of a chain), and that in the bulk, a chain occupies a practically constant surface σ, perpendicularly to the force. Finally, we shall neglect the influence of the surface energy.

Study the equilibrium of a small calotte of rubber (subjected to the elastic forces and to the pressure) and write down the relationship between the pressure p in the balloon and its radius ρ. Draw the graph of $p(\rho)$ and distinguish the phases φ_1, φ_2 and φ_3 on it. Compare with the results of question 17, in particular, for ρ_c (or rather ρ_c/ρ_0).

IV

Excluded volume constraint

We now introduce the 'excluded volume' constraint. That is, we study the consequences on the description in **I**, of the fact that two monomers i and j cannot be superimposed in space. Thus, we now have to introduce the interaction energies (repulsive) u_{ij} between pairs of monomers i and j. For a particular configuration (λ) of the chain of N monomers and path r, we set $U_{\lambda r} = \frac{1}{2} \sum_{ij} u_{ij}$.

19. *Verify that for this chain the partition function has the expression*

$$Z(r) = g(r)\langle\exp(-U_{\lambda r}/kT)\rangle \qquad (4)$$

where $g(r)$ is the function (6) of the Course (p. 91) and where $\langle\ \rangle$ denotes an average over all the configurations (λ) with a given path r.

20. Faced with the inextricable difficulty of evaluating the $U_{\lambda r}$, we shall now have recourse to an argument due to Flory and Fisher. Here, we liken the polymeric chain to a cloud of monomers and extend the simplification to the point of considering the density of this cloud to be *uniform in a sphere* taken here to be of size $\simeq r$, or *of volume r^3*, and *zero outside*[1]. Finally, for the energy u_{ij} we adopt an all-or-nothing model: u_{ij} is taken equal to ε (> 0) if j is contained in a small volume v around the monomer i (see diagram), and equal to zero if j lies outside this volume.

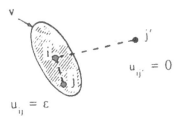

In this so-called 'mean-field' approximation (in which all the fluctuations of position, and thus of interaction, have been erased and it is assumed that all the monomers 'see' the same average environment) give the expression for $U_{\lambda r}$ as a function of N (large), ε and r.

21. Knowing, now, the partition function Z (which, in particular, involves the 'size' r of the previous sphere, the number of monomers N and the quantity $R^2 = Nb^2$), show that, at equilibrium, the size of the chain varies as $N^{3/5}$ (and not now as $N^{1/2}$).

N.B. Readers will have noted the additional approximation which, since question **19**, involves likening the chain to a sphere of size r (volume r^3) equal to the path r and thus seeking in question **21**, the function $F(N) = r$ (and not $\sqrt{\langle r^2\rangle}$).

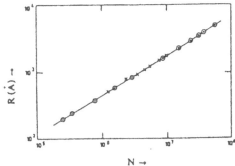

Figure 4 Size R of polystyrene molecules in solution in benzene, as a function of the degree of polymerisation, N. After **J P Cotton** J. Physique, **41**, (1980), L-231.

[1] In reality this density exhibits a Gaussian variation.

22.   If, after chains are dissolved (degree of polymerisation N as well defined as possible) in an appropriate solvent, one measures the diffusion of light, one can deduce the size R of these chains. Figure 4 shows a set of measurements of this type.

   Does the previous model give a good account of these measurements?

SOLUTION TO PROBLEM 19

I

1.

$$r^2 = Nb^2 + 2\sum \bar{b}_i \cdot \bar{b}_j \quad (j > i)$$

2.

$$\langle \bar{b}_i \cdot \bar{b}_{i+1} \rangle = \tfrac{1}{2}[b^2 \cos\alpha + b^2 \cos(-\alpha)] = b^2 \cos\alpha$$

($+\alpha$ and $-\alpha$ are equally likely).

$$\langle \bar{b}_i \cdot \bar{b}_{i+2} \rangle = \tfrac{1}{4}[b^2 \cos 2\alpha + b^2 + b^2 + b^2 \cos(-2\alpha)] = b^2 \cos^2\alpha.$$

   Suppose now that $\langle \bar{b}_i \cdot \bar{b}_j \rangle = b^2 \cos^{(j-i)}\alpha$. The 'average' monomer $\bar{b}_j$ makes an angle β with $\bar{b}_i$ such that $\cos\beta = \cos^{(j-i)}\alpha$. Thus, we have

$$\langle \bar{b}_i \cdot \bar{b}_{j+1} \rangle = \tfrac{1}{2}b^2[\cos(\beta + \alpha) + \cos(\beta - \alpha)] = b^2 \cos\beta \cos\alpha = b^2 \cos^{(j+1-i)}\alpha.$$

Thus, formula (2) is quite general.

3.   We set $\cos\alpha = \gamma$. We have

$$\frac{1}{2}\langle r^2 \rangle = \frac{Nb^2}{2} + \langle \bar{b}_1 \cdot \bar{b}_2 \rangle + \langle \bar{b}_1 \cdot \bar{b}_3 \rangle + \ldots + \langle \bar{b}_1 \cdot \bar{b}_N \rangle$$

$$+ \langle \bar{b}_2 \cdot \bar{b}_3 \rangle + \ldots$$

$$\vdots$$

$$+ \langle \bar{b}_{N-1} \cdot \bar{b}_N \rangle$$

$$\frac{1}{2b^2}\langle r^2 \rangle = \frac{N}{2} + \gamma(1 + \gamma + \ldots \gamma^{N-2}) + \gamma(1 + \gamma + \ldots \gamma^{N-3}) + \ldots$$

$$= \frac{N}{2} + \frac{\gamma}{1 - \gamma}\sum_0^{N-1}(1 - \gamma^p).$$

Whence:

$$\langle r^2 \rangle = Nb^2 + \frac{2\gamma b^2}{1 - \gamma}\left(N - \frac{1 - \gamma^N}{1 - \gamma}\right)$$

or, for large N,

$$\langle r^2 \rangle = Nb^2 + \frac{2\gamma Nb^2}{1 - \gamma} = Nb^2\frac{1 + \cos\alpha}{1 - \cos\alpha}. \tag{5}$$

4. The correlation between monomers (i) and (j) decreases as $\cos^{(j-i)}\alpha$ as $j-i$ increases. It becomes equal to η for $j-i=\nu$ such that $\cos^{\nu}\alpha=\eta$, in other words,

$$\nu = \ln\eta/\ln\cos\alpha.$$

For $\eta=0.1$ and $\cos\alpha=\sqrt{2}/2$, we find $\nu=6.6\simeq7$.

5. For the freely articulated chain (N,l), we have

$$\langle r^2\rangle = Nl^2.$$

This formula coincides with (5), provided we take

$$l = b\left(1+\frac{2\cos\alpha}{1-\cos\alpha}\right)^{1/2}$$

that is $l\simeq2.4b$ for $\alpha=45°$ and $l=3.7b$ for $\alpha=30°$.

6. Let us suppose, as in question 2, that formula (2) is valid up to order j. Setting $\cos\beta=\langle\cos(\bar{b}_i\cdot\bar{b}_j)\rangle=\cos^{(j-i)}\alpha$ and adopting an axis x_3 collinear with $\bar{b}_j$, we see that $\bar{b}_i=(b\sin\beta,0,b\cos\beta)$ and $\bar{b}_{j+1-i}=(b\sin\alpha\cos\phi,b\sin\alpha\sin\phi,b\cos\alpha)$, in other words $\bar{b}_i\cdot\bar{b}_{i+1}=b^2\sin\beta\sin\alpha\cos\phi+b^2\cos^{(j+1-i)}\alpha$, or again

$$\langle\bar{b}_i\cdot\bar{b}_{j+1}\rangle = b^2\cos^{(j+1-i)}\alpha$$

(since $\langle\cos\phi\rangle=0$).

7. We have $\alpha=180°-104°=76°$ (see Figure 3) and $\cos\alpha=0.242$, whence $R=nb^2\times1.64$.

8. When we rotate (rotation ϕ) the element $i+1$ around the axis $\Delta(C_{(i)}-C_{(i+1)})$, we see that, for sterical reasons, there exist three favourable positions, offset by $120°$. The first position (s_1) corresponds to the carbon atoms $(i-1)$, (i), $(i+1)$ and $(i+2)$ in the same plane. A view of this situation along the axis Δ is shown in the accompanying diagram. The two other positions (s_2) and (s_3) are obtained from (s_1) by a rotation of $\pm120°$ about the axis Δ. The energies E of (s_2) and (s_3) are clearly equal. Whence we have the two possibilities shown in the accompanying diagram. (**N.B.** The real situation is that in which (s_1) is more stable than (s_2) and (s_3)).

9. The result of question **6** is unchanged if the three positions (s_1), (s_2) and (s_3) are equiprobable. It does change if (s_1) is more stable than (s_2) and s_3) and the change is greater the greater the difference in stability is.

N.B. The positions (s_1), (s_2) and (s_3) are called *trans* (t), *left* + (l^+) and left $-$ (l^-), respectively. An arbitrary chain will be a random sequence of the type $l^+l^+tl^-tl^-l^+\ldots$. Under particular conditions, periodic chains may be developed, for example, helical chains: $tl^+l^-tl^+l^-\ldots$.

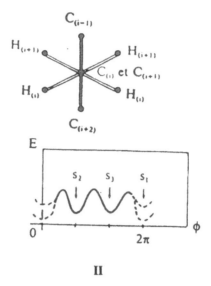

II

10. From (2) on p. 90, we obtain

$$S = S_0 - \frac{3kr^2}{2R^2}.$$

12. If the internal energy does not depend on r, we write

$$F = F_0 + \frac{3kTr^2}{2R^2}.$$

12. At equilibrium we have $f = \partial F / \partial r$ or

$$f = \frac{3kT}{R^2} r \qquad\qquad (6)$$

where r describes the elongation of the chain. If the force f is suppressed the chain returns to its most probable shape ($r = 0$). Thus, we are dealing with an elastic phenomenon, which is, moreover, of Hooke type since the elongation is proportional to the force applied. The modulus of elasticity, Y, defined here by *force* = Y × *elongation* is equal to $3kT/R^2$.

13. By decreasing R, the vulcanisation increases the modulus of elasticity of the rubber.

14. From (6), we see that, with the application of a constant force f, increasing T amounts to decreasing r: unlike most solids (metals, ionic crystals ...) stretched rubber has a *negative* coefficient of thermal expansion.

   When the rubber stretched by the action of the mass of the nut is cooled, we see that it actually becomes longer.

15. After violent (adiabatic) stretching of the elastic, we note a heating: $\Delta T > 0$.

16. The following classical relationship exists between the partial derivatives:

$$(\partial T/\partial r)_s = - \left(\frac{\partial S}{\partial r}\right)_T \Bigg/ \left(\frac{\partial S}{\partial T}\right)_r.$$

The definition of the specific heat $C_r = T(\partial S/\partial T)_r$, and the Maxwell equation $(\partial S/\partial r)_T = -(\partial f/\partial T)_r$ (itself deduced from the equation $dF = -SdT + fdl$) enables us to write down equation (3). It follows that

$$\Delta T = \int (\partial T/\partial r)_s \, dr = \frac{T}{C_r} \frac{3k}{R^2} \frac{(\Delta r)^2}{2}$$

which is positive, as required.

N.B. 1 Since the elongation is *adiabatic* and *reversible*, the total entropy does not vary (taking into account the fact that $dS = dQ/T$). The decrease in the *configuration entropy* calculated in question **10** must be compensated by an increase in another component of the entropy. Thus, it is the *vibration entropy* (that is, the molecular thermal agitation) which increases, as well as the temperature.

N.B. 2 The effect is at least as clear when it is realised in the other direction. When the elastic has been stretched, it is allowed to return room temperature, then released suddenly: the configuration entropy *increases* and the temperature *decreases* in a perceptible manner.

III

17. ρ_0 is equal to $\simeq 2$ cm. When blowing up the balloon we experience that the breath needed to create the pressure p ($> p_0$) seems to pass through a maximum for $\rho_c = 3.5$ cm, before decreasing (phase φ_3), while ρ continues to increase.

18. Let us consider a small circular element (radius l) on the balloon. It brings the tension of $2\pi lh/\sigma$ chains into play. Thus, the total elastic force is $A2\pi lh\Delta r/\sigma$, the radial projection of which is $F_\rho = A2\pi l^2h\Delta r/\sigma\rho$, in equilibrium with the force due to the pressure, namely $(p - p_0)\pi l^2$. Whence, $p - p_0 = 2Ah\Delta r/\sigma\rho$.

Δr is the elongation of a chain. If R is likened to the size of a chain, we write $2\pi(\rho - \rho_0) = \Delta r \times$ *number of chains on a circumference* $= \Delta r \times 2\pi\rho_0/R$. Whence,

$$p - p_0 = 2Ah(\rho - \rho_0)R/\sigma\rho$$

where $\rho_0^2 = \rho^2 h$.

Finally, let

$$p - p_0 = \frac{A}{2\pi} \frac{VR}{\sigma} \frac{\rho - \rho_0}{\rho_0\rho^3}.$$

The corresponding variation $p(\rho)$ (see diagram) leads to the appearance of a maximum for $\rho_c = \frac{3}{2}\rho_0$. As measured, the overpressure $p - p_0$ which is zero during φ_1, increases during φ_2 and decreases during φ_3. The observed values of ρ_0 and ρ_c are in approximate agreement with the above result.

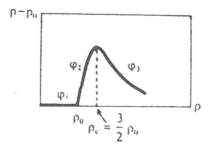

IV

19.   We write $Z = \sum_\lambda \exp(-U_{\lambda r}/kT)$. Since the number of configurations with 'path' r is $g(r)$, we have

$$Z(r) = g(r)\langle\exp(-U_{\lambda r}/kT)\rangle$$

where

$$g(r) \propto r^2 \exp\left(-\frac{3r^2}{2R^2}\right).$$

20.   Each of the N monomers interacts with the $N - 1$ ($\simeq N$) other monomers which are found within the interaction volume with probability v/r^3. Whence

$$U_{\lambda r} = \frac{N^2}{2}\frac{v}{r^3}\varepsilon. \tag{7}$$

21.   We then seek to maximise Z. Taking into account (4) and (7), we obtain

$$\frac{2}{3} + \frac{N^2 v\varepsilon}{kTr^3} = \frac{r^2}{Nb^2}.$$

If we postulate a law $r \propto N^d$ where $d > \frac{1}{2}$, then the term on the right-hand side diverges for N large, which necessarily makes the term $\frac{2}{3}$ negligible. It follows that

$$r = \left(\frac{v\varepsilon b^2}{kT}\right)^{1/5} N^{3/5}. \tag{8}$$

The measurements recorded on Figure 4 lead to the appearance of a law in $N^{0.586}$, which is in good agreement with the law (8).

N.B.   The pseudo Brownian motion in which the walker is not allowed to return to a point already visited and which, in this case, models the polymeric chain, is called a *self-avoiding walk* (SAW).

COMPLEMENTARY READING

Several books have already been indicated in the body of the text.
For a simple, but at the same time most complete overview of **materials**, see:

C KITTEL
Introduction to Solid State Physics
John Wiley (1971)

M F ASHBY and D R H JONES
Materials
Pergamon Press (1980)

A GUINIER and R JULIEN
La matière à l'état solide (in French)
Hachette, Paris (1987)

J M DORLOT, J P BAÏLON and J MASOUNAVE
Des matériaux (in French)
Edit. Ecole Polytechnique de Montréal (1986)

W KURZ, J P MERCIER and G ZAMBELLI
Introduction à la science des matériaux (in French)
Presses Polytechniques Romandes (1987)

and for a more detailed description:

N W ASHCROFT and N D MERMIN
Solid State Physics
Holt, Rinehard and Winston (1976)

W A HARRISON
Electronic Structures and the Properties of Solids
Dover Publ. (1989).

The following works relate to **specific chapters**:

Chapters III to VI

C JANOT, M GERL, J GRILHE and J CAISSO
Propriétés électroniques des métaux et alliages (in French)
Masson, Paris (1973)

C WEISBUCH and B VINTER
Quantum Semiconductor Structures
Academic Press (1991)

Chapters VII and VIII

C P FLYNN
Point Defects and Diffusion
Clarendon Press, Oxford (1972)

A STONEHAM
Theory of Defects in Solids
Clarendon Press (1975)

J PHILIBERT
Atom Movements
Diffusion and Mass Transport in Solids
Editions de Physique, Paris (1991)

M PRUTTON
Introduction to Surface Physics
Oxford Science Publ. (1995)

A P SUTTON and R W BALLUFFI
Interfaces in Crystalline Materials
Oxford Science Publ. (1995).

Chapters IX and X

J FRIEDEL
Dislocations
Pergamon Press (1964)

A H COTTRELL
The Mechanical Properties of Solids
John Wiley (1964)

J P DEN HARTOG
Strength of Materials
Dover Publ. (1977)

J P POIRIER
Creep of Crystals
Cambridge University Press (1985).

For general questions of **Metallurgy**, see:

R W CAHN and P HAASEN
Physical Metallurgy
Elsevier (1996).

For a list of **Exercises and Problems**, see:

J CAZAUX
Initiation à la physique du solide (in French)
Masson, Paris (1996).

SUBJECT INDEX

Pb refers to the Problems with Solutions

α particles   96, 157

absorption of light   26–27, 168, **Pb** 11
acceptor level   148
adsorption   160, **Pb** 10
allotropy   108
alloys   11, 186–200, 237, **Pb** 9
aluminium   79–81, 157, 222
amorphous materials   83, 149, 192
Anderson (localisation)   149
angular correlation   77
anharmonic vibrations   75
anharmonicity   75
Arrhenius (diagram)   178
atmospheric rarefaction   **Pb** 13
atomic planes   96
atomic rows   99
atomic units   36
atomic volume   68
attractive junctions   236
austenite   **Pb** 15
Azbel–Kaner (geometry)   144

band
        conduction   135, 199, **Pb** 3
        energy   48, 64, 135, 270
        forbidden   119, 126–129
        rigid   64
        valence   135, **Pb** 3
barium titanate   **Pb** 8
basis   98
B.C.S.   61, **Pb** 6
blende   3
Bloch
        (functions)   122–124, 259–260,
            267–268
        (theorem)   122, 259–260
body-centred cubic (bcc)   106–107, 125, 129
Boltzmann (equation)   261–273
Born–von Karman (conditions)   34, 44, 48,
    259
Bravais (lattice)   97, 100, **Pb** 5, **Pb** 6
Brillouin
        (condition)   124
        (planes)   124–126, 131

        (zones)   119, 124–126, 134, 270, **Pb** 2,
            Pb 3, **Pb** 4, **Pb** 6
brittle cast irons   249, 250
brittle–ductile transition   252
brittleness   221, 246–253
Brownian motion   89, 185
bulk modulus   73
Burgers (vector)   206, 226, 236, 271–273

carbides   163, 217
carotene   121, **Pb** 2
carotenoid pigments   121
carriers   148
catalysis   161
cavities   157, **Pb** 17
Cayley (tree)   95
cementite   **Pb** 15
channelling of particles   96
charge oscillations   159, 164, 183
cholesteric crystals   109
cleavage   155, 246
climb, of dislocations   215–217, 243, 244
close-packed hexagonal (hcp)   106
coherence length   62
cohesion   6, 67–76
cold work   224, 235
collision cascades   **Pb** 17
collisions   17, 56, 57, 261
colloids   102, 107
coloration   121, 138, 168, **Pb** 11
complex refractive index   26
composite materials   149, 219
compressibility   8–9, 73, 74, 80
conductivity
        complex   23
        electrical   12, 16–21, 55–59, 261–265
        thermal   11, 13, 21–22, 63, 80
configuration
        diagram   169
        entropy   162
Cooper (pairs)   62
coordinance   186
correlation coefficient   181
Cottrell
        (clouds)   238, 251, **Pb** 13
        (cracks)   249

Coulomb
    (attraction)   2–3, 68
    (energy)   4, 50, 68, 70, 72
creep   225, 239–245
critical field   60
crystal defects   58, 155–171, 201–222
crystal growth   213, Pb 10
crystalline texture   217
Curie (law)   Pb 7
Curie–Weiss (law)   Pb 7
cyclohexane   161
cyclotron
    resonance   144
    trajectory   144

Debye
    (layer)   167
    (temperature)   Pb 6
Debye–Waller (factor)   258, Pb 18
dendrites   197
density of states   35, 41, 45–49, 134
dielectric constant   23, 26, 199
diffraction   100, 117, 255–258, Pb 18
diffusion
    coefficient   174, 185
    equations   184–185, Pb 13, Pb 14
dipole moment   5
disinclinations   85, 207
dislocations   195, 201–222, 225, 271, Pb 13,
    Pb 16, Pb 17
    core   209
    density   210
    dissociation   214
    loops   211, 216, 227, 272, Pb 17
    networks   218, Pb 13
    sources   226, 227, 250
    width   229
disorder   83, 147, Pb 9
divacancy   162
donor level   148
Doppler (effect)   77
drift velocity   16, 19, 22, 177, 185, Pb 13,
    Pb 14
drink cans   250
Drude (model)   14–29, 53
dual space   100
Dulong and Petit (law)   29
Dupin (cyclids)   202
duralumin   173, 238

effective mass   145, 199, 265
    thermal   80

effective potential   69, 72
Einstein
    (formulae)   174, 177
    (frequency)   Pb 18
elastic balloon, inflation of   Pb 19
elastic energy   209
elastic stresses   205–209
    internal   212
    shear   203, 208
    true   224
elasticity   73, 208, Pb 13
electrical resistivity   59
electromagnetic induction   142
electron concentration   41, 47
electron core   69, 70
electron mean free path   57, 80
embryos   193
energy
    binding   4
    cohesive   4, 8, 72
    formation   162, 165, 200
    ionisation   4, 72
    migration   164, 165, 200
    self-diffusion   178
    surface   155, 156, 247
energy saving   27
epitaxy   107, 219
equilibrium diagrams   190–193
eutectic   191
Ewald (sphere)   101, 257
exchange   36, 39, 49, 50
exciton   170
excluded volume   92, Pb 19
expansion coefficient   75
exponential distribution   17

F centres   167, Pb 11
Fermi
    (energy)   40, 47
    (level)   40, 156, Pb 1
    (sphere)   46, 158
    (surface)   129, 144, 264
    (temperature)   47
    (velocity)   80
Fermi–Dirac (statistics)   29, 42–43, 65, 263
ferroelectricity   Pb 8
ferromagnetic metals   13, Pb 7
Fibonacci (sequences)   112
Fick (laws)   176, 178, 196, 240, Pb 13, Pb 14
fluctuating valency   62
fluidity   242
fluorescence   169

focal conics   202
forest of dislocations   235
Fourier (law)   11
fractal   93
fracture   246–253
Franck and **Condon** (principle)   **Pb** 11
Frank and **Read** (source)   226
free electrons   14
Frenkel (pairs)   170, 200
Friedel (oscillations)   164
frustration   85

gap   62, 133, 136, 137, 149
gas bubbles   157, **Pb** 16
Gauss (theorem)   **Pb** 5
gels   93
glass   87, 149, 153, 168, 192, 249, 250, 253
glide   201, 202–203, 215, 228–239, 243
    and climb   243
grain boundaries   218, 232
    melting   221
grains   200, 217, 218, 232, 240, 251
graphite   232, 249
Great Army   252
Griffith
    (cracks)   247
    (stresses)   248
growth spirals   213, **Pb** 10
Guinier–Preston (zones)   173, 183, 197
gyration, radius of   90

halides
    alkali   1, 7, **Pb** 11
    silver   198, 201
Hall and **Petch** (law)   233
hardening   232, 235, 237
    structural   238
hardness   224, 230
Hartree–Fock
    (effective mass)   52
    (equations)   50
heat equation   176
heterodiffusion   181–185
hole   136, 146–147, 170
Hooke (law)   203, 225, 246
hopping   150
Hubbard (bands)   151
Hurter and **Driffield** (graph)   198
hydrogen
    atom   **Pb** 12
    molecules   **Pb** 5

icosahedron   86
impurities   161, 220
incommensurable phases   110
independent electrons   36
industrial ceramics   231
insulators   133, 136
interatomic distance   67
intercalation   149
interfaces   193, 219
intergranular brittleness   221
interstellar dusts   **Pb** 12
interstitials   169, 199
    diffusion   199, **Pb** 17
    sites   104, 184
INVAR   76
ion implantation   107
ionic conduction   153
ionic solids   1
ionosphere   26
irradiation   169, 183, 239, **Pb** 16, **Pb** 17
isoprene   121
isotopes   179, 180
isotopic effect   62

Jahn–Teller (effect)   151, 168, 169, **Pb** 11
jogs   215, 216, 236
Joule (effect)   20, 57
Jupiter   150

Kinchin and **Pease** (formula)   **Pb** 17
kinks   160, 161
Kirchoff (law)   273

La_2CuO_4 **Pb** 6
Lamé (moduli)   208, 230
Laplace (force)   22, 142
LCAO   124, 268–270
Lennard–Jones (potential)   **Pb** 12
lever arm, rule of the   189
levitation   61
liberty ships   252
line tension   210
linear chain   31, 116, **Pb** 2, **Pb** 4
liquid crystals   109, 202
liquids   109, 190, 243
localisation   148, **Pb** 4
London (equation)   61
Lorenz (constant)   12, 63
Louvre pyramid   168
lycopene   121

machining   161

Madelung (constant)   2, 166
magnesia   102, 107
magnetic properties   **Pb** 7
maize   121
mass action, law of   136, 167
Matthiessen (law)   58
Maxwell (equations)   22, 61
Maxwell–Boltzmann (statistics)   16, 29, 42–43
mean field   **Pb** 19
mechanical tests   223, 225
Meissner (effect)   61
melting point   190, 221
metal   11, 31, 67, 135
metal–insulator transition   120, 150, **Pb** 5
metastability   192
microcracks   247, 249
microscopy
    electron   102, 157, 201, 213, 217, 221, 239
    tunnel   160
Miller (indices)   99
minerals   1
miscibility   187
    partial   190
    total   187, 192
moduli of elasticity   73, 203, 208, 230, 246
molecular beam epitaxy   219
Mollwo–Ivey (law)   168, **Pb** 11
molybdenum bisulphide   232
momentum operator, eigenfunctions of   34, 138
mother phase   193
Mott
    (conduction)   151
    (transition)   150
Mott and **Gurney** (model)   199

Nabarro–Herring (creep)   242
Navier (equation)   201, 207
nematic crystals   109
Newton
    (creep) 239
    (dynamics)   141
nodes   272
non-radiative transition   170
nuclear magnetic resonance   **Pb** 1
nuclear materials   **Pb** 16, **Pb** 17
nuclear reactors   183, **Pb** 16, **Pb** 17
nucleation and growth   192
    heterogeneous   195
    homogeneous   195

nuclei   193, 195

Ohm (law)   20
optical properties   22, 26
orbitals
    binding   **Pb** 3, **Pb** 5
    non-binding   **Pb** 3, **Pb** 5
order–disorder, transformation   **Pb** 9
organic conductors   31, 120
Orowan (formula)   225, 244
oxidation   **Pb** 14

pair distribution function   87–88
paramagnetism   13, 64
Pauli
    (hole)   39
    (paramagnetism)   13, 64
    (principle)   3, 46, 53, 73
Peach and **Koehler** (law)   212
Peierls
    (transition)   120
    (valley)   229, 230, 250
Peierls–Nabarro (stress)   229, 232
penetration distance   62
Penrose (tiling)   111
percolation   94
peritectic   192
perturbations   68, 116, 118
phonons   58, 80, **Pb** 6
photo-electrons   199
photographic process   198
pile-ups   232, 234, 244
pinning   226, 238
plasmons   24–25, 62
plastic deformation   223
plastic flow   225–226
plasticity   11, 223
point defects   155–171, 200, 239, **Pb** 17
Poisson
    (equation)   51
    (ratio)   209, 230
polarisability   5, 7, **Pb** 8
polarons   151
polycrystals   217, 233, 242
polymers   89, **Pb** 19
polytypes   109
Pooley (mechanism)   170
positrons, annihilation of   76–77
precipitates   195, 212, 237
precipitation   186–200
primitive cell   98, 126, **Pb** 6

quasi-crystals   113, 149
quasi-periodicity   111
quenching   165, 183
        chemical   251

radius
        atomic   68, 71
        electron   47
        ionic   71
rate
        accumulation   180
        deformation   224, 241
rational deformation   224
reciprocal lattice   100, 104, 106, 116, 122, 124, 256, **Pb** 18
recrystallisation   210, 235
redox potential   198
reduced zone   140
reflection   27, 132
relaxation time   17–19, 57, 262
replica   160, **Pb** 16
repulsion
        interionic   3
        kinetic   68
rubber   92, **Pb** 19
Russia, retreat from   252
rutile   219

saturation, by defects   **Pb** 17
Schmid and **Boas** (law)   203
Schottky (pairs)   166
Seeger (zones)   239
segregation at boundaries   221
self-diffusion   · 173, 178–181, 245
semi-metals   138
semiconductors   119, 133, 136, 148
shear   203, 208, 225
shear modulus   203, 225
short-range order (SRO)   85
silicon   134, 137, 138
singlet state   38, **Pb** 5
sinks   165
sintering   217, 231
sites
        octahedral   104
        tetrahedral   104
skin depth   144
Slater (determinant)   37, 50
smectic crystals   109
Snoek (effect)   **Pb** 15
soaps   109, 214, 232
sol–gel transition   93

solid solution   186, 187
solitons   **Pb** 4
Sommerfeld (model)   31
sources
        dislocations   226
        vacancies   165
specific heat   29, 54, 80, **Pb** 7, **Pb** 9
spin   37, 62, 64
spinodal transition   185
stacking faults   215, **Pb** 4
states
        localised   148, **Pb** 4
        stationary   32
steels
        ferritic   252, **Pb** 15, **Pb** 17
        martensitic   **Pb** 15
        stainless   239, 250
stoichiometry   62, 110, 163, 168
Stokes (shift)   169, **Pb** 11
Stokes–Einstein (formula)   243
structure
        crystalline   95
        CsCl   2, 7, 107
        diamond   107
        NaCl   2, 7, 105
sub-boundaries   217, 253
sublimation   4, 72, 221
sulphur   138
Sun   26
superalloys   220
superconductivity   60, **Pb** 6
supersaturation   165, 183, 196, 221
surface dipole   159
surface steps   160, **Pb** 10
surface traps   199
surfaces   43, 155–171, **Pb** 10
susceptibility, magnetic   13, 64, **Pb** 6, **Pb** 7
swelling   **Pb** 16
sylvine   1

talc   232
tensile test   202, 223–224
tensor
        conductivity   266
        effective mass   145, 265
        strain   208
terraces   161
thermal cycling   76
thermal etching   221
thermal expansion   74
thermal vibrations   74, 95, 255, **Pb** 18

tight binding   124, 267–270, **Pb** 2, **Pb** 3,
    Pb 4, **Pb** 5, **Pb** 6
tiling   88, 110
time of flight   18, 20, 140
TMTSF–DMTCNQ   120
Tolman (experiment)   15
tomatoes   121
torsion test   223
transformation-related hysteresis   194
transparency   27, 137, 149
triplet states   38, **Pb** 5
TTF–TCNQ   31
tunnel effect   151, **Pb** 12
    micrography   160

ultrasound   246

vacancies   163–169, 174, 182, 240, **Pb** 10,
    Pb 11, **Pb** 16
Van der Waals (forces)   5–7, **Pb** 12

Van Hove (singularity)   **Pb** 6
viscosity   242
Volterra (dislocation)   205
von Laue (experiment)   101
vulcanisation   92

wave packet   139
Weertman (equation)   245
Weiss (molecular field)   **Pb** 7
Wiedemann–Franz (law)   12, 21, 63
Wigner–Seitz (cell)   72, 88
window panes   27–28
wurtzite   3
wüstite   110

$YBa_2Cu_3O_7$ 61, **Pb** 6
yield stress   223, 250
Young (modulus)   246, **Pb** 15

zero creep method   157

AUTHOR INDEX

Pb refers to the Problems with Solutions

Allain M   214
Alloul H   x, Pb 5
Amelinckx S   Pb 13
Anderson P W   149
Ashby M F   471
Ashcroft N W   471
Averback R S   Pb 17

Baïlon J-P   471
Balian R   ix, x, 141, Pb 10
Balluffi R W   472
Barbier L   Pb 10
Barbu A   83
Bardeen J   61
Barton J   168
Basdevant J-L   ix, x, Pb 12
Bednorz J G   60
Bellon P   Pb 17
Benedek R   Pb 17
Bennet H E   132
Béranger G   Pb 14
Bernard M   x, Pb 2, Pb 7, Pb 9
Bethge H   160, 214
Beuneu F   171
Blakely D W   161
Blandin A   183
Blech I   113
Bois P   171
Bok J   Pb 6
Boos J Y   221
Boulanger L   239
Bourret A   213
Brézin E   ix
Brillouin L   115
Bullough R   Pb 17
Bunting M G   220
Burton W K   Pb 10

Cabaud B   86
Cabrera N   Pb 10
Cahn J W   113
Cahn R W   472
Caisso J   471
Caput M   157
Carel Cl   110
Casimir H B G   61

Cazaux J   472
Chazalviel J-N   x, Pb 3, Pb 4, Pb 6
Colliex C   25
Cooper L N   61
Cotton J P   Pb 19
Cottrell A H   230, 249, 472
Cousty J   Pb 10

Daguer C   110
Dahmen U   219
Dammak H   83
Dautreppe D   Pb 9
Den Hartog J P   472
Dimitrov O   59
Dorlot J M   471
Drude P   14, 27
Duneau M   111, 112
Dunlop A   83
Duplantier B   92
Dupouy G   218
Duwez P   83

Fincher C R   Pb 4
Finney J L   89
Fisher   Pb 19
Flinn P A   Pb 18
Flory P   Pb 19
Flynn C P   471
Forro L   120
Frank F C   226, Pb 10
Friedel G   202
Friedel J   164, 183, 234, 249, 472

Galvari J R   110
Gebbart W   Pb 11
de Gennes P G   62, 89
Gerl M   471
Gorter C J   61
Goux C   221
Gratias D   113
Grilhé J   471
Guéron M   x, Pb 2
Guinier A   173, 471, Pb 1

Haasen P   472
Harris I A   86

Harrison W A   471
Haüy R J   96
Hermann C   x, 134
Hirsch P B   201
Hoareau A   86
Hogan D W   220
Holcomb D F   Pb 5
Holland W A   Pb 16
Honda K   64

Iwao O   222

Jahn H A   Pb 11
Janot C   471
Jones D R H   471
Jones R B   242
Jouffrey B   201
Jousset J C   Pb 17
Julien R   471

Katz A   111, 112
Keller K W   160, 214
Kidwell R S   86
Kittel C   471
Kléman M   85, 202
Kormann R   152
Kuhnert A   Pb 11
Kurz W   471

Labbé J   Pb 6
Lacombe P   Pb 13
Lambert C M   28
Lambert M   Pb 1
Laugier J   Pb 9
Lesueur D   83–88
London F   6, 61

Mandelbrot B   93
Manenc J   189
Manuel A A   78
Martin G   Pb 17
Masounave J   471
Masson L   Pb 10
Mazières C   Pb 1
McLean D   234
Mélinon P   86
Mercier J P   471
Mermin N D   471
Mikailoff H   Pb 16
Mönch W   Pb 4
Moret R   Pb 6
Morin F   137

Mory J   97
Mott N F   152, 199, 249
Müller K A   60
Müller W   Pb 4
Mundy J N   180

Néel L   Pb 9
Nienhuis B   92
Noguéra C   Pb 6
Northby J A   86

Omnès R   x, Pb 8

Paciornik S   219
Pascard H   61
Perrier F   218
Perrin J   Pb 13
Petiau P   x, Pb 12
Philibert J   472
Pierre L   110
Poirier J P   472
Pouget J-P   Pb 6
Prutton M   472

Quéré Y   Pb 17

Read   226
Reynolds F W   59
Rosenbaum T F   Pb 5

Sadoc J F   85
Salençon J   207
Sapoval B   x, 93, 134
Schneck J   110
Schrieffer J R   61
Schwoebel R   Pb 14
Sébilleau F   246
Seeger A   239
Seidman D N   Pb 17
Shechtman D   113
Slifkin L   198
Solomon I   x, Pb 1
Solorzano I G   219
Somorjai G A   161
Stauffer D   93
Stoneham A   472
Stoto T   217
Sutton A P   472

Takahashi J   97
Taupin C   Pb 1
Teller E   Pb 11

Tessman J R   7
Tolédano J C   110
Toulouse G   85
Trebbia P   25
Treilleux M   102

Vandersande J B   219
Veysseyre R   110
Vidal-Madjar A   x, Pb 12
Vinter B   471
von Laue M   100

Weertman J   245
Weigel D   110
Weisbuch C   471
Wenzl H   Pb 17

Zachariasen W H   87
Zambelli G   471
Zener C   27, 249
Zuppiroli L   120, 152, 217

TABLE OF CONSTANTS

		S.I.	a.u. (see p. ??)
Avogadro's number		6.02×10^{23}	
Planck's constant	h	6.63×10^{-34} J s	
	$\hbar = h/2\pi$	1.06×10^{-34} J s	1
Boltzmann's constant	k	1.38×10^{-23} J K^{-1}	
Velocity of light	c	2.99×10^{8} m s^{-1}	137
Electron mass	m	9.11×10^{-31} kg	1
Proton mass	$1836m$		1836
Electron charge	$\|e\|$	1.60×10^{-19} C	1
Rydberg's constant		2.18×10^{-18} J	0.5
1 eV		1.60×10^{-19} J	3.7×10^{-2}
Bohr's radius	a_H	5.29×10^{-11} m	1
Bohr's magneton	μ_B	9.27×10^{-24} J T^{-1}	

ENERGY EQUIVALENCIES

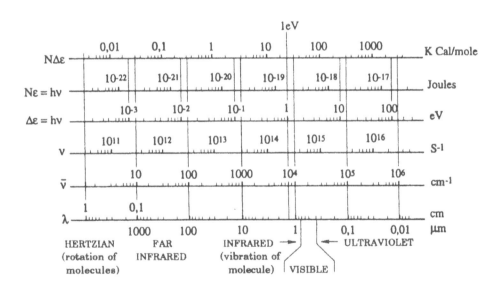